Advances in

Heterocyclic
Chemistry

Volume 94

Advances in

HETEROCYCLIC CHEMISTRY

Edited by

ALAN R. KATRITZKY, FRS

Kenan Professor of Chemistry
Department of Chemistry
University of Florida
Gainesville, Florida

Volume 94

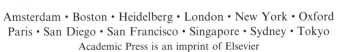
Amsterdam • Boston • Heidelberg • London • New York • Oxford
Paris • San Diego • San Francisco • Singapore • Sydney • Tokyo
Academic Press is an imprint of Elsevier

ACADEMIC
PRESS

ELSEVIER

Academic Press is an imprint of Elsevier
84 Theobald's Road, London WCIX 8RR, UK
Radarweg 29, PO Box 211, 1000 AE Amsterdam, The Netherlands
Linacre House, Jordan Hill, Oxford OX2 8DP, UK
30 Corporate Drive, Suite 400, Burlington, MA 01803, USA
525 B Street, Suite 1900, San Diego, CA 92101-4495, USA

First edition 2007

ISBN: 978-0-12-373963-6
ISSN: 0065-2725

For information on all Academic Press publications
visit our website at books.elsevier.com

Printed and bound in USA

07 08 09 10 11 10 9 8 7 6 5 4 3 2 1

Contents

The Synthesis of Heterocycles Using Cascade Chemistry

SCOTT K. BUR, and ALBERT PADWA

Organometallic Chemistry of Polypyridine Ligands II

ALEXANDER P. SADIMENKO

Aminoisoxazoles: Preparations and Utility in the Synthesis of Condensed Systems

V.P. KISLYI, E.B. DANILOVA, and V.V. SEMENOV

Isothiazolium Salts and Their Use as Components for the Synthesis of S,N-Heterocycles

J. WOLF, and B. SCHULZE

Contributors

Numbers in parentheses indicate the pages on which the author's contribution begins.

Scott K. Bur (1) Department of Chemistry, Gustavus Adolphus College, 800 West College Avenue, Saint Peter, MN 56082, USA

E.B. Danilova (173) N.D. Zelinsky Institute of Organic Chemistry, Russian Academy of Sciences, 47 Leninsky prosp., 117913, Moscow, Russian Federation

V.P. Kislyi (173) N.D. Zelinsky Institute of Organic Chemistry, Russian Academy of Sciences, 47 Leninsky prosp., 117913, Moscow, Russian Federation

Albert Padwa (1) Department of Chemistry, Emory University, 1515 Dickey Drive, Atlanta, GA 30322, USA

Alexander P. Sadimenko (107) Department of Chemistry, University of Fort Hare, Alice 5701, Republic of South Africa

B. Schulze (215) Department of Organic Chemistry, University of Leipzig, Johannisallee 29, D-04103 Leipzig, Germany

V.V. Semenov (173) N.D. Zelinsky Institute of Organic Chemistry, Russian Academy of Sciences, 47 Leninsky prosp., 117913, Moscow, Russian Federation

J. Wolf (215) Department of Organic Chemistry, University of Leipzig, Johannisallee 29, D-04103 Leipzig, Germany

Preface

Volume 94 of *Advances in Heterocyclic Chemistry* commences with a review of cascade reactions in heterocyclic synthesis by S.K. Bur (Gustavus Adolphus College, Minnesota) and A. Padwa (Emory University, Atlanta). The chapter presents a fascinating array of complex sequences, which provide efficient routes to a wide variety of heterocyclic systems.

The second chapter is the twelfth in the series on the organic chemistry of heterocyclic ligands in metallic complexes by A.P. Sadimenko of University of Fort Hare (Republic of South Africa). The present contribution deals with the chemistry of polypyridine ligands in organomanganese and organorhenium complexes. Its current importance can be measured by the fact that, of the nearly 700 references, approximately half date from the last 10 years.

V.P. Kislyi, E.B. Danilova, and V.V. Semenov of Zelinsky Institute (Moscow) discuss the preparation of aminoisoxazoles and their utility in the synthesis of condensed systems.

In the final chapter, J. Wolf and B. Schulze (University of Leipzig, Germany) review isothiazolium salts and their use in synthesis. Many condensed S,N-heterocyclic systems are described in this, the first review dedicated to this topic.

Alan R. Katritzky
Gainesville, Florida

The Synthesis of Heterocycles Using Cascade Chemistry

Scott K. Bur[a], and Albert Padwa[b]

[a]Department of Chemistry, Gustavus Adolphus College, 800 West College Avenue, Saint Peter, MN 56082, USA
[b]Department of Chemistry, Emory University, 1515 Dickey Drive, Atlanta, GA 30322, USA

I. Introduction

Molecules containing heterocyclic substructures continue to be attractive targets for synthesis since they often exhibit diverse and important biological properties (96CHC1). Accordingly, novel strategies for the stereoselective synthesis of hetero-polycyclic ring systems continue to receive considerable attention in the field of synthetic organic chemistry (81JOC2002, 83T3707, 94PAC2095, 91TL1787, 93JA3030, 83JOC2685, 85HCA745, 89JOC1236, 83JA4750, 84ACR35, 93JOC4473, 93T10219, 95JA7834, 96T10569, 98JOC5587, 90JOC1624, 91TL5183, 93JA2064, 93JA7904, 95JOC8044, 96JA6210, 95JOC6258, 92CL443, 94JOC5633, 97JOC3263). The efficiency with which heterocycles can be constructed is important not only because it affects the production costs for the desired material, but also the environmental impact associated with waste disposal, conservation of source ma-terials like petroleum stocks, and energy consumption. The rate of increase in mo-lecular intricacy as one progresses from simple starting materials to the final product can also serve as a measure of efficiency (01T6855). On one end of the continuum, a single synthetic step could convert an inexpensive material into a highly complex heterocyclic product. On the other end lies a linear series of transformations, wherein a single atom or group is added in each step to build complexity. As a prerequisite for an ideally proceeding one-pot sequential transformation, the reactivity pattern of all participating components has to be such, that each building block is involved in the reaction only when required to do so. The reality of chemical synthesis is somewhere between these extremes, with the one-step process held as the ideal.

Domino reactions (reactions in which several bonds are formed in one sequence without the isolation of intermediates, the changing of reaction conditions, or the addition of reagents) (96CR115), multi-component reactions, and the so-called "telescoping" of reactions (the sequencing of multiple transformations in a single reaction vessel through the changing of conditions and/or adding of reagents at appropriate times) allow for a rapid increase in molecular complexity in a single chemical operation. The terms "*tandem*" and "*cascade*" have been applied to all three of these reaction types and are thus used as general descriptors in this work (93AGE131, 96CR115, 92TOR13103, 95CR195, 91COS, 95DROSH193, 91COS779, 96CR1). Because of the rate at which they increase molecular intricacy, cascade reactions have received considerable attention from the synthetic organic commu-nity. The development of sequences that combine transformations of differing fundamental mechanism broadens the scope of such procedures in synthetic chemistry.

This review contains a representative sampling from the last 15 years on the kinds of reactions that have been sequenced into cascades to produce heterocyclic mol-ecules. The fact that multiple reactions give rise to a cascade sequence makes the categorization of these processes difficult. The structure we have imposed, therefore, is somewhat arbitrary but is loosely based on what, in our judgment, is the key reaction in the cascade sequence. This mini-review is not intended to be a critical or comprehensive coverage, but rather provides an overview of the field and thus some cascade processes are covered in more detail than others.

II. [4π + 2π]-Cycloadditions

A. [1,3]-Dipolar Cycloadditions

1. *Metallo-Carbenoid-Initiated Cascades*

Many different examples of cascade processes that employ 1,3-dipoles as reactive intermediates have been described in the literature. The transition metal-catalyzed decomposition of diazoimides results in the formation of isomünchnone dipoles, a class of mesoionic betaines (03SA681, 99JOC2049, 00JOC2368), which are known to readily undergo 1,3-dipolar cycloaddition chemistry. Ibata and Hamaguchi were the first to report that diazoimide **1** formed isomünchnone **2** on heating in the presence of Cu$_2$(acac)$_2$ (74TL4475), and that this reactive dipole could be trapped with various dipolarophiles such as **3** to give the oxabicyclic product **4** (Scheme 1) (75CL499).

Rhodium(II) catalysts also initiate a similar reaction (91JOC820). By tethering the dipolarophile moiety within the diazoimide, complex polycyclic frameworks can be formed in a single step (88TL1677). Thus, heating compounds such as **5** with catalytic Rh$_2$(OAc)$_4$ produced cycloadduct **6** as a single diastereomer in 73–91% yield (Scheme 2). Over a period of years, Padwa and coworkers demonstrated that this cascade sequence is quite general. Diazoimides with the general structure **7** ($n = 1, 2, 3$; $m = 1, 2$) were readily converted to the corresponding polycyclic system **8** (92TL4731).

Compounds of type **6** and **8** contain an *N,O*-acetal functional group and have been used as precursors to *N*-acyliminium ions. This method was exploited for the synthesis of B-ring homologs of the erythrinane family of alkaloids (97JOC67). In these studies, diazoimides of type **9** were exposed to catalytic amounts of rhodium(II) and oxabicycles **10** are formed in 90–98% yield (Scheme 3). Treating **10** with BF$_3$•OEt$_2$ provided the ring opened products **11** in 85–95% yield as single diastereomers.

Isomünchnone dipoles generated by the cyclization of rhodium carbenoid intermediates with adjacent amido groups have also been shown to undergo cycloaddition with both electron-rich and certain heteroaromatic π-bonds. For example, the catalytic decomposition of diazoimide **12** provided dipole **13** which subsequently added across the indole π-bond to give a cycloadduct possessing the aspidosperma alkaloid skeleton (Scheme 4) (95JOC6258).

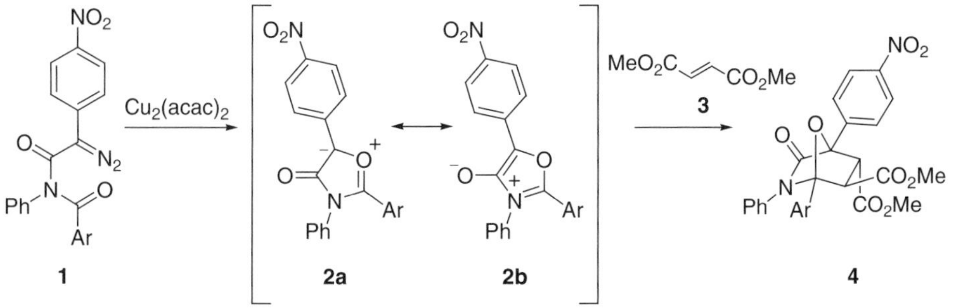

Scheme 1

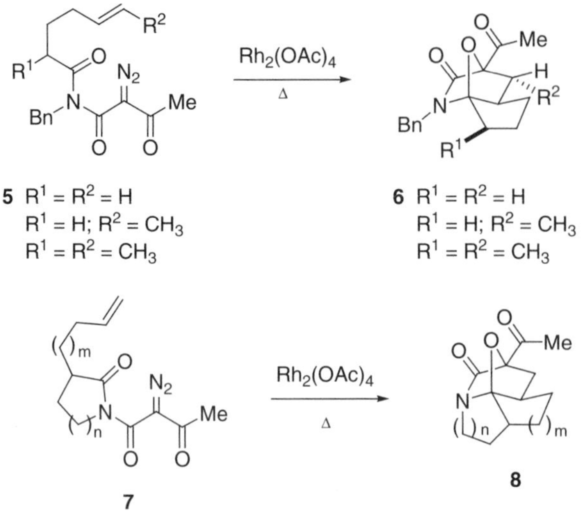

Rh2(OAc)4
Δ
5 R1 = R2 = H
R1 = H; R2 = CH3
R1 = R2 = CH3
6 R1 = R2 = H
R1 = H; R2 = CH3
R1 = R2 = CH3
7
8

Scheme 2

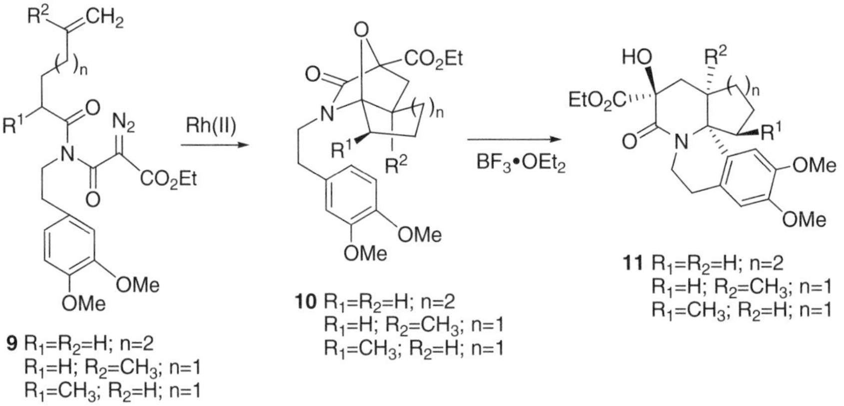

Rh(II)
BF3•OEt2
9 R1=R2=H; n=2
R1=H; R2=CH3; n=1
R1=CH3; R2=H; n=1
10 R1=R2=H; n=2
R1=H; R2=CH3; n=1
R1=CH3; R2=H; n=1
11 R1=R2=H; n=2
R1=H; R2=CH3; n=1
R1=CH3; R2=H; n=1

Scheme 3

This sequence was recently used in the synthesis of the alkaloid (±)-aspidophytine (**19**). The key sequence of reactions was initiated by treatment of diazo ketoester **15** with $Rh_2(OAc)_4$ to generate a transient metallocarbene that reacted with the proximal imido carbonyl group to form dipole **16** (06OL3275). A subsequent 1,3-dipolar cycloaddition across the tethered indole π-bond gave cycloadduct **17** in 97% yield. Oxabicycle **17** was then converted into **18** by the action of $BF_3 \cdot OEt_2$ in 70% yield and this compound was eventually converted into (±)-aspidophytine (**19**).

Dauben used a related cyclization–cycloaddition approach for the synthesis of tigliane **22**. Carbonyl ylide **21**, derived from the diazo ketoester **20**, underwent intramolecular cycloaddition to form **22**, a molecule which contains the C(6), C(9)-oxido bridge of the tigliane ring system (Scheme 5) (93JOC7635).

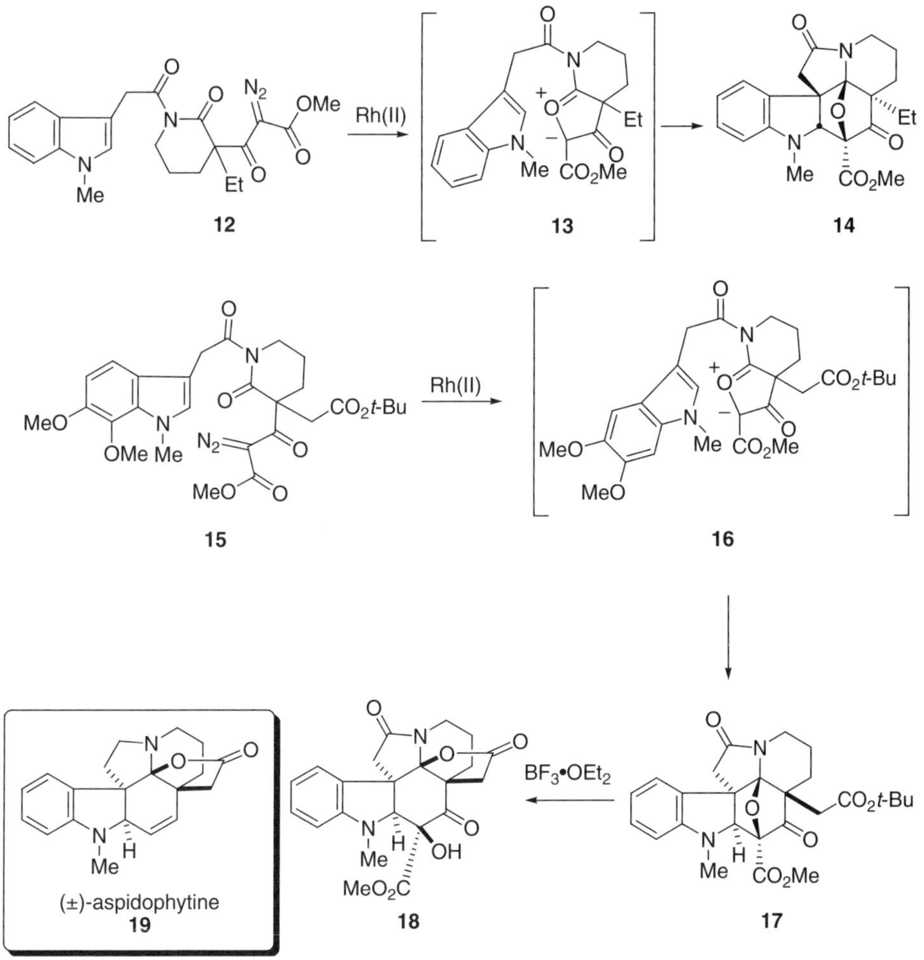

Scheme 4

2. *Pummerer-Initiated Cascade*

A Pummerer-initiated cascade reaction has also been used as a method for generating isomünchnones for further use in cycloaddition chemistry. For example, treatment of sulfoxide **23** with acetic anhydride first resulted in the formation of a reactive thionium ion that reacted with the distal amide carbonyl group to produce isomünchnone **24** (Scheme 6) (99JOC2038). Exposure of **24** to a dipolarophile, such as *N*-phenylmaleimide, resulted in 1,3-dipolar cycloaddition to give **25** as a single diastereomer in 85% yield.

The specific conditions required to successfully effect this transformation were important and warrant comment. The initial attempts to form the isomünchnone intermediate, employing TFAA to promote the cyclization and Et₃N to deprotonate the oxonium ion intermediate, failed to produce the isomünchnone dipole. Rather, cyclic

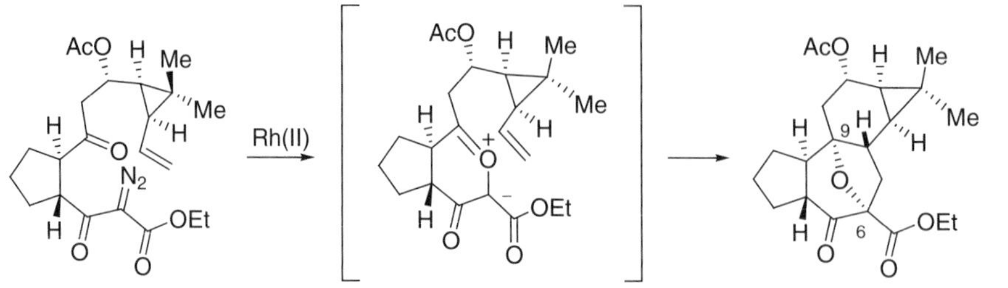

20 21 22

Scheme 5

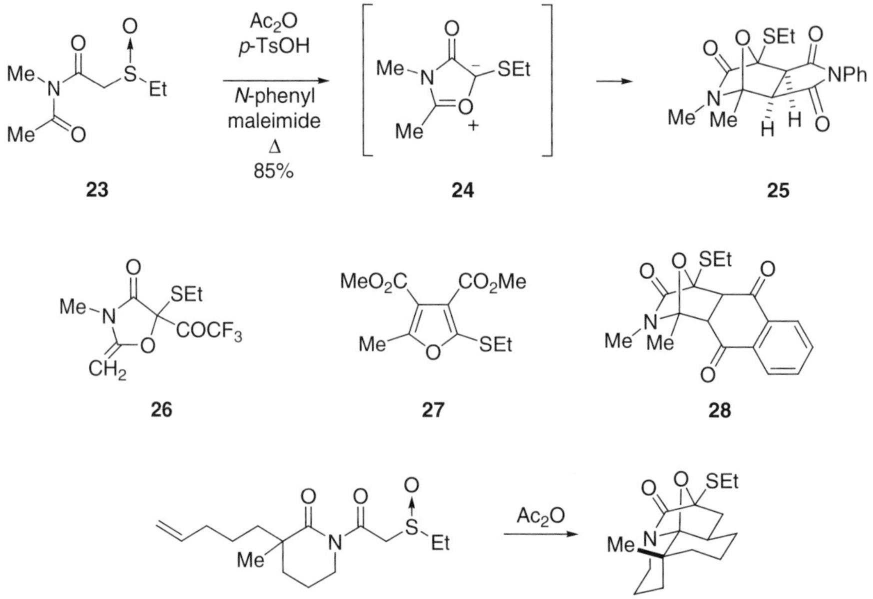

23 24 25

26 27 28

29 30

Scheme 6

ketene acetal **26** was obtained. After considerable experimentation, it was found that the slow addition of **23** to a mixture of acetic anhydride, a catalytic amount of *p*-TsOH, and the appropriate dipolarophile at 120 °C in toluene gave consistently good yields of the 1,3-cycloadduct. A variety of dipolarophiles were found to participate in these cycloadditions. When **24** was allowed to react with DMAD, the initially formed oxa-bicycle underwent rapid fragmentation to produce furan **27** (41% yield) and methyl isocyanate. The reaction of isomünchnone **24** with 1,4-naphthoquinone afforded cycloadduct **28** in 73% yield. Other suitable dipolarophiles include vinyl sulfones, maleic anhydride, and acrylate derivatives. Unactivated olefins also participated in the cycloaddition reaction when they were tethered to the isomünchnone dipole.

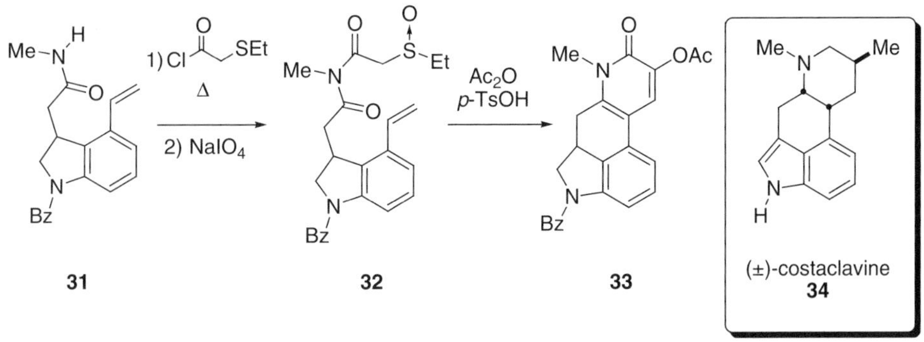

Scheme 7

For example, when sulfoxide **29** was treated with acetic anhydride, azapolycycle **30** was isolated as a single diastereomer in 73% yield.

The synthesis of the ergot alkaloid (±)-costaclavine **34** demonstrates the utility of this methodology for the synthesis of natural products (Scheme 7) (00JOC2368). Construction of the ergot skeleton began by acylation of the methyl amide functionality of **31** with (ethylsulfanyl)acetyl chloride and this was followed by a subsequent oxidation of the sulfide with $NaIO_4$ to provide sulfoxide **32**. A tandem Pummerer cyclization/cycloaddition cascade, initiated by exposing **32** to acetic anhydride and catalytic p-TsOH, gave tetracycle **33** in 64% yield. Several functional group interconversions of **33** then delivered (±)-costaclavine **34**. The synthesis of several other alkaloids, including onychine, dielsquinone, (±)-lupinine, and pumiliotoxin C, was also accomplished using this methodology (00JOC2368).

3. Nitrones

The intramolecular 1,3-dipolar cycloaddition of nitrones is a well-precedented reaction for the formation of cyclic isooxazolidines. An interesting method that has been used for the generation of N–H nitrones from readily available starting materials is through the 1,2-prototropic shift of oximes. Although it is unusual to observe cycloadditions using these N–H nitrones, a few examples have been reported. For example, while studying the synthesis of a series of Amaryllicaceae alkaloids, Wildman observed that the reaction of 6-hydroxybuphandidrine (**35**) with hydroxylamine produced cycloadduct **38** in good yield (Scheme 8) (71JOC3202). This reaction presumably occurs by formation of the intermediate oxime **36** that undergoes a 1,2-prototropic shift to give nitrone **37**. Cycloaddition of the nitrone dipole across the adjacent alkene moiety furnishes cycloadduct **38**.

Christy and coworkers observed a similar cyclization on condensing ketone **39** with hydroxylamine in hot ethanol to provide compound **41** (Scheme 9) (79JOC3117). Oxime **40** could be prepared under milder conditions, and was found to undergo a 1,2-prototropic shift followed by intramolecular cycloaddition to provide **41** on warming to 75 °C in toluene.

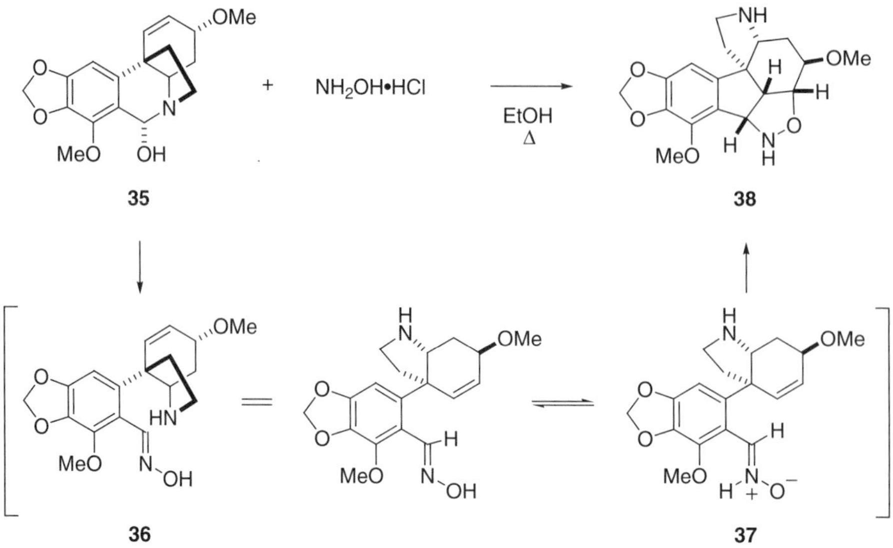

Scheme 8

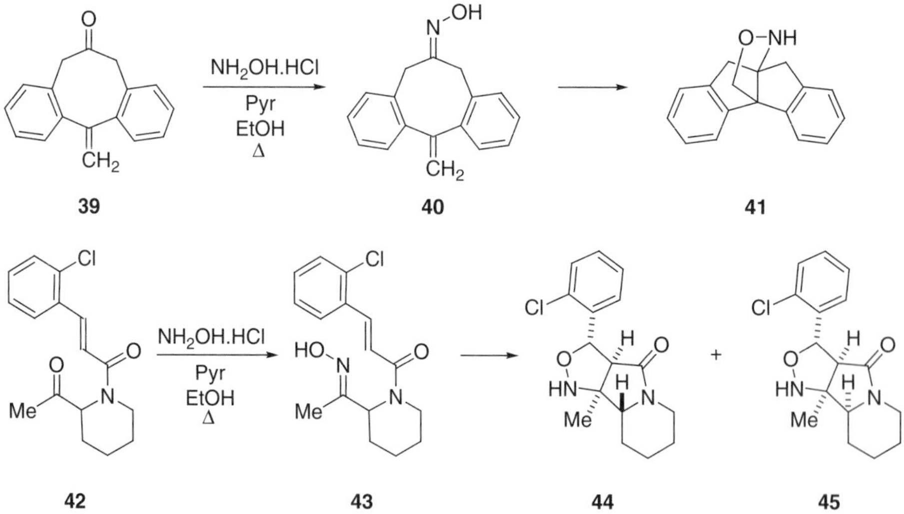

Scheme 9

In a related study, Heathcock reported that ketone **42** reacted with hydroxylamine hydrochloride under similar conditions to produce **44** (87JOC226). The intermediate oxime **43** that was first formed could be isolated under milder conditions. Whereas heating oxime **43** at reflux in acetonitrile for 30 h gave **44** in 92% yield, heating **43** in DMF formed a 1:1-mixture of the diastereomeric cycloadducts **44** and **45**.

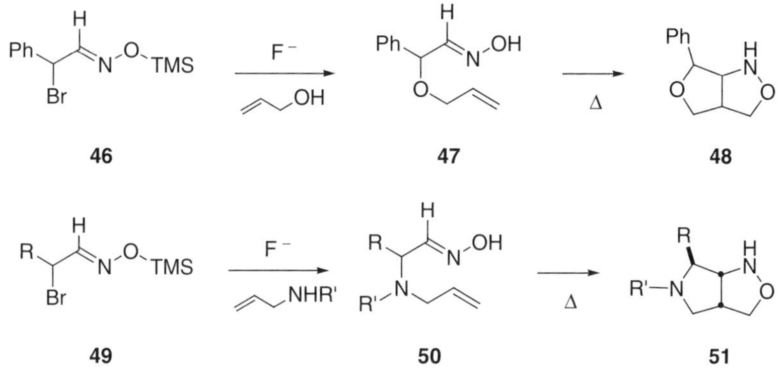

Scheme 10

While exploring the chemistry of α-brominated aldoxime derivatives, Padwa and Hassner observed the cycloaddition of oximes that contained pendant olefins. For example, the reaction of **46** with fluoride ion in the presence of allyl alcohol produced oxime **47** (Scheme 10) (88TL4169). Heating a benzene solution of **47** at 80 °C led to the formation of **48** as a single diastereomer, though only in 25% yield. The Hassner group later expanded the method to include the fluoride-mediated reaction of aldoxime **49** with various allyl amines to generate oximes **50** in 70–80% yield (88TL5313). Heating toluene solutions of these oximes at reflux temperatures led to the formation of pyrrolidines **51**.

The cascade sequence was also used to synthesize indolizidine, pyrrolizidine, and quinolizidine structures. Thus, heating oximes **52** at 180 °C in a sealed tube provided cycloadducts **53** or **54** in 60–76% yields (equation (1)) (89TL2289). Each of the products were isolated as single diastereomers. When five-membered rings were obtained from the cycloaddition, *cis-anti* isomers (i.e. **53a,b**) were formed, whereas formation of a six-membered ring led only to the *cis-syn* isomer (i.e. **54a,b**).

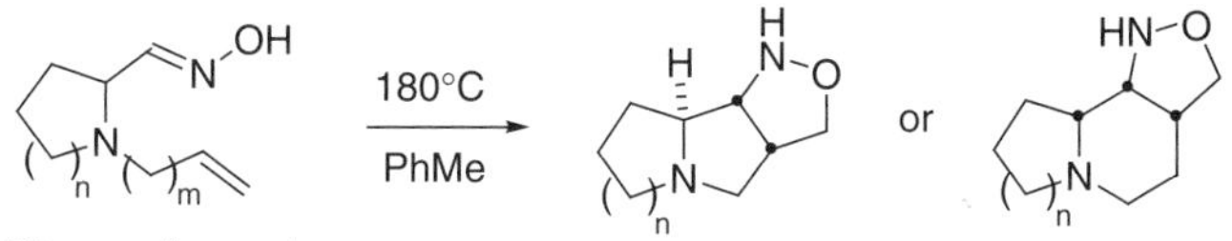

52a n = 1, m = 1
52b n = 1, m = 2
52c n = 2, m = 1
52d n = 2, m = 2

53a n = 1
53b n = 2

54a n = 1
54b n = 2

(1)

The Grigg group also studied the tautomerization of oximes to N–H nitrones followed by a dipolar cycloaddition reaction. The well-known H-bonding dimeric association of oximes, in both solution and the solid state, allows for a concerted proton transfer to occur and provides nitrone **56** (Scheme 11) (91TL4007). Another possible pathway involves tautomerization of the oxime to an ene-hydroxylamine (i.e. **57**) followed by a 1,4-hydride shift to give nitrone **58**. To probe the ene-hydroxylamine mechanism, deuterated oxime **59** was prepared and heated at 140 °C in xylene. The physical characteristics of the isolated product, however, were consistent with compound **60**, suggesting that the 1,2-prototropic reactions does not proceed

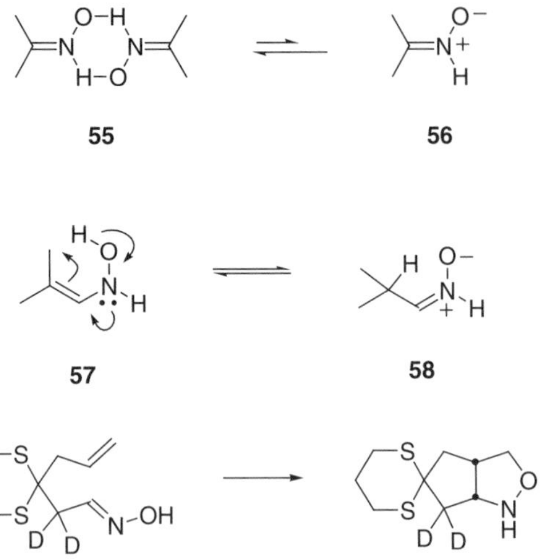

Scheme 11

via the ene-hydroxylamine. Grigg postulated that while the tautomerization between an oxime and a N–H nitrone is facile, dipolar cycloadditions involving these types of nitrones are relatively rare because (a) unactivated or electron rich dipolarophiles have too large a HOMO/LUMO gap, although intramolecular cycloadditions to form five-membered rings can overcome this gap, and (b) electron deficient dipolarophiles preferentially undergo Michael-type reactions with oximes.

Another interesting cascade involving nitrones is the copper catalyzed reaction with alkynes to produce β-lactams that was originally reported by Kinugasa (72CC466). Stoichiometric amounts of copper(I) phenylacetylide (**61**) react with various aryl nitrones **62** in pyridine to give β-lactams **63** in 50–60% yield (equation (2)). In each case, only the *cis*-lactams were isolated.

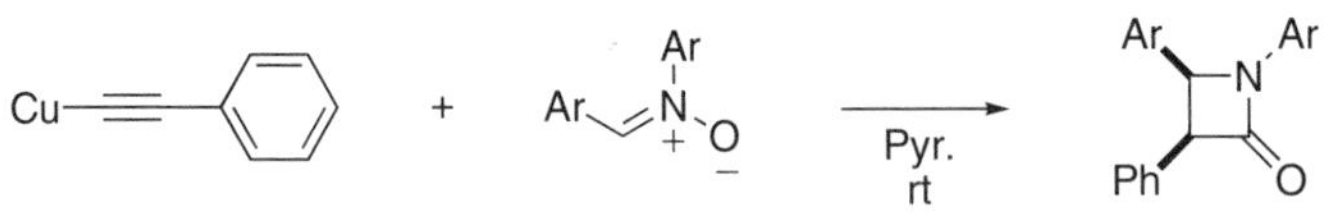

(2)

Miura and coworkers showed that the reaction could also be carried out using catalytic amounts of CuI in the presence of pyridine (95JOC4999). Asymmetric reactions were reported to occur with chiral bisoxazoline ligands producing β-lactams with moderate (40–68%) enantiomeric excesses. Use of an oxazolidinone with a chiral auxiliary appended to the alkyne also provided enantiomerically pure products (02TL5499). In all of these latter reports, mixtures of *cis* and *trans* lactam isomers were obtained in which the *trans*-product predominates. It was also shown that the *cis*-isomer could easily be converted to the *trans*-product when exposed to base.

The Fu group recently reported the use of C_2-symmetric planar-chiral *bis*(azaferrocene) ligands for the catalytic enantioselective Kinugasa reaction. A variety of terminal alkynes **64** (R^1 = Ar, Bn, 1-cyclohexenyl) were allowed to react with nitrones **65** (R^2 = Ar, Cy, PhCO; R^3 = Ar) in the presence of catalytic amounts of the CuCl•**67** complex to give diastereomeric mixtures (>90:10) predominating in *cis*-substituted β-lactams **66** in moderate to good yields (45–90%) and with good enantiomeric excesses (67–92%; Scheme 12) (02JA4521). With regard to the R^3

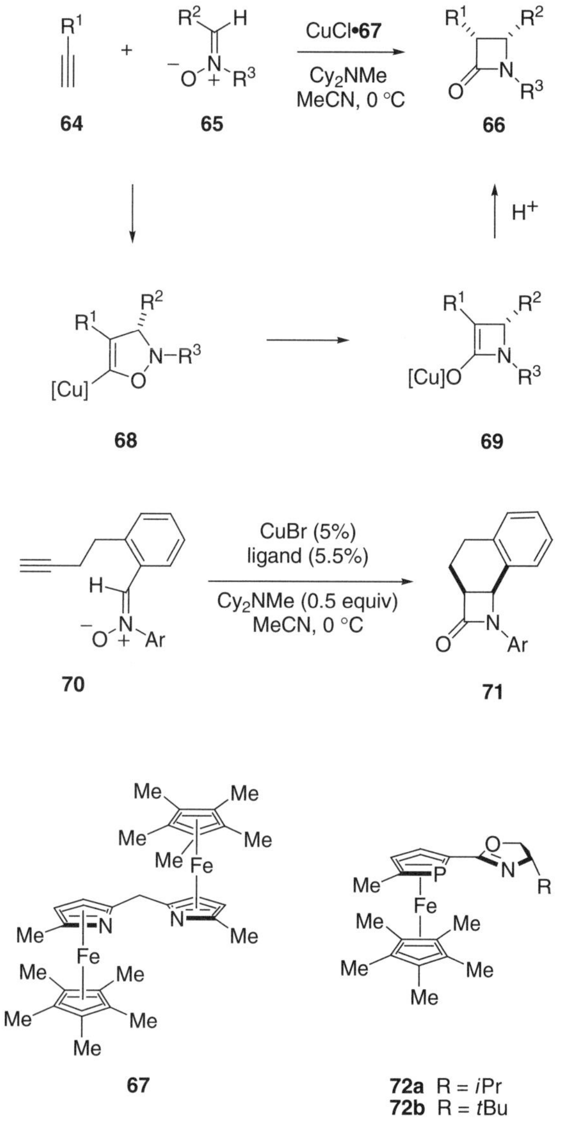

Scheme 12

group on nitrone **65**, electron-rich aromatic groups increased the enantioselectivity, although the yields were somewhat lower.

An intramolecular variant of this catalytic enantioselective process was recently reported. Nitrone **70** was converted to azetidinone **71** in the presence of the CuBr•**72a** complex in 74% yield and with 88% *ee* (03AG4082). Ligand **72b** was also quite effective, providing **71** with 90% *ee*, although the yield was only 47%.

The mechanism for the Kinusaga reaction is thought to involve a [3 + 2]-cyclo-addition of the nitrone with the copper-acetylide to give isoxazolidine **68**. Rearrangement of **68** then provides the copper enolate of the corresponding β-lactam (i.e. **69**), which is subsequently protonated to provide the observed product. The proton source for this last step is likely to be the conjugate acid of the base used to generate the copper-acetylide. Through considerable experimentation, the Fu group developed conditions that allowed for the reaction of the enolate with added electrophiles. Thus, exposing **73** to CuBr•**72a** in the presence of KOAc, allyl iodide, and the silyl enol ether of acetophenone gave rise to β-lactam **74** in 70% yield and with 90% *ee* (equation (3)) (03AG4082).

Brandi and coworkers developed an interesting tandem cycloaddition/thermal rearrangement cascade involving nitrones and methylenecyclopropane derivatives to produce 4-pyridones. For example, heating nitrone **75** with **76** at 110 °C for 7 days afforded pyridone **78** in 63% yield (Scheme 13) (92JOC5666). The intermediate isoxazolidine **77** was suggested to undergo homolytic cleavage of the weak N–O bond to give diradical **79** which underwent further fragmentation to provide di-radical **80** (93SL1). Cyclization of **80** then gave rise to 4-pyridone **78**.

Nitrone **81** underwent a dipolar cycloaddition reaction with bicyclopropylidene **82** at 60 °C over a 30-day period to give isoxazolidine **83** in 93% yield (Scheme 14) (96JOC1665). Further heating of **83** in toluene at reflux for 5 days produced pyri-done **84** in 63% yield. When a toluene solution of **81** and **82** was heated at reflux for several days, pyridone **84** was isolated in 61% yield. Nitrone **85** produced indolizi-dine **86** when heated with **82** in benzene at reflux temperatures for 7 days.

4. Nitrile Oxides

Brandi also examined the subsequent thermal behavior of the dipolar cyclo-adducts that arise from the reaction of nitrile oxides with methylenecyclopropanes. In general, the isoxazoline intermediates require much higher temperatures than their isoxazolidine counterparts for the rearrangement to occur, and the yields ob-tained from the single pot cascades are somewhat low. In part, this is because some

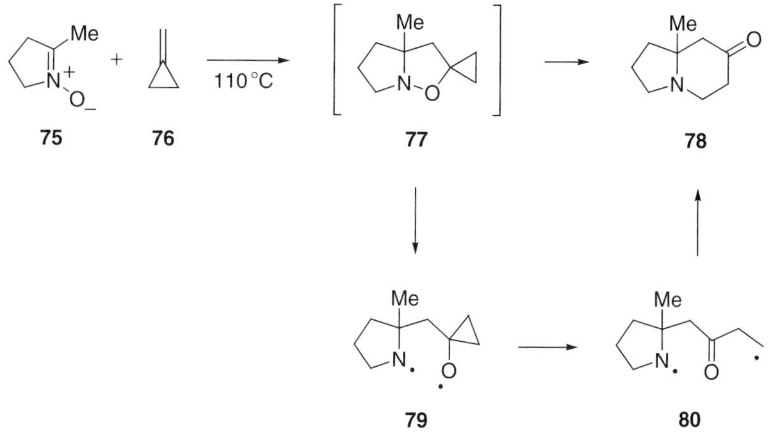

Scheme 13

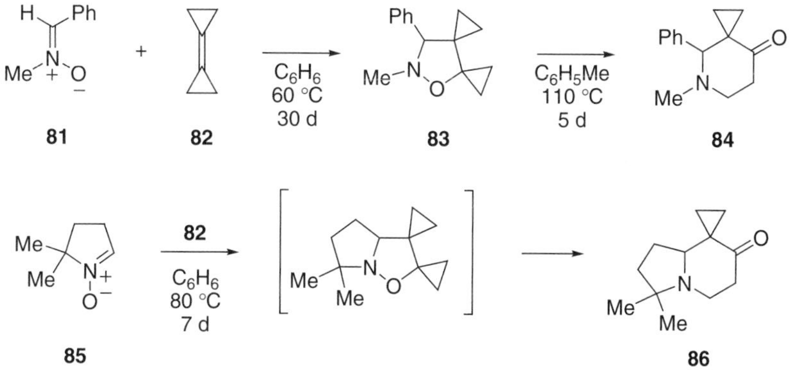

Scheme 14

nitrile oxides rearrange to the corresponding isocyanates at elevated temperatures. Another complication is that the intermediate isoxazolines can act as dipolarophiles in [3 + 2]-cycloadditions with the starting nitrile oxides. For example, nitrile oxide **87a** undergoes the cycloaddition/rearrangement cascade with **82** at 170 °C over 5 days to provide dihydrofuropyridine **88a** in only 7% yield (equation (4)) (96JOC1665). Similarly, nitrile oxide **87b** reacted with **82** under the same conditions to give **88b** in 21% yield.

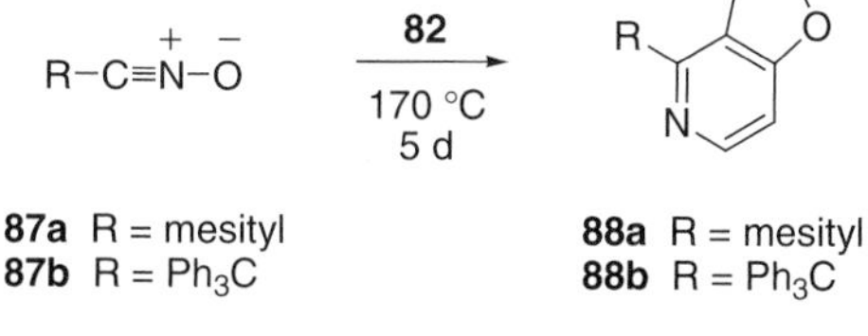

(4)

87a R = mesityl
87b R = Ph₃C

88a R = mesityl
88b R = Ph₃C

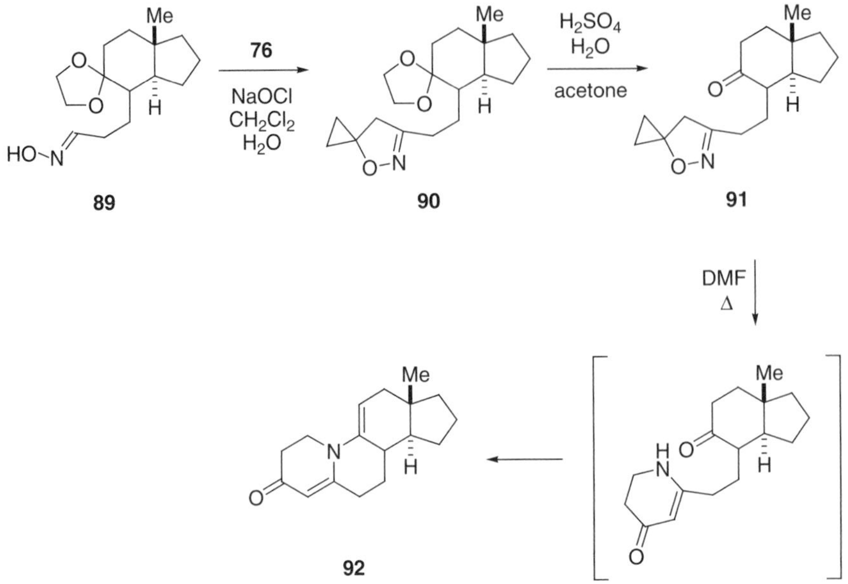

Scheme 15

Other cascade sequences have also been observed to occur from the thermolysis of isoxazolines, thereby increasing the utility of the nitrile oxide cycloaddition reaction. For example, in the context of synthesizing testosterone derivatives, Guarna and coworkers reported that the reaction of a nitrile oxide derived from oxime **89** with **76** gave isoxazoline **90** (Scheme 15) (91TL6395). Hydrolysis of the ketal moiety provided cycloadduct **91**, which was heated at reflux in DMF to furnish **92** in 30% yield.

5. Azides

The Pearson group studied a synthetically useful cascade in which azides undergo dipolar-cycloaddition with dienes followed by a thermal rearrangement to produce pyrrolidine containing products. Thus, heating azido diene **92** to 100 °C in CHCl$_3$ for 15 h afforded the pyrrolizidine derivative **93** in 90% yield (Scheme 16) (90JOC5719). The phenylsulfanyl substituent was critical both in terms of the products isolated and the rates of the reaction. The mechanism for this cascade was suggested to involve an initial dipolar cycloaddition to provide an intermediate triazole **94** in the rate-determining step. Triazole **94** then fragments by loss of nitrogen to either produce a diradical or a zwitterionic intermediate **95**. The resulting intermediate could cyclize either to **93** directly or alternatively give vinylaziridine **96**. The formation of **96** would likely be reversible, and some products isolated with other substrates suggest its involvement as an intermediate. Formally, this process can be considered as a [4 + 1]-cycloaddition of a nitrene with a 1,3-diene.

The stereoselectivity of the azide–diene cascade was examined using substrates bearing chiral centers. Choosing azido-dienes that could lead to natural product precursors, Pearson's group prepared azides **97, 99**, and **101** (Scheme 17) (90JOC5719).

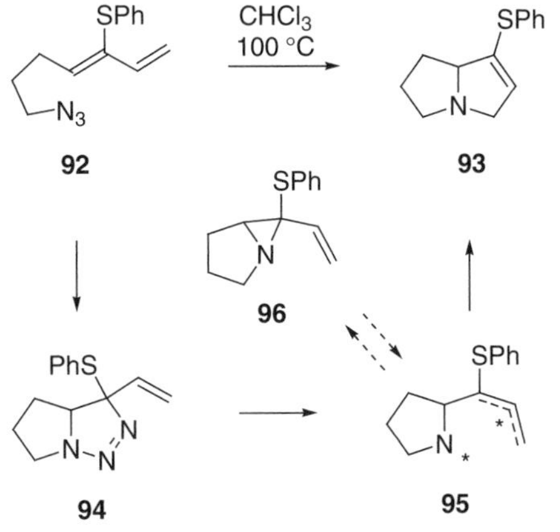

Scheme 16

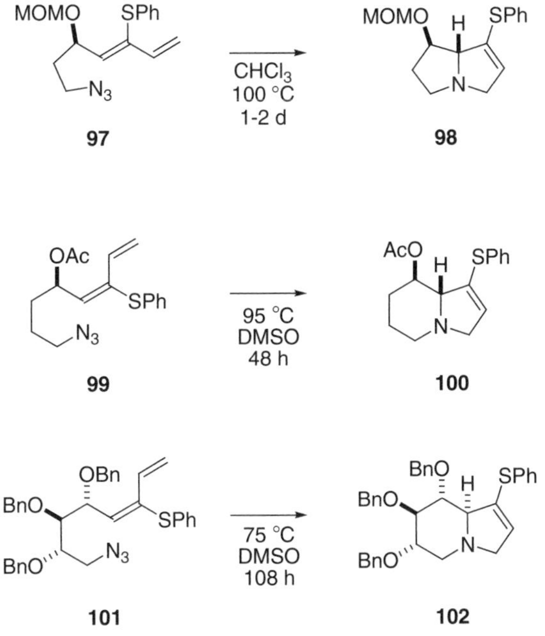

Scheme 17

Thermolysis of **97** at 100 °C in CHCl₃ for 2 days provided **98** in 74% yield. Diene **99** produced **100** in 62% yield on heating at 95 °C in DMSO for 48 h, and the poly-oxygenated azide **101** furnished **102** in 55% yield on heating at 75 °C in DMSO.

An unusual tandem Wittig/[3 + 2]-cycloaddition sequence of an azide was used for the synthesis of azasugars. Thus, acetal **103** was allowed to react with Ph₃P = CHCO₂Et to provide diazoamine **106** (Scheme 18) (96T14745, 99EJOC1407).

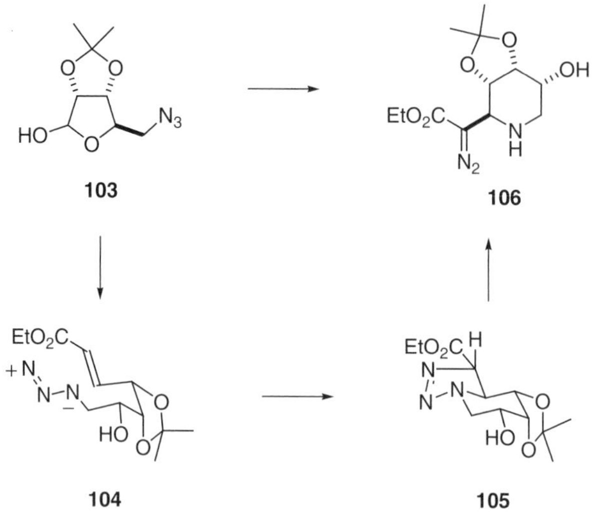

Scheme 18

The Wittig reagent presumably underwent reaction with the ring-opened form of ketal **103** to provide the intermediate enoate **104**. Dipolar cycloaddition of the pendant azide across the π-bond then produces triazole **105** that undergoes a subsequent dipolar cycloreversion reaction to give diazoamine **106**.

B. DIELS–ALDER AND RELATED PROCESSES

1. Diels–Alder

Diels–Alder cycloaddition chemistry has also been extensively exploited for many cascade reactions. A tandem Diels–Alder approach (92TL5649, 91TL2549) toward tricyclic molecules was used by Markó's group as an approach toward gibbarellic acid. Enone **107** was reacted with 2-pyrone (**108**) under high pressure to provide the Diels–Alder adduct **109** in 33% yield (Scheme 19) (93TL7305). Exposure of **109** to $TiCl_4$ effected the extrusion of CO_2 to give the intermediate cyclohexadiene **110** that underwent cycloaddition to afford **111** in 69% yield. Dihydroxylation followed by a periodate-mediated ring cleavage produced **112**, which contains the core of gibberellic acid, in 80% yield.

A tandem Claisen/Diels–Alder sequence was recently used to construct the tricyclic structure found in a series of *Garcinia* natural products, represented by morellin. On heating at 140 °C, acrylate ester **113** underwent an initial [3,3]-sigmatropic rearrangement to provide intermediate **114** (Scheme 20) (02OL909). A subsequent intramolecular Diels–Alder cycloaddition then produced **115** in 92% yield.

Several reaction cascades where amidofurans act as 4π components in Diels–Alder chemistry have recently been examined as a strategy for alkaloid synthesis (97JOC4088, 98JOC3986, 99JOC4617). For example, the thermolysis of amidofuran

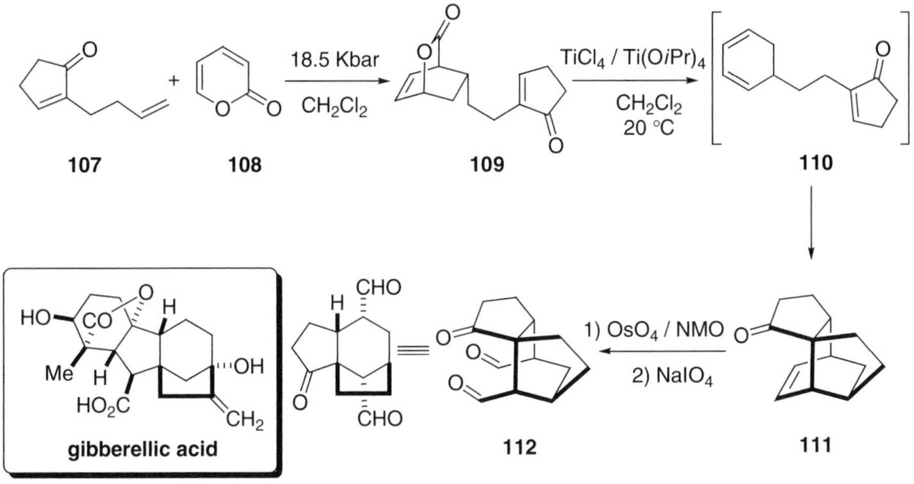

Scheme 19

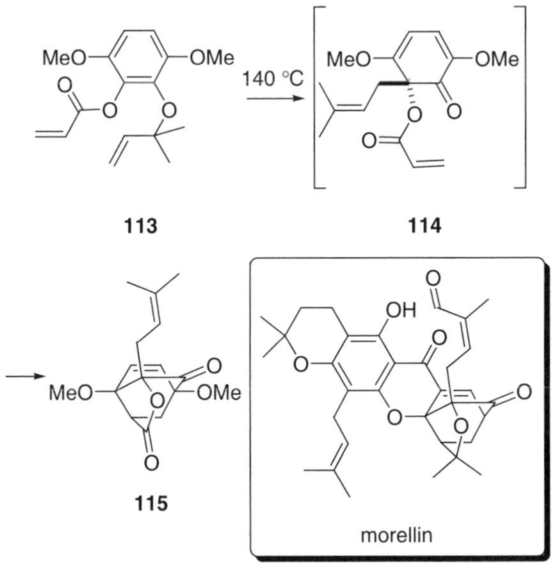

Scheme 20

116 led to the formation of **119** in 71% yield (Scheme 21) (98JOC5304). In this reaction, an intramolecular [4 + 2]-cycloaddition of **116** first provides oxabicycle **117**. Nitrogen-assisted opening of the oxygen bridge then leads to zwitterionic **118**. A 1,2-hydride shift of **118**, driven by the formation of a strong C = O double bond, results in the formation of **119**. Interestingly, if the 2π reaction partner was not geminally substituted (as in **120**), a deprotonation/dehydration cascade proceeds at a faster rate than the 1,2-hydride shift. This reaction sequence constitutes a *de novo* synthesis of the carbocyclic ring of an indole.

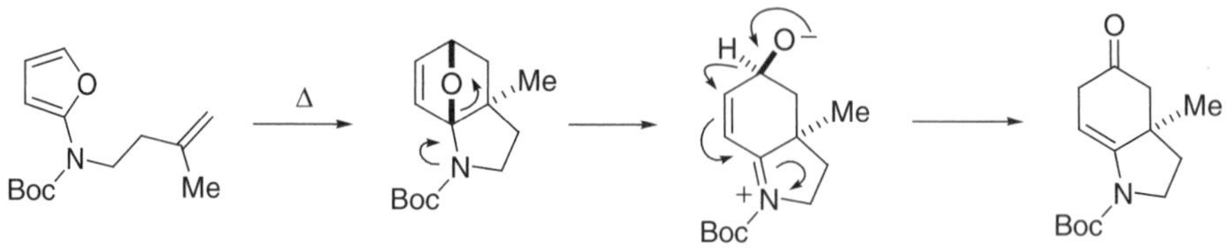

116   **117**   **118**   **119**

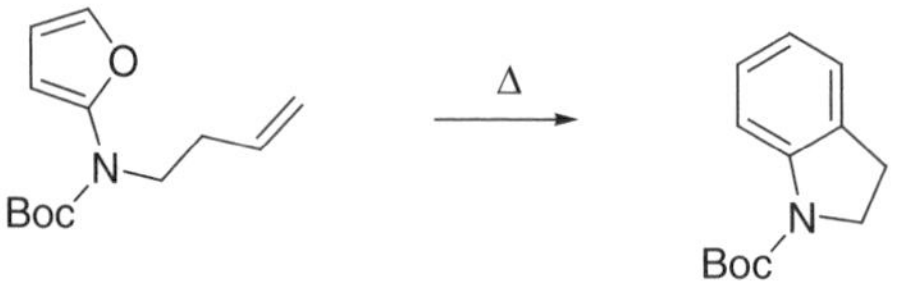

120   **121**

Scheme 21

Coupling of this indole methodology with Rawal's azadiene work (99JOC3039, 00JOC9059) led to a synthesis of Kornfeld ketone analogues with substitution patterns that are difficult to otherwise obtain. Heating a mixture of **122a** and Rawal's diene (**123**) in CH$_3$CN at reflux for 2 h furnished a 2:1-mixture of diastereomeric amines **124** that was immediately treated with HF at room temperature (rt) to unmask the enone **125** (Scheme 22) (02OL4135, 05JOC6833). The crude reaction mixture was then heated at reflux in toluene for 30 min to effect an intramolecular Diels–Alder cycloaddition. Ring opening of the resulting cycloadduct followed by dehydration provides the tricyclic ketone **126a** in 60% yield from **122a**. In a similar manner, amidofurans **122b,c** were converted to dihydroindoles **126b,c**.

Furans are also useful 4π components for tandem Ugi condensation/intra-molecular Diels–Alder cascade reactions. For example, stirring a methanolic mixture of compounds **127–129** and benzylamine at rt provided the Ugi condensation product **130** that underwent a subsequent intramolecular Diels–Alder cycloaddition to furnish **131** in 70–90% yield (Scheme 23) (99TL1851). This methodology also allowed for a solid phase synthesis by using an ArgoGel-Rink resin as the amine component, providing cycloadducts **131** (after cleavage from the resin) in *ca.* 90–95% yields.

In a related sequence, pyrrole was found to act as a 4π reaction partner leading to the formation of aza-bridged derivatives. Propionic acid (**132**) was used in the Ugi condensation with **128** and **133** to provide alkyne **134** (Scheme 24) (04JOC1207). Heating **134** at reflux temperature in toluene promoted a somewhat rare intramolecular Diels–Alder reaction of a pyrrole, giving rise to the formation of intermediate **135**. Ring opening of the nitrogen bridge in **135** produced isoindolone **136** in 65% yield.

A novel tandem Pictet–Spengler/intramolecular Diels–Alder sequence has been used to prepare carboline derivatives. Reaction of imine **137** with maleic anhydride in CH$_2$Cl$_2$ provided cycloadduct **140** in 60–80% yields (Scheme 25) (02TL203). The

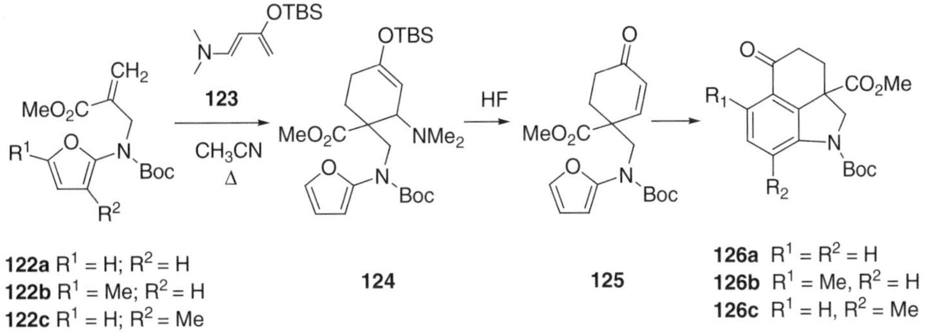
OTBS
123
CH3CN
Δ
HF
OTBS
NMe2
MeO2C
CH2
Boc
R1
R2
O
CO2Me
R1
R2
122a R1 = H; R2 = H
122b R1 = Me; R2 = H
122c R1 = H; R2 = Me
124
125
126a R1 = R2 = H
126b R1 = Me, R2 = H
126c R1 = H, R2 = Me

Scheme 22

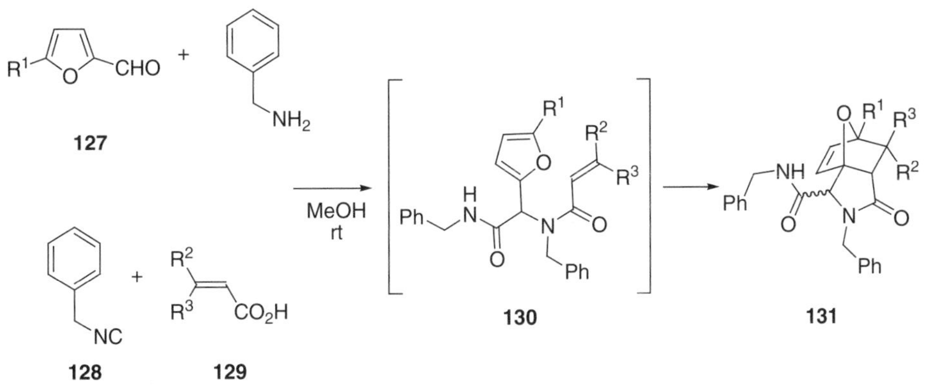
CHO
NH2
127
NC
CO2H
128
129
MeOH
rt
Ph
130
131

Scheme 23

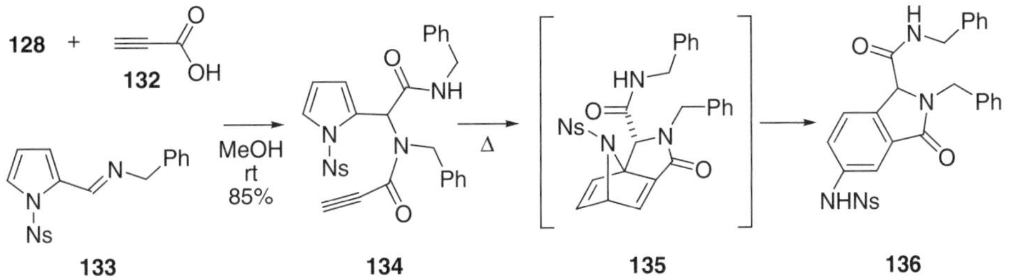
128
132
OH
Ns
Ph
MeOH
rt
85%
Δ
NHNs
133
134
135
136

Scheme 24

reaction proceeds by acylation of the imine with the available anhydride to first produce iminium ion **138** which then cyclizes with the indole ring to give **139**. An intramolecular Diels–Alder reaction of the furan with the proximal π-bond ultimately provides **140**.

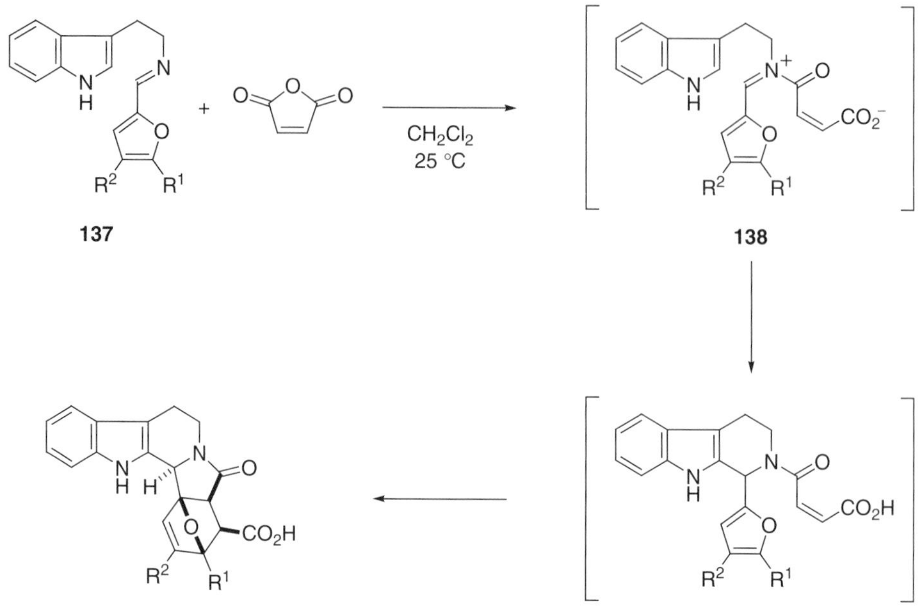

Scheme 25

In another example of multi-component reactions involving Diels–Alder cyclo-adducts, Zhu and coworkers found that a mixture of an amine, an aldehyde, and isonitrile **141** led to oxazole **142** when the reaction was carried out in the presence of a mild acid catalyst (Scheme 26) (02JA2560). Further reaction of **142** with a variety of α,β-unsaturated acid chlorides produced Diels–Alder substrates **143** that under-went cyclization to give bridged ethers **144**. Ring opening with concomitant loss of morpholine afforded **145** that rapidly tautomerized to give **146** in 32–75% yield.

Taylor and Raw recently designed a tethered imine–enamine cascade sequence that converts 1,2,4-triazenes into substituted pyridines. In the presence of molecular sieves, N-methylethylenediamine (**147**) underwent condensation with excess cyclic ketone **148** ($n = 1$–4) to give imine–enamine **150** (04CC508). The enamine portion of the molecule then participated in an inverse-demand Diels–Alder cycloaddition reaction with **149** to provide intermediate **151**. Cycloreversion of **151** with loss of N_2 then gave **152** in which the tertiary amino group underwent addition to the adjacent imine functionality to afford zwiterionic **153**. Finally, an intramolecular Cope elimination produced **154** in 74–100% yield. Several other triazines were also shown to participate in this novel cascade (Scheme 27).

2. *Hetero Diels–Alder*

A domino Knoevenagel/hetero-Diels–Alder cycloaddition cascade was developed by Tietze (96CR115) and has continued to attract considerable attention. For

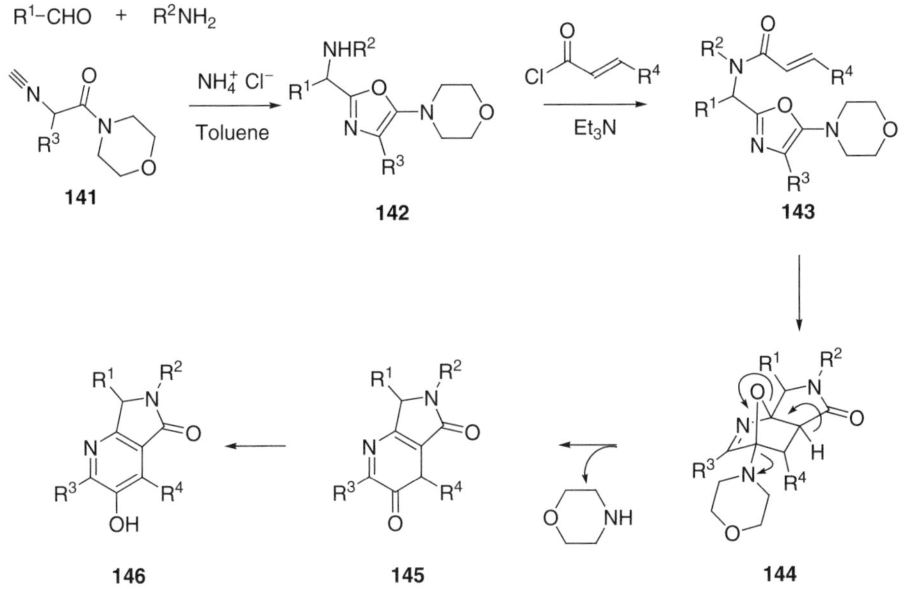

Scheme 26

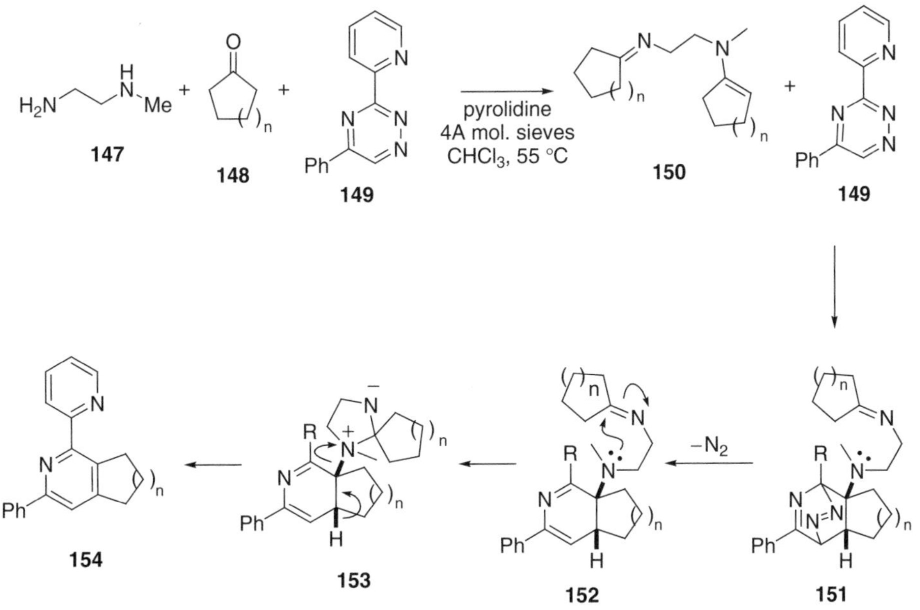

Scheme 27

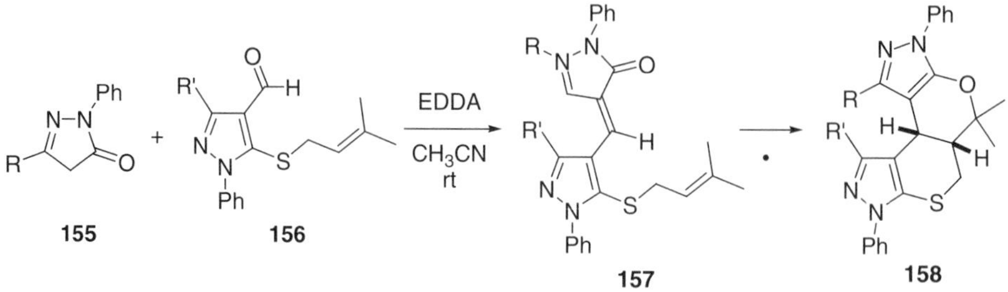

Scheme 28

example, variously substituted pyrazolones **155** and thio substituted heterocycles of type **156** were condensed to furnish novel heterocyclic structures (Scheme 28) (02T531). The reaction of **155** and **156** in the presence of EDDA (ethylene diammonium diacetate) at rt in CH_3CN gave **157**. On heating the reaction mixture at reflux, hetero-Diels–Alder cycloadducts such as **158** could be isolated in good yields (81–87%).

A related domino process was used for the synthesis of coumarin derivatives that contain sugar-fused moieties. In the presence of NaOAc and HOAc, the reaction of coumarin **159** with prenylated sugar aldehyde **160** produced **161** in 82% yield (equation (5)) (04TL3493). A variety of 1,3-dicarbonyl compounds also participated in the reaction and provided tandem condensation/cycloaddition products in good yields (70–80%).

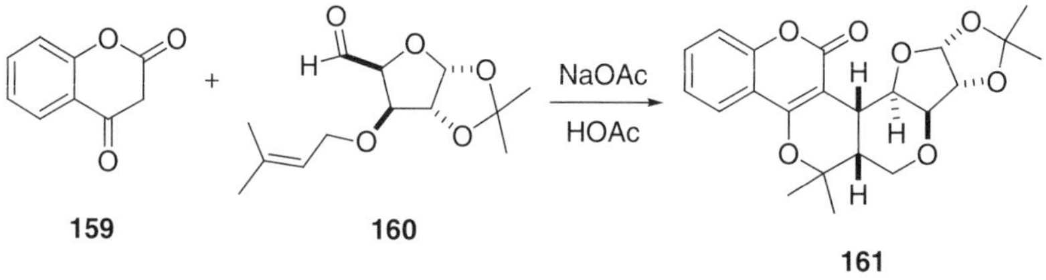

$$(5)$$

An interesting example of a formal [4 + 2]-cycloaddition has been found to occur on condensing N-substituted anilines with ω-unsaturated aldehydes in the presence of Lewis Acids. In this study, N-phenylamines **162** underwent condensation with **163** to provide acridine products **165** in ca. 60–75% yields (Scheme 29) (96CC1213). The intermediate iminium ions **164** that are first formed either participate in a concerted [4 + 2]-cycloaddition (followed by proton transfer) or else undergo polar addition to the pendant alkene by addition of the resultant benzylic carbocation onto the aniline ring.

3. Nitroalkene [4+2]/[3+2]-Cycloadditions

The Denmark laboratory has developed an elegant tandem [4 + 2]/[3 + 2]-cyclo-addition strategy for the synthesis of a variety of alkaloid natural products (96CR137). Nitroethylene (**166**) readily undergoes a Lewis acid promoted cyclo-addition with vinyl ethers that contain a chiral auxiliary group to give nitronates **168**

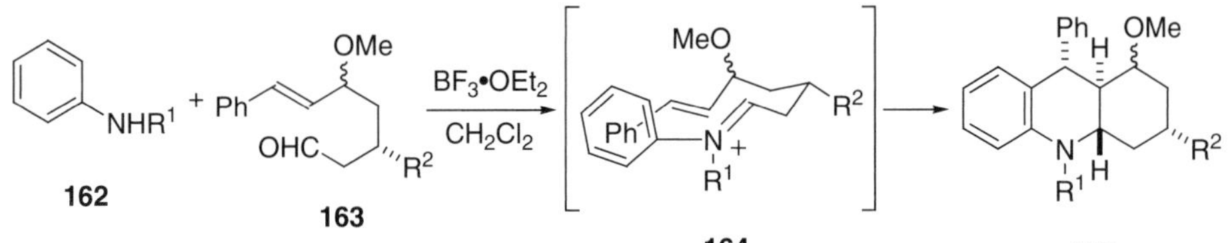

Scheme 29

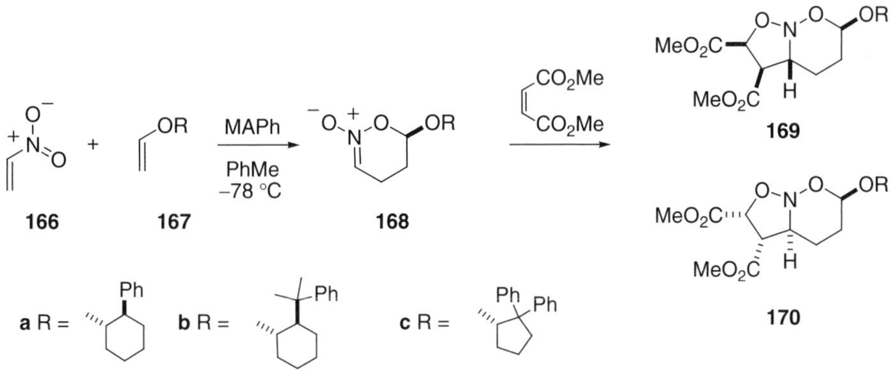

MAPh = methylaluminum bis(2,5-diphenoxide)

Scheme 30

with good stereoselectivity (Scheme 30) (98JOC3045). For example, vinyl ether **167a** provided **168a** with a 20:11 diastereoselectivity, whereas **167c** afforded **168c** with > 50:1 selectivity. The initially formed nitronates **168** were unstable to silica gel chromatography, but the crude products underwent a ready [3 + 2]-cycloaddition reaction with electron deficient dipolarophiles. In these reactions, dimethyl maleate reacted with **168** to provide 6:1-mixtures of **169** and **170** in 84–89% yield.

This tandem intermolecular [4 + 2]/intermolecular [3 + 2]-cycloaddition strategy was successfully applied to the synthesis of (+)-casuarine. In this synthesis, nitroalkene **171** was allowed to react with enol ether **172** in the presence of SnCl₄ at –78 °C to give intermediate nitronate **173** (Scheme 31) (00JOC2875). A dipolar cycloaddition of **173** with **174** provided **175** in 76% yield as a mixture predominating in the stereoisomer shown in Scheme 31. Stereoselective reduction of the ketone moiety in **175**, followed by conversion to the corresponding mesylate, gave **176** in 84% yield. Exposure of **176** to Raney nickel under high pressure afforded pyrrolizidine **177** in 64% yield and with 98% ee. Oxidative removal of the silyl group produced (+)-casuarine (**178**) in 84% yield.

The Denmark group has also developed several interesting variants of this sequence. For example, the intermolecular [4 + 2]/intramolecular [3 + 2]-cycloaddition cascade (98JOC1604, 97JA125, 97JOC7086) was used to construct several natural

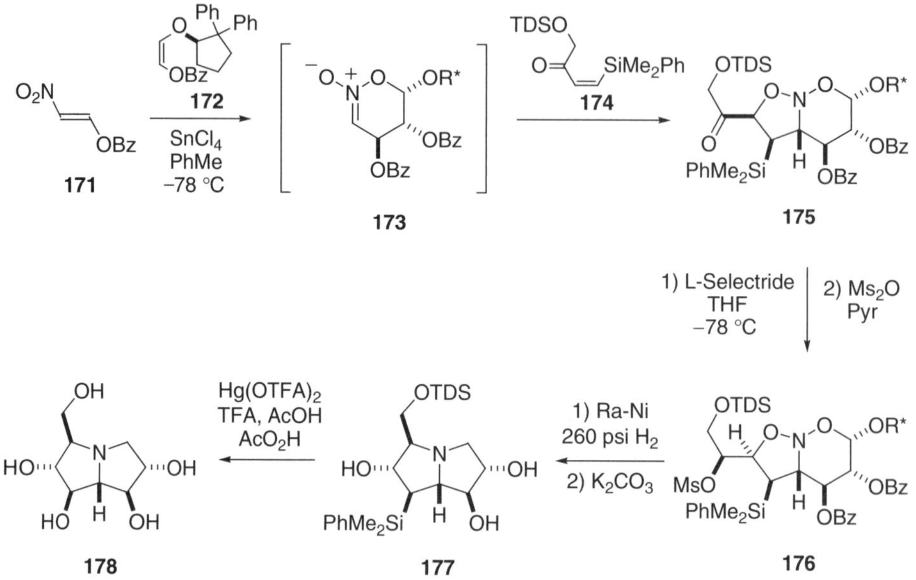

Scheme 31

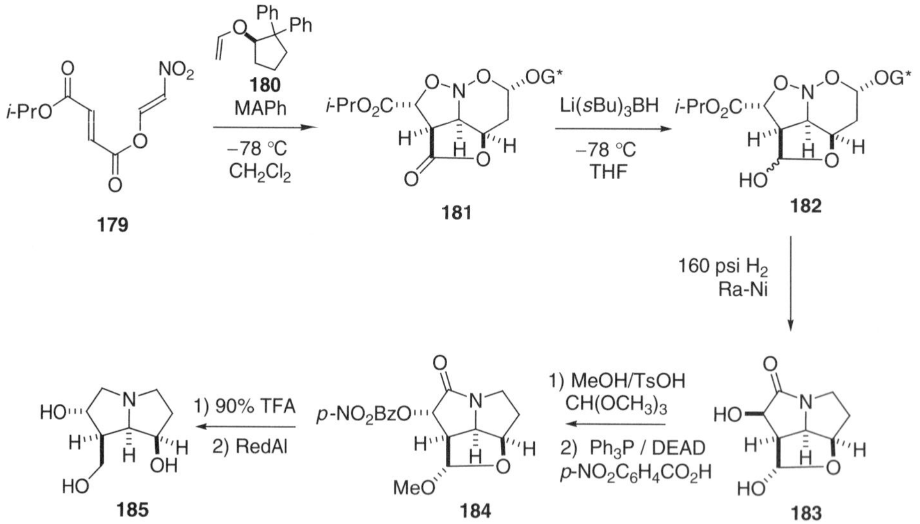

Scheme 32

products, such as (–)-rosmarinecine (96JA8266). For this particular natural product, the Lewis acid promoted reaction of nitroalkene **179** with chiral enol ether **180** produced nitrosoacetal **181** in 94% yield and with excellent stereoselectivity (25:1— exo:endo) as shown in Scheme 32. Reduction of the lactone moiety afforded lactol

182 in 91% yield. Exposing **182** to Raney nickel under H_2 (160 psi) gave the bicyclic lactam **183** in 64% yield. The chiral auxiliary could be recovered in 98% yield. Protection of the lactol followed by reaction with p-nitrobenzoic acid, Ph_3P, and DEAD provided benzoate ester **184** in 69% yield from **183**. Finally, deprotection of the lactol in compound **184** followed by exposure to RedAl produced (−)-rosmarinecine (**185**) in 57% yield for the two-step procedure.

More recently, the Denmark group reported on the tandem intramolecular [4 + 2]/intramolecular [3 + 2]-cycloaddition of nitroalkenes. Exposure of nitrone **186a,b** to $SnCl_4$ produced nitronate **187a,b** (Scheme 33) (01OL2907). Warming the crude reaction mixture containing **187a** in toluene at 80 °C for 90 min afforded **188a** as a single diastereomer in 82% overall yield. Nitronate **187b** required heating at 100 °C in toluene for 3 days in order to give **188b** as a single diastereomer, though in only 44% yield (along with 40% of **187b**). Reduction of **188a,b** with Raney nickel under a hydrogen atmosphere (160 psi) provided the fused tricycles **189a,b** in 71 and 78% yield, respectively. The selectivity of this tandem sequence is remarkable in that compounds **189a,b** each contain six contiguous stereogenic centers. Similarly, nitroalkene **190** produced **191** in 87% yield when exposed to a Lewis acid. The reaction of **191** with Raney nickel in the presence of hydrogen provided the bridged tricycle **192** in 81% yield.

In these cited examples, Denmark employed a Lewis acid (often in twofold to threefold excess) to effect the tandem cycloaddition reaction. In an alternate approach, Scheeren promoted the tandem [4 + 2]/[3 + 2]-cycloadditions by using high pressure. For example, nitroalkene **193a** reacted with methyl acrylate and ethyl vinyl ether under 15 kbar pressure to produce the bicyclic nitroso acetals **195a** and **196a** in 17 and 45% yield, respectively, after heating for 1 h (equation (6)) (97T11929). Nitrone **193b** reacted under similar conditions to produce **194b**, **195b**, and **196b** in 29,

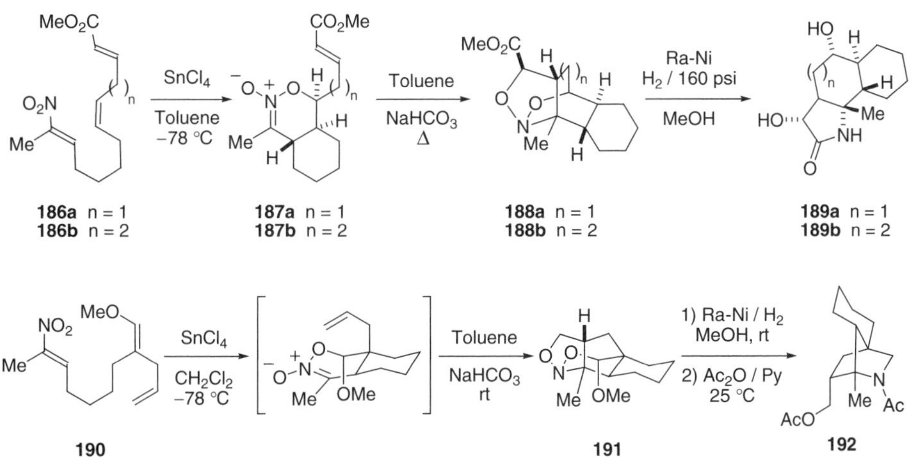

Scheme 33

18, and 29% yields, respectively.

(6)

Heteroaromatic substituted nitroalkenes also participate in this high-pressure reaction sequence. For example, the reaction of **198a–c** with **197** and methyl acrylate afforded diastereomeric mixtures of **199a–c** in 53–74% yields (Scheme 34) (01EJOC553). In contrast, **198a** reacted with **197** and *N*-phenyl maleimide to provide **200** as a single diastereomer.

A tandem cycloaddition sequence involving nitroalkenes derived from carbohydrates was recently investigated. In this study, nitroalkene **201** reacted with ethyl vinyl ether in EtOH at 25 °C to produce **202** as a single diastereomer in 89% yield (Scheme 35) (96JOC1880). A subsequent reaction of **201** with ethyl vinyl ether and 1,4-benzoquinone gave rise to a single diastereomer, whose structure was tentatively assigned as **203**, in 41% yield (99JOC1494).

4. *[4+3]-Cycloadditions*

Based on the Harmata group's earlier work using alkoxyallyl sulfone (88JOC6154, 90TL5981) and vinyl sulfoxide (91JA9861) substrates, Bai and coworkers applied a Pummerer rearrangement/intramolecular [4 + 3]-cycloaddition cascade toward the synthesis of pseudolaric acid A (Scheme 36) (99TL545, 99T13999). In their studies, sulfoxide **204** was allowed to react with TFAA in the presence of 2,6-lutidine to give cycloadduct **205** in 50% yield and with a remarkably high diastereoselectivity (>95% de). Hydrolysis of the trifluoroacetyl group delivered an advanced intermediate (**206**) that was used for the synthesis of pseudolaric acid A.

III. Rearrangements and Electrocyclizations

A. [2,3]-Sigmatropic Shifts

The [2,3]-sigmatropic rearrangement of ammonium ylides can lead to interesting heterocycles. Although it has been known for some time that the Simmons–Smith

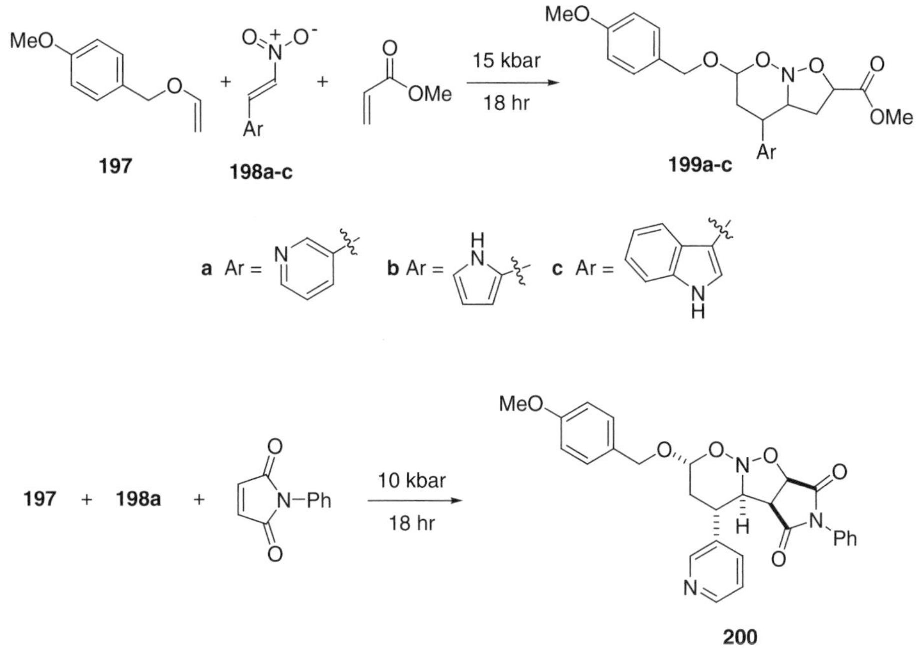

Scheme 34

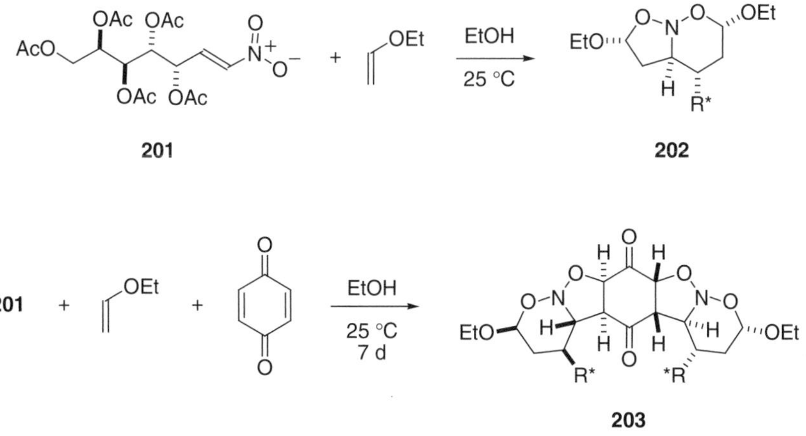

Scheme 35

reagent $(ClCH_2)_2Zn$ reacts with tertiary amines to provide quaternary ammonium salts, the chemistry of the intermediate ammonium ylide had received little attention. Recently, Aggarwal reported that the reaction of $(ICH_2)_2Zn$ with allyl amine **207a** produced the unreactive ylide **208** (03OL1757). Treatment of **208** with *n*-BuLi,

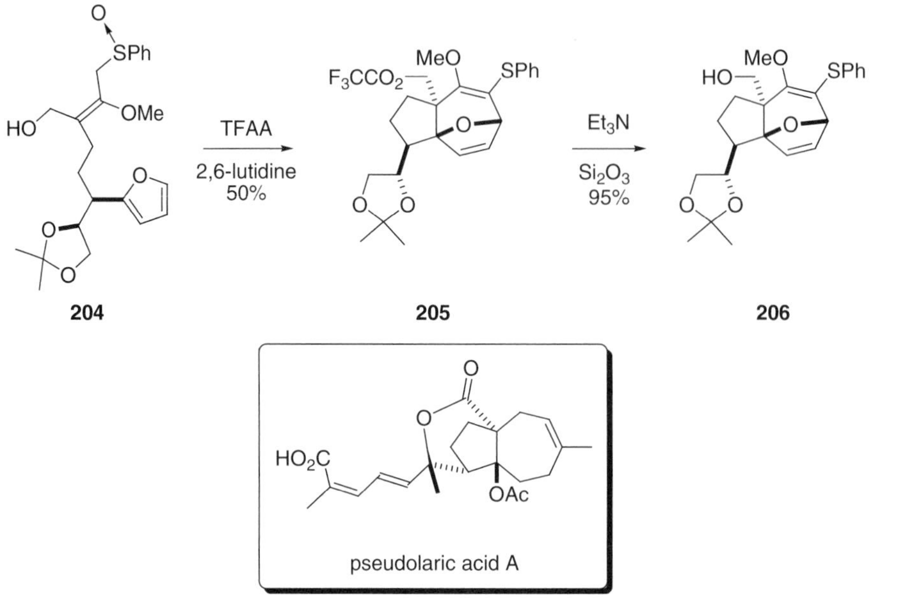

Scheme 36

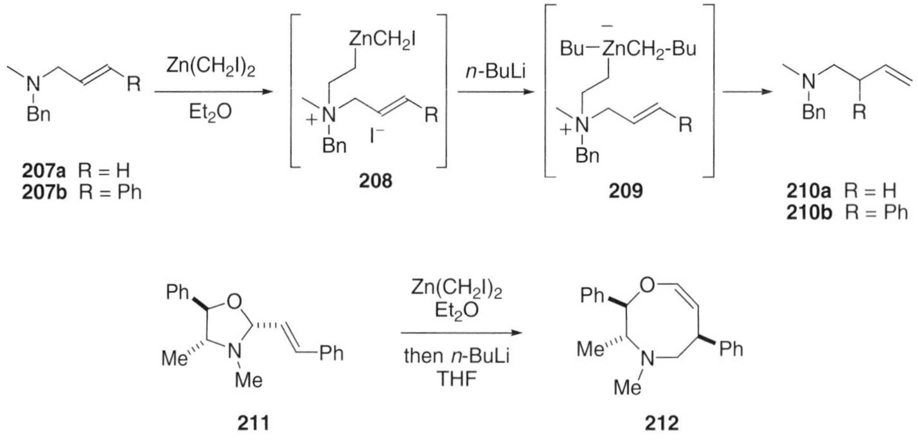

Scheme 37

however, generated an activated zincate complex **209** that rearranged to give homoallyl amine **210a** in 70% yield (Scheme 37). That a [2,3]-sigmatropic rearrangement occurs, as opposed to a Stevens rearrangement, was established by treating **207b** with (ICH₂)₂Zn followed by *n*-BuLi to produce **210b** in 76% yield. This reaction was also applied to oxazolidine **211**, furnishing the eight-membered ring **212** in 72% yield and with a >98% diastereoselectivity.

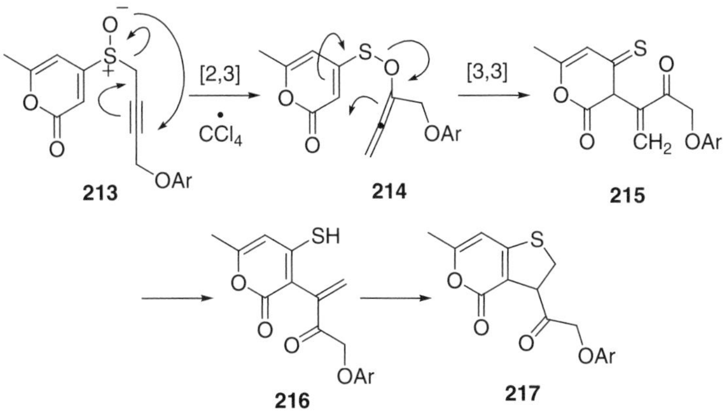

Scheme 38

A novel cascade sequence was encountered during a study of the thermolysis of propargylic sulfoxide **213** which gave the rearranged structure **217** in 60–70% yield (Scheme 38) (02T10309). The cascade was initiated by a [2,3]-rearrangement of the sulfoxide, which first produced the allene intermediate **214**. A subsequent [3,3]-rearrangement of the transient allene then gave enone **215**. Tautomerization of the thione functionality afforded **216**, and this was followed by intramolecular Michael addition to give the observed product **217**.

B. [3,3]-Sigmatropic Rearrangements

Of the various heterocycle-forming cascade reactions involving [3,3]-rearrangements, Overman's use of the Aza-Cope rearrangement/Mannich cyclization sequence certainly represents the best-known example of this methodology (88JA4329, 92ACR352). Condensation of a secondary homoallylic amine containing an allylic alcohol or ether such as **218** with aldehydes produces the intermediate iminium ion **219** (Scheme 39) (79JA1310). A Cope rearrangement then provides a new iminium ion (**220**) that contains a transient enol, which attacks the cationic center in a Mannich fashion to deliver pyrrolidines of type **221** (79CB1913).

This methodology has been the strategic core of several clever synthetic endeavors carried out by the Overman group. For example, amine **222** was converted into the pentacyclic core of the aspidosperma alkaloid family (Scheme 40) (91JA2598). Condensation of **222** with paraformaldehyde produced oxazoline **223**. Heating **223** with excess camphorsulfonic acid (CSA) effected the Aza-Cope/Mannich cyclization cascade to furnish **224** in nearly quantitative yield.

An efficient synthesis of the strychnos alkaloid skeleton was also achieved using this novel cascade process. The key transformation in this sequence occurs by heating bicyclic amine **225** with formaldehyde and CSA in CH_3CN to give **226** as a single diastereomer in 88% yield (Scheme 41) (91JA5085). Hydrolysis of the amide and subsequent condensation of the ketone with the aniline derivative provided dehydrotubifoline (**227**).

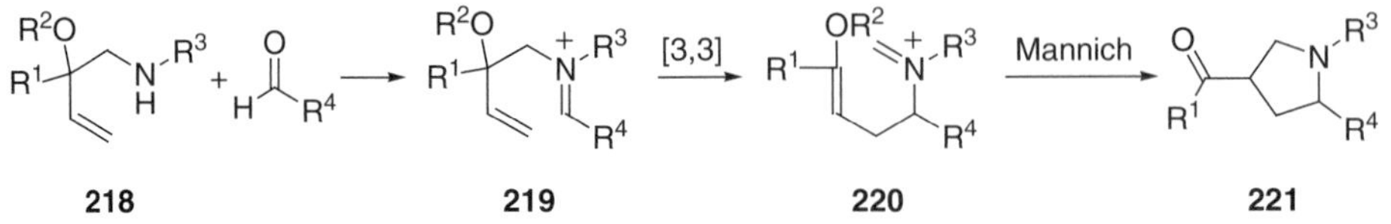

Scheme 39

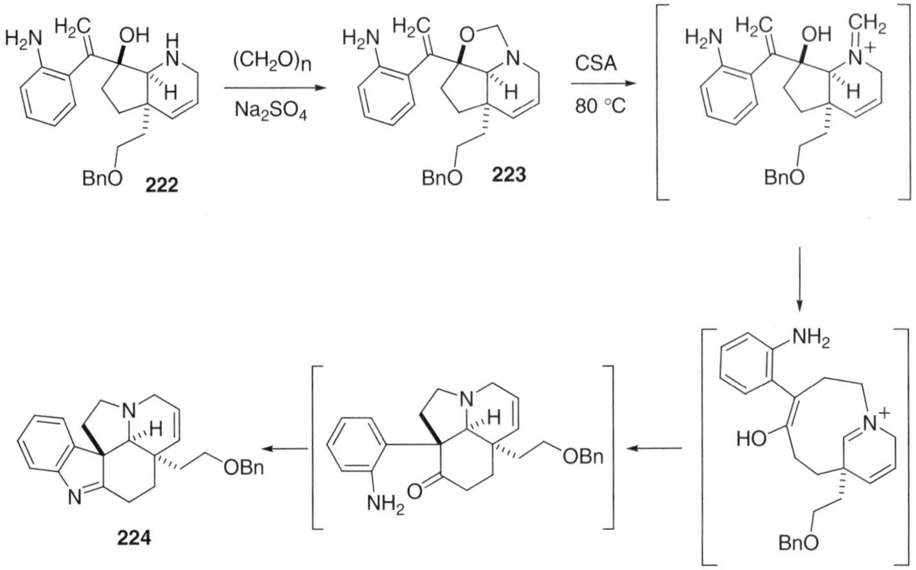

Scheme 40

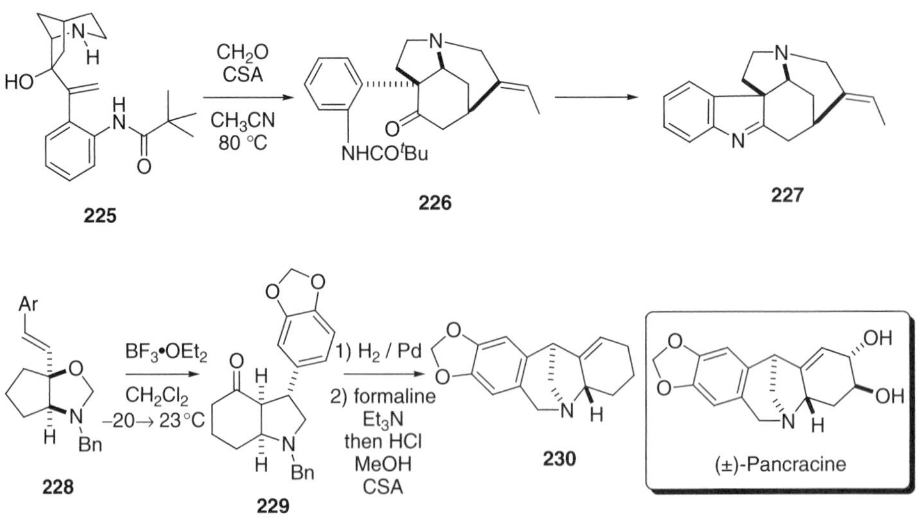

Scheme 41

Overman also used his Aza-Cope/Mannich cascade for a total synthesis of (±)-pancracine. In this particular synthesis, *N,O*-acetal **228** was allowed to react with BF$_3$•OEt$_2$, which resulted in the eventual formation of amine **229** in 97% yield (91JOC5005). Hydrogenolysis removed the *N*-benzyl group, and the resulting amine was then heated with formaline in the presence of catalytic amounts of CSA to effect a Pictet–Spengler reaction, the product of which (**230**) contains the pancracine skeleton.

More recently, Overman designed a variant of this process for the construction of angularly substituted bicyclic amines. Heating ketal **231** with TFA and dimedone (**232**) resulted in condensation with the pendant amine group to give iminium ion **233** (Scheme 42) (05OL913). The [3,3]-rearrangement resulted in the formation of a second iminium ion **234** that was intercepted by enol **232** to give the Mannich adduct **235**. Finally, elimination of the α-methylene 1,3-dione afforded amine **236**. For ease of isolation, the crude reaction mixtures were subjected to the action of benzyl chloroformate. Several examples demonstrated the versatility of this sequence in that the original ring size could be varied (*m* = 1–3) as well as the annulated ring size (*n* = 1, 2) to produce predominantly *cis*-fused bicycles **237** in *ca.* 65–95% yields.

A novel [3,3]-sigmatropic process involving an additive-Pummerer reaction that produces γ-butyrolactones by the reaction of dichloroketene with vinyl sulfoxides was developed by the Marino group (81JA7687). The oxygen atom of vinyl sulfoxide **238** first attacks dichloroketene to produce an internal salt **239** (Scheme 43). The resulting enolate present in **239** then undergoes a [3,3]-sigmatropic rearrangement to provide thionium ion intermediate **240**. Finally, the resulting carboxylate adds to the neighboring thionium ion to furnish butyrolactone **241** whose stereochemistry depends on the geometry of the starting olefin. The use of chiral sulfoxides **238** led to the enantiospecific formation of butyrolactones **241** (87S1088, 85TL5381, 94TA641).

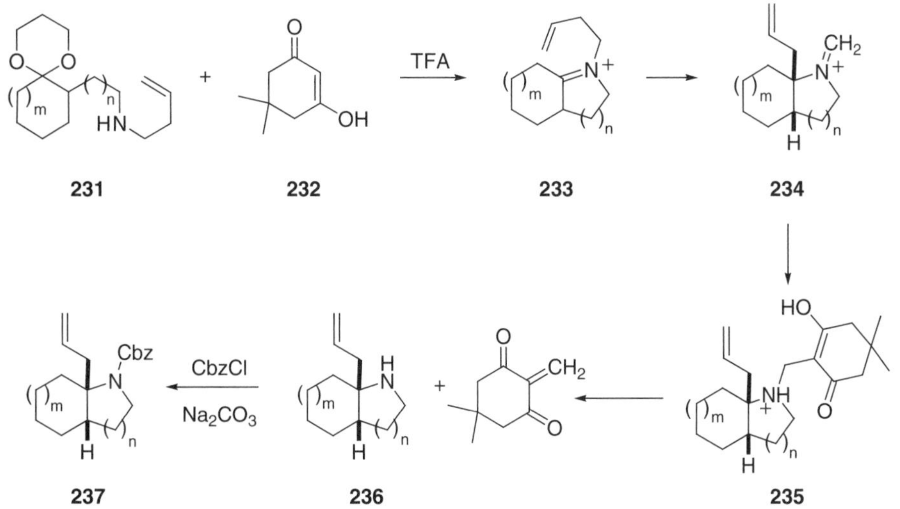

Scheme 42

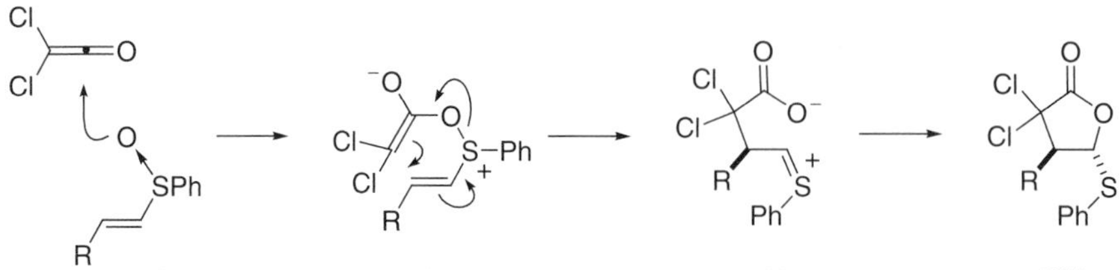

Scheme 43

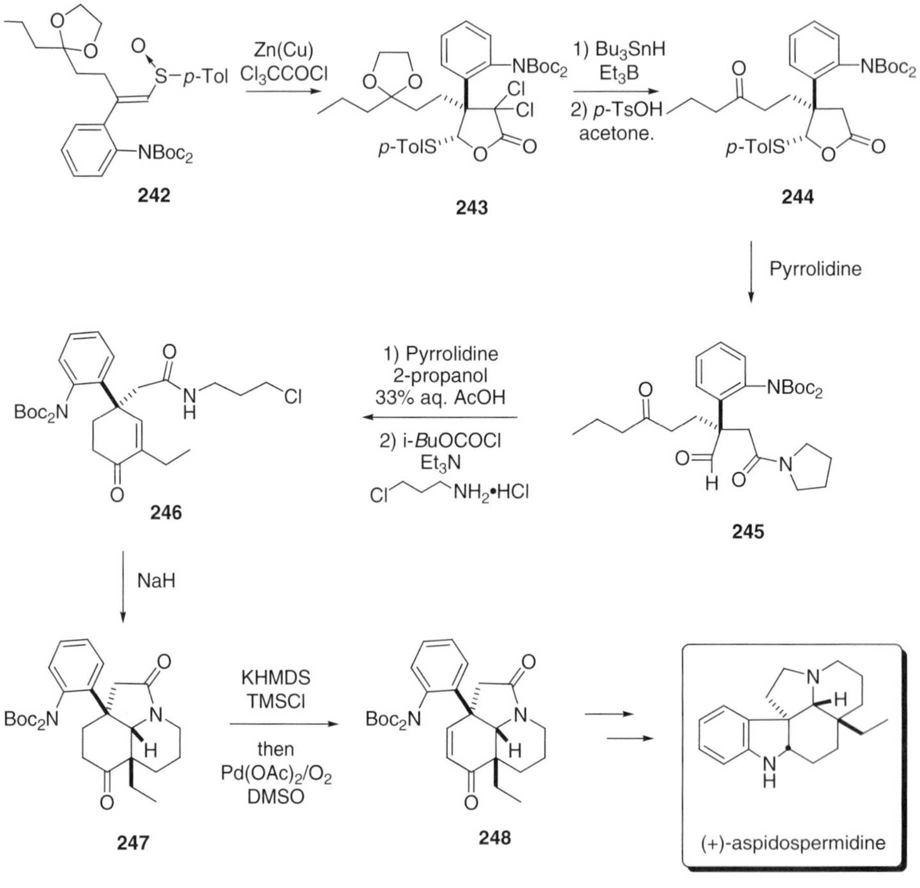

Scheme 44

This novel strategy was applied to a synthesis of (+)-aspidospermidine. In this approach, enantiomerically pure sulfoxide **242** was treated with trichloroacetyl chloride in the presence of zinc–copper couple (Zn–Cu) to give lactone **243** in 78% yield (Scheme 44) (02JA13398). Removal of the chloro substituents followed by the deprotection of the ketal afforded **244** in 96% yield. Reaction of **244** with pyrrolidine effected an *O*- to *N*-transacylation with a subsequent elimination of thiolate to furnish the amido aldehyde **245** in 86% yield. Further exposure of **245** to pyrrolidine in the presence of 33% aqueous AcOH and 2-propanol promoted an intramolecular aldol reaction and simultaneously hydrolyzed the amide group to furnish an inter-mediate carboxylic acid. Conversion of the carboxylic acid to a mixed anhydride followed by the addition of 3-chloropropylamine gave **246** in 64% yield from **245**. Exposure of **246** to NaH initiated a tandem intramolecular conjugate addition/ alkylation to provide **247** in 86% yield. Subjection of the silyl enol ether of **247** to modified Segusa oxidation conditions delivered **248** (85%), which was subsequently carried onto (+)-aspidospermidine.

Kawasaki and Sakamoto developed a [3,3]-sigmatropic rearrangement cascade to introduce angular substituents found in several indole alkaloids. In one of the cases studied, the Claisen rearrangement was first preceded by a Horner–Emmons olefi-nation of indolinone **249** to give **250** (Scheme 45) (96TL7525). Isomerization of **250** provided indole **251** that then underwent a [3,3]-rearrangement to furnish **252** in 73% yield. Reduction of the nitrile by the action of Red-Al gave **253** in 89% yield. Methylation of the imine nitrogen in the presence of NaHCO$_3$ then afforded flustramine C (**254**) in 38% yield. A more complete study on the scope of this cascade sequence has been reported (05JOC2957), and the application of domino Wittig-pericyclic reactions to bioactive heterocycles has recently been reviewed (02COC1181).

The stereoselective formation of imidazolidine thiones *via* the rearrangement of chiral thiocyanates has recently been reported. Heating allylic thiocyanates such as **255** at 80 °C produced 1:1-mixtures of diastereomeric isothiocyanates **256a** in 92%

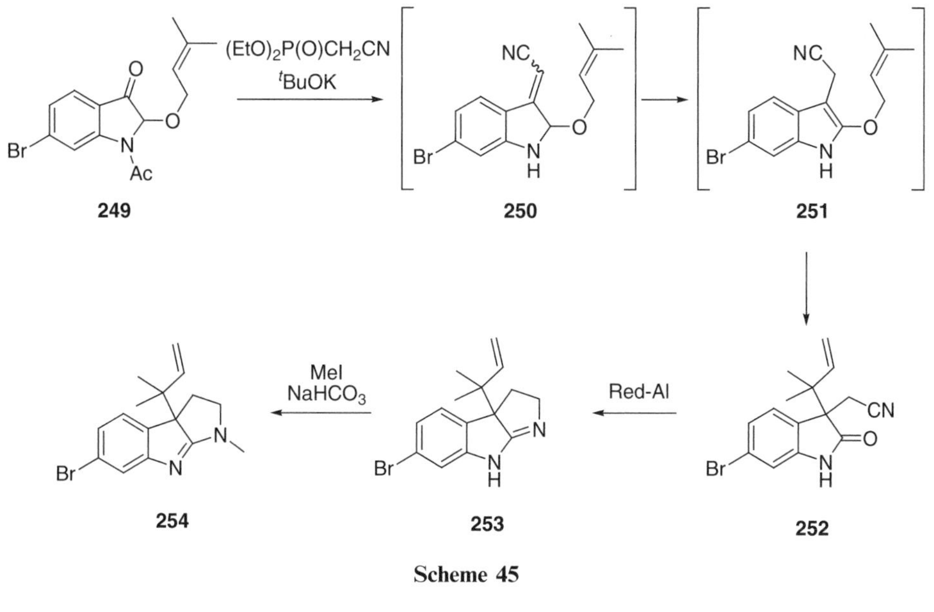

Scheme 45

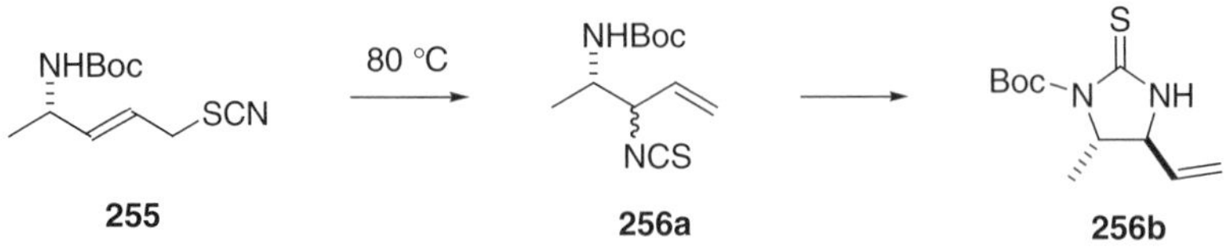

Scheme 46

yield (Scheme 46) (97TL875, 02T1611). Prolonged heating, however, led to the isolation of the cyclic thiourea **256b** as a single stereoisomer in 89% yield. Several other examples, differing in the nature of the alkyl substituent, were also reported.

C. OTHER REARRANGEMENTS

In the context of developing rapid access to thioaurone structures, De and coworkers observed an interesting 6π-electrocyclization/isomerization cascade. The reaction of sulfanyl amide **257** with an excess of LDA and cinnamaldehyde produced thioaurone **258** in 83% yield (Scheme 47) (03SL1479). On heating at 210 °C, compound **258** isomerized to give **259**, which underwent a subsequent electrocyclization reaction to produce **260**. A formal [1,3]-hydride shift then furnished the observed product **261**.

A tandem Wolff rearrangement/cyclization process has been used to synthesize benzopyran derivatives. In this sequence, α-diazo ketones **262a,b** were heated to effect a Wolff rearrangement, giving rise to ketenes **263** (Scheme 48) (98T6457). The authors propose that a [1,5]-hydride shift then provided **264**, and a subsequent cyclization gave **265a,b** in 88 and 75% yields, respectively. While thiophene derivative **266** was also found to rearrange to **267** in 75% yield, the reaction of a related furan derivative led to extensive decomposition.

A clever synthetic approach toward the synthesis of cephalotaxine relies on an asymmetric Beckman rearrangement/allyl silane-terminated cation-cyclization cascade. In these studies, Schinzer and coworkers found that the reaction of racemic **268** with DIBAL-H produced **269** in 36% yield (Scheme 49) (94SL375). The racemic oxime ether **270a** was converted into **271** in 23% yield under similar conditions. By changing the size of the silicon group (i.e. **270b**), the yield was increased to 41%. Non-racemic (S)-**270b** was synthesized using a chiral chromium–arene complex and afforded (S,R)-**271** in 55% yield and with 81% ee on exposure to excess DIBAL-H (97SL632).

A tandem anionic cyclization/Dimroth rearrangement was employed for the preparation of γ-lactams containing alkylidene substituents (02EJOC221). In this cascade sequence, the dianion of ethyl acetoacetate (**272**) reacted with **273** to provide furan derivative **274** (Scheme 50) which underwent a subsequent rearrangement to give **275** in 56% yield (04EJOC1897).

The rearrangement of *bis*-allenyl disulfides provides an interesting route to prepare fused thieno[3,4-c]thiophenes. Thus, Braverman reported that **276** reacted with lithium methoxide to give **280** in 70% yield (Scheme 51) (90T5759). Presumably,

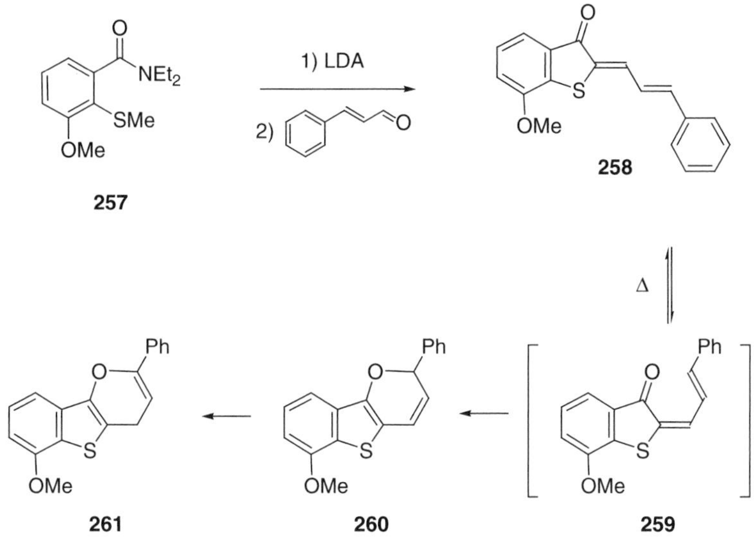

Scheme 47

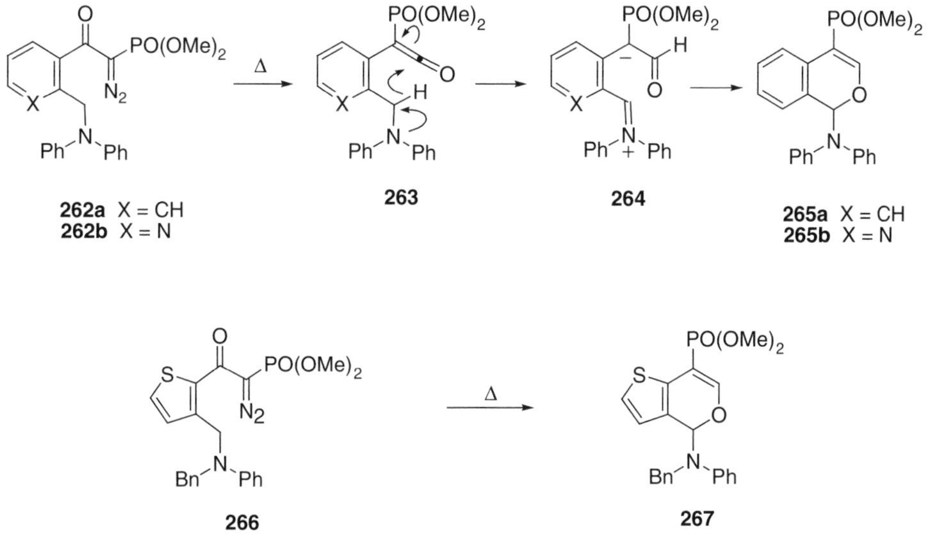

Scheme 48

allene **276** first dimerized under the reaction conditions to generate disulfide **277**. Cyclization of **277** would then produce a diradical species **278** that fragments into **279a**. A further cyclization of the *E*-isomer **279b** nicely accounts for the formation of **280**. The analogous diselenide underwent a related reaction to give the corresponding selenophene derivative.

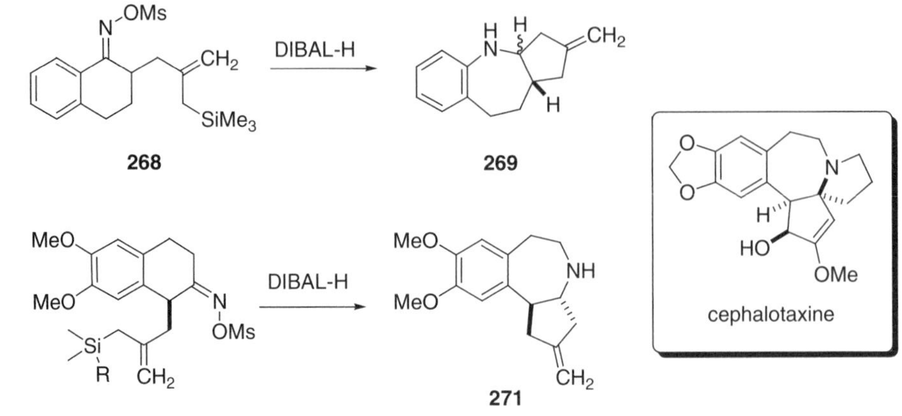

268

269

270a R = Me
270b R = Ph

271

Scheme 49

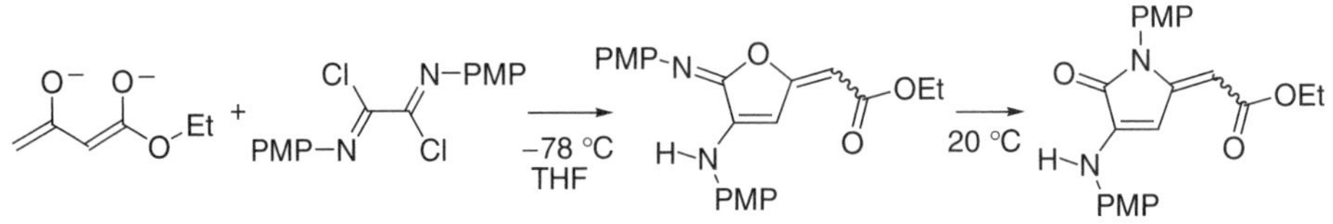

272 273 274 275

Scheme 50

IV. Cation-Promoted Cyclization Cascades

A. Nitrogen Stabilized Carbocations

The Mannich reaction is a very common process that occurs in many tandem reaction sequences. For example, the Overman Aza-Cope cascade sequence is terminated by a Mannich reaction (*cf.* Scheme 35). Several groups have used variants of the Mannich reaction to initiate cascades that lead to the formation of heterocyclic molecules. For example, the Lewis acid-catalyzed intermolecular vinylogous Mannich reaction (01T3221) of silyloxy furan **281** with nitrone **282** produced a diastereomeric mixture (49:3:42:6) of azabicycles **284a–d** in 97% combined yield (Scheme 52) (96TA1059). These products arose from an intramolecular Michael addition of the initially formed oxonium ion **283**.

A Mannich/Michael reaction sequence was used by Waldman for the formation of several piperidone derivatives. The reaction of **285** with **286** in the presence of a

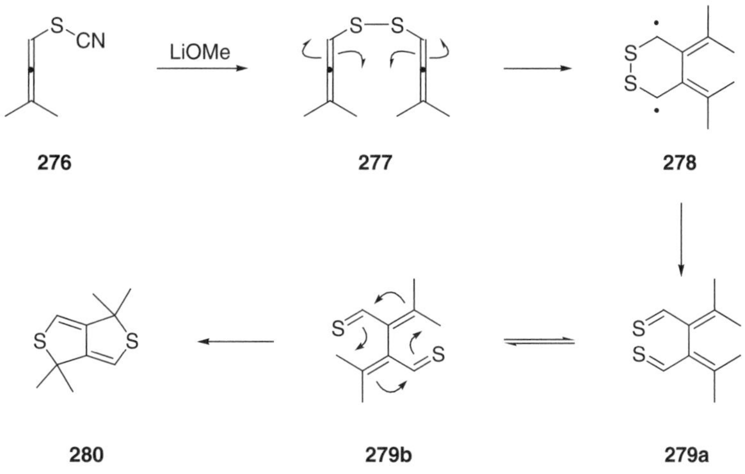

Scheme 51

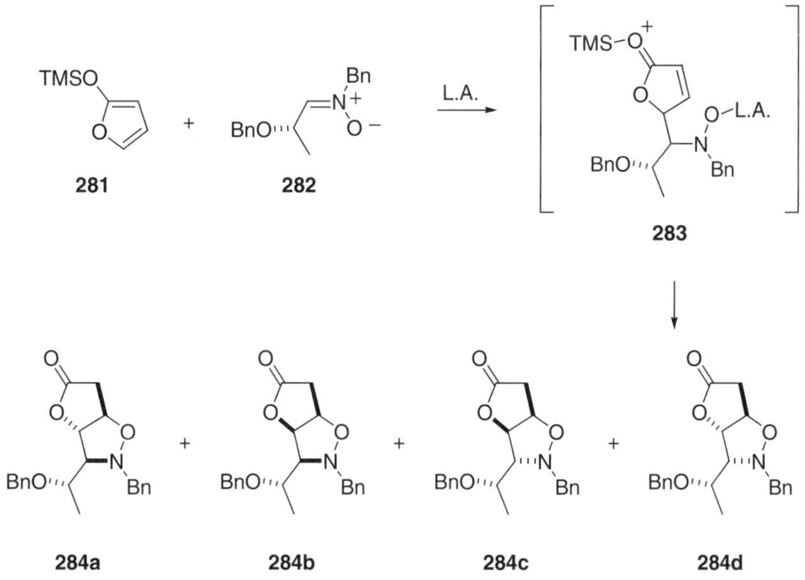

Scheme 52

variety of Lewis acids produced mixtures of **287a,b** in 84% yield (Scheme 53) (91TA1231, 97CEJ143).

Using diene **288** and imine **289**, the tandem Mannich/Michael reaction sequence afforded the vinylogous amide **290** in 66% yield (97TL2829). Imines derived from other aldehydes were also studied, providing derivatives of **290** in moderate yields

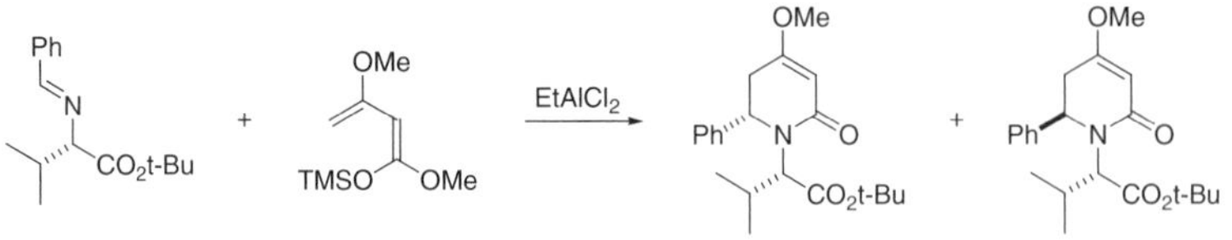

Scheme 53

(*ca.* 40–65%). The palladium catalyzed cyclization of **290** furnished tricyclic benzoquinolizine **291** in 76% yield.

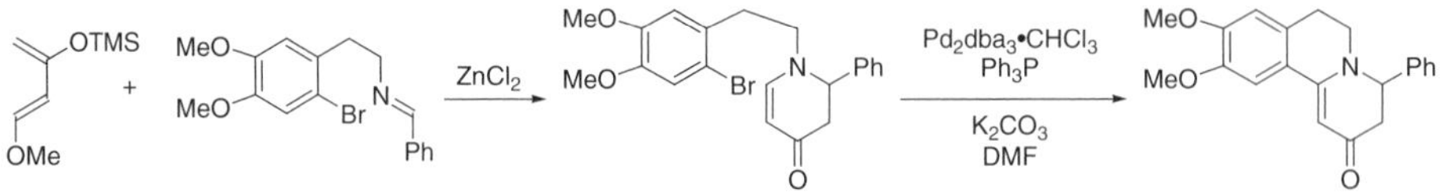

In addition to their use in Mannich (and variant) reactions, iminium ions are useful for other cationic type cyclizations. Corey employed a novel tandem iminium ion cyclization as part of an elegant cascade used for the synthesis of aspidophytine. The reaction of tryptamine **292** and dialdehyde **293** in CH_3CN at ambient temperature afforded the pentacyclic skeleton of the alkaloid (**296**; Scheme 54) (99JA6771). Condensation of the free amino functionality of **292** with the dialdehyde produced a dihydropyridinium intermediate **294** that then cyclized onto the indole π-bond to give **295**. The iminium ion so produced underwent a second cyclization with the tethered allylsilane moiety to give **296**. Protonation of the enamine in **296** provided still another iminium ion (**297**) that was then reduced with $NaCNBH_3$ to furnish **298** in 66% yield. All of the above reactions could be made to occur in a single pot.

B. Pummerer Cascade Reactions

The combination of a Pummerer-based reaction (04CR2401) followed by an *N*-acyliminium ion cyclization in tandem to form pyrrolidine-containing ring systems represents a unique method to synthesize heterocycles. In a typical example from the Padwa laboratory, enamide **299** was treated with *p*-TsOH in boiling benzene to produce thionium ion **300**. A subsequent Nazarov-like ring closure of **300** furnished iminium ion **301**. Finally, an intramolecular Pictet–Spengler reaction with the pendant aromatic ring of **301** provided **302** as a single diastereomer in 78% yield (Scheme 56) (98JOC6778, 02JOC5928). The stereochemistry of **302** was established by X-ray crystallographic analysis and is compatible with a *conrotatory* ring closure.

Other π-bonds were also found to efficiently participate in the Pummerer/Mannich cascade. For example, allylsilane **303** gave bicycle **304** in 61% yield when heated with *p*-TsOH (equation (7)). The terminal alkene present in **305** cyclized to give **306**,

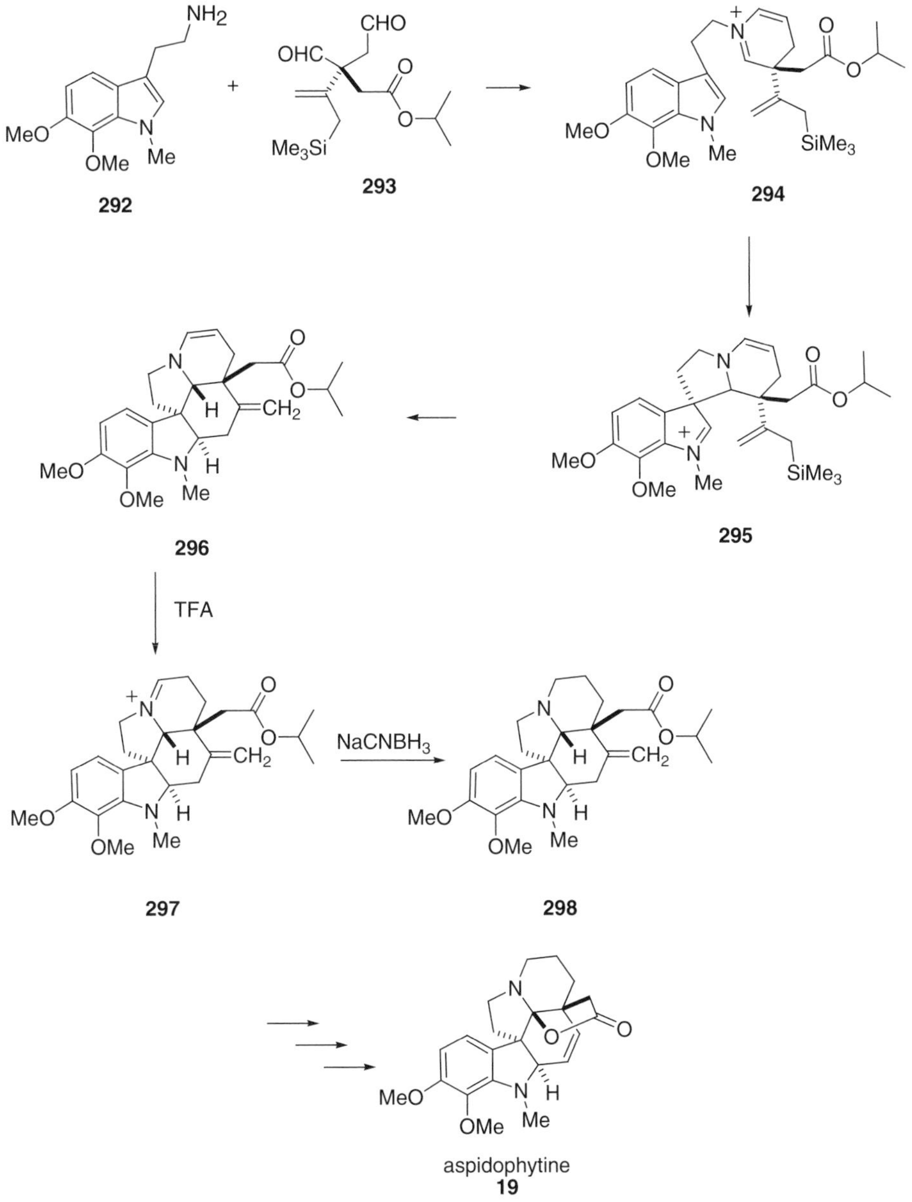

Scheme 54

wherein the resultant secondary carbocation was captured by the sulfonate anion in 80% yield (equation (8)). In each case, only one diastereomer was isolated, suggesting that a concerted 4π-electrocyclization reaction occurs from the intermediate thionium ion.

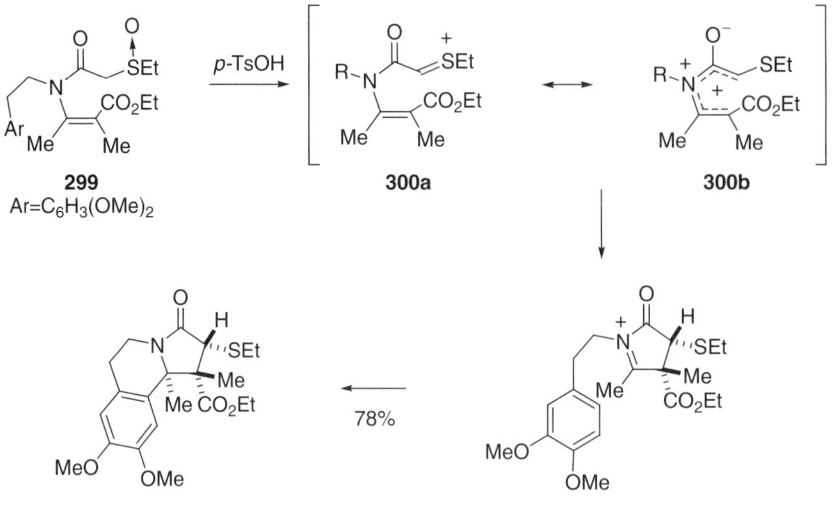

299
Ar=C$_6$H$_3$(OMe)$_2$

300a

300b

302

78%

301

Scheme 56

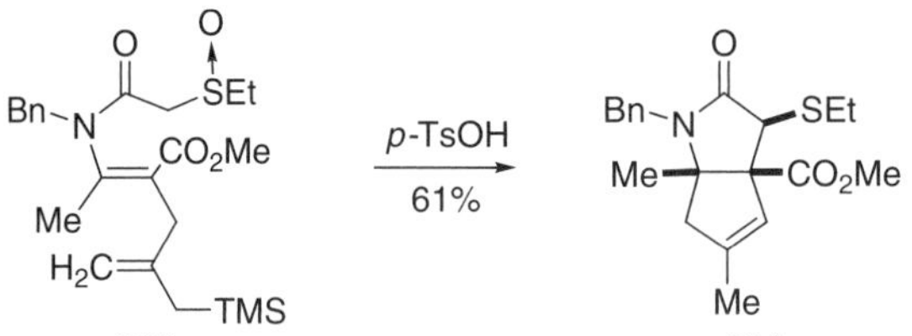

303

p-TsOH
61%

304

(7)

305

p-TsOH
80%

306

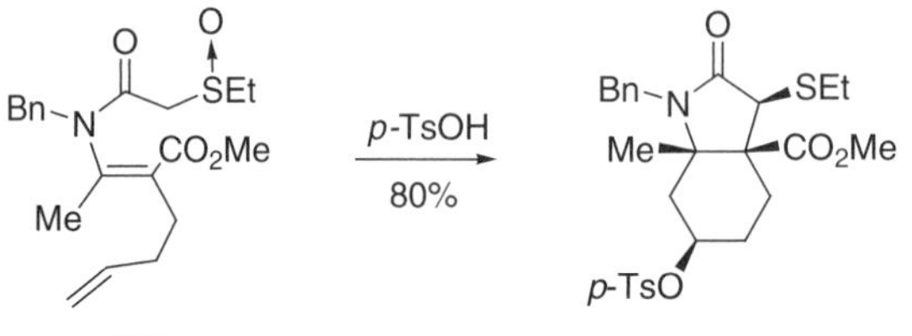

(8)

This methodology was employed for the synthesis of the reported structure of the alkaloid jamtine (02OL715, 02JOC929). The key sulfoxide intermediate **307** was heated with CSA to produce several tricyclic products (98% yield) as a mixture (5:2:1:1) of diastereomers in which **308** predominated (Scheme 57). The stereochemistry of **308** was secured by X-ray crystallographic analysis and is consistent with a Nazarov-type *conrotatory* 4π-electrocyclization followed by attack of the nucleophilically disposed aromatic ring from the least hindered side of the intermediate iminium ion. Reaction of α-ethylthio amide **308** with NaH effected an intramolecular alkylation to provide tetracycle **309**.

As part of their investigations dealing with *N,S*-fused polycyclic ring systems, Daich and coworkers reported the use of a tandem Pummerer/*N*-acyliminium ion

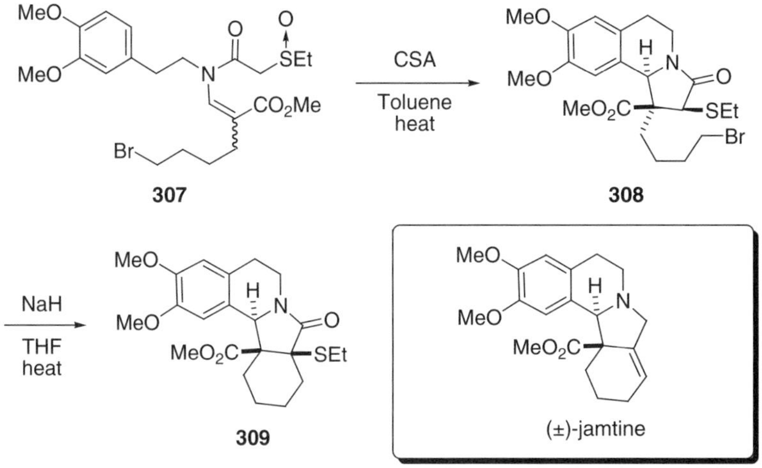

Scheme 57

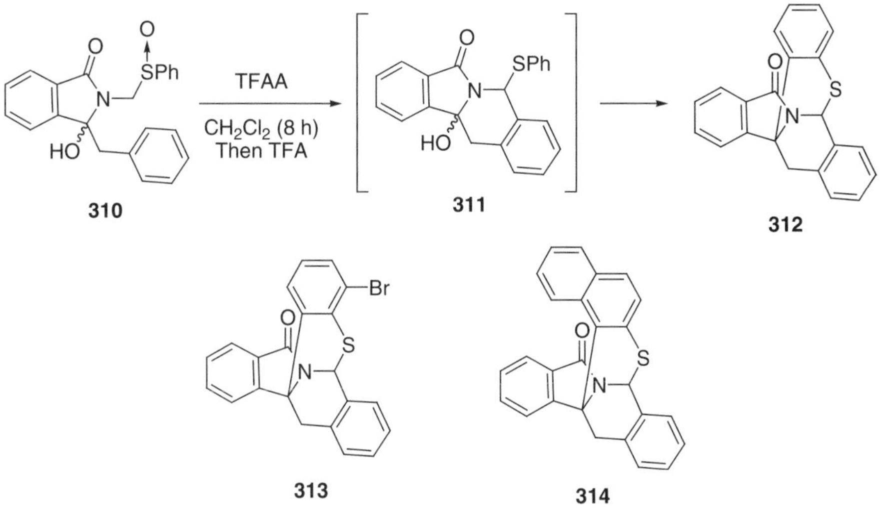

Scheme 58

cyclization to construct interesting isoquinolinone structures. Thus, treatment of sulfoxide **310** with TFAA in CH$_2$Cl$_2$ at rt for 8 h followed by the addition of TFA produced **312** in 42% yield through the intermediacy of **311** (Scheme 58) (00OL1201). By conducting the reaction under buffered conditions (TFAA and pyridine), compound **311** could be isolated in 56% yield. An N-acyliminium ion intermediate was then generated by treating **311** with neat TFA, and a subsequent cyclization of the resulting cationic intermediate gave **312** in 58% yield. Other arylthio groups were also studied, with compounds **313** and **314** being obtained from the TFAA/TFA conditions in 62 and 41% yields, respectively.

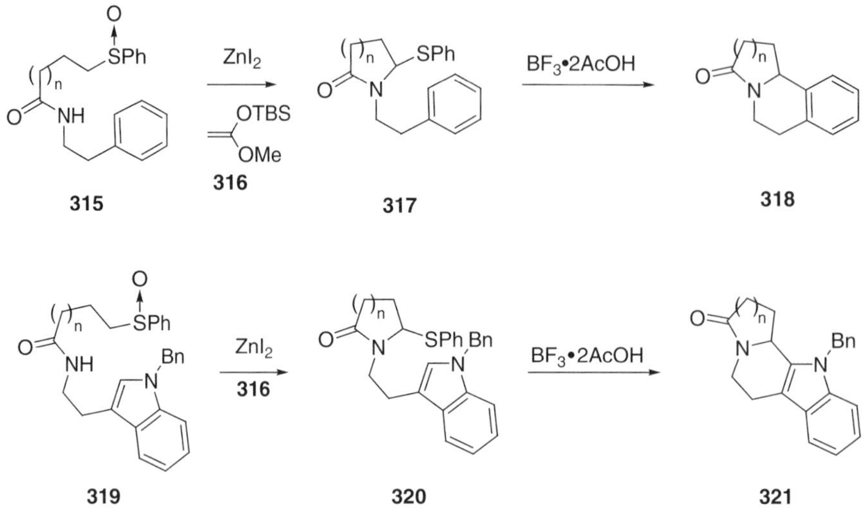

Scheme 59

α-Thiophenylamides were also employed as precursors for the formation of N-acyliminium ions, which were then used as intermediates for subsequent cyclization chemistry. For example, treatment of amido sulfoxide **315** with silylketene acetal **316** in the presence of ZnI$_2$ gave lactam **317** in excellent yield (>90%, Scheme 59) (00T10159, 98TL8585). The action of BF$_3$•2AcOH on **317** led to further ionization of the phenylthio group and cyclization of the resultant iminium ion onto the aromatic ring furnished **318** in 98% ($n = 1$) and 79% ($n = 2$) yield, respectively. The indole-substituted amido sulfoxide **319** gave compound **321** *via* the intermediacy of **320** in good overall yield when subjected to these reaction conditions.

The above tandem Pummerer/Mannich cyclization cascade was modified to allow the use of dithioketals rather than sulfoxides as thionium ion precursors (00JOC235). This change in thionium ion precursor allowed the Pummerer cyclization to produce the requisite iminium ion in a single reaction vessel. An efficient synthesis of the erythrina alkaloid core demonstrated the utility of this cascade. Keto acid **322** was transformed into the thioketal **323** (Scheme 60). Coupling of **323** with 3,4-di-methoxyphenethylamine using carbonyl diimidazole (CDI) gave **324**. Treatment of **324** with dimethyl(methylthio)-sulfonium tetrafluoroborate (DMTSF) in CH$_2$Cl$_2$ at refluxing temperatures delivered the indoloisoquinoline **325** in 71% yield.

The Padwa group has also made extensive use of a Pummerer-based cyclization cascade for the formation of amidofurans (95TL3495, 95JOC3938, 96JOC3706, 96JOC6166). For example, the lithium enolate of cyclic amides such as **326** added cleanly to *bis*-(methylsulfanyl)acetaldehyde (**327**) to furnish aldol products of type **328** (Scheme 61) (00TL9387). Reaction of **328** with DMTSF triggered a Pummerer cascade process by first inducing the loss of a methylthio group in **328** which provided a reactive thionium ion intermediate. This cation reacts with the proximal carbonyl group to give the dihydrofuran derivative **329**. Elimination of acetic acid under the reaction conditions furnished amidofurans **330** in 70–80% isolated yields.

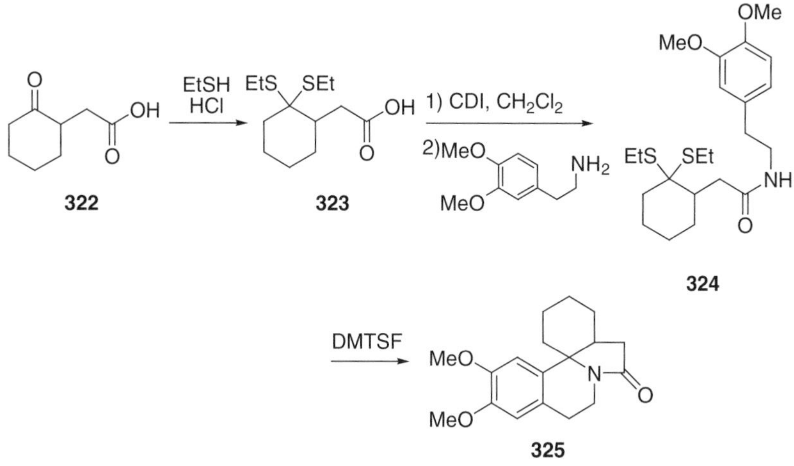

Scheme 60

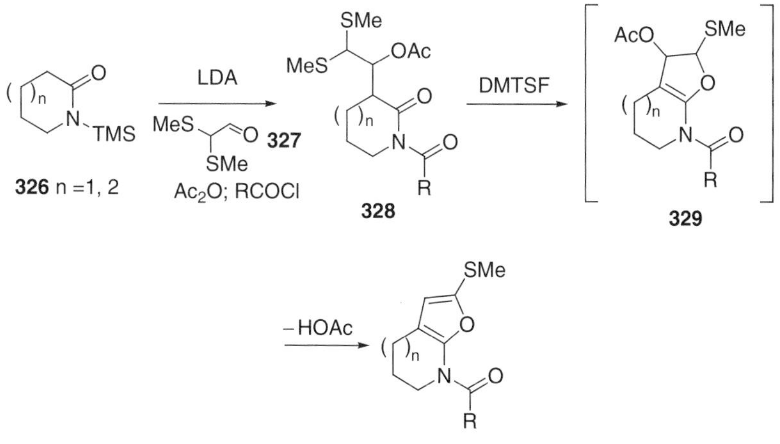

Scheme 61

A variety of 2-methylthio-5-amidofuran systems containing a tethered π-bond on the amido nitrogen were prepared and utilized for a subsequent intramolecular Diels–Alder reaction (02JOC3412). Thus, exposure of imides **331** to DMTSF resulted in the formation of furans **332** in 40–70% yields (Scheme 62). Thermolysis of these furans in toluene at reflux initiated an intramolecular Diels–Alder reaction to first produce an intermediate oxabicyclo adduct. A subsequent fragmentation of the intermediate cycloadduct followed by a 1,2-thio shift provided the bicyclic amides **333** in good yields (*ca.* 70%). In an analogous manner, the cycloaddition chemistry of amidofurans **334** provided the azatricyclic products **335**. Apparently, the rate of the 1,2-thio shift of the initially formed cycloadduct is much faster than the deprotonation/dehydration pathway previously described in Scheme 21.

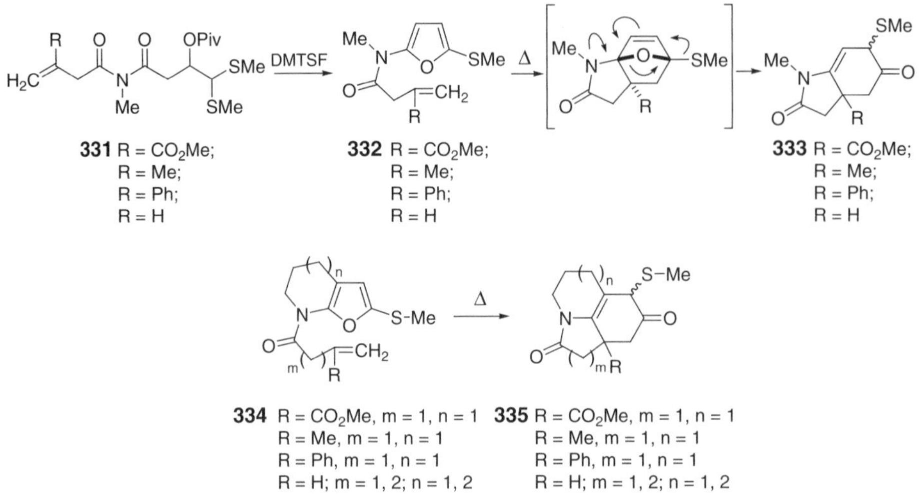

331 R = CO₂Me;
  R = Me;
  R = Ph;
  R = H

332 R = CO₂Me;
  R = Me;
  R = Ph;
  R = H

333 R = CO₂Me;
  R = Me;
  R = Ph;
  R = H

334 R = CO₂Me, m = 1, n = 1
  R = Me, m = 1, n = 1
  R = Ph, m = 1, n = 1
  R = H; m = 1, 2; n = 1, 2

335 R = CO₂Me, m = 1, n = 1
  R = Me, m = 1, n = 1
  R = Ph, m = 1, n = 1
  R = H; m = 1, 2; n = 1, 2

Scheme 62

An interesting example of a Wagner–Merwein-type rearrangement that triggers a subsequent Pummerer cyclization has recently been reported (02OL2565). Phenylsulfanyl-cyclopropane **336** was heated with *p*-TsOH in dry benzene at reflux. Ionization of the hydroxyl group occurred with concomitant ring expansion to give the transient cyclobutyl thionium **337** ion that was subsequently captured by the pendant aryl group to furnish **338** in 77% yield (Scheme 63). Other aryl groups, such as those containing a *p*-Me or a *p*-Cl substituent, also participated in this reaction, as did the unsubstituted analog (67–80% yield). Chromene **338** could be converted into the core structure of the radulanins by treatment with *m*-CPBA, which gave sulfoxide **339** in 70% yield. Thermolysis of **339** in toluene resulted in the elimination of PhSOH producing **340** in 83% yield. Further exposure of **340** to *m*-CPBA induced a ring contraction reaction. This reaction presumably proceeds through the intermediacy of epoxide **341** and provides **342** whose carbon skeleton is found in the radulanin family of natural products.

C. Prins-Pinacol Cascades

The Overman group has made effective use of a pinacol-terminated Prins cyclization cascade for the synthesis of oxygen-containing heterocycles (92ACR352, 02JOC7143). His synthetic strategy for the synthesis of several *Laurencia* sesquiterpenes, such as kumausyne and kumausallene, focused on the acid-mediated reaction of 1-vinylcyclopentane-1,2-diol with 2-(benzyloxy)acetaldehyde. This reaction led to tetrahydrofuran **343**, which contains the requisite stereochemistry for these natural products (Scheme 64) (91JA5378, 93JOC2468). In this reaction, the *p*-TsOH-mediated condensation first generated oxonium ion **344**. A Prins cyclization then afforded carbocation **345**, which underwent a pinacol rearrangement to furnish racemic **343** in 69% yield. Enantiomerically enriched starting (1*S*, 2*R*)-diol (84% *ee*) gave (–)-**343** in 57% yield under similar conditions.

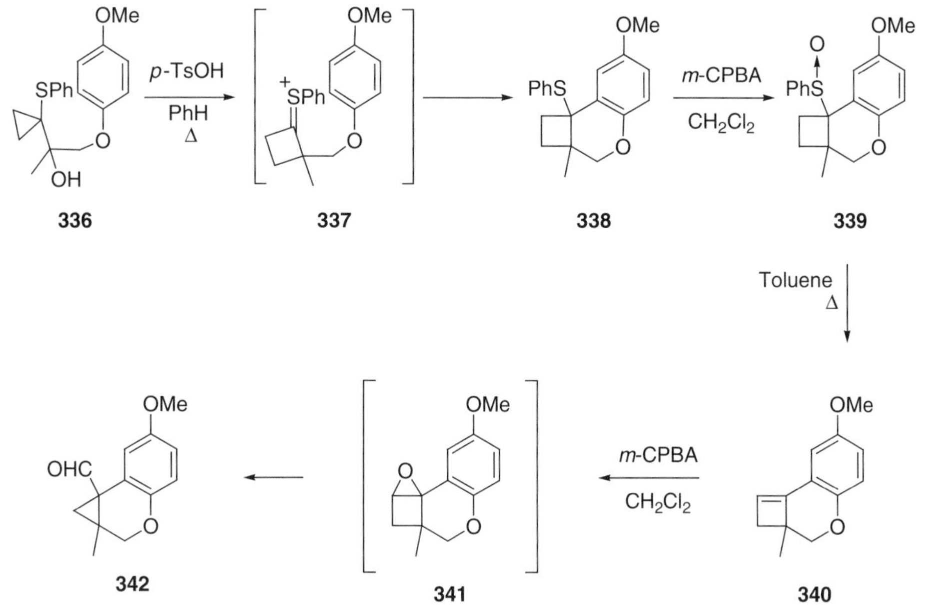

Scheme 63

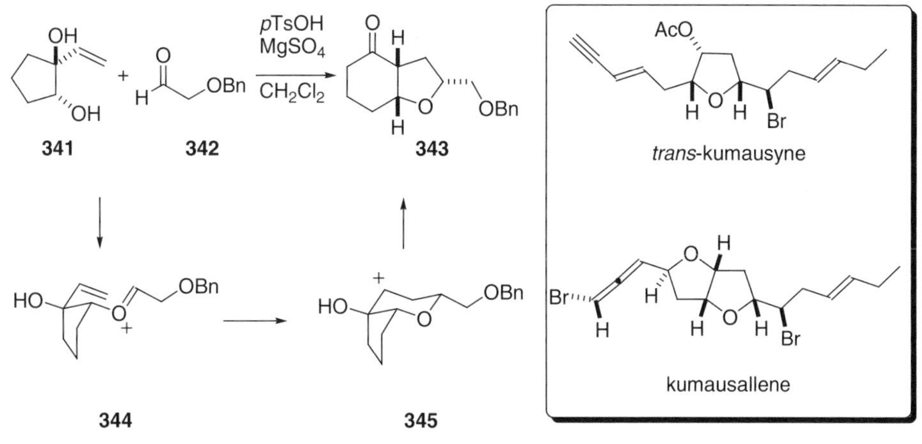

Scheme 64

Application of the Prins-pinacol strategy also led to the synthesis of several cembranoid diterpenes. In these syntheses, $BF_3 \cdot OEt_2$ promoted the condensation of aldehyde **346** with diol **347** that generated oxonium ion **348** that underwent a subsequent Prins cyclization to provide **349** (Scheme 65) (95JA10391, 01JA9033). The Pinacol rearrangement of **349** then afforded tetrahydrofuran **350** in 79% yield. This compound was employed for the construction of several natural products, including sclerophytin A (01OL135).

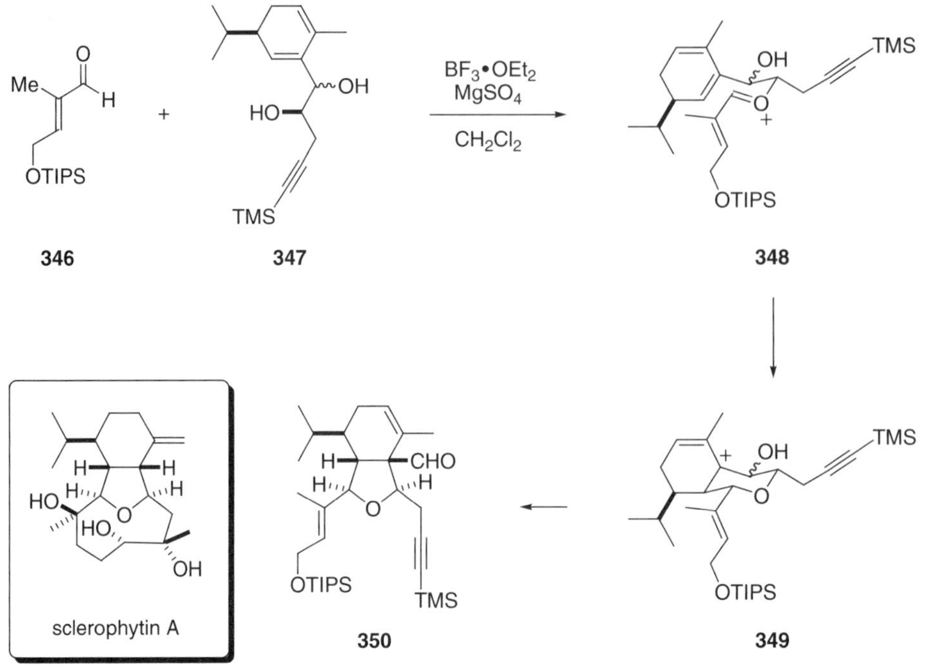

Scheme 65

D. Other Cationic Cyclizations

A tandem Wagner–Merwein rearrangement/carbocation cyclization was used to synthesize several fenchone-derived systems (97TL2159). Heating a mixture of HCl and amide **351** at reflux temperature in aqueous ethanol for 24 h produced the indole derivative **352** in 60% yield (Scheme 66). Presumably, this reaction involves hydrolysis of amide **351** to initially produce compound **353**. Solvolysis of **353** then provided carbocation **354**, which undergoes a rearrangement to give **355**. Carbocation capture by the adjacent nitrogen ultimately affords the ammonium salt of **352**.

When stirred in 85% H_3PO_4, the triptophan derived α-amino nitrile **356** underwent a stereospecific cyclization cascade to give **357** in nearly quantitative yield (Scheme 67) (04OL2641). The formation of tetracyclic **357** is interesting because this compound incorporates both the tetrahydropyrrolo[2,3-*b*]indole structure, which is found in physostigmine and related alkaloids, and the tetrahydroimidazo[1,2-*a*] indole skeleton, which is present in asperlicin and related natural products.

V. 1,4-Additions

A. Michael Addition-Initiated Sequences

Dihydropyridine derivatives can be formed from a cascade sequence wherein conjugate addition of enaminoesters occur with α,β-unsaturated carbonyl compounds

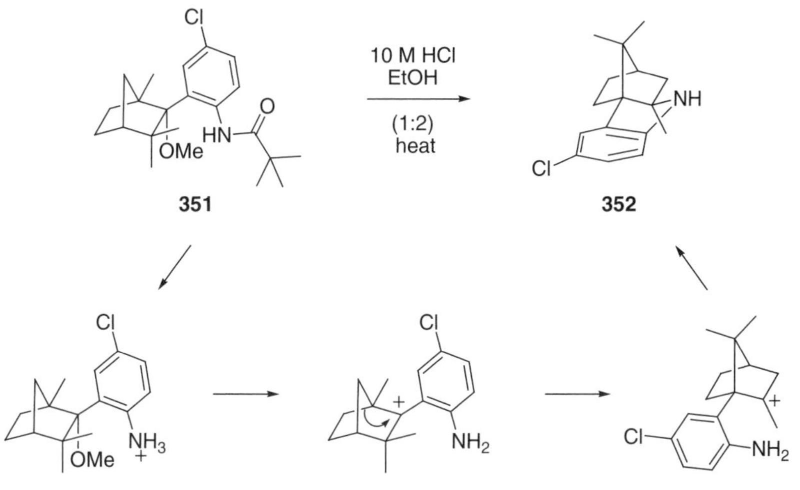

Scheme 66

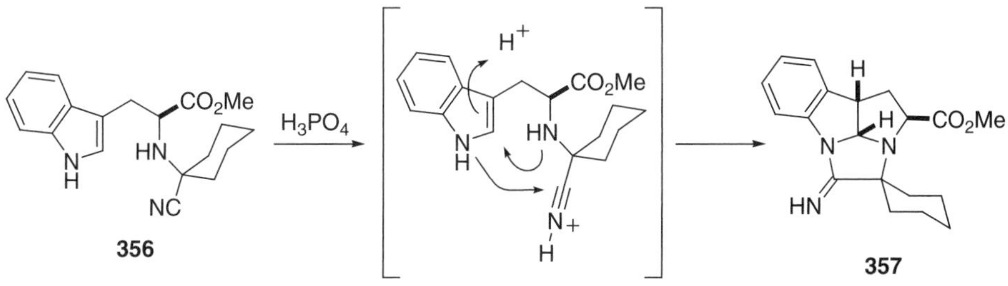

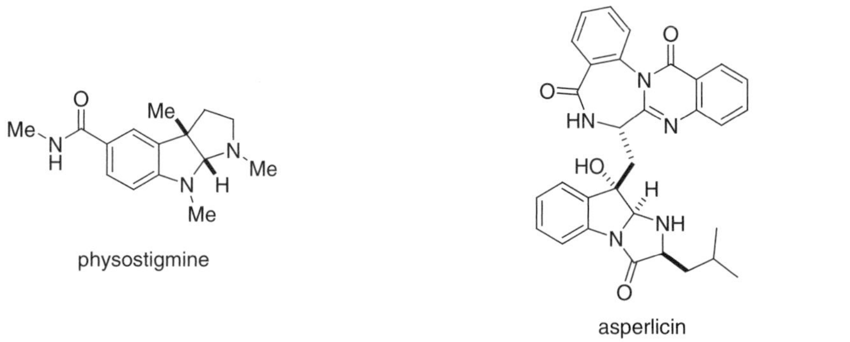

Scheme 67

followed by a subsequent condensation reaction. For example, Caballero and coworkers showed that enaminoester **358** underwent reaction with enone **359** to form **360** in 48% yield (Scheme 68) (98TL455). Ester **358** was shown to be in equilibrium with **361**, presumably through the intermediacy of enol **362**. The conjugate addition

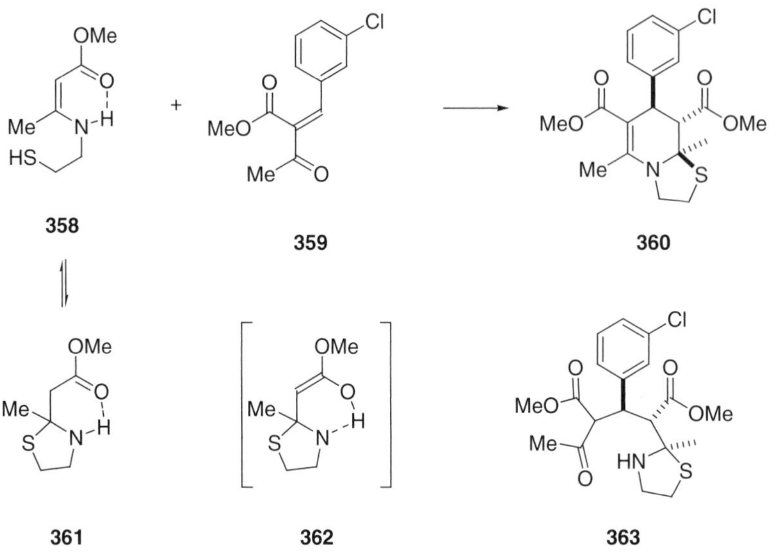

Scheme 68

of **362** with **359** led to compound **363**, which then reacted with the pendant keto group to produce **360**.

Pyrrolidine derivatives can also be prepared from a 1,4-addition of mixed organocopper/zinc reagents to appropriately substituted enones. Thus, Chemla and coworkers showed that the addition of the organometallic reagent **364** to enone **365** in the presence of added ZnBr (3 equiv.) produced a 9:1-diastereomeric mixture of **366a,b** in approximately 80% yield (Scheme 69) (02EJOC3536). The mixed copper/zinc reagents **367a–c** selectively produced substituted pyrrolidines **368a–c** in 50–60% yield under similar experimental conditions. Quenching the reaction with a reactive electrophile resulted in the formation of a more complex pyrrolidine structure. For example, addition of allyl bromide to the reaction mixture after the reaction of **365** with the copper/zinc reagent **367a** was complete provided **369** as a single diastereomer in 57% yield.

A tandem intramolecular Michael/S_N2 reaction was also used to produce pyrrolidone derivatives. Heating a mixture of KF, dichloroacetamides **370a,b** and acetonitrile at reflux temperature produced bicyclic amides **372a,b** in 85 and 30% yield, respectively (Scheme 70) (94T9943). In this reaction, the choice of fluoride ion as a base was critical. Also, the isolation of the conjugate acid of intermediate **371** provided confirmation that the reaction occurred by the tandem cyclization sequence, rather than simple cyclization of a carbene intermediate.

B. CYCLIZATION FOLLOWED BY A MICHAEL ADDITION REACTION

Bunce and coworkers explored the tandem S_N2/Michael addition cascade of 6- and 7-halo-2-alkenoate esters using both primary amines and thiourea to form

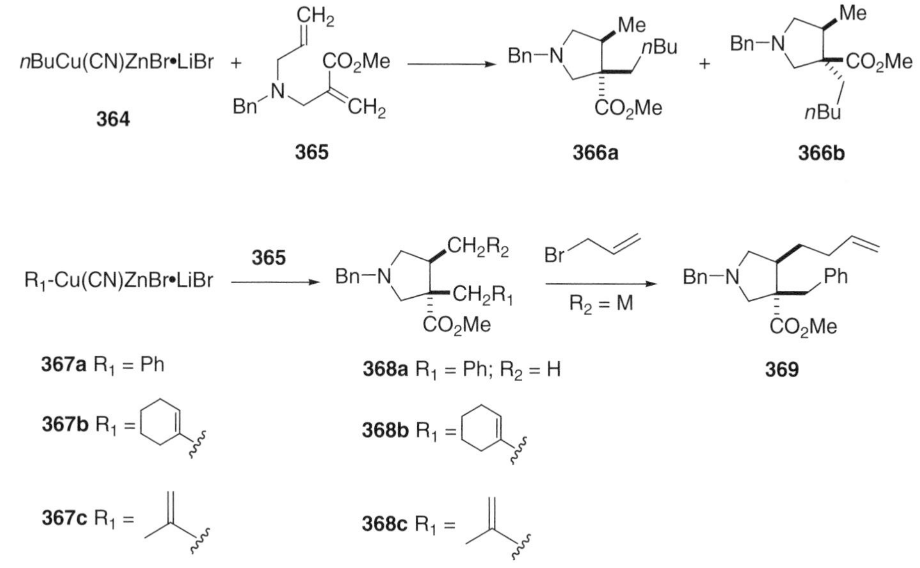

Scheme 69

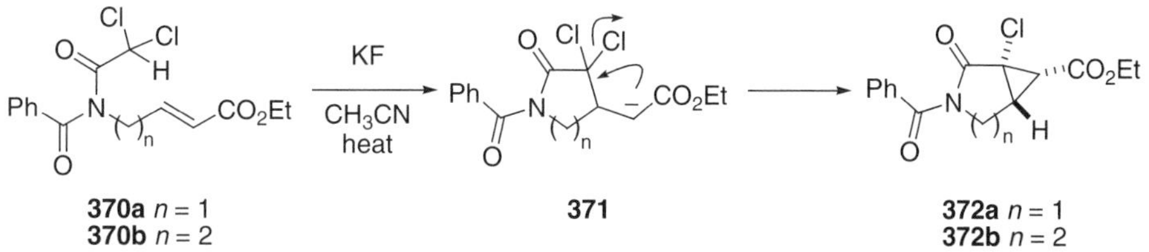

Scheme 70

5- and 6-membered ring heterocycles. Mixing enoate **373a,b**, BnNH$_2$, and Et$_3$N in ethanol produced pyrrolidine **374a** as well as piperidine **374b** in 59–63% yield (Scheme 71) (92JOC1727). Variously substituted halo-enoates participated in this substitution/cyclization cascade, giving rise to nitrogen heterocycles **375–378** in 60–70% yield, though enoate bromides reacted at a slower rate. While the reactions were quite successful with several different primary amines, attempts to fashion azepine derivatives *via* this cascade failed.

In similar experiments, heating thiourea with **373a,b** in ethanol at reflux temperatures produced isothiouronium salts **379a,b**. Exposure of these salts to an aqueous KOH solution followed by an acid work-up afforded the 5- and 6-membered ring thio-heterocycles **380a,b** in 60 and 69% yield, respectively. Although both reactions could be conducted in a single vessel, the intermediate isothiouronium salts could be isolated and completely characterized if so desired. Using this method, sulfur-containing heterocycles analogous to compounds **375–378** were produced in 60–70% yield.

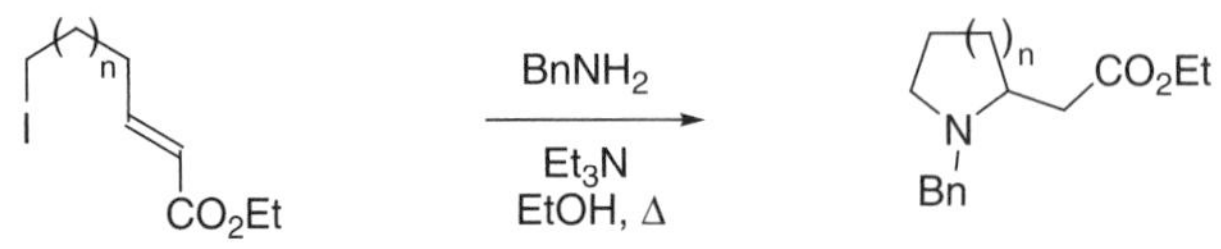

373a *n* = 1
373b *n* = 2

374a *n* = 1
374b *n* = 2

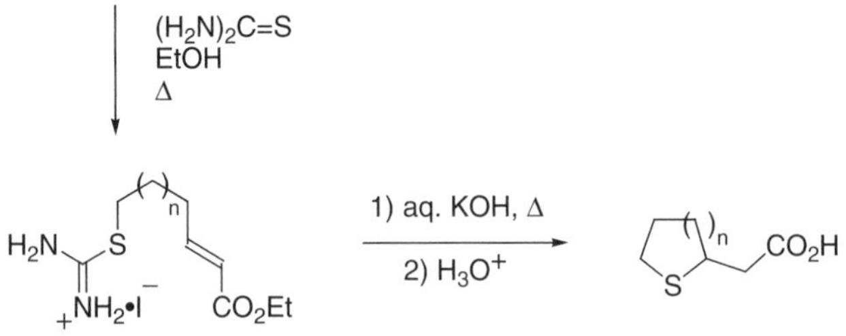

379a *n* = 1
379b *n* = 2

380a *n* = 1
380b *n* = 2

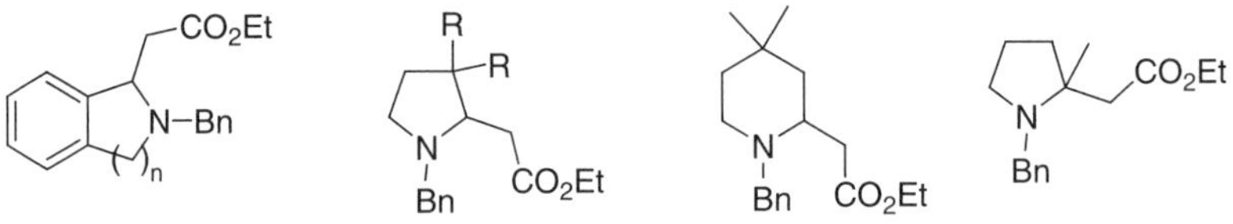

375a *n* = 1
375b *n* = 2

376a R, R = Me
376b R, R = –(CH$_2$)$_5$–

377

378

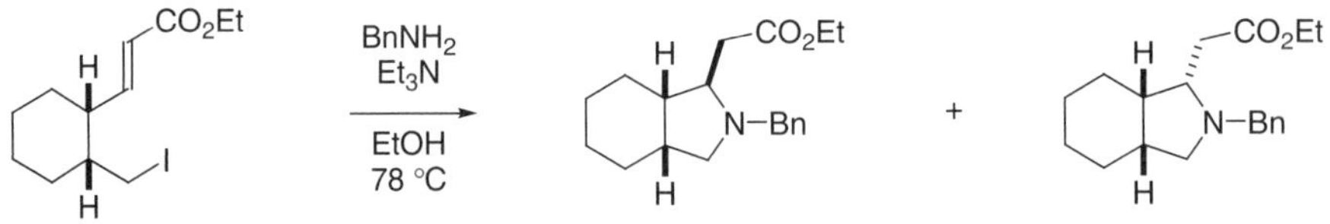

381

382

383

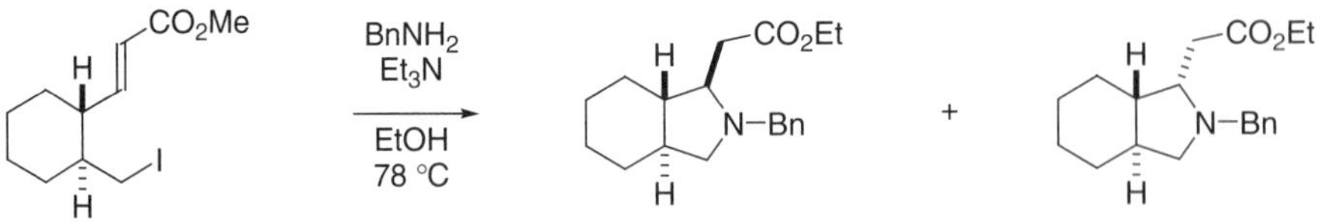

384

385

386

Scheme 71

More recently, Bunce and coworkers examined the stereoselectivity of the substitution/Michael addition sequence for the formation of bicyclic systems. For example, the *cis*-disubstituted cyclohexane derivative **381** underwent reaction with BnNH$_2$ and Et$_3$N when heated in EtOH to produce a mixture of **382** and **383** in 80 and 5%, respectively (02JHC1049). The *trans*-derivative **384**, when subjected to the same conditions, was less selective and provided **385** and **386** in 70 and 12% yields, respectively.

One of the themes to emerge from Bunce's research is the use of cascade reactions to produce an internal nucleophile that is subsequently intercepted by an enone to deliver heterocyclic architectures. Thus, nitro aromatics of type **387** undergo an iron-mediated reduction to furnish benzo-fused heterocycles **388** in 88–98% yield (Scheme 72) (00JOC2847). Exposure of unsaturated esters like **389** to NaOEt

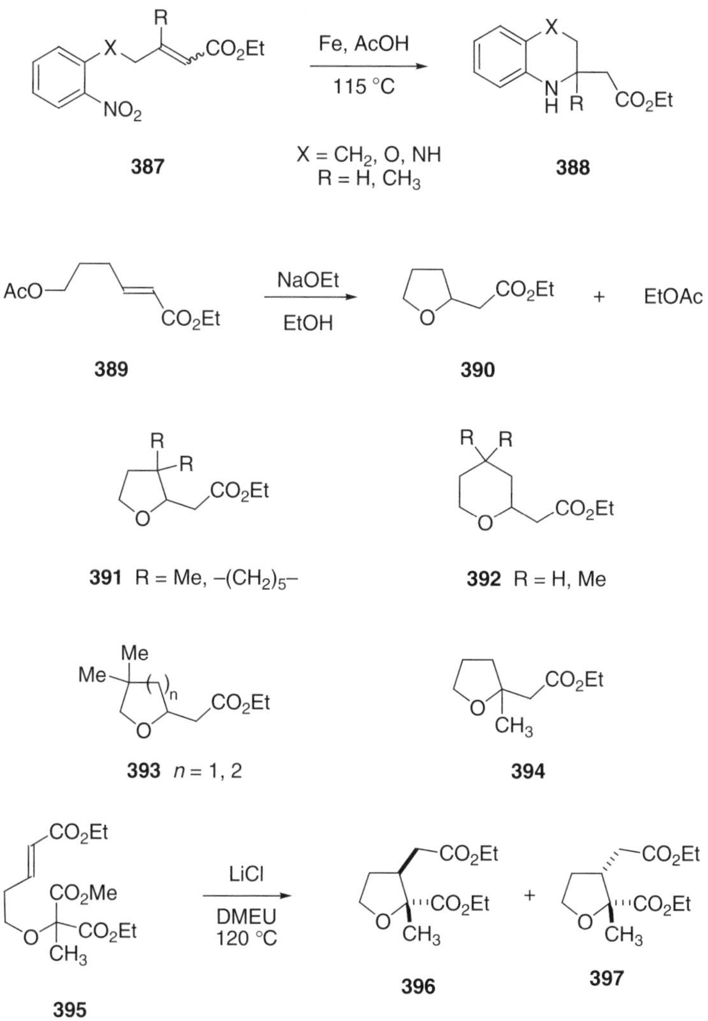

Scheme 72

triggered a cyclization reaction to produce tetrahydrofuran derivatives **390–394** in 70–90% yields (93SC1009). A variant of the Krapcho decarbalkoxylation (73TL957) linked to a Michael reaction of the intermediate enolate was reported to occur onto a tethered enoate moiety (98JOC144). Thus, the reaction of **395** with LiCl in dimethyl-2-imidazolidinone (DMEU) at 120 °C selectively excised the methyl ester and the resulting enolate then underwent addition to the neighboring enone moiety to furnish a 3:1-mixture of the substituted tetrahydrofurans **396** and **397** in 67% combined yield. As with the other examples of cascade reactions terminated by a Michael cyclization, substitution about the alkyl tether was found to afford a variety of related derivatives in 60–90% yield.

Several other groups have explored reactions that work in tandem with Michael additions to form molecules of much higher complexity. For example, Saito showed that carbodiimide **398** underwent reaction with either alcohol or amine nucleophiles to give addition/cyclization products (Scheme 73) (96TL209). Specifically, **398** was found to react with methanol to give **399** in 62% yield. When **398** was treated with n-BuNH$_2$, it produced a mixture of regioisomers **400** and **401** in 54 and 32% yield, respectively. Thiols also participated in this addition sequence as demonstrated by the addition of p-thiocresol with **398** to give **402** in 70% yield.

A tellurium ion-triggered cascade was reported by Dittmer as a method to synthesize isobenzofuran derivatives. Exposing epoxide **403** to Te^{-2} under phase transfer conditions afforded dihydroisobenzofuran **404** in 90% yield (Scheme 74) (00TL6001). The mechanism of this reaction presumably begins by S$_N$2

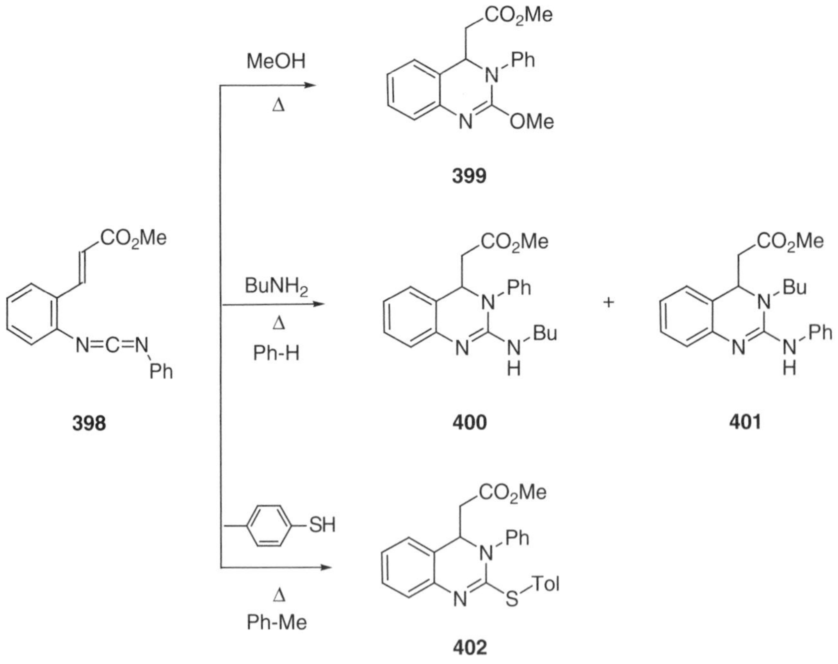

Scheme 73

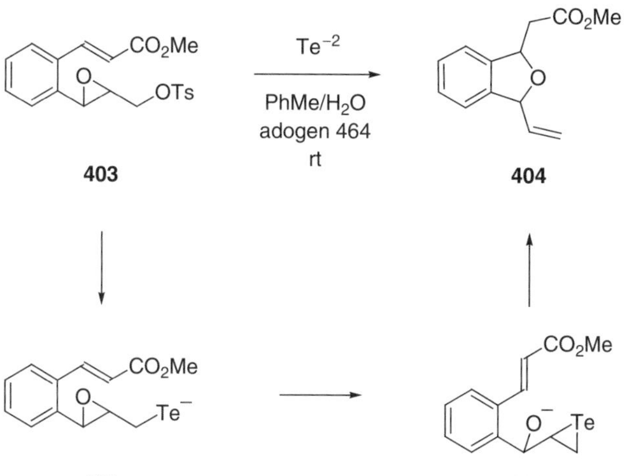

Scheme 74

displacement of the tosylate group with Te^{-2} to give **405**. Rearrangement of **405** furnished the epitelluride **406**, whose alkoxide anion is properly positioned for conjugate addition to the acrylate π-bond followed by loss of tellurium to provide **404**. It is unclear if **405** is a discrete reactive intermediate or whether a double displacement directly gives rise to compound **406**.

A phosphine-catalyzed reaction involving electron deficient allenes and hydrogen-donor nucleophiles has been reported by Lu. Cyclohexanedione underwent a double addition to allene **407** when the reaction was carried out in the presence of catalytic amounts of PPh_3 to give compound **408** in 68–84% yield (Scheme 75) (02OL4677). The mechanism proposed for this transformation begins with addition of phosphine to the allene, and this step is followed by a proton transfer from the nucleophile to produce the vinyl phosphonium ion **409**. Addition of the conjugate base of cyclohexanedione to **409** followed by elimination of phosphine gives compound **410**. 1,4-Addition of the enolic hydroxyl group present in **410** then leads to the observed product **408**.

Alkynes can also undergo this double addition cascade as shown by the reaction of **411** with cyclohexanedione that also gave **408** in 71–92% yield. The reaction of alkyne **411** with **412** provided **413** in 81–93% yield.

Tandem Michael reactions are also useful for the construction of heterocycles containing per- and polyfluorinated alkyl substituents. For example, the fluorinated alkynoic acid **414a** reacted with ethylenediamine in aqueous EtOH at rt to produce **415** (as a 1:1 salt with ethylenediamine) in 97% yield (94BCSJ3021). On heating at reflux, decarboxylation occurred producing **416** in 88% yield. Fluorinated derivatives **414b,c** reacted in a similar fashion to give analogous products. 2-Aminoethane thiol underwent reaction with **414a** at rt, although this reaction stopped after the first Michael addition and gave compound **417** in 87% yield. On heating **417** in hot aqueous EtOH, cyclization and decarboxylation then proceeded to provide **418** in 74% yield. The longer chain fluoroalkane derivative **414c** did not provide double

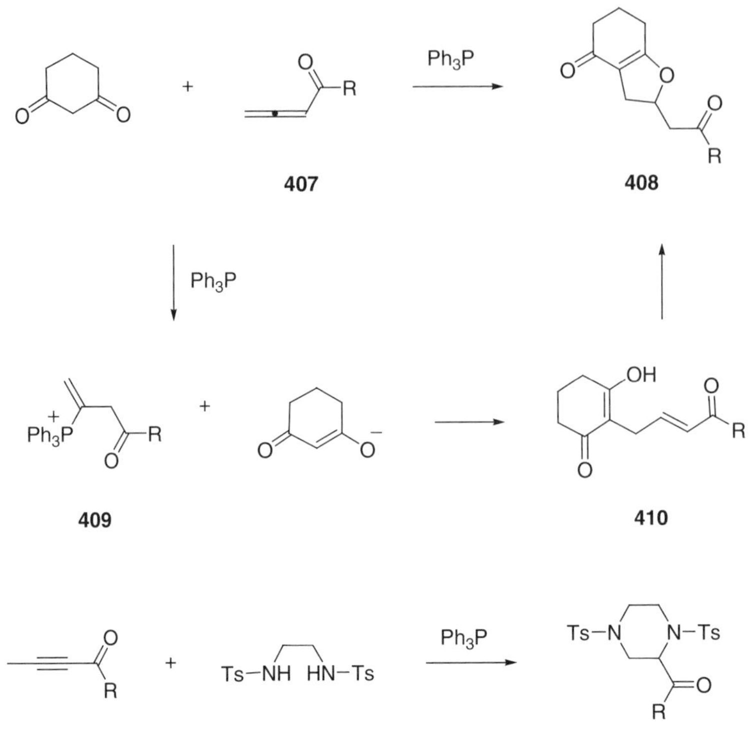

Scheme 75

addition products, even at elevated temperatures; rather, a single-addition product similar to **417** was isolated in 89% yield. The reaction of **414a** with 2-aminoethanol at ambient temperature provided **419** in 75% yield, whereas conducting the reaction at reflux temperature afforded **420** in 74% yield (Scheme 76).

Michael acceptors such as nitroethylene have been found to be useful substrates for tandem Michael reactions. In a typical example, exposure of **421** to the nitro-ethylene precursor **422** at rt produced pyrrolidine **423** in good yield. Likewise, compound **424** reacted with **422** to give piperidine derivative **425** (Scheme 77) (90TL3039). This reaction presumably occurs by initial addition of the amine to nitroethylene followed by a subsequent addition of the resulting anionic carbon to the α,β-unsaturated carbonyl group. An extension of this methodology was applied toward the synthesis of (−)-merquinene. The nitroalkene precursor **426** was stirred with **427** to give a 4:1-diastereomeric mixture of **428** (94T2583). Both of the di-astereomers formed were converted to (−)-merquinene **429**.

α,β-Unsaturated amides have been used to initiate double Michael additions to form piperidones. Thus, *N*-benzyl-*trans*-cinnamamide **430** undergoes reaction with methyl acrylate in the presence of *tert*-butyldimethylsilyl trifluoromethanesulfonate (TBSOTf) and Et$_3$N to produce a 1:1-diastereomeric mixture of piperidinone **431** in 80% isolated yield (Scheme 78) (03TL7429). Compound **431a** could easily be

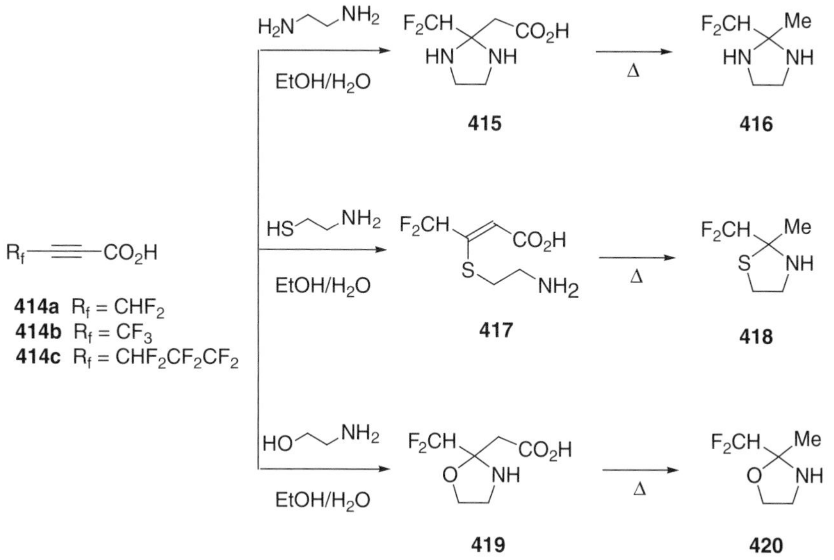

Scheme 76

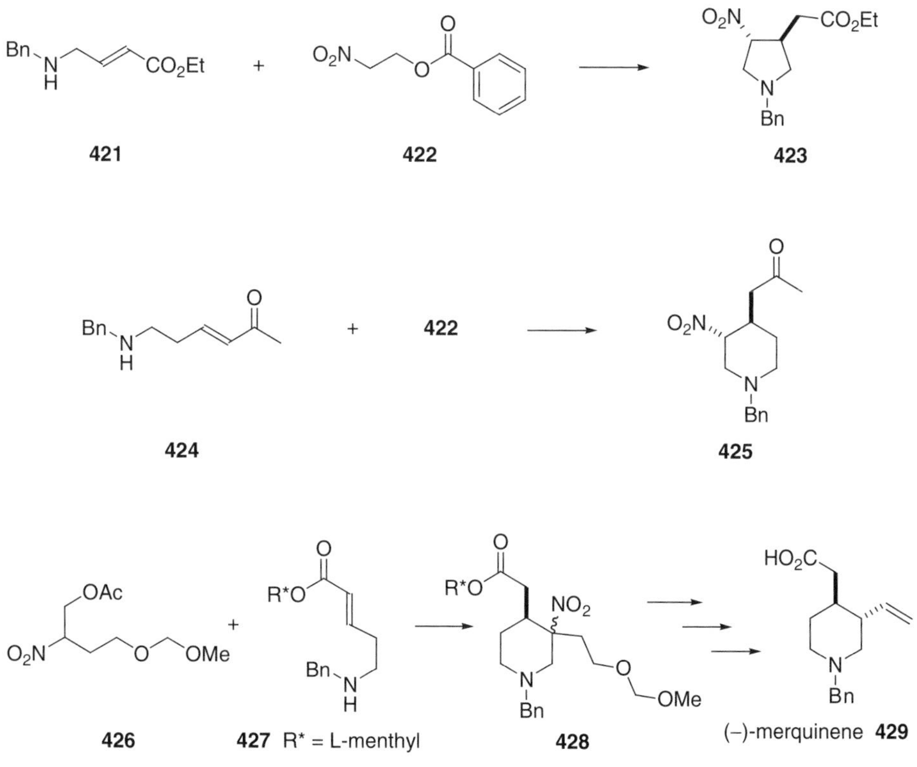

Scheme 77

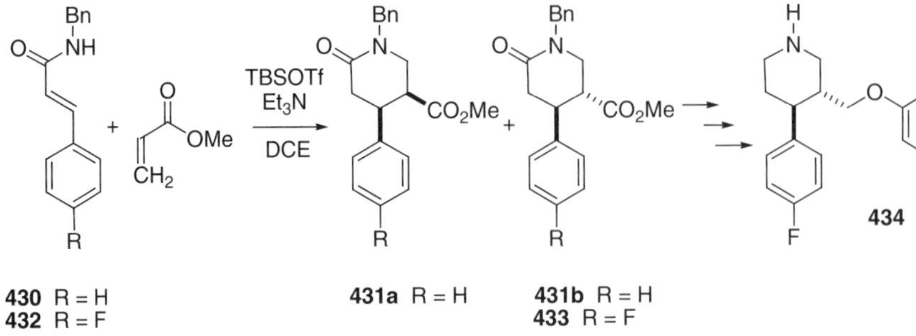

Scheme 78

converted into diastereomer **431b** by exposure to NaOMe. A variety of unsaturated amides and unsaturated carbonyl compounds react under similar conditions to provide variously substituted piperidones. To demonstrate the synthetic utility of the Michael addition cascade process, racemic paroxetine (Paxil®) **434** was synthesized by this method. Specifically, compound **432** was reacted with methyl acrylate and this was followed by equilibration of the resulting mixture of diastereomers to give **433** in 58% isolated yield. Piperidone **433** was then converted to paroxetine **434** in several additional steps.

Another double Michael addition sequence was reported by Carreño, and in this work catalytic amounts of $TiCl_2(Oi\text{-}Pr)_2$ were used as the Lewis acid. Thus, treatment of *p*-quinamine **435** with cyclopentenone in the presence of $TiCl_2(Oi\text{-}Pr)_2$ gave **436** in 60% yield (equation (9)) (02AG2753). Cyclohexenone and ethyl vinyl ketone also reacted with **435** under similar conditions to give compounds related to **436** in 60 and 51% yield, respectively. The stereoselectivity of the reaction is quite remarkable and is thought to arise form a tightly coordinated titanium complex.

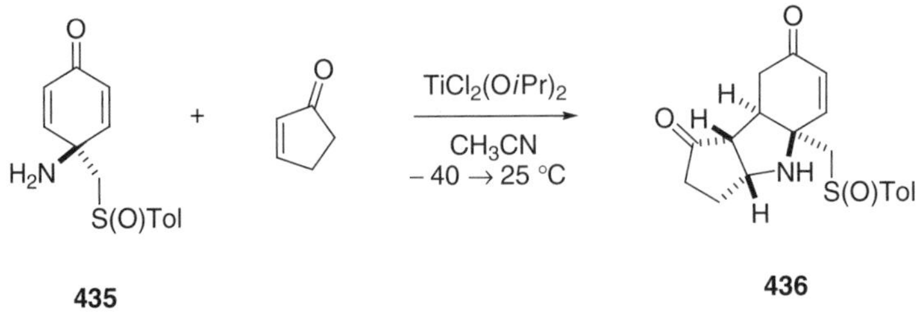

(9)

VI. Carbanion and Carbanion-Like Processes

A carbolithiation cascade was used by Taylor for the synthesis of silacyclopentanes. Addition of *n*-BuLi to vinyl silane **437** afforded a 2:1-diastereomeric mixture of **438** (70%), in which the *trans*-diastereomer predominated (equation (10)) (03TL7143). Interestingly, addition of *n*-BuLi to the alkynyl tethered vinyl silane **439** produced a 3:1-mixture containing predominantly the *Z*-isomer **440a** in 67% yield. Addition of *t*-BuLi to the same substrate provided a 10:1-mixture of **440b** (62%) in which the *Z*-isomer was also the dominant product formed (equation (11)).

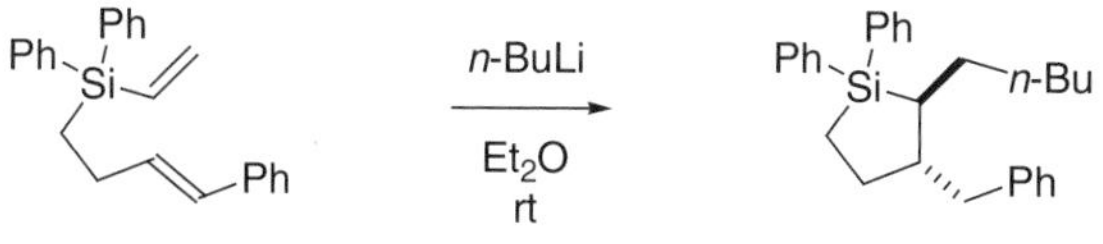

$$(10)$$

437   **438**

$$(11)$$

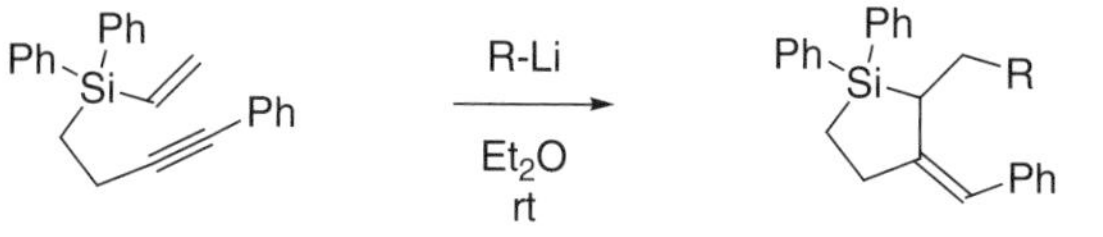

439   **440a** R = *n*-Bu
                440b R = *t*-Bu

A tandem *C–O* alkylation process was used to generate vinylidene substituted tetrahydrofurans. Thus, the base promoted reaction of **441** with dibromoalkyne **442** gave the bicyclic allene derivative **444** in 92% yield, presumably *via* the intermediacy of **443** (Scheme 79) (97JOC3787). Acyclic ketone **445** also participated in this tandem alkylation reaction to give the related compound **446** in 95% yield. The initially formed dihydrofuran isomerized to vinylfuran **447** when subjected to silica gel chromatography. Other substrates were shown to undergo a related tandem alkylation/cyclization sequence in high yield.

Interesting benzimidazoles structures were obtained when *o*-nitroanilines were sequentially treated with base and alkylating agent combinations. Thus, alkylation of aniline **448** followed by the addition of a second base and another alkylating reagent produced *N*-alkoxybenzimidazole **449** in approximately 90% overall yield (equation (12)) (01TL5109).

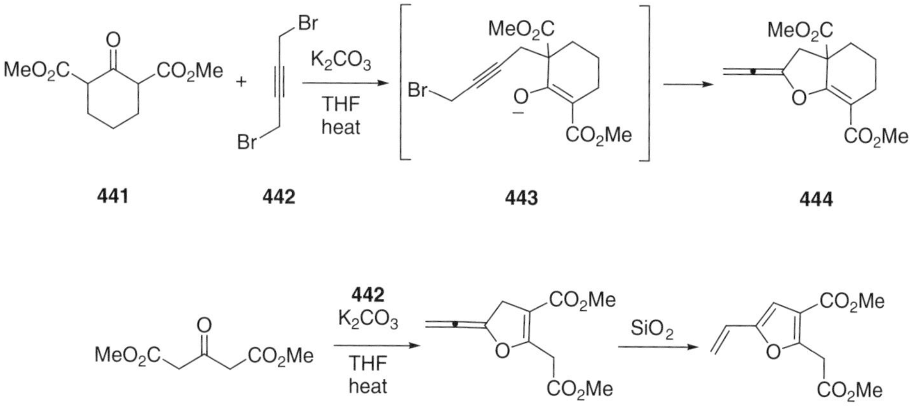

441   **442**   **443**   **444**

445   **446**   **447**

Scheme 79

$$(12)$$

VII. Radical Cyclizations

A. SIMPLE HETEROCYCLES

Radical reactions represent a well-established method for the formation of cyclic molecules. The sequencing of a radical addition followed by a radical cyclization provides the opportunity of increasing molecular complexity in a single reaction. In recent years, attention has focused on novel ways to initiate the radical chain reaction for the synthesis of heterocyclic ring systems.

Heterocycles containing fluorinated substituents were readily available by radical addition of fluorinated hydrocarbons to an alkene followed by cyclization of the resulting radical onto a pendant alkenyl group. Sodium dithionite–sodium bicarbonate was employed by Lu to initiate such a cascade wherein iodotrifluoromethane, **451a** reacts with **450** to form lactam **452a** in 72% yield (Scheme 80) (95T2639). Similarly, compound **451b** reacted with **450** to give **452b** in 83% yield.

Naito described the use of indium metal to initiate a radical addition/cyclization cascade in aqueous media. Stirring a mixture of isopropyl iodide, **453**, and indium powder in water at 20 °C for 2 h produced lactam **454** in 63% yield as a 3:1-mixture of *trans*- and *cis*-isomers (02OL3835). The reaction of sulfonamide **455** under similar conditions led to the isolation of sultam **456** as a 1:1.5-mixture of isomers in 81% yield.

Oximes and hydrazones also participate in these cyclization cascades. For example, oxime **457** underwent reaction with isopropyl iodide in the presence of Et$_3$B to give lactam **458** in 67% yield as a 3:1-mixture of diastereomers (Scheme 81) (03JOC5618). Interestingly, hydrazone **459** produced a 3.2:1-mixture of **460** and **461** in 76% combined yield without having to add a Lewis acid, while the addition of Zn(OTf)$_2$ produced only compound **460** in 50% yield. In contrast, sulfonamide **462** bearing a hydrazone group reacted with indium metal and isopropyl iodide in aqueous methanol to provide **463** in 93% yield as a nearly equimolar mixture of *cis*- and *trans*-diastereomers (02OL3835). Lactones could also be formed under Et$_3$B radical-initiated conditions. Exposure of compound **464** to Et$_3$B in hot toluene produced **465** in 70% yield, along with minor amounts of three other diastereomers (03JOC5618).

Sibi and coworkers examined some of the factors that control the radical addition/cyclization reactions for the formation of cyclic ethers in terms of the stereoselectivity of the reaction as well as the size of the cyclic ether that is formed.

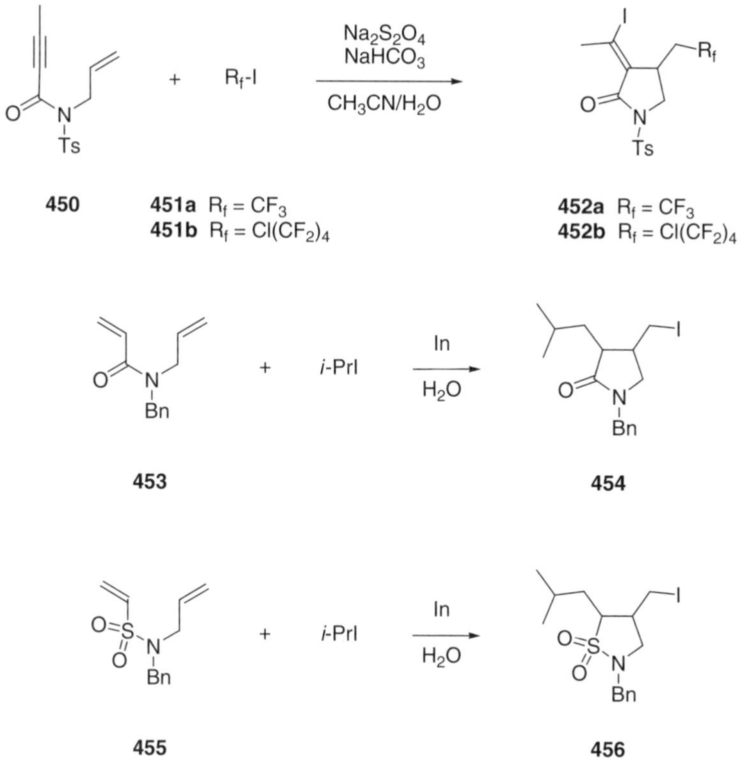

Scheme 80

The Et$_3$B-initiated radical reaction of **466** with alkyl halides in the presence of both Lewis acids and a hydride donor provided cyclic ethers such as **467** *via* a 5-exo-trig pathway (Scheme 82) (04JOC372). Without a Lewis acid, only **467a** was obtained in 39% yield, though the diastereoselectivity of the cyclization was >50:1. In the presence of Yb(OTf)$_3$, however, the yield increased to 72%, while the diastereo-selectivity still remained high. Other Lewis acids produced **467a** in slightly higher yields, but the diastereoselectivity fell dramatically. Secondary alkyl halides and acyl halides also can be used for this cascade, producing compound **467b,c** in 70 and 78% yield, respectively. Ether **467b** formed with excellent diastereoselectivity (>50:1), whereas ketone **467c** was obtained as a 6:1-mixture of diastereomers. The disub-stituted olefin **468** reacted in a similar fashion and gave **469** in 70% yield, though the diastereoselectivity (5:1) of the cyclization was significantly diminished.

Substrates for which the rate of the 5-exo-trig cyclization is greatly reduced, such as **470**, underwent a 6-endo-trig cyclization to give pyran **471** in 68% yield, although with poor (2.5:1) diastereoselectivity. Other alkyl halides that were examined produced related pyrans but in better yield (70–85%) and with similar (1.2–6:1) diastereoselectivities. Using similar substituent placement to control relative rates, 7- and 8-membered rings were also constructed in 50–86% yields.

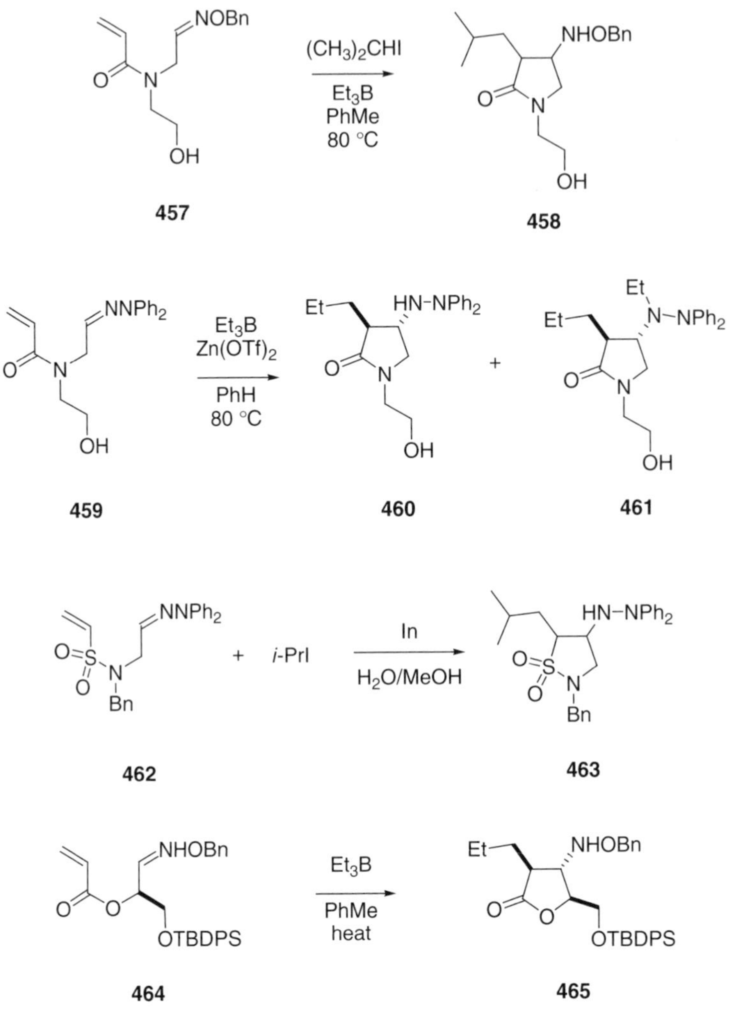

Scheme 81

In all of these cases (Schemes 80–82), the cascade begins with radical addition of a group onto the π-bond so as to position the system for a subsequent cyclization. A complimentary route to substituted heterocycles that has also been used involves an S$_{RN}$1 reaction. For example, indole and benzofuran derivatives have been prepared using a sequential radical ring closure/S$_{RN}$1 strategy. With this approach, the order of the reaction steps has been reversed. Irradiation of **472** in the presence of tri-methylstannyl anion produced dihydrobenzofuran **473** in 87% yield (Scheme 83) (02JOC8500). The reaction course involves cyclization of the initially formed radical **474** to give **475**. Combination of the available nucleophile with radical **475** provides radical anion **476**. Dihydroindoles such as **477a,b** (97% yield) and dihydronaptho-furans **478** (84% yield) were constructed in this manner. Other nucleophiles such as

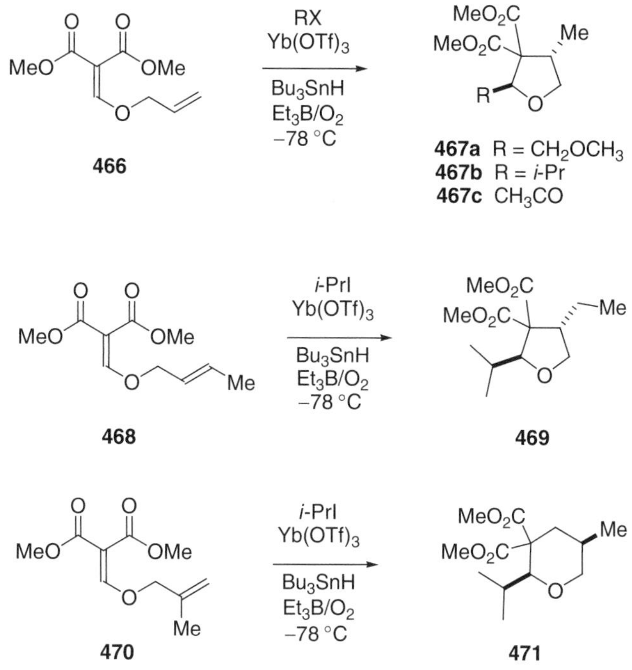

Scheme 82

the anions derived from diphenylphosphine and nitromethane have also been used in these reactions and produce products such as **479** in 60–97% yields.

B. Polycyclic Cascades

Polyethers are readily accessible by tandem radical cyclizations. For example, *bis*-allylether **480a,b** reacts with a trimethyl tin radical and then undergoes a sequential radical cyclization to provide **481a,b** in 86 and 85% yield, respectively (equation (13)) (92JA3115). A ceric ammonium nitrate oxidation of **481** was carried out in methanol and converted the stannyl moiety into the corresponding dimethylacetal.

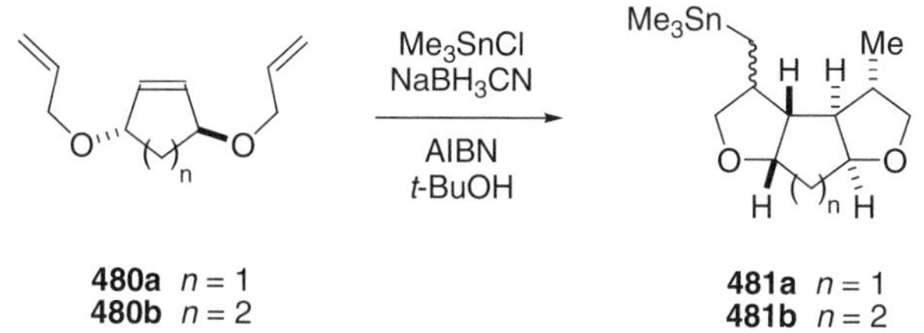

(13)

Several groups have reported the use of a radical cyclization cascade to form nitrogenous polycyclic structures. In one example, Parsons treated enamide **482** with

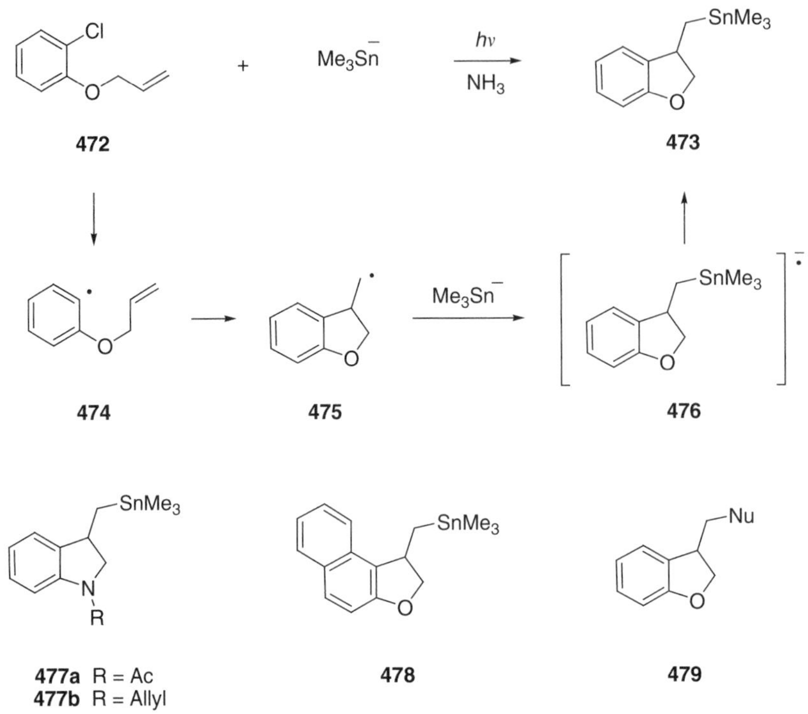

Scheme 83

Ph$_3$SnH in the presence of AIBN to produce **484** in 40% yield (Scheme 84) (98TL7197, 99JJCS(P1)427). In this case, cyclization of the intermediate α-amino ester radical proceeded through a 6-endo-trig pathway rather than the typically more rapid 5-exo-trig closure. The isolation of the 6-endo-trig product most likely reflects the reversibility of the ring closure step, thereby allowing thermodynamic product stability to dictate the course of the reaction. When subjected to the same conditions, **485a** produced **486a** as a single diastereomer. Unfortunately, the incorporation of a menthol chiral auxiliary onto the ester group (i.e. **485b**) led to **486b** as mixture of six diastereomers in 38% yield, suggesting that this is not a suitable way to control stereoselectivity in these cyclization reactions.

The pyrrolizidinone ring can also be generated using this methodology if the intermediate α-amino ester radical undergoes cyclization onto an appropriately tethered electron-poor double bond. For example, enamide **487** reacted with Ph$_3$SnH in the presence of AIBN to produce **488** in 52% yield as a 1.6:1-mixture of diastereomers, where the *cis*-isomer predominates (99JJCS(P1)427). By incorporating a radical stabilizing group onto the π-bond, the reversibility of the 5-exo ring closure was reduced, thereby allowing isolation of the kinetically controlled product.

Ynamides also participate in radical cascade reactions. The Bu$_3$SnH-mediated cyclization of **489** afforded tricylic amide **491** in 70% yield *via* the intermediacy of radical **490** (Scheme 85) (03OL5095). Similarly, subjection of ynamide **492**, in which the carbonyl group is no longer part of the radical acceptor, to the same

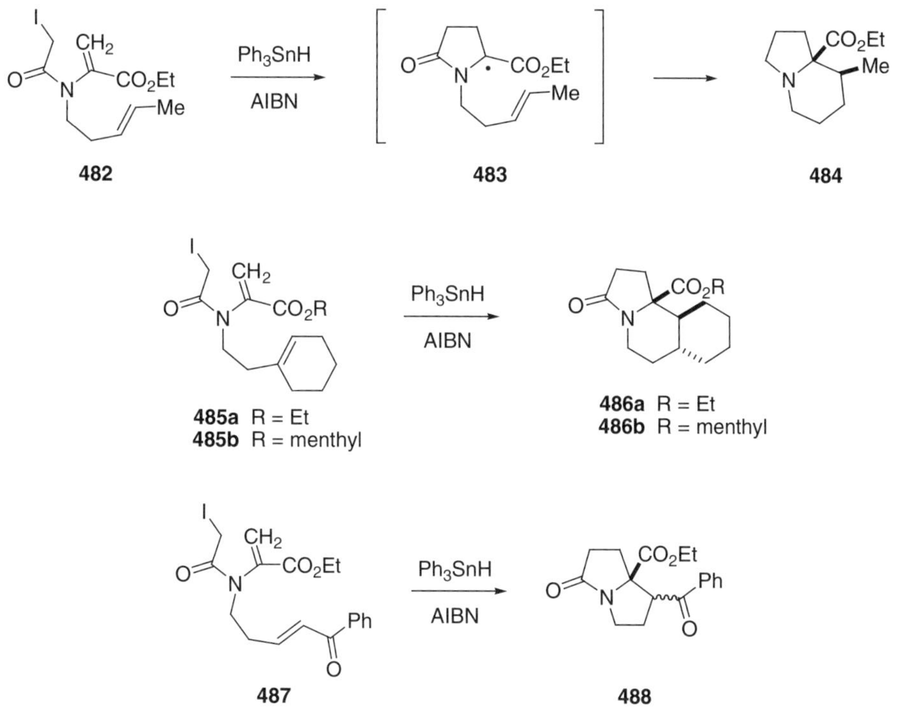

Scheme 84

experimental conditions gave **493** in 90% yield. Ynamide **494**, which contains a benzoyl radical acceptor, produced compound **495** in 67% yield, whereas **496** gave only pyrrolidinone **497** in 57% yield under identical conditions. Photolysis of $(Bu_3Sn)_2$, however, promoted the conversion of **496** into **498** in 46% yield.

The Curran group has examined the use of thiocarbonyl derivatives for the radical cyclization cascade and employed this as a method to form quinoline derivatives. Thus, thiocarbamate **499** was allowed to react with *tris*-trimethylsilyl silane (TTMSH) under irradiation (UV) conditions, and this resulted in the formation of **500** in 67% yield (Scheme 86) (03OL1765). Tin reagents failed to mediate this reaction. Variously substituted analogs of **499** also participated in this cyclization cascade, affording quinoline derivatives (i.e. **500**) in 44–88% yield. The mechanism of the cyclization is believed to involve addition of the radical derived from TTMSH onto the sulfur atom to generate an α-thioamino radical **501** which undergoes a subsequent cyclization onto the pendant alkyne to give vinyl radical **502**. A second cyclization onto the aryl ring then provides **500** (Scheme 86). Substrates possessing substituents in the *meta*-position of the aryl ring afforded 1:1-mixtures of regioisomeric products.

Thioamides and thioureas also undergo the silyl-mediated cascade reaction. For example, compound **503** gave **504** in 67% yield, while structural variants generally afforded related cyclized products in 50–87% yield. Thiourea **505** provided **506** in 64% yield, although a related substrate whose alkyne tether was conformationally more flexible failed to produce any cyclized product.

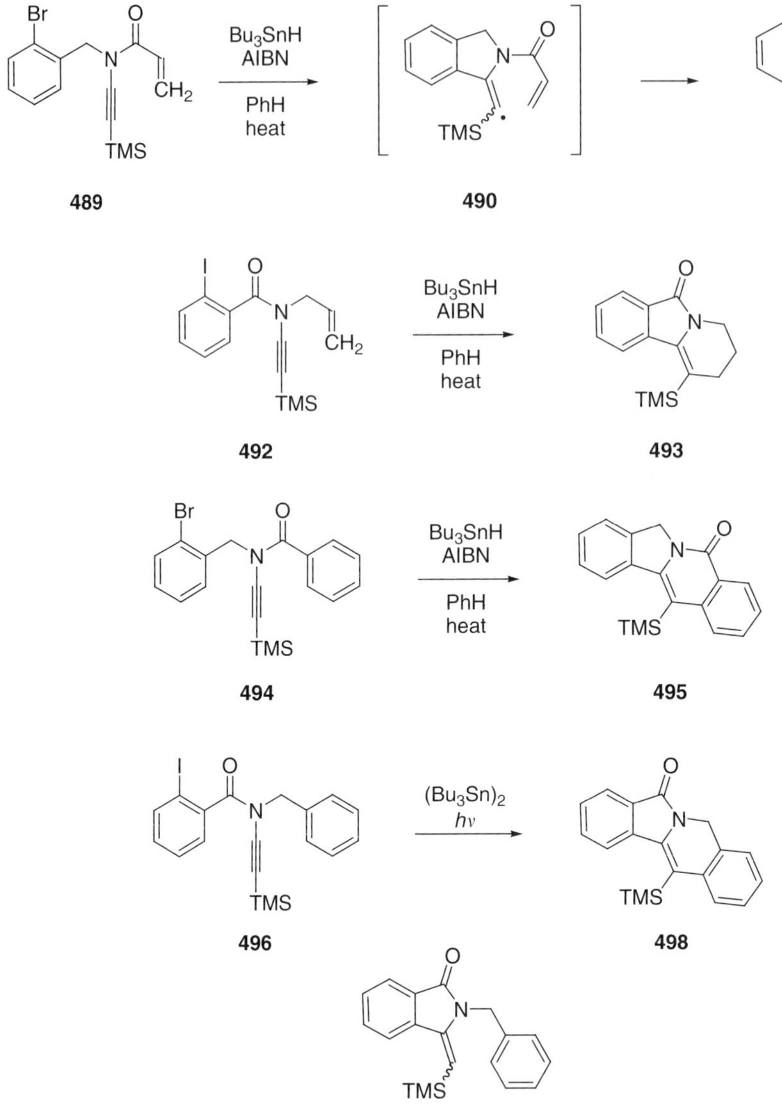

Scheme 85

A novel application of the radical cascade for construction of the indolizidinone skeleton focused on the initial formation of O-stannyl ketyls. The tributyl tin radical was found to react with the carbonyl group of **507** to give ketyl **508** (Scheme 87) (04T8181). Consecutive 6-endo- and 5-exo-trig cyclizations then furnished stannyl enol ether **510**. Eventual hydrolysis of the enol ether provided indolizidinone **511** in 36% yield as a 1:1-mixture of diastereomers. Again, the predominant isolation of the thermodynamic favored products derived from a 6-endo-trig cyclization can be

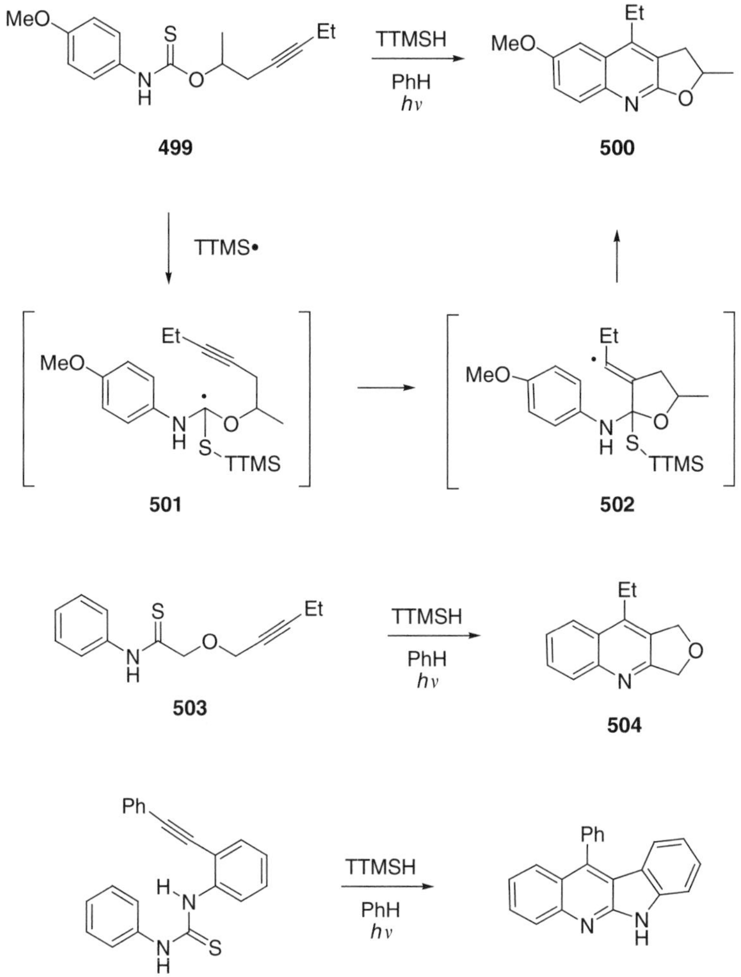

Scheme 86

attributed to the stability of **508b**, suggesting that the cyclization to **509** is a reversible process. Without the stabilizing phenyl group, the conditions required to effect the first cyclization were much harsher, and a 5-exo-trig product was isolated.

Nitrogen centered radicals have received considerable attention in recent years. In particular, amidyl radicals have been shown to enter into cascade reactions to form pyrrolizidinone and indolizidinone derivatives. Thus, heating the *O*-benzoyl hydroxamic acid derivative **512** with Bu$_3$SnH in the presence of AIBN produced **513** as a 3:2-mixture of diastereomers (Scheme 88) (99SL441). Separation of compound **513** from the tin residues was difficult, and the isolated yield (17%) was consequently low. When **514** was subjected to identical conditions, a 2:1-mixture of indolizidine **515** and pyrrolizidine **516** was isolated in 42% yield, along with the monocyclic

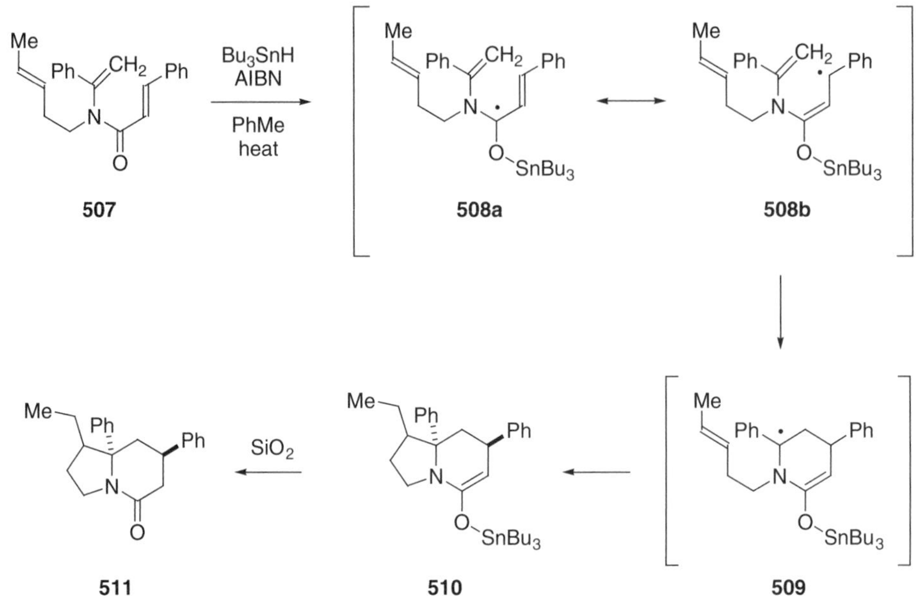

Scheme 87

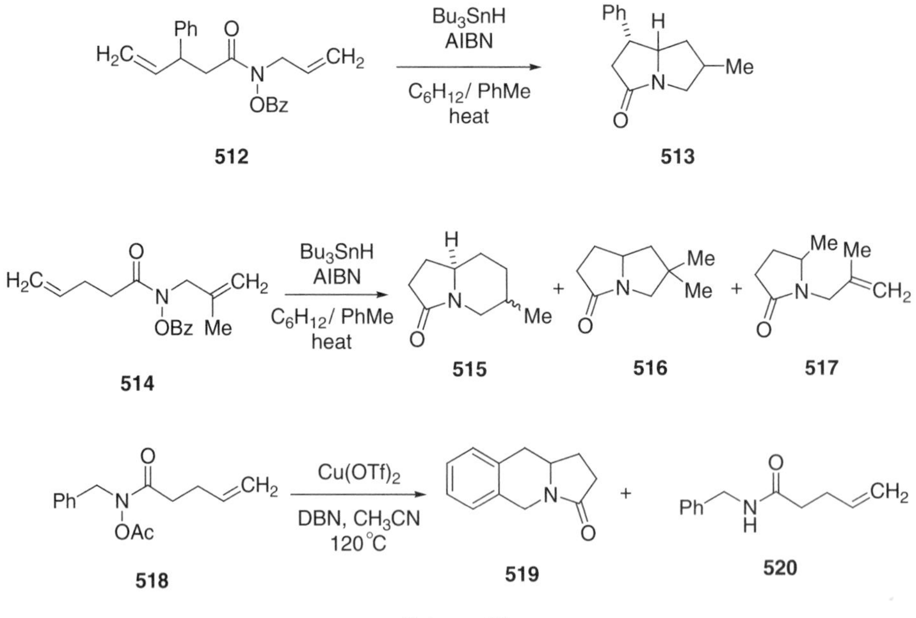

Scheme 88

product **517** in 5% yield. Attempts to induce addition of a radical intermediate onto an aromatic ring and thereby form molecules like **519** failed. However, by adding Cu^{+2} salts to the reaction mixture, this permitted the tandem radical cyclization to occur. It was suggested that the intermediate carbon centered radical was first oxidized to a carbocation and this was followed by a Friedel–Craft type reaction. Thus, under high dilution conditions in CH₃CN, compound **518** was converted into **519** in 53% yield. Some reduced starting material (i.e. **520**) was also produced in 40% yield.

The Bowman group investigated different ways to use aminyl radicals for cyclization so as to produce azacycles. Aminyl radicals generally do not react well with alkenes. Bowman found, however, that these radicals will cyclize onto alkenes that are "activated." For example, sulfenamide **521** when reacted with Bu₃SnH and AIBN in THF at reflux temperatures delivered pyrrolizidine **523** as a mixture of three diastereomers in 49% yield, as well as indolizidine **524** in 14% yield (Scheme 89) (92TL4993). Although the 5-exo-trig cyclization pathway is kinetically favored, the 6-endo-trig pathway does lead to the thermodynamically more stable radical. The formation of **524** suggests that cyclization of the intermediate radical **522** onto the tethered π-bond is a reversible process.

The difficulty associated with cyclization of the aminyl radical onto a π-bond is probably related to a competition between the rate of cyclization versus hydrogen abstraction from the tin hydride. Since the 5-exo-trig cyclization of the endo-2-(bicyclo[2.2.1]hept-2-en-5-yl)ethyl system is one of the fastest radical reactions

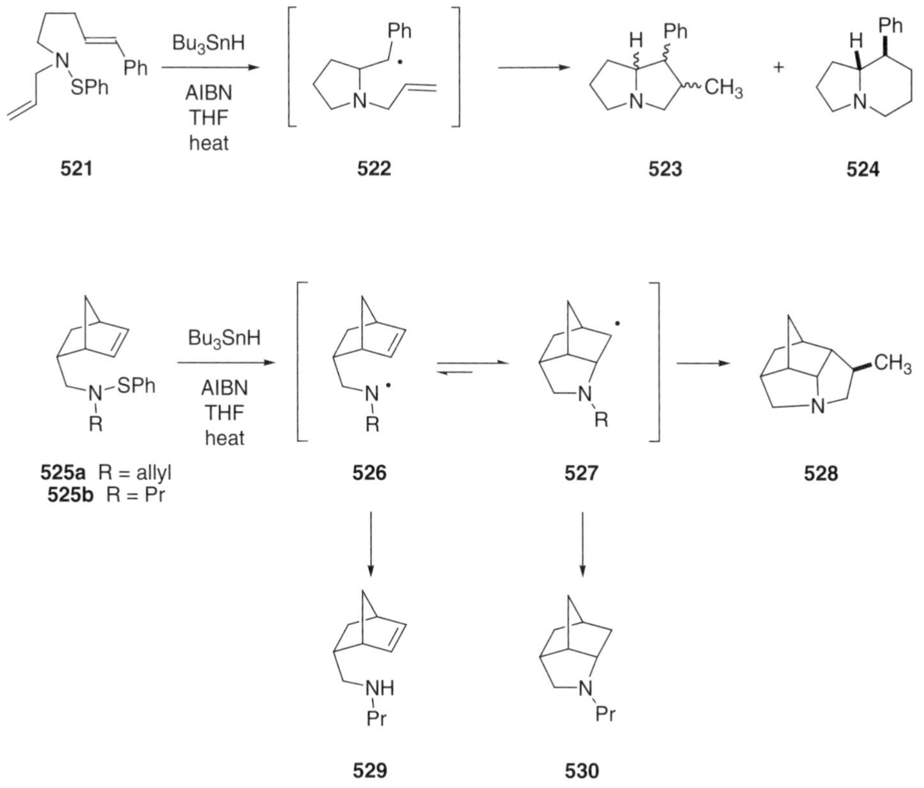

Scheme 89

known, sulfenamides **525a,b** were constructed and then subjected to the cyclization conditions. The course of the reaction was found to depend on the choice of the substituent group on nitrogen. When **525a** was treated with Bu_3SnH and AIBN, the *N*-allyl group acted as an internal radical trap and gave rise to **528** in 90% yield. Without the presence of the π-bond, products such as **529** were isolated in approximately 6% yield along with the cyclization product **530** (29% yield). The results were interpreted in terms of a fast but reversible cyclization of **526** to give **527**. In the absence of the internal allyl group, hydrogen abstraction by radical **526** becomes competitive with hydrogen abstraction by radical **527**.

Having established that aminyl radicals can undergo cyclization, several different modes of reaction were explored. Allyl sulfenamide **531** was found to participate in a tandem radical cyclization reaction to produce a 2:2:1-mixture of hexahydro-indolines **532** in 30% yield (Scheme 90) (94T1295). The cyclization failed when a related substrate lacking the allyl group was subjected to the same reaction conditions. Sulfenamide **533** underwent a tandem 5-exo/6-endo-trig cyclization to give indolizidine **534** in 64% yield. Substrates that would require a 6-endo/5-exo cascade to form the indolizidine skeleton (i.e. **535**) failed to cyclize under the radical conditions.

Spirocyclic amines can also be formed by using cyclization cascades that involve aminyl radicals. The AIBN promoted reaction of ketimines **536a,b** with Bu_3SnH in toluene provided **537a,b** in 34 and 30% yields, respectively (95TL5623). Ketimines **536c,d** failed to produce spirocycles under these conditions. However, in the presence of $MgBr_2 \cdot Et_2O$, compounds **536c,d** underwent the tandem cyclization reaction to give **537c,d** in 24 and 33% yield, respectively. Aldimines **538a–d** reacted similarly to afford bicyclic amines **539a–d** in 40, 58, 27, and 33% yield. Again, substrates containing an aromatic group, which provide stabilization for the intermediate radical produced from the cyclization, gave higher yields of bicyclic products. Adding $MgBr_2 \cdot Et_2O$ to the reaction mixture increased the yield of **539c** to 35% yield.

Crich and coworkers developed a novel radical–cation cascade for the construction of azapolycycles. Thermolysis of compound **540** with Bu_3SnH and AIBN in benzene produced pyrrolizidine **541** as a mixture of diastereomers in 85% overall yield (Scheme 91) (03JA7942). The reaction was thought to proceed *via* the formation of radical cation **542**, which was generated by homolytic cleavage of the nitro group and ionization of the phosphate ester moiety. Intramolecular addition of the nitrogen onto the cationic center would then provide radical **543** that could cyclize to give the final product. Depending on the substitution pattern of the starting nitro compound, a variety of fused and bridged bicyclic amines are possible. For example, allylamine **544** produced **545** as a 1:1-mixture of diastereomers in 78% yield, whereas **546** afforded **547** as a 2:1-mixture of diastereomers in 78% yield.

VIII. Aza-Wittig Cyclizations

Iminophosporanes (phosphazenes) are easily formed by a Staudinger reaction of azides with triphenylphosphine or a Kirsanov reaction, which is a process that takes place between an amine and dichlorotriphenylphosphorane. The ready availability of

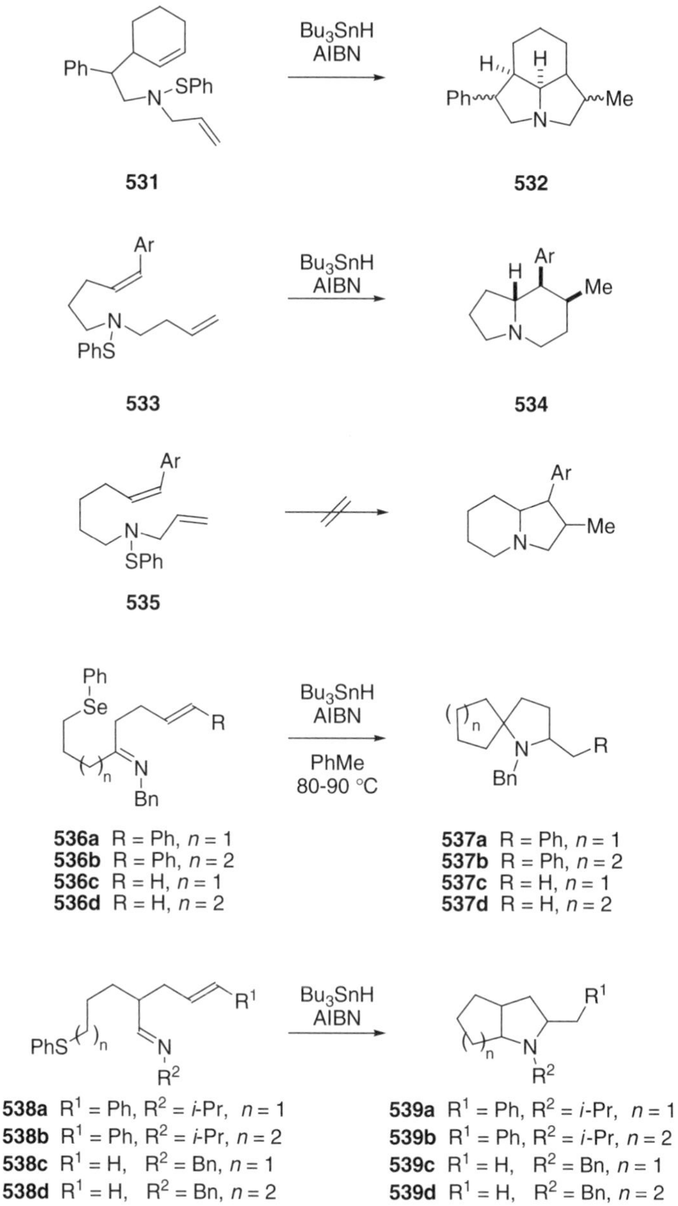
Bu3SnH
AIBN
531
532
533
534
535
PhMe
80-90 °C
536a R = Ph, n = 1
536b R = Ph, n = 2
536c R = H, n = 1
536d R = H, n = 2
537a R = Ph, n = 1
537b R = Ph, n = 2
537c R = H, n = 1
537d R = H, n = 2
538a R1 = Ph, R2 = i-Pr, n = 1
538b R1 = Ph, R2 = i-Pr, n = 2
538c R1 = H, R2 = Bn, n = 1
538d R1 = H, R2 = Bn, n = 2
539a R1 = Ph, R2 = i-Pr, n = 1
539b R1 = Ph, R2 = i-Pr, n = 2
539c R1 = H, R2 = Bn, n = 1
539d R1 = H, R2 = Bn, n = 2

Scheme 90

these reagents has made the Aza-Wittig a key reaction for the synthesis of a variety of natural products (04SL1, 05ARK98). Molina and coworkers have developed a tandem Aza-Wittig/heterocumulene-mediated annulation reaction as a unique method for the synthesis of fused heterocycles. For example, iminophosporane **548** reacted with phenyl isocyanate, carbon dioxide, and carbon disulfide to give

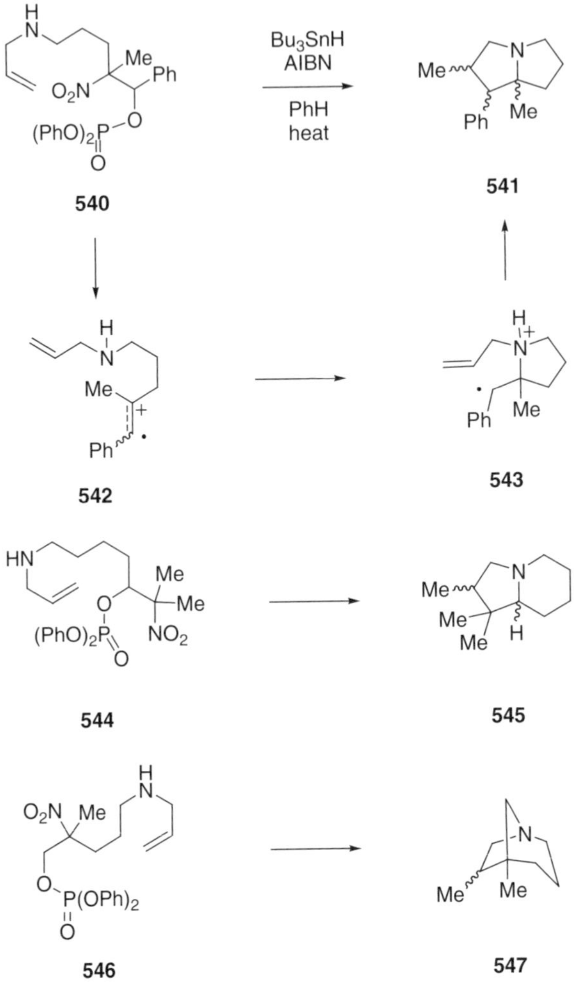

Scheme 91

quinazoline derivatives **549** (96% yield), **550** (96% yield), and **551** (93% yield) as shown in Scheme 92 (88TL3849).

An interesting extension of this methodology involves the sequential use of two phosphazene cascades. Heterocumulene **553** is derived from the reaction of **552** with phenyl isocyanate and gave **554** when treated with hydrazine (04EJOC3872). An intramolecular *O*- to *N*-acyl transfer then produced compound **555** in 81% yield. Exposing **555** to Ph₃P and hexachloroethane in the presence of Et₃N produced iminophosphorane **556** in 85% yield. The reaction of **556** with phenyl isocyanate eventually gave **557** in 92% yield.

A similar cascade was employed for the formation of nitrogen and sulfur containing polycyclic systems. Thus, imino-phosphoranes of the general structure **557**

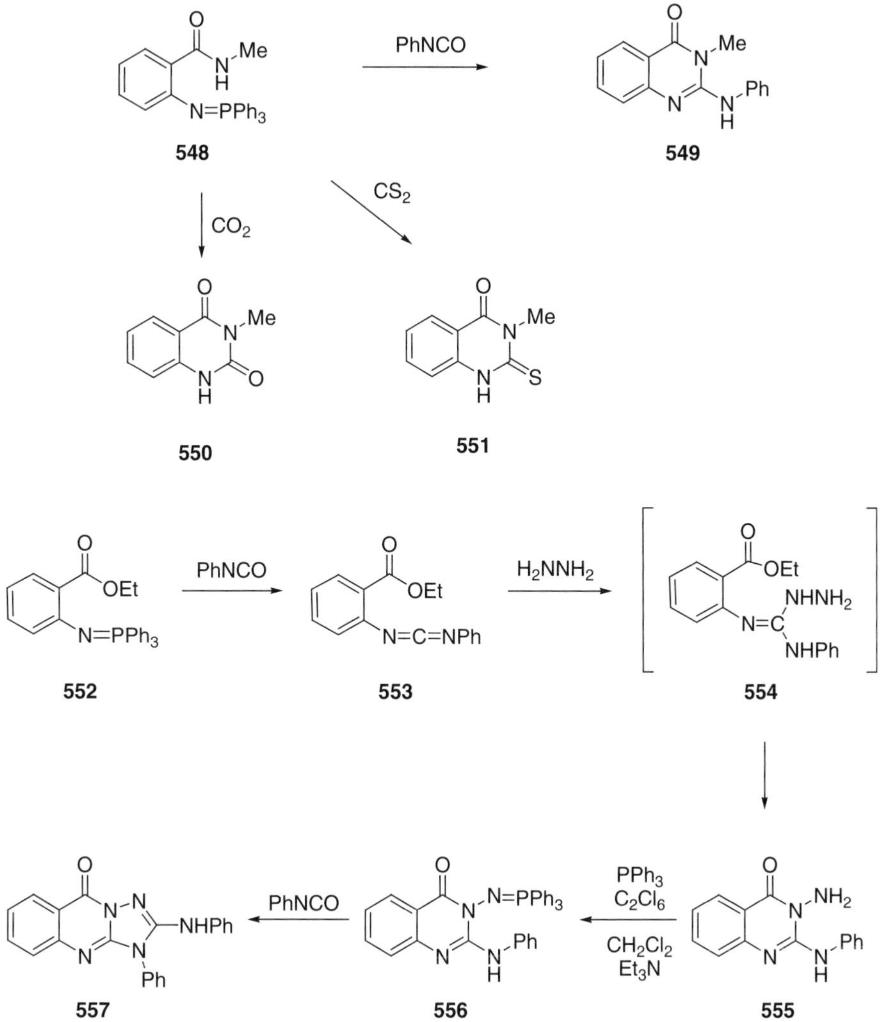

Scheme 92

reacted sequentially with various isocyanates and amines to give compounds **558** in 25–89% yield (equation (14)) (04T275).

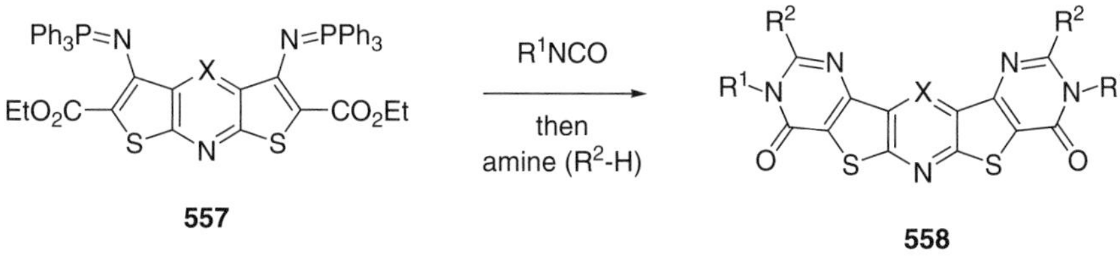

(14)

X = C; N

R^1 = C$_6$H$_6$; 4-MeOC$_6$H$_4$; 4-Cl-C$_6$H$_4$

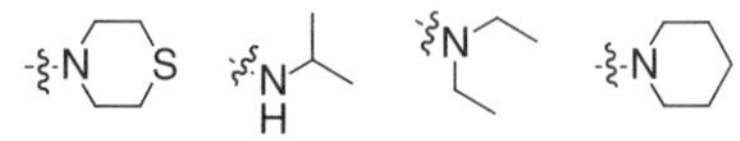

Another example of a tandem Aza-Wittig/heterocumulene-mediated annulation cascade makes use of a subsequent electrocyclization to form a fused heterocycle. Thus, imine **559** reacted with ethyl isocyanate to give the intermediate heterocumulene **560** that underwent a subsequent 6π-electrocyclic ring closure to afford pyrazolopyrimidine **561** in 63% yield (Scheme 93) (88JOC4654). Other iso-cyanates also entered into this cascade producing pyrimidine derivatives with yields generally higher than 60%. Derivatives of **559** containing thiazole and triazole scaffolds react with isocyanates in an analogous way. Treatment of **559** with carbon dioxide provided **562** in 50% yield, and reaction with CS$_2$ produced **563** in 97% yield. A similar strategy led to the construction of both pyridine and pyridone derivatives (89CB307).

Vinyl phosphazenes have been found to undergo tandem Aza-Witting/1,6-elect-rocyclization as illustrated by the reaction of **564** with tricarbonyl compound **565** to give **566** (Scheme 94) (95T3683). The reaction of **564** with aliphatic, heteroaryl, or aryl aldehydes first produced an intermediate imine **567** that ultimately cyclized to produce imidazol[1,5-*a*]pyridines **568** in good (65–82%) yields.

The recognition that a tandem Staudinger/Aza-Wittig reaction would provide the critical imine component necessary for an Ugi reaction prompted another series of investigations. Thus, the Me$_3$P promoted reaction of azide **569** with the pendant aldehyde group produced imine **570** (Scheme 95) (05TA177). Further exposure of **570** to benzoic acid and cyclohexyl isocyanide gave **571** in 36% yield. Other carbo-xylic acids and isocyanide combinations produced several derivatives of **571** in

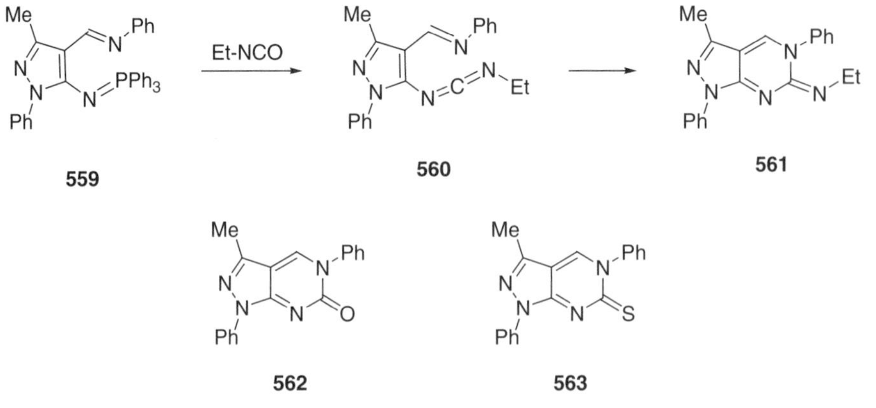

Scheme 93

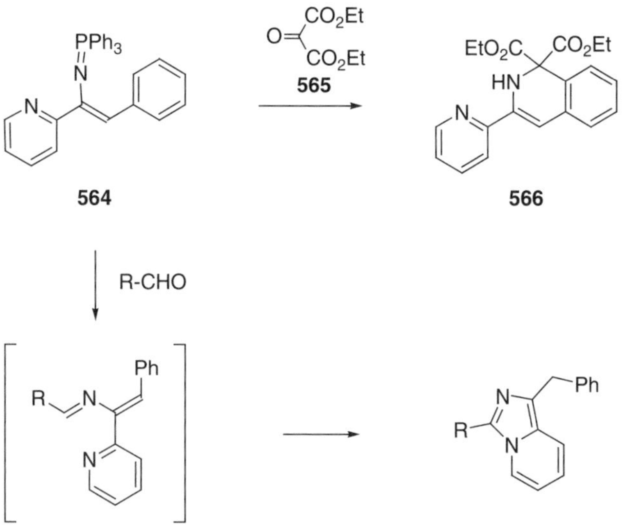

Scheme 94

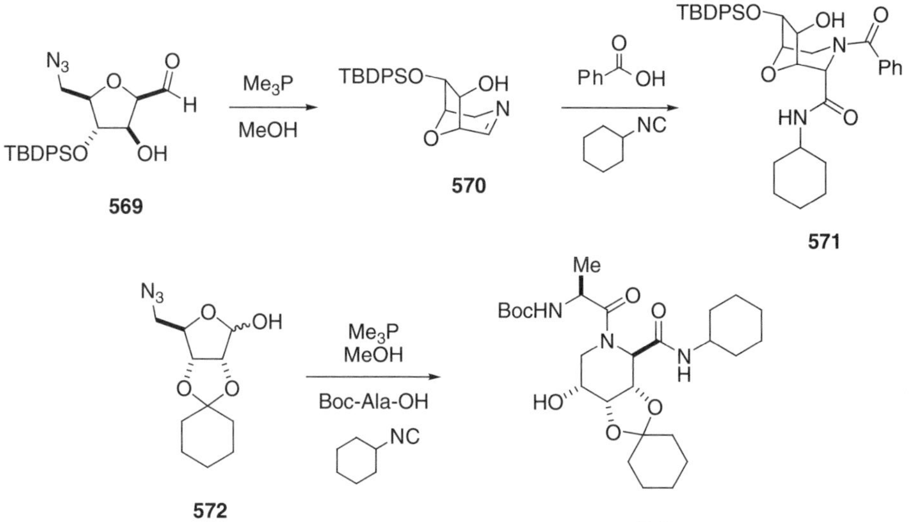

Scheme 95

36–61% yields. Likewise, treatment of azide **572** with Me₃P followed by reaction with Boc-protected alanine and cyclohexyl isocyanide gave **573** in 34% yield. Derivatives of **573** could also be formed by judicious choice of carboxylic acid and isocyanide in 25–75% yield.

IX. Oxidation Reactions

Oxidation reactions often produce reactive functional groups that can engage in further reactions with various nucleophiles. For example, the allylic epoxidation of **574** using VO(acac)$_2$/t-BuOOH in toluene provided a mixture of **575** and **576** in 40 and 44% yields, respectively (Scheme 96) (04TL6349). Formation of **576** occurred by the spontaneous cyclization of the initially formed epoxide **577**. Cyclization of **575** was initiated by the action of TFA to give **578** in 82% yield. The epimeric allylic alcohol **579** reacted in a similar fashion.

The Taylor group has been interested in tandem oxidation processes, and they examined the MnO$_2$ promoted oxidation of alcohols followed by a subsequent condensation of the resulting carbonyl group with amines for the construction of nitrogen-containing heterocycles. Thus, heating a mixture of **580** and **581** in CH$_2$Cl$_2$ with MnO$_2$ and molecular sieves produced quinoxaline **582** in 79% yield (Scheme 97) (03CC2286, 04OBC788). The reaction of diamine **583** with **580** gave **584** in 64% yield under the same experimental conditions. A variety of α-hydroxy ketones and 1,2-diamines participated in this sequence, giving quinoxaline and dihydropyrazine products in 50–80% yield.

By adding NaBH$_4$ to the reaction mixture of **580** and **583**, piperazine **585** was produced in 75% yield. Dilution of the mixture with methanolic KOH, after the consumption of starting materials, facilitated an oxidative aromatization to eventually give **586** in 66% yield. Again, a variety of α-hydroxy ketones and 1,2-diamines participated in these tandem processes.

As an extension of this methodology, a series of benzamidazoles and related heterocycles were prepared starting from simple alcohols. For example, benzyl alcohol was oxidized with MnO$_2$ in order to generate benzaldehyde *in situ*. In the presence of diamine **581**, HCl, and molecular sieves, compound **587** was formed and was then further oxidized under the reaction conditions to give **588a** in 90% yield (Scheme 98) (04SL1628). Monosubstituted 1,2-phenylenediamines such as **589** could also be used, giving rise to *N*-substituted benzimidazole **588b** in good yield. By

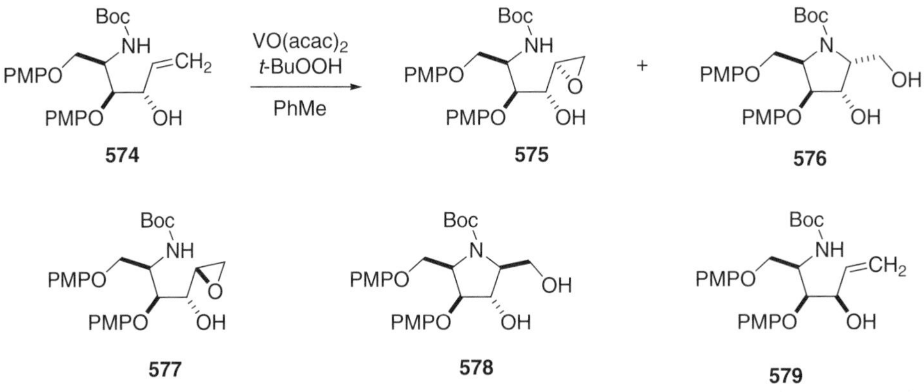

Scheme 96

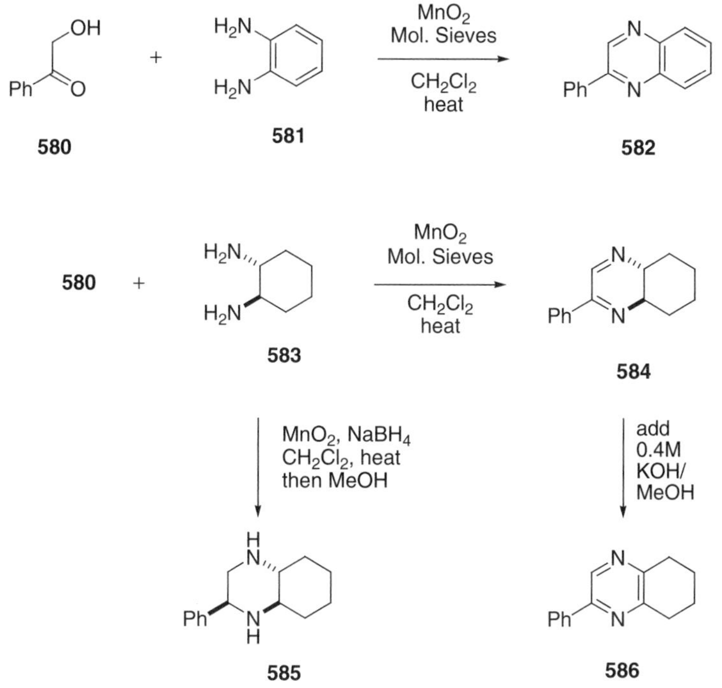

Scheme 97

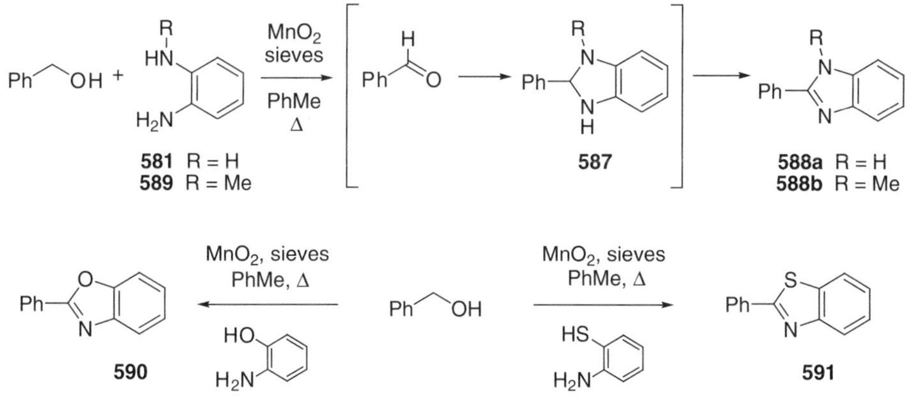

Scheme 98

replacing the 1,2-phenylene diamine with either 2-aminophenol or 2-aminothiophenol, benzoxazole **590** and benzothiazole **591** were formed in 73 and 66% yield, respectively. Other alcohols also participated in this cascade sequence, although yields were much lower (30%) for non-benzylic alcohols.

Shibata developed an interesting one-pot cascade sequence for the formation of trifluoromethyl-substituted pyrimidines. The combination of amidine **592** with enone **593**

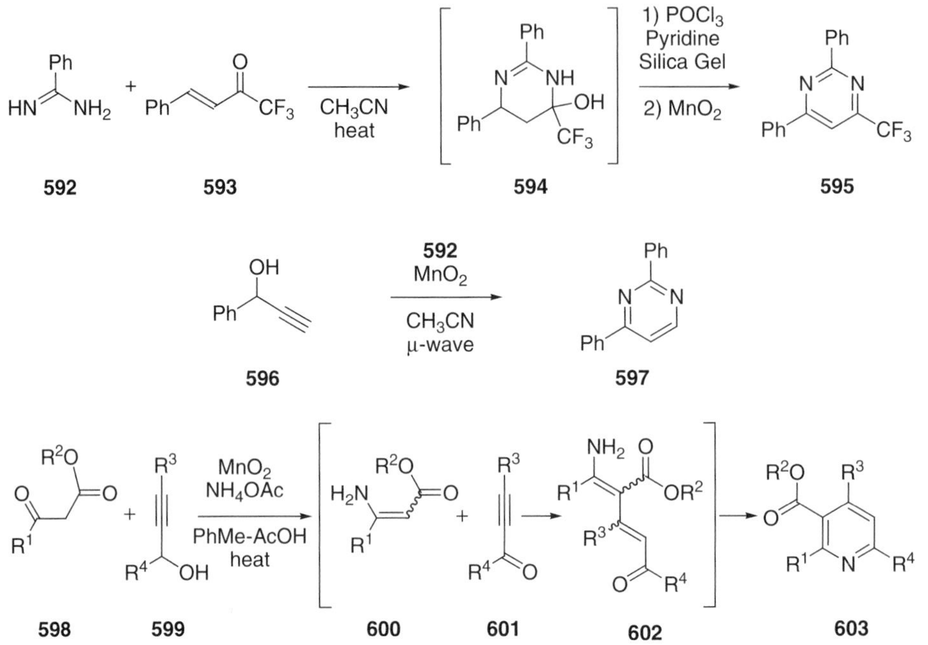

Scheme 99

provided intermediate **594** (Scheme 99) (99SL756). Dehydration with POCl₃ followed by MnO₂ oxidation gave pyrimidine **595** in 86% yield. A variety of aryl-substituted amidines and enones were used to provide pyrimidine derivatives in good yield.

More recently, Bagley and coworkers demonstrated that the oxidation of propargylic alcohols in the presence of amidines gave pyrimidines in good yield using microwave irradiation conditions. For example, the microwave irradiation of a mixture containing propargyl alcohol **596**, amidine **592**, and MnO₂ provided pyrimidine **597** in 84% yield (03SL1443). *o*-Iodoxybenzoic acid (IBX) also mediated the same transformation in 80% yield. An extension of this methodology enabled the synthesis of highly substituted pyridine derivatives from ketones, ammonium acetate, and propargylic alcohols in the presence of MnO₂. In this sequence, an acid catalyzed condensation of ammonia and β-ketoesters **598** produced enamine **600**, while alcohols of type **599** were oxidized to the corresponding ketones. Michael addition of the enamine onto the ynone produced dienone **602** as an intermediate that cyclized to give pyridines **603** in good to excellent yields (60–96%).

X. Transition Metal-Mediated Processes

A. Palladium

Palladium-mediated cyclizations have lately become a central theme for the synthesis of many different types of heterocycles. It is not surprising, therefore, that palladium catalyzed reactions have appeared in numerous cascade reactions (95SL1, 99JOMC42, 02H(56)613, 02H(58)667). Given the diversity of cyclization modes (fused, bridged, or spiro), catalyst ligands, and reacting functionality (alkynes and alkenes, aryl and vinyl halides), a large number of cascade processes can be imagined. The following are just a representative example of the various possibilities.

As mentioned previously, Brandi and coworkers have thoroughly examined the thermal rearrangement of 5-spirocyclopropane isoxazolidines (*cf.* Schemes 13 and 14). As an extension of the earlier thermal work, this group also studied the palladium-mediated rearrangement of the amino alcohols derived from the hydrogenolysis of the isoxazolidine ring. Exposure of isoxazolidine **604** to hydrogen gas in the presence of Pearlman's catalyst gave amino alcohol **605** in 70% yield (Scheme 100) (03SL2305). Interestingly, the reaction of **605** with Pd(dba)$_2$ effected ring opening of the cyclopropyl alcohol to furnish enone **606**, and this step was followed by an intramolecular Michael addition to give **607** in 30% yield. On the other hand, exposure of **605** to PdCl$_2$ and pyridine with an air atmosphere provided **608** in 59% yield. Although the mechanism of this oxidative cyclization process is not well understood, a comparison to the Wacker oxidation was proposed. Accordingly, when **605** was subjected to Wacker-like conditions (Pd(OAc)$_2$, LiOAc, and Cu(OAc)$_2$ as a stoichiometric oxidant), compound **608** was formed in 70% yield. Alternatively, the reaction of **609** with Pd(dba)$_2$ afforded a 5:1-mixture of **610** and **611** in 90% combined yield. Whereas the exposure of **609** to PdCl$_2$, pyridine, and atmospheric oxygen provided **611** in 50% yield, the Wacker-like conditions gave a 5:1-mixture of **610** and **611** in 75% yield.

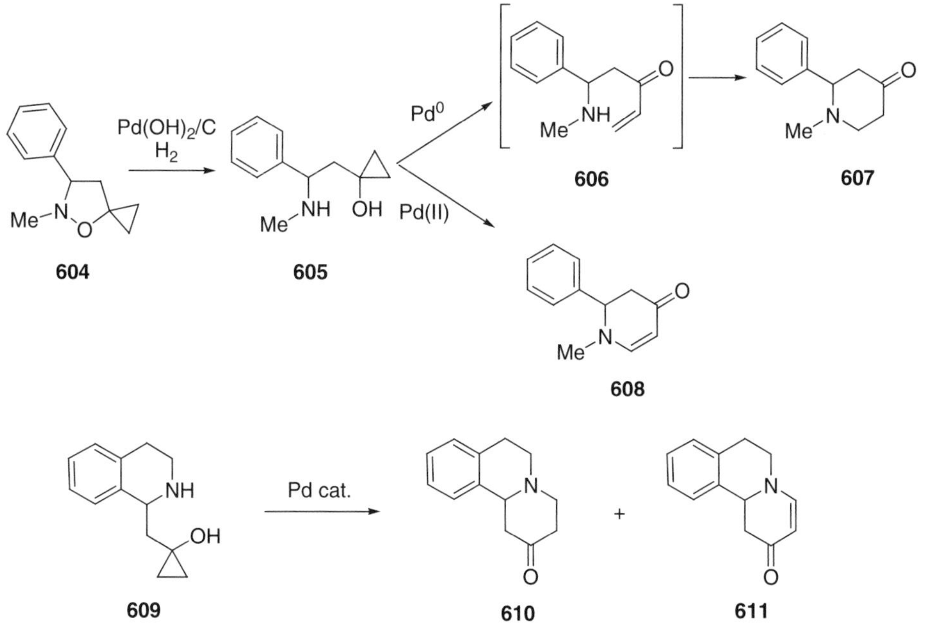

Scheme 100

A linked Pd(0) deprotection–Pd(II) cyclization cascade was used for the synthesis of benzofuran derivatives. Heating a mixture of **612**, Pd(PPh$_3$)$_4$, PdI$_2$, and MeOH under a CO atmosphere provided benzofuran **613** in 91% yield (Scheme 101) (05CC271). This cascade presumably involves a Pd(0)-mediated deprotection of the phenol to give phenolate anion **614**, the substrate needed for the second reaction in the cascade process. Nucleophilic addition of the oxygen anion onto the pendant Pd(II)-activated alkyne followed by a methoxy carbonylation reaction affords **615** and H-Pd-I. Elimination of H$_2$O from **615** by reaction with H-Pd-I then produces a π-allyl palladium species **616** that is regioselectively protonated to ultimately give **613**.

Arylidene butyrolactones were also accessible by addition of nucleophiles onto a Pd(II)-activated alkyne. Thus, the oxidative addition of Pd(0) to **617** in the presence of a base gave **618** in 70% yield (Scheme 102) (96T11463). The proposed mechanism of this reaction was suggested to involve an oxidative insertion of Pd(0) into the Ar–I bond to give the organopalladium species **619**. A subsequent addition of the pendant carboxylate to the activated alkyne provides metallocycle **620** that undergoes reductive elimination to furnish **618**. Similarly, subjecting **621** to the same reaction conditions produced **622** in 77% yield.

The Trost group coupled the palladium-mediated alkyne heterocyclization with a subsequent palladium-catalyzed addition to a terminal alkyne to produce ynoates. Thus, the reaction of homopropargylic alcohol **623** with alkyne **624** in the presence of Pd(OAc)$_2$ (10 mol.%) and tris-(2,6-dimethoxyphenyl)phosphine (TDMPP, 4 mol.%) at rt in benzene afforded compound **626** in 61% yield (Scheme 103) (00JA11727). In this case, formation of the intermediate enyne **625** was fast compared to the heterocyclization process. When attempts were made to reduce the reaction time by heating the mixture, the formation of lactone **627** became competitive, and the yields of **626** were significantly diminished. Addition of Pd(O$_2$CCF$_3$)$_2$ to the mixture after the alkyne addition was complete, increased the cyclization rate. In some cases, particularly those involving terminal alkynes whose

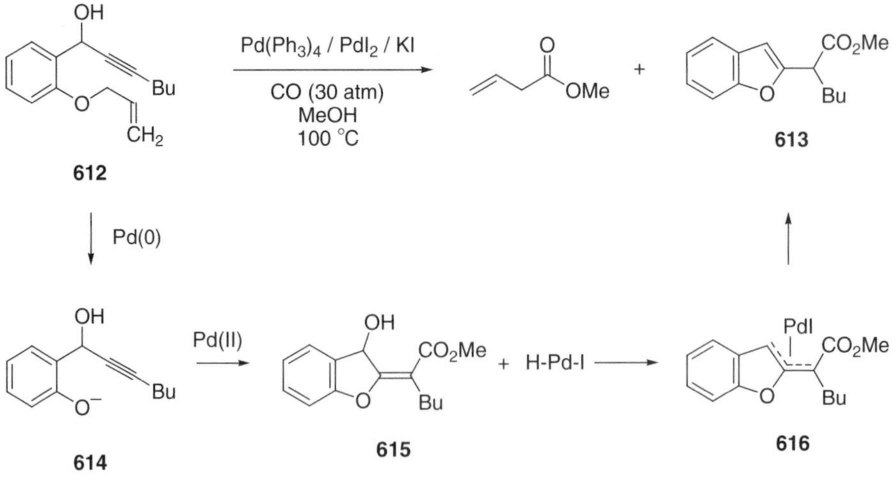

Scheme 101

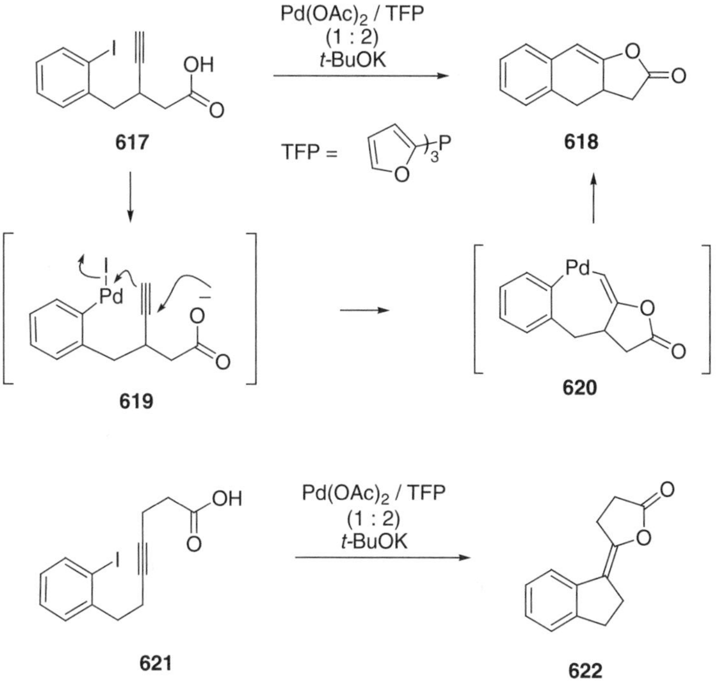

Scheme 102

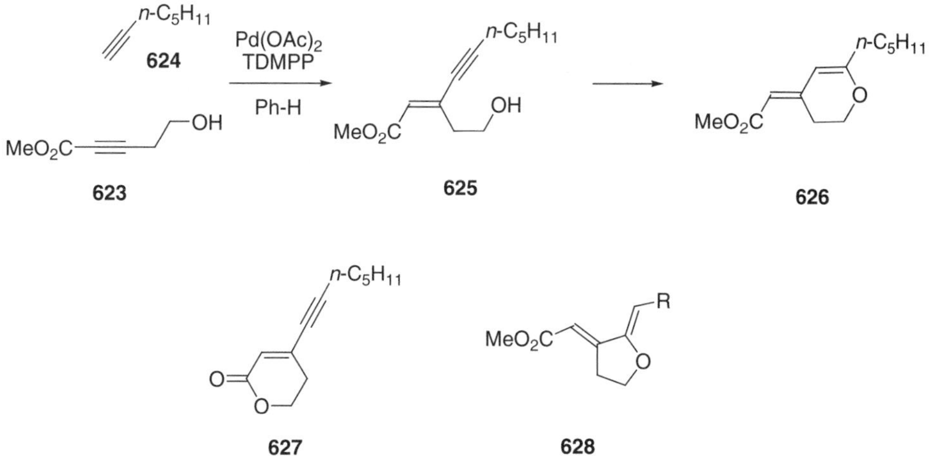

Scheme 103

substituents disfavored the buildup of positive charge at the proximal alkyne carbon, products arising from a 5-exo-dig cyclization (i.e. **628**) were obtained. The stereo-chemical control of the exocyclic double bond with these cases is most unusual since stereochemistry is often difficult to control using other methods.

Pd(II)-mediated cascade cyclizations have also been employed for the construction of other types of heterocyclic systems. For example, Bäckvall reported that the oxidative cyclization of dienyl amides occurred to give pyrrolizidines and indolizidines. Thus, treatment of **629a,b** with Pd(OAc)$_2$ and CuCl$_2$/O$_2$ provided **630a,b** in 90 and 85% yield, respectively (Scheme 104) (92JA8696).

More recently, a tandem Heck/allylic substitution cascade was used as a route toward various lactams. The reaction of **631a,b** with 2-isopropenyl bromide in the presence of Pd(OAc)$_2$ and (*o*-tolyl)$_3$P gave the Heck adduct **632** that reacted further with H-Pd-Br to give the π-allylpalladium complex **633**. Intramolecular attack of the amide anion on the π-allyl complex followed by reductive elimination of Pd(0) afforded **634a,b** in 77 and 75% yield, respectively (03OL259). Other vinyl bromides were also examined and found to give substituted pyrrolidinones and piperidones in moderate to good yields (50–82%).

A tandem Buchwald–Hartwig cross-coupling/Heck cyclization cascade was used to synthesize indole derivatives. In this cascade, 1,2-dibromobenzene reacted with the vinylogous amide **635** in the presence of Pd$_2$(dba)$_3$, ligand **636** and Cs$_2$CO$_3$ to form the intermediate *N*-aryl vinylogous amide **637** (Scheme 105) (00OL1109). By adding a second portion of palladium and the ligand, indole **638** was produced in 61% yield. Interestingly, when *o*-bromobenzaldehyde **639** was allowed to react with **635** under similar conditions, quinoline **640** was isolated in virtually quantitative yield.

Enolates are useful cross-coupling partners and have been used to construct isochromene derivatives. The tandem arylation/cyclization cascade is demonstrated by the reaction of **641** with pinacolone **642** to give **644** in 71% yield (Scheme 106) (01JOC3284). This reaction was optimized for solvent and palladium ligand by using combinational protocols and a variety of ketones were screened. Simple alkyl ketones participated efficiently in this tandem sequence, though several aryl ketones also afforded isochromene products. The most effective ligands for palladium seemed to be DPPF and *o*-biaryl-*tert*-butylphosphine. Other ligands tended to catalyze the arylation reaction (i.e. formation of **643**) but produced less of the cyclic products. The mechanism of the reaction seemingly involves arylation of the enolate derived from **642** to produce intermediate **643** followed by nucleophilic attack on the

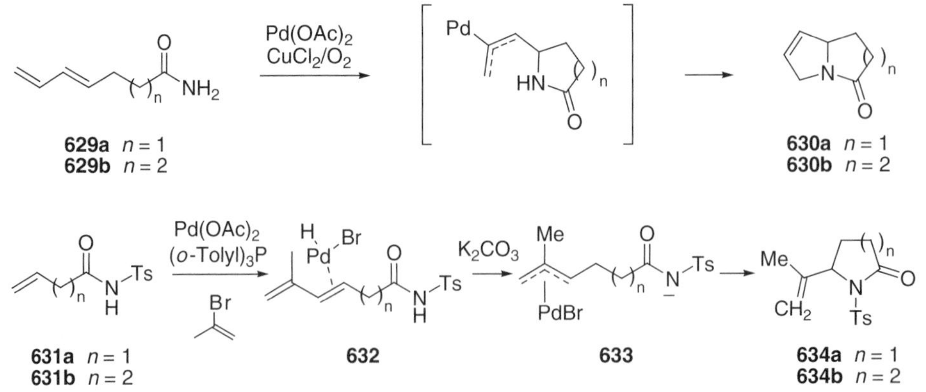

Scheme 104

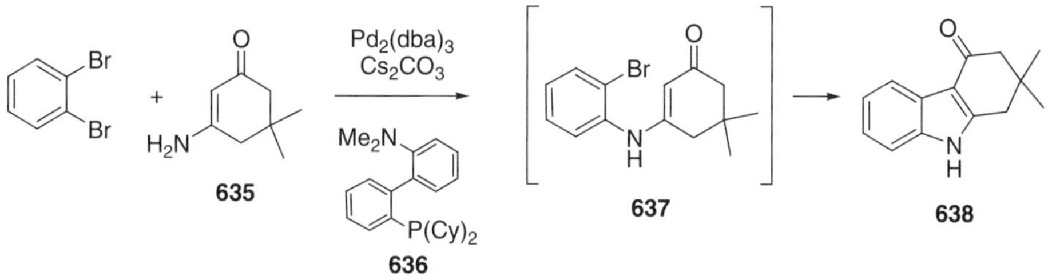

Scheme 105

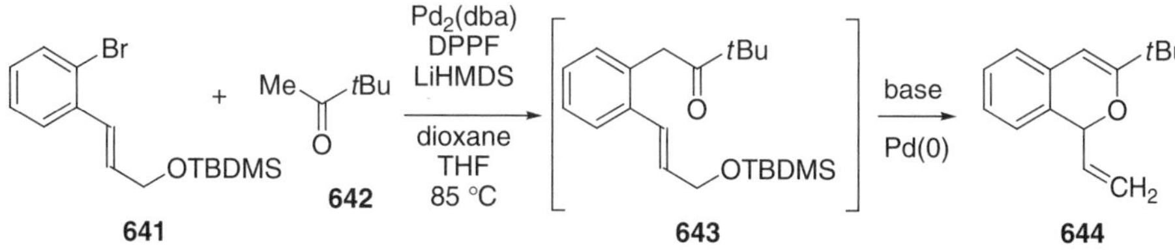

Scheme 106

π-allyl palladium complex by the oxygen atom of the enolate anion derived from **643**. It is interesting to note that the requisite π-allyl palladium complex could be generated from an allylic silyl ether.

Grigg's group has studied palladium-mediated cascade reactions for quite some time (98PAC1047, 99JOMC65). In some of the early work, a tandem alkyne-arene vinylation/alkylation sequence that was terminated by a formal Friedel–Crafts alkylation was used to form polyfused heterocycles. Thus, the reaction of **645** with Pd(0) produced compound **648** in good yield (Scheme 107) (96TL3399). In this sequence, Pd(0) underwent an oxidative insertion into the Ar–I bond and this was followed by carbopalladation of the proximal alkyne to give the vinyl palladium species **646**. A second cyclization generated intermediate **647**. A formal Friedel Crafts alkylation of **647** with the neighboring aromatic ring gave **648**.

Another of Grigg's approaches to heterocyclic synthesis involves the "*anion-capture*" termination of the palladium cascade (91TL2545). In a typical example, a 1:1-mixture of benzyl halide **649** was reacted with sodium tetraphenylborate in the presence of Pd(0) to give compound **650** in 69% yield. Likewise, allyl acetate **651** reacted under similar conditions to afford **652** in 80% yield.

This methodology was used to construct several indole and benzofuran derivatives (00T7541). Oxidative insertion of Pd(0) into the aryl-I bond of **653a,b** followed by carbopalladation onto the pendant olefin afforded an intermediate organopalladium

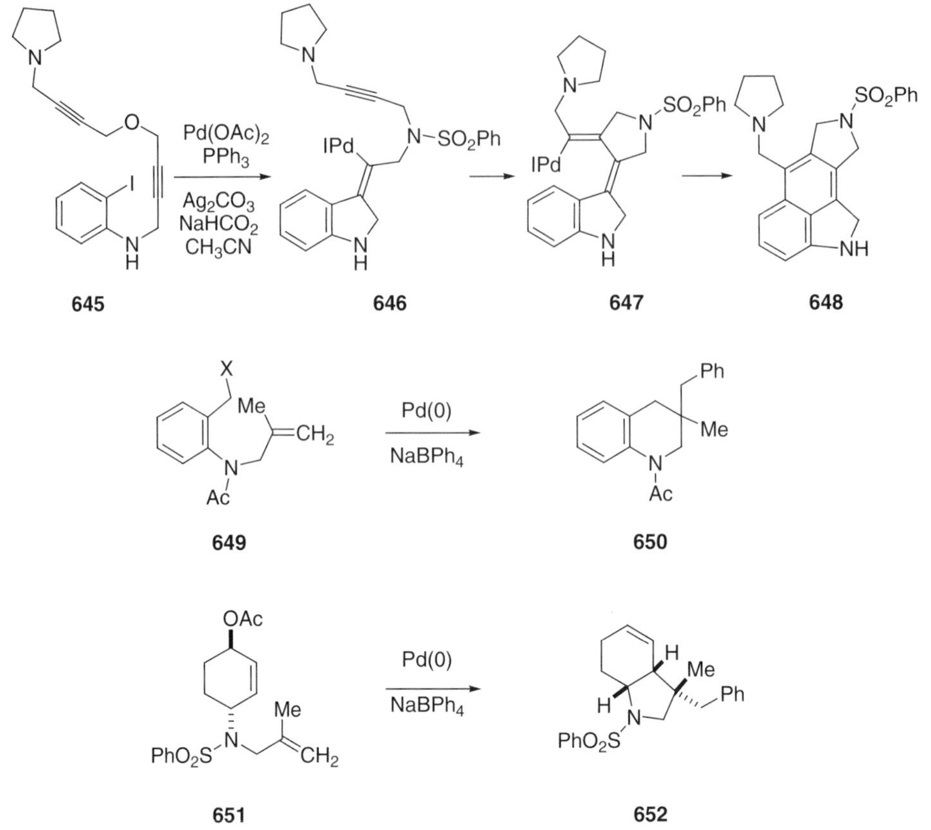

Scheme 107

species **654** (Scheme 108) (00T7525). This organometallic species then reacted with an appropriate stannane (R = alkynyl, vinyl, allyl, heteroaryl) to give **655** or **656** in good yield. By incorporating a second carbopalladation site on the starting aryl halide, heterocycles of much greater complexity were obtained. For example, aryl iodide **657** cyclized to give intermediate **658** when exposed to Pd(0). A second cyclization then produced the organopalladium species **659** that was captured by tributyl-(2-furyl)stannane to give a 5:1-diastereomeric mixture of **660** in 74% yield. Aryl iodide **661** underwent a related cyclization to produce spirocyclic **662** as a single diastereomer in 48% yield.

Manipulation of the temperature of the reaction mixture was also exploited to incorporate a hydrostannylation reaction that ultimately produced an internal stannane terminating group. When held between 0 °C and rt, a toluene solution of compound **663**, HSnBu$_3$, and a palladium (0) catalyst only gave vinyl stannane **664** (Scheme 109) (00T7541). However, on heating the solution containing **664** to 100°C, compound **665** was produced in 56% yield. Macrocycles could also be prepared using this thermally switched cascade reaction. Thus, low-temperature hydrostannylation of **666** followed by heating the intermediate vinyl stannane **667** to 110 °C furnished **668** ($n = 2$–8) in 38–53% isolated yield.

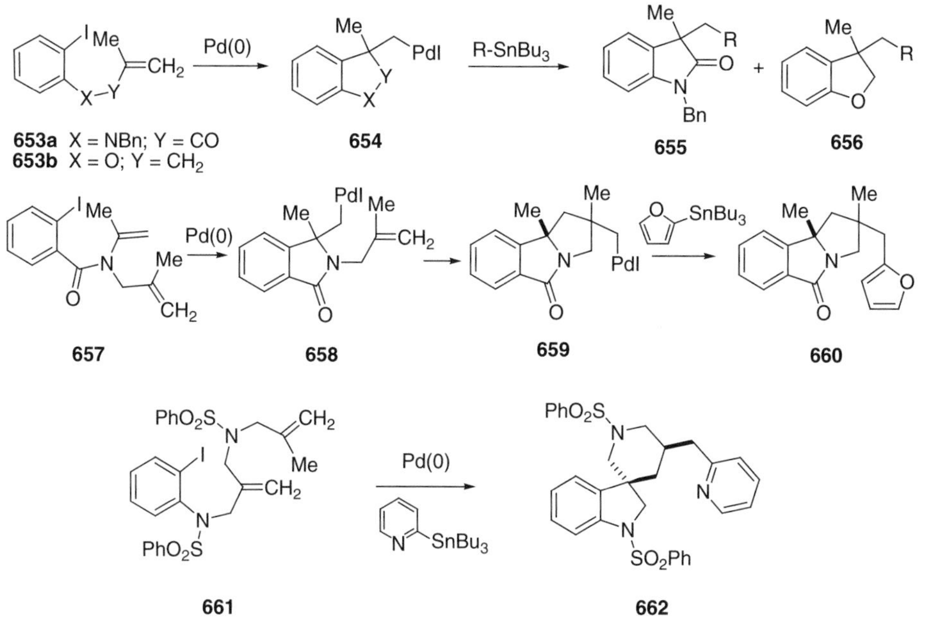

653a X = NBn; Y = CO
653b X = O; Y = CH2
Pd(0)
R-SnBu3
654
655
656
657
658
659
660
661
662
Pd(0)
SnBu3

Scheme 108

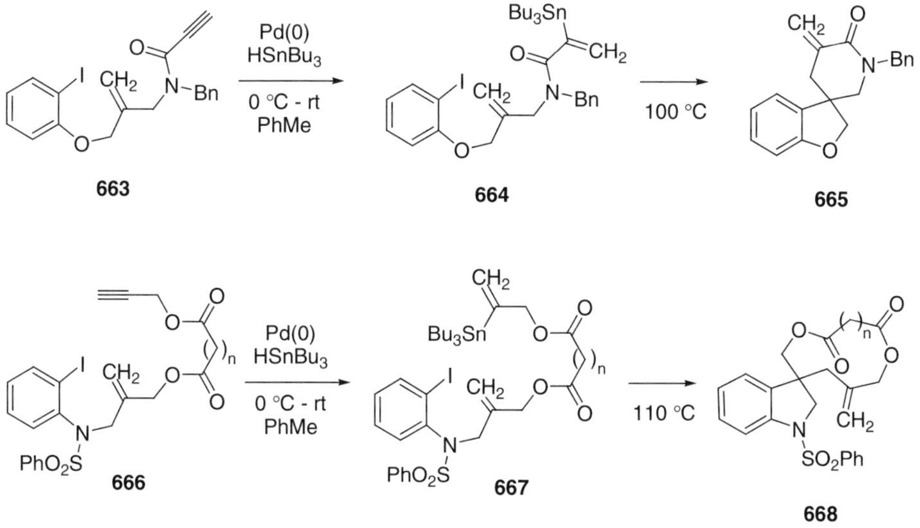

Pd(0)
HSnBu3
0 °C - rt
PhMe
100 °C
663
664
665
666
667
668
110 °C

Scheme 109

B. Copper

The Buchwald group reported an interesting copper-catalyzed lactam *N*-arylation/ring-expansion cascade wherein compound **669** reacted with 2-azetidinone **670** in the presence of CuI and K_2CO_3 to give **671** in 96% yield (Scheme 110) (04JA3529).

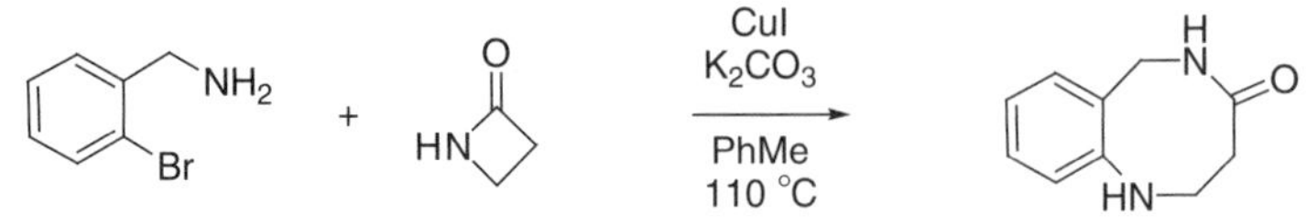

669   **670**   **671**

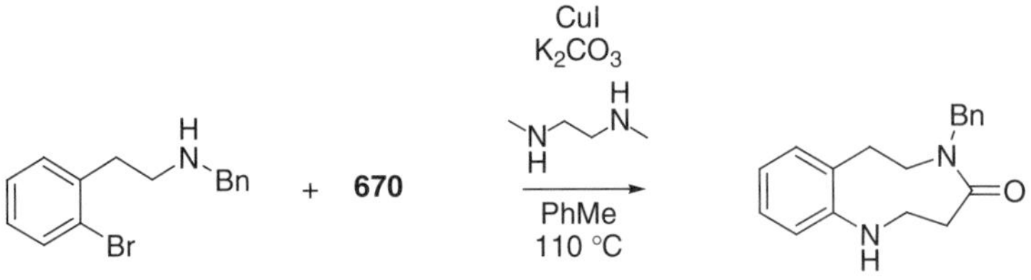

672   **673**

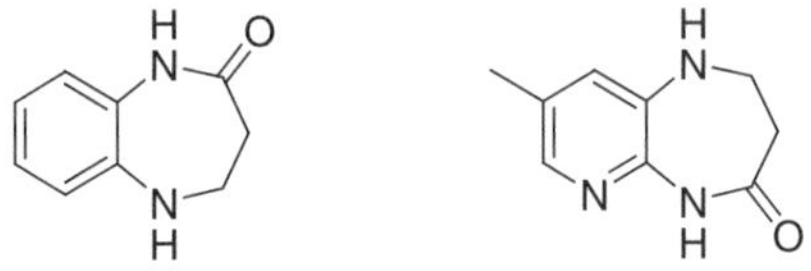

674   **675**

Scheme 110

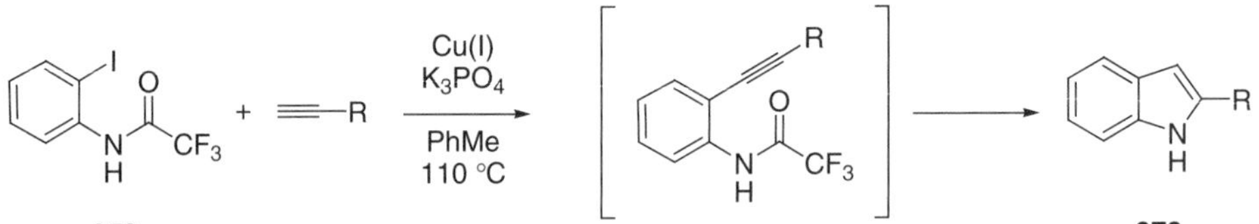

676   **677**   **678**

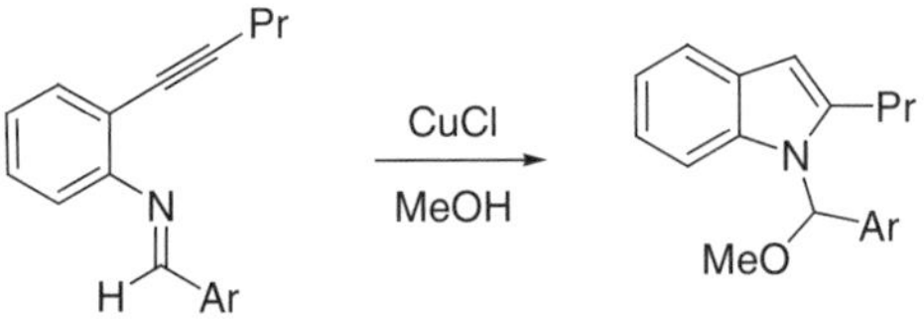

679   **680**

Scheme 111

Several substituted β-lactams and benzyl amine derivatives participated in this coupling reaction, though the ring expansion was slower when *N*-substituted benzyl amines were used. Exposure of the intermediate *N*-aryl lactam to AcOH in THF, however, overcame the sluggish transamidation. In some examples, the addition of *N,N*′-dimethylethylene-diamine facilitated the reaction, though it was not required in all cases. Phenethyl amine **672** reacted with azetidinone **670** to give **673** in 78% yield.

Bromo-substituted anilines underwent the copper-catalyzed coupling reaction, but required a second transamidation step with Ti(O*i*Pr)₄, to produce benzazapinones such as **674** and **675**.

Indoles were also formed by addition of nitrogen nucleophiles to copper-activated alkynes. For example, trifluoroacetanilide **676** underwent reaction with a variety of aryl and heteroaryl alkynes in the presence of a copper catalyst to give 2-substituted indoles **678** in 60–96% yields (Scheme 111) (03OL3843). Use of CuI with (±)-1, 2-*trans*-cyclohexanediamine or PPh₃ ligands effectively facilitated the cyclization of the intermediate Sonogashira product **677**. With indole **678**, however, use of 15 mol.% CuI was required to induce a high yield reaction. The process proceeded more efficiently when 10 mol.% [Cu(phen)(PPh₃)₂]NO₃ was used as the catalyst.

Similarly, the aniline-derived imine **679** bearing an appropriately placed alkyne, underwent the CuCl-mediated cyclization with capture of methanol to give indoles **680** in moderate (53%) to good (74%) yields (04TL35). Several alcohols also participated in the reaction, including phenol, as did aryl alkynes.

C. Cobalt

The Pauson–Khand [2 + 2 + 1]-cycloaddition, is undoubtedly the most common reaction mediated by cobalt and has been incorporated into several cascade sequences. For example, diynes **681a,b** undergo the Pauson–Khand cyclization in the presence of Co₂(CO)₈ to first give a cyclopentadienone that then participates in a subsequent [2 + 2 + 2]-cycloaddition reaction to furnish **682a,b** in 35 and 89% yield, respectively (equation (15)) (99CC2099). In the case of diyne **681c**, a 4.2:1-mixture of regioisomers **682c** and **683** were isolated in 80% combined yield.

681a R = H, X = O
681b R = H, X = NTs
681c R = Ph, X = O

682a R = H, X = O
682b R = H, X = NTs
682c R = Ph, X = O

683

(15)

A less common cobalt-mediated cascade, which involves a radical cyclization and subsequent cross coupling, was employed to form tetrahydrofuran and pyrrolidine derivatives. For example, catalytic Co₂Cl₂(dppe) and an excess of PhMgBr reacted with **684** to give tetrahydrofuran **685** in 80% yield (Scheme 112) (01JA5374). Likewise, toluidine **686** produced pyrrolidine **687** in 81% yield. Both the stereochemical outcome and the required stoichiometry of the Grignard reagent suggest a radical mechanism. It was proposed that excess Grignard reagent converted the cobalt catalyst to the 17-electron-ate complex [Co(0)Ph₂(dppe)](MgBr)₂ along with biphenyl, and that this electron-rich zero-valent-ate species acted as an electron transfer reagent. The reaction of alkyl halide **686** with this Co(0) species gave rise to a new

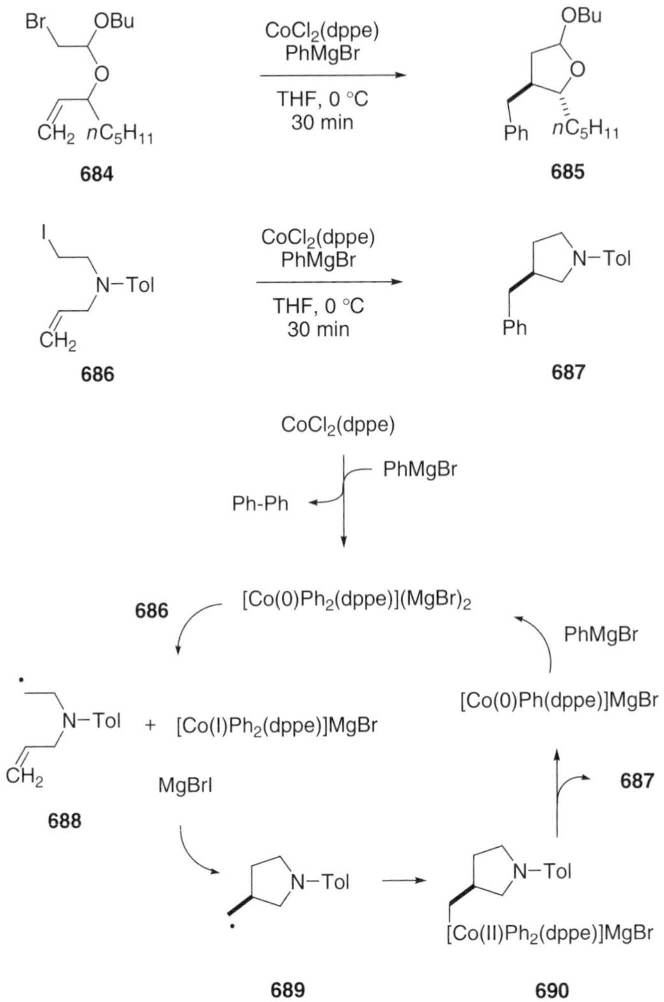

Scheme 112

Co(I) species and radical **688**. Cyclization of **688** would lead to the formation of radical **689** that then combines with the Co(I) species to afford the organometallic species **690**. Reductive elimination produced **687** and regenerated the Co(0) species. Whereas a variety of aryl and heteroaryl Grignard reagents readily participated in this cascade, *o*-substituted aryl Grignards appeared to be too sterically encumbered to efficiently react.

D. TITANIUM AND RHODIUM

The Kelly group showed that TiCl₄ could be used to remove a triphenylmethyl (trityl, Tr) protecting group from cysteine residues and effect a subsequent cyclodehydration to give thiazolines. For example, exposure of **691** to 3 equiv. of TiCl$_4$ at 0 °C for 4 h produced thiazoline **692** in 96% yield and with 99% *ee* (equation (16)) (00OL3289). Other examples of this tandem deprotection/cyclodehydration cascade demonstrated that this process is a general reaction of trityl-protected cysteine *N*-amide derivatives. A variety of thiazolines were formed in moderate to high yields, but some racemization can occur during the cyclodehydration step.

(16)

691   **692**

Hydroformylation reactions that are mediated by rhodium catalysts can also be incorporated into cascade sequences. The zwitterionic rhodium complex **694** promotes a tandem cyclohydrocarbonylation/CO insertion reaction producing pyrrolinone derivatives that contain an aldehyde functional group in good yields (01JA10214). In one example, exposure of α-imino alkyne **693** to catalytic quantities of **694** and (PhO)$_3$P under an atmosphere of CO and H$_2$ at 100 °C produced pyrrolinone **695** in 82% yield (Scheme 113). A variety of alkyl substitutents can be tolerated in this reaction.

Tandem hydroformylation/acetylization reactions have also been examined. Thus, heating a mixture of **696a,b** with [Rh(cod)Cl]$_2$, Ph$_3$P, and CH$_2$Cl$_2$ under a CO/H$_2$ atmosphere afforded **697a,b** in 72 and 55% yield, respectively (02OL289). Six-membered rings were also accessible as shown by the conversion of **698** into **699** under similar conditions.

E. MIXED METAL SYSTEMS

In work that compliments the palladium-mediated alkyne heterocyclization cascade (*cf.* Scheme 103), the Trost group sequenced a ruthenium-catalyzed ene–yne coupling with a palladium catalyzed allylic substitution reaction for the formation of five- and six-membered aza or oxa-heterocycles. For example, alkyne **700a**, which contains a *p*-nitrobenzenesulfonamide group, reacted with nitrophenyl ether **701** in the presence of a ruthenium catalyst to give the intermediate ene–yne **702** (Scheme 114) (02AG4693). Addition of a chiral palladium catalyst to the mixture effected a subsequent cyclization to give pyrrolidine **703a** in 90% yield and with 91% *ee*. Under these same conditions, compound **700b** reacted with **701** to provide piperidine **703b** in 83% yield with 88% *ee*. Solvent choice was important: the ene–yne coupling was more efficient in acetone, whereas the allylic alkylation reaction was more effective in CH$_2$Cl$_2$. Also, the nitrophenyl ether was selected as the allylic leaving group since it was not strong enough a nucleofuge to undergo ionization under the ruthenium catalyzed conditions, yet ionized readily with palladium. Tetrahydrofurans and

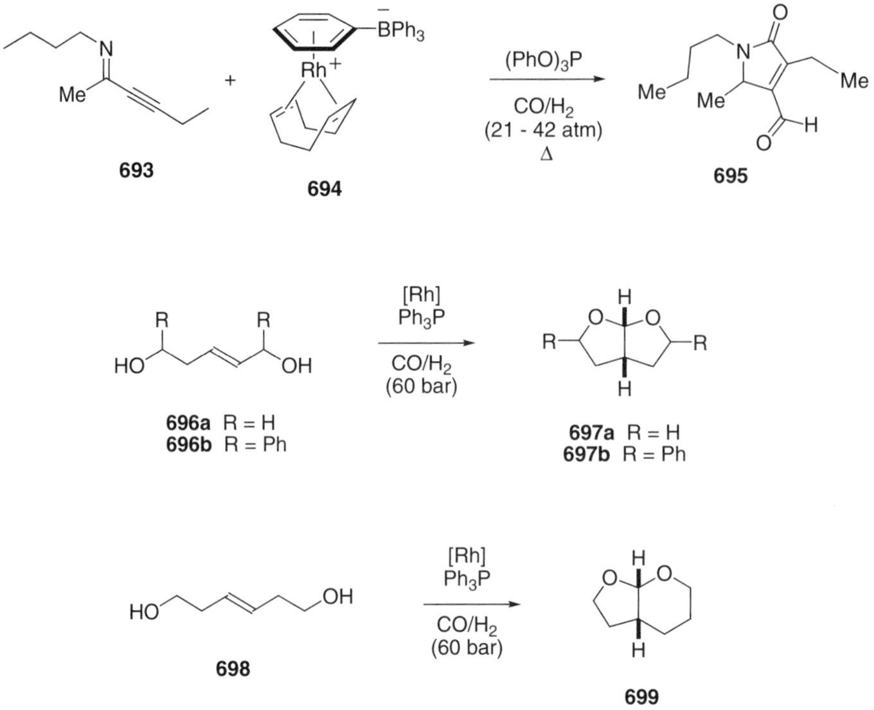
693
694
695
(PhO)3P
CO/H2
(21 - 42 atm)
Δ
[Rh]
Ph3P
CO/H2
(60 bar)
696a R = H
696b R = Ph
697a R = H
697b R = Ph
698
699

Scheme 113

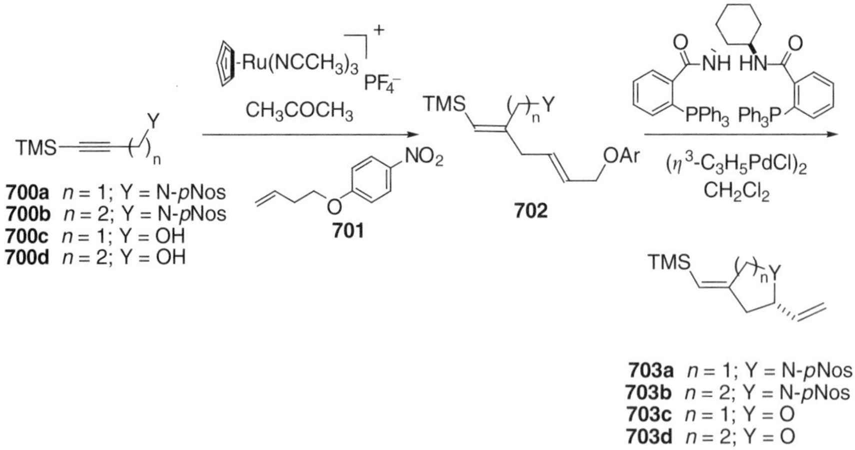
Ru(NCCH3)3 PF4−
CH3COCH3
700a n = 1; Y = N-pNos
700b n = 2; Y = N-pNos
700c n = 1; Y = OH
700d n = 2; Y = OH
701
702
(η3-C3H5PdCl)2
CH2Cl2
703a n = 1; Y = N-pNos
703b n = 2; Y = N-pNos
703c n = 1; Y = O
703d n = 2; Y = O

Scheme 114

pyrans were similarly constructed. Thus, the reaction of alkynyl alcohols **700c** or **700d** with **701** gave **703c** in 84% yield (76% *ee*) and **703d** in 80% yield (94% *ee*).

The use of two different metal catalysts for tandem reactions is often accomplished by the sequential addition of the appropriate catalyst at an intermediate stage of the

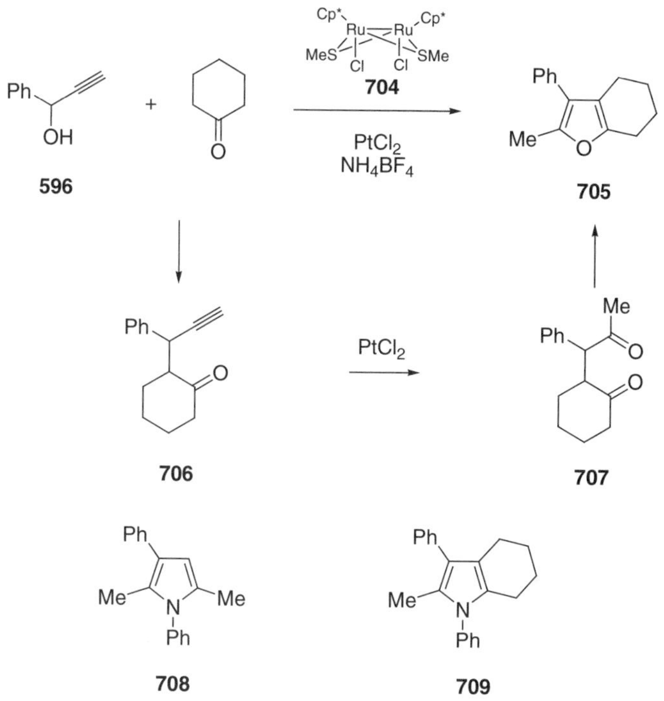

Scheme 115

reaction (*cf.* Scheme 114). This is particularly necessary if the catalysts react independently with the starting substrates at competitive rates. Heterobimetallic catalyst systems are less common. The sulfur bridged ruthenium catalyst **704** works in concert with PtCl$_2$ to produce furans from proparylic alcohols and ketones. Specifically, heating the aryl-substituted propargylic alcohol **596** and cyclohexanone with catalyst **704** provided furan **705** in 75% yield (Scheme 115) (03AG2681). Acetone, methyl ethyl ketone, and diethyl ketone all underwent this transformation in good yields (72–85% yield) as did cycloheptanone. The mechanism proposed involved an initial ruthenium catalyzed propargylic substitution reaction with the ketone to produce the intermediate keto-alkyne **706**. A subsequent platinum catalyzed hydration of the alkyne gave dione **707** and this was followed by an intramolecular condensation to provide furan **705**. The addition of aniline to the reaction mixture afforded pyrroles **708** and **709** in 50 and 56% yield, respectively.

XI. Metathesis

In recent years, the ring-closing olefin metathesis (RCM) reaction has become one of the more powerful tools for the synthesis of heterocycles (95ACR446, 04CR2199), particularly medium-sized rings that are hard to form by other methods. Of the cascade reactions involving RCM, many are multiple sequential ring-closures or involve a tandem ring-opening/ring-closing reaction. Appropriately arrayed polyenes easily undergo multiple ring-closing metathesis reactions. For example, Harrity and coworkers demonstrated that the Grubb's first-generation catalyst **711** catalyzed the conversion of tetraene **710** into the spirocyclic **712** in 90% yield (Scheme 116) (99TL3247). None of the corresponding seven-membered ring product was isolated under these conditions. RCM reactions are known to be sensitive to conformational preferences within a molecule as well as substitution on the participating alkene. For example, diester **713** afforded **714** in 50% yield and macrocycle **715** in 19% yield, along with other dimeric material, when exposed to catalytic quantities of **711** at rt in CH_2Cl_2 (02TL7851). Use of the second-generation Grubb's catalyst **717** increased the yield of **715** (up to 45% yield), but no detectable amount of **716** was observed. Diether **718a** underwent reaction with **711** to give mixtures that contained spirocycle **719**, but only in 21% yield. Formation of five-membered rings continue to be the dominant pathway. To retard the formation of the five-membered ring products, diether **718b** was synthesized and underwent reaction with **711** to produce spirocyclic **719** in 46% yield.

The Grubb's group has made extensive use of the tandem RCM cascade with tethered alkynes to produce heterocyclic compounds. For example, acyclic ester **720** reacted in the presence of 5 mol.% of catalyst **717** in CH_2Cl_2 at 40 °C to furnish

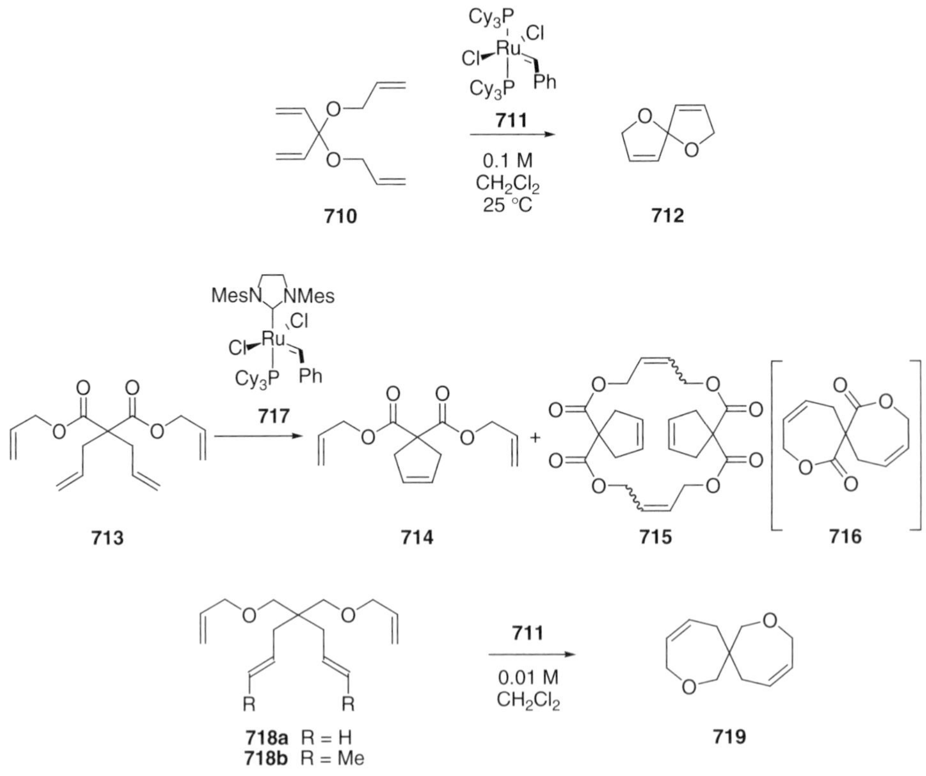

Scheme 116

bicyclic lactone **721** in 95% yield (Scheme 117) (01CC2648). The cyclic alkyne **722** reacted under the same conditions to give **723** in 74% yield.

Strained cycloalkenes that contain appropriately tethered olefins can also undergo the tandem ring-opening metathesis/RCM cascade. Using the first-generation catalyst **711**, Grubbs transformed a series of diallyloxy-cycloalkenes **724** into *bis*-furan derivatives **725** in moderate to good yields (57–90%) (96JA6634). Likewise, cyclopentenes **726a,b** reacted in the presence of catalyst **717** to afford **727a,b** in 81 and 89% yield, respectively (01CC2648). Bicycloalkene **728** was converted into tricyclic **729** in 47% yield.

The Blechert group exploited the ring-opening/ring-closing cascade for the synthesis of several natural products. In a synthesis of (–)swainsonine (**735b**), sulfonamide **730** was heated at 40 °C with 5 mol.% Grubbs catalyst **711** under an atmosphere of ethene to provide pyrroline derivative **731** in 98% yield (02JOC4325).

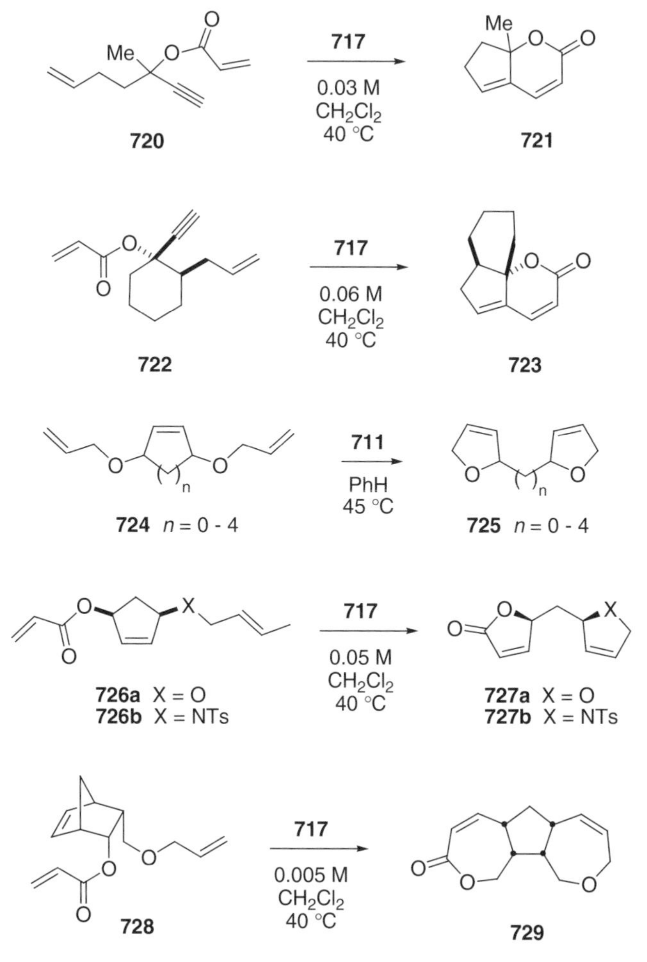

Scheme 117

Hydroboration of the terminal alkene in **731** with 9-BBN followed by an oxidative workup under alkaline conditions gave alcohol **732** in 83% yield. The sulfonamide was cleaved by the action of Na/Hg. Acylation of the resulting free amino nitrogen with allyl chloroformate followed by mesylation afforded **733** in 87% yield from **732**. Palladium-mediated removal of the allyl carbamate group liberated a basic nitrogen that subsequently displaced the mesylate group to produce **734** in 95% yield. An asymmetric dihydroxylation of **734** produced an inseparable 20:1-mixture of diastereomeric diols. Separation of these diastereomers required removal of the silyl ether and conversion of the resulting triol into the triacetate occurred in 68% yield and with isomer **735a** predominating. A base promoted hydrolysis then yielded the desired target **735b** in 96% yield (Scheme 118).

Extending this approach, Blechert's group reported a particularly efficient synthesis of indolizidine 167B (**740**) (Scheme 119). The cycloheptene derivative **736** was initially converted into silacycle **737** by the action of catalyst **711** (02TL6739). A subsequent addition of TBAF to the reaction mixture promoted cleavage of the silicon group to give **738** in 92% yield from **736**. Oxidation of the alcohol using the Dess–Martin periodane reagent afforded ketone **739** in 73% yield. A palladium-mediated hydrogenolysis of the benzyl carbamate was accompanied by intramolecular reductive amination of the ketone and this was followed by reduction of the alkene to provide **740** in 79% yield.

Synthesis of the alkaloid cuscohygrine (**744**) also came about from this methodology (Scheme 120). Exposing *bis*-carbamate **741** to catalyst **711** in hot CH$_2$Cl$_2$ produced a *bis*-pyrroline derivative whose double bonds proved to be unstable (02JOC6456). Accordingly, the crude metathesis reaction mixture was treated with palladium in the presence of a hydrogen atmosphere to give **742a** in 72% yield from

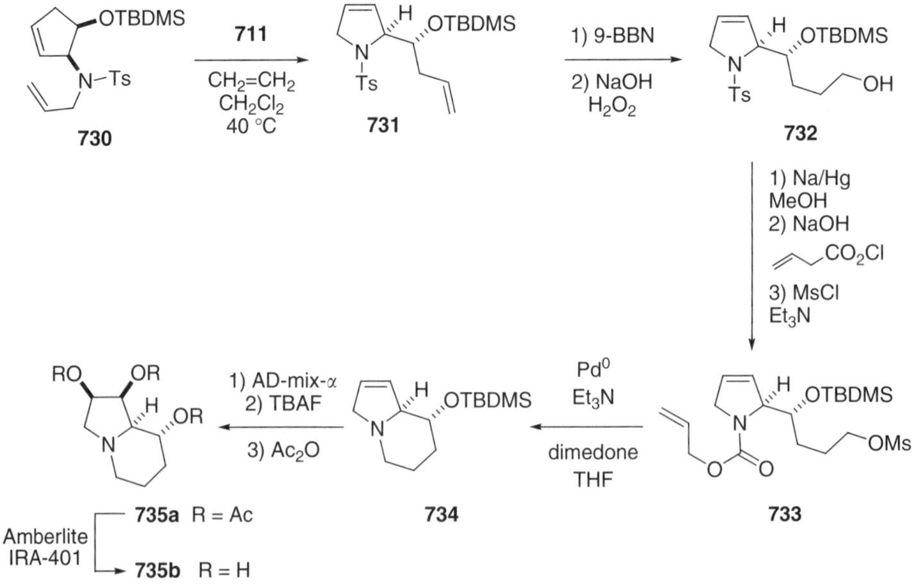

Scheme 118

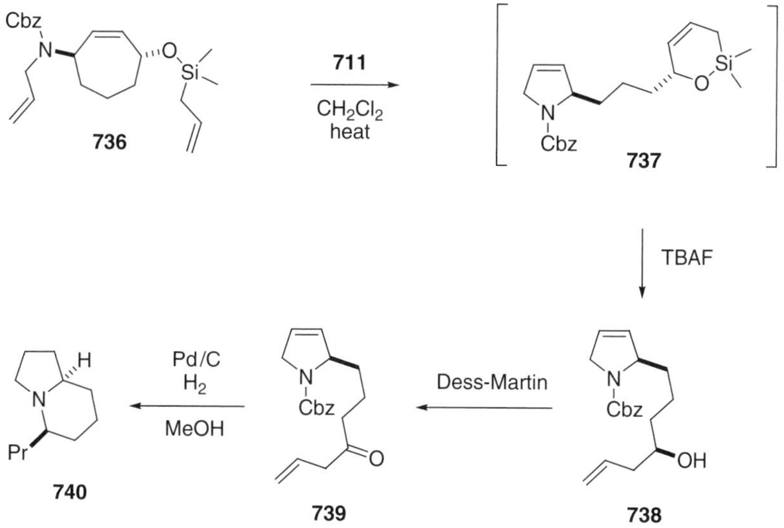

Scheme 119

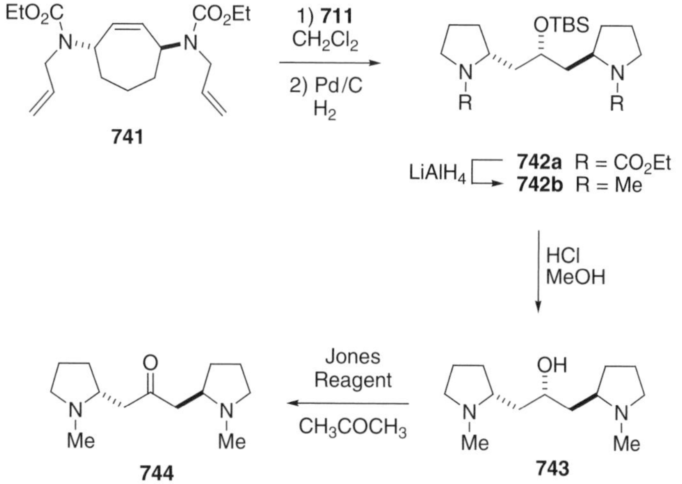

Scheme 120

741. Reduction of the carbamate-protecting group by the action of LiAlH₄ afforded **742b** in 92% yield. An acid-mediated hydrolysis of the silyl ether provided (+) -dehydrocuscohygrine (**743**) in 89% yield. Further oxidation of the alcohol under Jones conditions produced the desired target **744** in 73% yield.

Recently, Hoveyda and Schrock have developed molybdenum-based catalysts (i.e. **744a,b**) for an asymmetric ring-opening metathesis (AROM) reaction (Scheme 121) (98JA4041, 98JA9720, 99JA11603). Using a tandem AROM/RCM sequence, they examined the asymmetric synthesis of several heterocyclic compounds. For example,

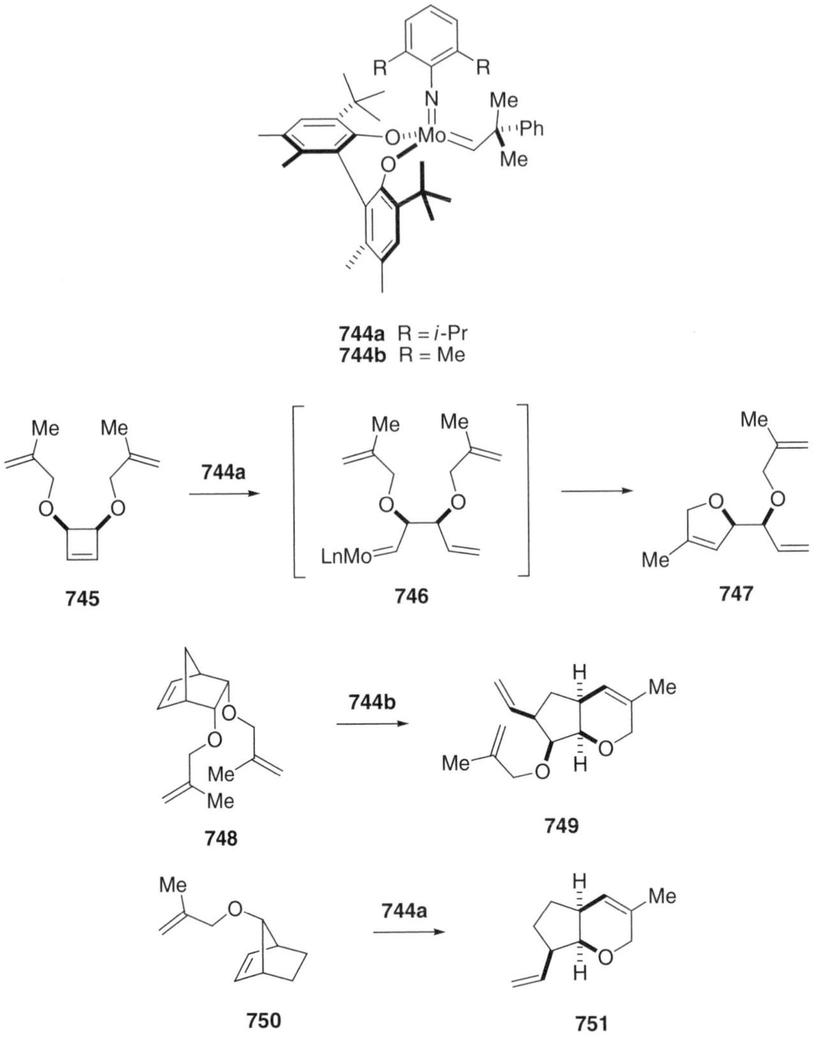

744a R = *i*-Pr
744b R = Me

Scheme 121

meso *bis*-ether **745** delivered **747** in 69% yield and with a 92% *ee* on exposure to 5 mol.% of **744a** (00JA1828). In this case, ring opening by the catalyst was faster than the metathesis reaction with one of the pendant alkenes, and alkylidene carbene **746** was enantiospecifically produced as an intermediate. The less sterically hindered catalyst **744b** mediated the transformation of **748** into **749** in 84% yield and with greater than 98% *ee*. Interestingly, **750** could be converted into **751** by the action of catalyst **744b** in 60% yield and with 72% *ee*, whereas catalyst **744a** did not promote the reaction. By adding diallyl ether to the reaction mixture, **744a** did catalyze the reaction and provided **751** in 54% yield but with 92% *ee*.

Building on their earlier tandem ring-opening/cross-metathesis studies (03EJOC611), Arjona and Plumet have applied the ring-opening/ring-closing/cross-metathesis cascade

(98SL169) to the synthesis of various nitrogenous heterocycles. For example, bicyclic amides **752a,b** reacted with ethene in the presence of Grubbs catalyst **711** to give **753a,b** in 65 and 60% yields, respectively (Scheme 122) (02JOC1380). Bicyclic lactams **753** were the expected products of a ring-opening/ring-closing metathesis cascade without the initial cross-metathesis with ethylene. To demonstrate the actual cross-metathesis reaction, allyl acetate was used as the partner. In this case, compound **752a** reacted with catalyst **711** and allyl acetate and gave **754** in 40% isolated yield along with **753a** in 30% yield. By changing to catalyst **717**, the yield of **754** was increased to 65% yield, though **753a** was again produced in 30% yield.

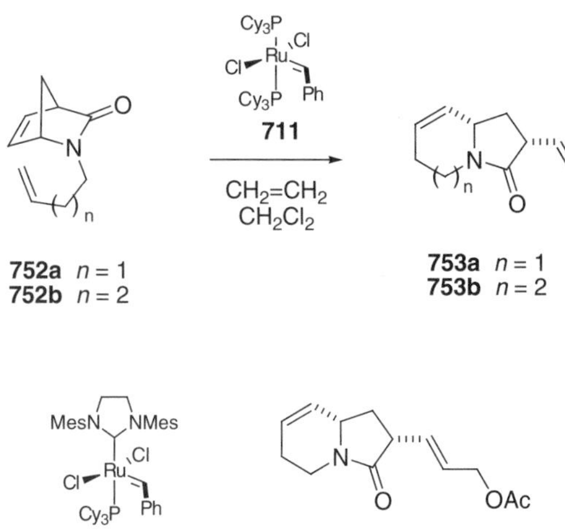

752a *n* = 1
752b *n* = 2

753a *n* = 1
753b *n* = 2

711

$CH_2=CH_2$
CH_2Cl_2

717

754

Scheme 122

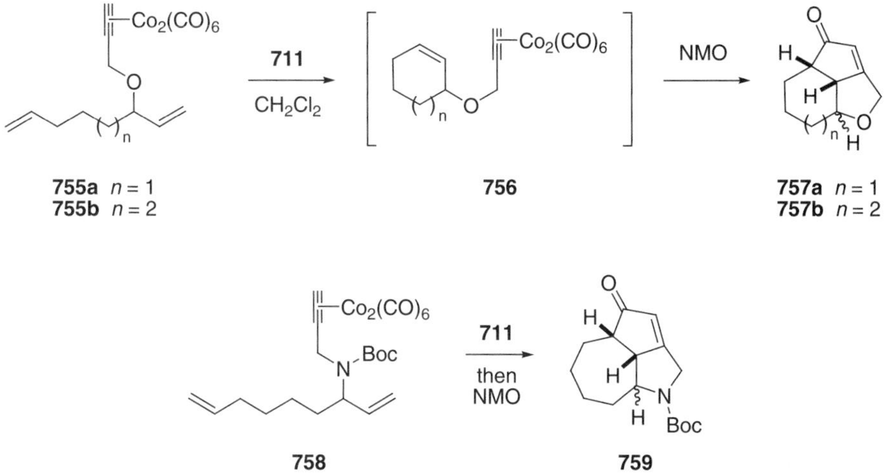

755a *n* = 1
755b *n* = 2

756

757a *n* = 1
757b *n* = 2

711

CH_2Cl_2

NMO

758

759

711

then
NMO

Scheme 123

A one-pot sequential RCM/Pauson–Khand reaction sequence has been used to synthesize nitrogen- and oxygen-containing polycycles. The cobalt–alkyne complexes **755a,b** reacted in the presence of Grubbs catalyst **711** to give **756** (Scheme 123) (03OBC1450). The addition of NMO to the crude reaction mixture promoted the Pauson–Khand reaction and provided **757a,b** in 81 and 70% yield as mixtures of diastereomers. Similarly, **758** underwent the tandem RCM/Pauson–Khand to produce **759** in 67% yield.

XII. Concluding Remarks

From the selective sampling of cascade reactions for the synthesis of heterocyclic molecules that has been outlined in this chapter, it is clear that virtually any reaction can be incorporated into a tandem sequence. Some cascade sequences increase molecular complexity more than others, but each seems to provide complex hetero-cyclic structures in a more efficient manner than the corresponding chemistry wherein each intermediate is isolated. Indeed, many of these cascades rapidly construct hetero-polycyclic systems that are difficult to produce in other ways.

Several cascade sequences for heterocyclic synthesis have been well explored; Padwa's rhodium carbene-initiated dipolar cycloadditions, Denmark's nitroalkene [4 + 2]/[3 + 2]-cycloadditions, Overman's Aza-Cope/Mannich cascade, Bunce's conjugate addition strategy, Molina's Aza-Wittig/heterocumulene cyclization reactions, and Grigg's use of relays and switches in palladium-mediated cascades have matured into significant synthetic tools. Familiar multi-component reactions, such as the Ugi reaction, are being used in interesting ways. Others sequences show tremendous promise; Fu's asymmetric Kinugasa reaction, indium-initiated radical cascades, Buchwald's copper catalyzed *N*-arylation reactions, Trost's alkyne heterocyclization, and the Hoveyda and Schrock tandem AROM/RCM reactions all provide improvements in stereoselectivity and involve the use of environmentally benign reagents. Continued development of these cascade reactions will have a significant impact on the processes used to make heterocyclic compounds on an industrial scale.

Acknowledgment

AP wishes to acknowledge the research support of our Cascade program in heterocyclic chemistry by the National Institutes of Health (GM 0539384) and the National Science Foundation (CHE-0450779).

REFERENCES

71JOC3202   M. R. Slabaugh and W. C. Wildman, *J. Org. Chem.*, **36**, 3202–3207 (1971).
72CC466   M. Kinugasa and S. Hashimoto, *J. Chem. Soc., Chem. Commun.*, (8), 466–467 (1972).

73TL957   A. P. Krapcho and A. J. Lovey, *Tetrahedron Lett*, **14**(12), 957–960 (1973).

74TL4475   M. Hamaguchi and T. Ibata, *Tetrahedron Lett*, **15**(51–52), 4475–4476 (1974).

75CL499   M. Hamaguchi and T. Ibata, *Chem. Lett.*, **4**, 499 (1975).

79CB1913   For an early example involving an Aza-Cope/vinylogous Mannich cascade in the synthesis of yohimbone, see W. Benson and E. Winterfeldt, *Chem. Ber.*, **112**, 1913–1915 (1979).

79JA1310   L. E. Overman and M. -A. Kakimoto, *J. Am. Chem. Soc.*, **101**, 1310–1312 (1979).

79JOC3117   M. E. Christy, P. S. Anderson, S. F. Britcher, C. D. Colton, B. E. Evans, D. C. Remy, and E. L. Engelhardt, *J. Org. Chem.*, **44**, 3117–3127 (1979).

81JA7687   J. P. Marino and M. Neisser, *J. Am. Chem. Soc.*, **103**, 7687–7689 (1981).

83JA4750   T. Gallagher, P. Magnus, and J. C. Huffman, *J. Am. Chem. Soc.*, **105**, 4750 (1983).

83JOC2685   L. E. Overman, M. Sworin, and R. M. Burk, *J. Org. Chem.*, **48**, 2685 (1983).

83T3707   M. E. Kuehne and W. G. Earley, *Tetrahedron*, **39**, 3707 (1983).

84ACR35   P. Magnus, T. Gallagher, P. Brown, and P. Pappalardo, *Acc. Chem. Res.*, **17**, 35 (1984).

85HCA745   L. E. Overman and S. Sugai, *Helv. Chim. Acta.*, **68**, 745 (1985).

85TL5381   J. P. Marino and R. F. de la Pradilla, *Tetrahedron Lett.*, **26**, 5381–5384 (1985).

87JOC226   M. H. Norman and C. H. Heathcock, *J. Org. Chem.*, **52**, 226–235 (1987).

87S1088   J. P. Marino, R. F. de la Pradilla, and E. Laborde, *Synthesis*, (12), 1088–1091 (1987).

88JA4329   E. J. Jacobsen, J. Levin, and L. E. Overman, *J. Am. Chem. Soc.*, **110**, 4329–4336 (1988).

88JOC4654   P. Molina, A. Arques, and M. V. Vinader, *J. Org. Chem.*, **53**, 4654–4663 (1988).

88JOC6154   M. Harmata and C. B. Gamlath, *J. Org. Chem.*, **53**, 6154–6156 (1988).

88TL1677   M. E. Maier and K. Evertz, *Tetrahedron Lett.*, **29**, 1677–1680 (1988).

88TL3849   P. Molina and A. Vidal, *Tetrahedron Lett.*, **29**, 3849–3852 (1988).

88TL4169   A. Padwa, U. Chiacchio, D. C. Dean, A. M. Schoffstall, A. Hassner, and K. S. K. Murthy, *Tetrahedron Lett.*, **29**, 4169–4172 (1988).

89CB307   P. Molina, A. Arques, P. M. Fresneda, M. V. Vinader, and M. d. l. C. Foces-Foces, *Chem. Ber.*, **122**, 307–313 (1989).

89JOC1236   L. E. Overman, G. Robertson, and A. J. Robichaud, *J. Org. Chem.*, **54**, 1236 (1989).

89TL2289   A. Hassner and R. Maurya, *Tetrahedron Lett.*, **30**, 2289–2292 (1989).

90JOC1624   G. A. Kraus, P. J. Thomas, D. Bougie, and L. Chen, *J. Org. Chem.*, **55**, 1624 (1990).

90JOC5719   W. H. Pearson, S. C. Bergmeier, S. Degan, K. -C. Lin, Y. -F. Poon, J. M. Schkeryantz, and J. P. Williams, *J. Org. Chem.*, **55**, 5719–5738 (1990).

90T5759   S. Braverman and M. Freund, *Tetrahedron*, **46**, 5759–5776 (1990).

90TL3039   A. Barco, S. Benetti, A. Casolari, G. P. Pollini, and G. Spalluto, *Tetrahedron Lett.*, **31**, 3039–3042 (1990).

90TL5981   M. Harmata, C. B. Gamlath, and C. L. Barnes, *Tetrahedron Lett.*, **31**, 5981–5984 (1990).

91COS   F. E. Ziegler in "*Comprehensive Organic Synthesis, Combining C–C π-Bonds*" (L. A. Paquette, ed.), Vol. 5, Chap. 7.3, Pergamon Press, Oxford (1991).

91COS779   D. P. Curran, in "*Comprehensive Organic Synthesis*" (B. M. Trost and I. Fleming, eds.), Vol. 4, p. 779, Pergamon, Oxford (1991).

91JA2598   L. E. Overman, G. M. Robertson, and A. J. Robichaud, *J. Am. Chem. Soc.*, **113**, 2598–2610 (1991).

91JA5085   J. M. Fevig, R. W. Marquis, and L. E. Overman, *J. Am. Chem. Soc.*, **113**, 5085–5086 (1991).

91JA5378   M. J. Brown, T. Harrison, and L. E. Overman, *J. Am. Chem. Soc.*, **113**, 5378–5384 (1991).

91JOC5005   L. E. Overman and J. Shim, *J. Org. Chem.*, **56**, 5005–5007 (1991).

91JOC820   M. P. Doyle, R. J. Pieters, J. Tauton, H. Q. Pho, A. Padwa, D. L. Hertzog, and L. Precedo, *J. Org. Chem.*, **56**, 820 (1991).

91TA1231   H. Waldmann, M. Braun, and M. Dräger, *Tetrahedron Asymm.*, **2**, 1231–1246 (1991).

91TL1787   J. Nkiliza and J. Vercauteren, *Tetrahedron Lett.*, **32**, 1787 (1991).

91TL2545   R. Grigg, S. Sukirthalingam, and V. Sridharan, *Tetrahedron Lett.*, **32**, 2545–2548 (1991).

91TL2549   T. M. Swarbrick, I. E. Markó, and L. Kennard, *Tetrahedron Lett.*, **32**, 2549–2552 (1991).

91TL5183   D. Sole and J. Bonjoch, *Tetrahedron Lett.*, **32**, 5183 (1991).

91TL6395   C. Belle, A. Cardelli, and A. Guarna, *Tetrahedron Lett.*, **32**, 6395–6398 (1991).

92ACR352   L. E. Overman, *Acc. Chem. Res.*, **25**, 352–359 (1992).

92CL443   S. Takano, K. Inomata, and K. Ogasawara, *Chem. Lett.*, **21**(3), 443 (1992).

92JA3115   S. Hanessian and R. Léger, *J. Am. Chem. Soc.*, **114**, 3115–3117 (1992).

92JA8696   P. G. Andersson and J. E. Bäckvall, *J. Am. Chem. Soc.*, **114**, 8696–8698 (1992).

92JOC1727   R. A. Bunce, C. J. Peeples, and P. B. Jones, *J. Org. Chem.*, **57**, 1727–1733 (1992).

92JOC5666   A. Brandi, Y. Dürüst, F. M. Cordero, and F. De Sarlo, *J. Org. Chem.*, **57**, 5666–5670 (1992).

92TL4731   D. L. Hertzog, D. J. Austin, W. R. Nadler, and A. Padwa, *Tetrahedron Lett.*, **33**, 4731–4734 (1992).

92TL4993   W. R. Bowman, D. N. Clark, and R. J. Marmon, *Tetrahedron Lett.*, **33**, 4993–4994 (1992).

92TL5649   I. E. Markó, P. Seres, T. M. Swarbrick, I. Staton, and H. Adams, *Tetrahedron Lett.*, **33**, 5649–5652 (1992).

92TOR13103   T. L. Ho, in "*Tandem Organic Reactions*" Wiley, New York (1992); R. A. Bunce, *Tetrahedron*, **51**, 13103 (1995).

93AGE131   L. F. Tietze and U. Beifuss, *Angew. Chem., Int. Ed. Engl.*, **32**, 131 (1993).

93JA2064   J. Bonjoch, D. Sole, and J. Bosch, *J. Am. Chem. Soc.*, **115**, 2064 (1993).

93JA3030   V. H. Rawal, C. Michoud, and R. F. Monestel, *J. Am. Chem. Soc.*, **46**, 3030 (1993).

93JA7904   A. G. Schultz, M. A. Holoboski, and M. S. Smyth, *J. Am. Chem. Soc.*, **115**, 7904 (1993).

93JOC2468   T. A. Grese, K. D. Hutchinson, and L. E. Overman, *J. Org. Chem.*, **58**, 2468–2477 (1993).

93JOC4473   J. H. Rigby and M. H. Qabar, *J. Org. Chem.*, **58**, 4473 (1993).

93JOC7635   W. G. Dauben, J. Dinges, and T. C. Smith, *J. Org. Chem.*, **58**, 7635–7637 (1993).

93SC1009   R. A. Bunce and M. J. Bennett, *Synth. Commun.*, **23**, 1009–1020 (1993).

93SL1   A. Brandi, F. M. Cordero, F. De Sarlo, A. Goti, and A. Guarna, *Synlett*, (1), 1–8 (1993).

93T10219   J. H. Rigby, M. Qabar, G. Ahmed, and R. C. Hughes, *Tetrahedron*, **49**, 10219 (1993).

93TL7305   I. E. Markó, P. Seres, G. R. Evans, and T. M. Swarbrick, *Tetrahedron Lett.*, **34**, 7305–7308 (1993).

94BCSJ3021   K. Funabiki, I. Tamura, and H. Yamanaka, *Bull. Chem. Soc. Jpn.*, **67**, 3021–3029 (1994).

94JOC5633   N. Uesaka, F. Saitoh, M. Mori, M. Shibasaki, K. Okamura, and T. Date, *J. Org. Chem.*, **59**, 5633 (1994).

94PAC2095   M. E. Kuehne, C. S. Brook, F. Xu, and R. Parsons, *Pure Appl. Chem.*, **66**, 2095 (1994).

94SL375   D. Schinzer and E. Langkopf, *Synlett*, (6), 375–377 (1994).

94T1295   W. R. Bowman, D. N. Clark, and R. J. Marmon, *Tetrahedron*, **50**, 1295–1310 (1994).

94T2583   A. Barco, S. Benetti, C. De Risi, G. P. Pollini, R. Romagnoli, G. Spalluto, and V. Zanirato, *Tetrahedron*, **50**, 2583–2590 (1994).

94T9943   S. Chan and T. F. Braish, *Tetrahedron*, **50**, 9943–9950 (1994).

94TA641   J. T. B. Ferreira, J. A. Marques, and J. P. Marino, *Tetrahedron Asymm.*, **5**, 641–648 (1994).

95ACR446   R. H. Grubbs, S. J. Miller, and G. C. Fu, *Acc. Chem. Res.*, **28**, 446–452 (1995).

95CR195   P. J. Parsons, C. S. Penkett, and A. J. Shell, *Chem. Rev.*, **95**, 195 (1995).

95DROSH193   H. Waldmann, in "*Domino Reaction in Organic Synthesis Highlight II*" (H. Waldmann ed.), pp. 193–202, VCH, Weinheim (1995).

95JA7834   J. H. Rigby, R. C. Hughes, and M. J. Heeg, *J. Am. Chem. Soc.*, **117**, 7834 (1995).

95JA10391   D. W. C. MacMillan and L. E. Overman, *J. Am. Chem. Soc.*, **117**, 10391–10392 (1995).

95JOC3938   J. E. Cochran and A. Padwa, *J. Org. Chem.*, **60**, 3938–3939 (1995).

95JOC4999   M. Miura, M. Enna, K. Okura, and M. Nomura, *J. Org. Chem.*, **60**, 4999–5004 (1995).

95JOC6258   A. Padwa and A. T. Price, *J. Org. Chem.*, **60**, 6258–6259 (1995).

95JOC8044   A. G. Schultz, P. R. Guzzo, and D. M. Nowak, *J. Org. Chem.*, **60**, 8044 (1995).

95SL1   J. L. van der Baan, A. J. van der Heide, J. van der Louw, and G. W. Klump, *Synlett*, (1), 1 (1995).

95T2639   Z. Wang and X. Lu, *Tetrahedron*, **51**, 2639–2658 (1995).

95T3683   F. Palacios, C. Alonso, and G. Rubiales, *Tetrahedron*, **51**, 3683–3690 (1995).

95TL3495   J. E. Cochran and A. Padwa, *Tetrahedron Lett.*, **36**, 3495–3498 (1995).

95TL5623   W. R. Bowman, P. T. Stephenson, and A. R. Young, *Tetrahedron Lett.*, **36**, 5623–5626 (1995).

96CC1213   U. Beifuss, A. Herde, and S. Ledderhose, *Chem. Commun.*, (10), 1213–1214 (1996).

96CHC1   A. Katritzky, C. W. Rees, and E. F. Scriven (eds.), "*Comprehensive Heterocyclic Chemistry*" Elsevier Science, Oxford, U.K. (1996).

96CR1   P. A. Wender (ed.), *Chem. Rev.*, **96**, 1–600 (1996).

96CR115   L. F. Tietze, *Chem. Rev.*, **96**, 115–136 (1996).

96CR137   S. E. Denmark and A. Thorarensen, *Chem. Rev.*, **96**, 137–165 (1996).

96JA6210   A. G. Schultz, M. A. Holoboski, and M. S. Smyth, *J. Am. Chem. Soc.*, **118**, 6210 (1996).

96JA6634   W. J. Zuercher, M. Hashimoto, and R. H. Grubbs, *J. Am. Chem. Soc.*, **118**, 6634–6640 (1996).

96JA8266   S. E. Denmark, A. Thorarensen, and D. S. Middleton, *J. Am. Chem. Soc.*, **118**, 8266–8277 (1996).

96JOC1665   A. Goti, B. Anichini, and A. Brandi, *J. Org. Chem.*, **61**, 1665–1172 (1996).

96JOC1880   M. Avalos, R. Babiabo, P. Cintas, F. J. Higes, J. L. Jiménez, J. C. Palacios, and M. A. Silva, *J. Org. Chem.*, **61**, 1880–1882 (1996).

96JOC3706   A. Padwa, J. E. Cochran, and C. O. Kappe, *J. Org. Chem.*, **61**, 3706–3714 (1996).

96JOC6166   C. O. Kappe and A. Padwa, *J. Org. Chem.*, **61**, 6166–6174 (1996).

96T10569   J. H. Rigby and M. E. Mateo, *Tetrahedron*, **52**, 10569 (1996).

96T11463   M. Cavicchioli, S. Decortiat, D. Bouyssi, J. Goré, and G. Balme, *Tetrahedron*, **52**, 11463–11478 (1996).

96T14745   C. Herdeis and T. Schiffer, *Tetrahedron*, **52**, 14745–14756 (1996).

96TA1059   C. Castellari, M. Lombardo, G. Pietropaolo, and C. Trombini, *Tetrahedron Asymm.*, **7**, 1059–1068 (1996).

96TL209   T. Saito, K. Tsuda, and Y. Saito, *Tetrahedron Lett.*, **37**, 209–212 (1996).

96TL3399   R. Grigg, V. Loganathan, and V. Sridharan, *Tetrahedron Lett.*, **37**, 3399–3402 (1996).

96TL7525   T. Kawaskaki, R. Terashima, K. -E. Sakaguchi, H. Sekiguchi, and M. Sakamoto, *Tetrahedron Lett.*, **37**, 7525–7528 (1996).

97CEJ143   R. Lock and H. Waldmann, *Chem. Eur. J.*, **3**, 143–151 (1997).

97JA125   S. E. Denmark and A. Thorarensen, *J. Am. Chem. Soc.*, **119**, 125–137 (1997).

97JOC3263   M. Mori, S. Kuroda, C. Zhang, and Y. Sato, *J. Org. Chem.*, **62**, 3263 (1997).

97JOC3787   T. Lavoisier-Gallo and J. Rodriguez, *J. Org. Chem.*, **62**, 3787–3788 (1997).

97JOC4088   A. Padwa, M. Dimitroff, A. G. Waterson, and T. Wu, *J. Org. Chem.*, **62**, 4088–4096 (1997).

97JOC67   A. Padwa, M. A. Brodney, J. P. J. Marino, M. H. Osterhout, and A. T. Price, *J. Org. Chem.*, **62**, 67–77 (1997).

97JOC7086   S. E. Denmark and J. A. Dixon, *J. Org. Chem.*, **62**, 7086–7087 (1997).

97SL632   D. Schinzer, U. Abel, and P. G. Jones, *Synlett*, (5), 632–634 (1997).

97T11929   R. M. Uittenbogaard, J. -P. G. Seerden, and H. W. Scheeren, *Tetrahedron*, **53**, 11929–11936 (1997).

97TL2159   S. M. Starling and S. C. Vonwiller, *Tetrahedron Lett.*, **38**, 2159–2162 (1997).

97TL2829   S. Kirschbaum and H. Waldmann, *Tetrahedron Lett.*, **38**, 2829–2832 (1997).

97TL875   M. Martinková and J. Gonda, *Tetrahedron Lett.*, **38**, 875–878 (1997).

98JA4041   J. B. Alexander, D. S. La, D. R. Cefalo, A. H. Hoveyda, and R. R. Schrock, *J. Am. Chem. Soc.*, **120**, 4041–4042 (1998).

98JA9720   D. S. La, J. B. Alexander, D. R. Cefalo, D. D. Graf, A. H. Hoveyda, and R. R. Schrock, *J. Am. Chem. Soc.*, **120**, 9720–9721 (1998).

98JOC144   R. A. Bunce, E. D. Dowdy, R. S. Chidress, and P. B. Jones, *J. Org. Chem.*, **63**, 144–151 (1998).

98JOC1604   S. E. Denmark and D. S. Middleton, *J. Org. Chem.*, **63**, 1604–1618 (1998).

98JOC3045   S. E. Denmark and A. R. Hurd, *J. Org. Chem.*, **63**, 3045–3050 (1998).

98JOC3986   A. Padwa, M. Dimitroff, A. G. Waterson, and T. Wu, *J. Org. Chem.*, **63**, 3986–3997 (1998).

98JOC5304   A. Padwa, M. A. Brodney, and M. Dimitroff, *J. Org. Chem.*, **63**, 5304–5305 (1998).

98JOC5587   J. H. Rigby, S. Laurent, A. Cavezza, and M. J. Heeg, *J. Org. Chem.*, **63**, 5587 (1998).

98JOC6778   A. Padwa, T. M. Heidelbaugh, J. T. Kuethe, and M. S. McClure, *J. Org. Chem.*, **63**, 6778–6779 (1998).

98PAC1047   R. Grigg and V. Sridharan, *Pure Appl. Chem.*, **70**, 1047–1057 (1998).

98SL169   R. Stragies and S. Blechert, *Synlett*, (2), 169–170 (1998).

98T6457   F. Léost, B. Chantegrel, and C. Deshayes, *Tetrahedron*, **54**, 6457–6474 (1998).

98TL455   E. Caballero, P. Puebla, M. Medarde, Z. Honores, P. Sastre, J. L. López, and A. San Feliciano, *Tetrahedron Lett.*, **39**, 455–458 (1998).

98TL7197   S. R. Baker, A. F. Parsons, J. -F. Pons, and M. Wilson, *Tetrahedron Lett.*, **39**, 7197–7200 (1998).

98TL8585   A. Padwa and A. G. Waterson, *Tetrahedron Lett.*, **39**, 8585–8588 (1998).

99CC2099   S. H. Hong, J. W. Kim, D. S. Choi, Y. K. Chunk, and S.-G. Lee, *Chem. Commun.*, (20), 2099–2100 (1999).

99JA11603   D. S. La, J. G. Ford, E. S. Sattely, R. R. Schrock, and A. H. Hoveyda, *J. Am. Chem. Soc.*, **121**, 11603–11604 (1999).

99JA6771   F. He, Y. Bo, J. D. Altom, and E. J. Corey, *J. Am. Chem. Soc.*, **121**, 6771–6772 (1999).

99EJOC1407   C. Herdeis and J. Tesler, *Eur. J. Org. Chem.*, (6), 1407–1414 (1999).

99JOC1494   M. Avalos, R. Babiabo, P. Cintas, F. J. Higes, J. L. Jiménez, J. C. Palacios, and M. A. Silva, *J. Org. Chem.*, **64**, 1494–1502 (1999).

99JOC2038   A. Padwa, T. M. Heidelbaugh, and J. T. Kuethe, *J. Org. Chem.*, **64**, 2038–2049 (1999).

99JOC3039   S. A. Kozmin, J. M. Janey, and V. H. Rawal, *J. Org. Chem.*, **64**, 3039–3052 (1999).

99JOC4617   A. Padwa, M. A. Brodney, K. Satake, and C. S. Straub, *J. Org. Chem.*, **64**, 4617–4626 (1999).

99JOMC65   R. Grigg and V. Sridharan, *J. Organomet. Chem.*, **576**, 65–87 (1999).

99SL441   A. J. Clark, R. P. Filik, J. L. Peacock, and G. H. Thomas, *Synlett*, (4), 441–443 (1999).

99SL756   K. Funabiki, H. Nakamura, M. Matsui, and K. Shibata, *Synlett*, (6), 756–758 (1999).

99T13999   L. Ou, Y. Hu, G. Song, and D. Bai, *Tetrahedron*, **55**, 13999–14004 (1999).

99TL1851   K. Paulvannan, *Tetrahedron Lett.*, **40**, 1851–1854 (1999).

99TL3247   M. J. Bassindale, P. Hamley, A. Leitner, and J. P. A. Harrity, *Tetrahedron Lett.*, **40**, 3247–3250 (1999).

99TL545   Y. Hu, L. Ou, and D. Bai, *Tetrahedron Lett.*, **40**, 545–548 (1999).

00JA11727   B. M. Trost and A. J. Frontier, *J. Am. Chem. Soc.*, **122**, 11727–11728 (2000).

00JA1828   G. S. Weatherhead, J. G. Ford, E. J. Alexanian, R. R. Schrock, and A. H. Hoveyda, *J. Am. Chem. Soc.*, **122**, 1828–1829 (2000).

00JOC235   A. Padwa and A. G. Waterson, *J. Org. Chem.*, **65**, 235–244 (2000).

00JOC2368   A. Padwa, T. M. Heidelbaugh, and M. E. Kuehne, *J. Org. Chem.*, **65**, 2368–2378 (2000).

00JOC2847   R. A. Bunce, D. M. Herron, and M. L. Ackerman, *J. Org. Chem.*, **65**, 2847–2850 (2000).

00JOC2875   S. E. Denmark and A. R. Hurd, *J. Org. Chem.*, **65**, 2875–2886 (2000).

00JOC9059   J. M. Janey, T. Iwama, S. A. Kozmin, and V. H. Rawal, *J. Org. Chem.*, **65**, 9059–9068 (2000).

00OL1109	S. D. Edmonson, A. Mastracchio, and E. R. Parmee, *Org. Lett.*, **2**, 1109–1112 (2000).
00OL1201	N. Hucher, A. Daich, and B. Decroix, *Org. Lett.*, **2**, 1201–1204 (2000).
00OL3289	P. Raman, H. Razavi, and J. W. Kelly, *Org. Lett.*, **2**, 3289–3292 (2000).
00T10159	A. Padwa and A. G. Waterson, *Tetrahedron*, **56**, 10159–10173 (2000).
00T7525	P. Fretwell, R. Grigg, J. M. Sansano, V. Sridharan, S. Sukirthalingam, D. Wilson, and J. Redpath, *Tetrahedron*, **56**, 7525–7539 (2000).
00T7541	A. Casaschi, R. Grigg, J. M. Sansano, D. Wilson, and J. Redpath, *Tetrahedron*, **56**, 7541–7551 (2000).
00TL6001	B. Chao and D. C. Dittmer, *Tetrahedron Lett.*, **41**, 6001–6004 (2000).
00TL9387	J. D. Ginn, S. M. Lynch, and A. Padwa, *Tetrahedron Lett.*, **41**, 9387–9391 (2000).
01CC2648	T. -L. Choi and R. H. Grubbs, *Chem. Commun.*, (24), 2648–2649 (2001).
01JA10214	B. G. Van den Hoven and H. Alper, *J. Am. Chem. Soc.*, **123**, 10214–10220 (2001).
01JA5374	K. Wakabayashi, H. Yorimitsu, and K. Oshima, *J. Am. Chem. Soc.*, **123**, 5374–5375 (2001).
01JA9033	D. W. C. MacMillan, L. E. Overman, and L. D. Pennington, *J. Am. Chem. Soc.*, **123**, 9033–9044 (2001).
01JOC3284	R. Mutter, I. B. Campbell, E. M. Martin de la Nava, A. T. Merritt, and M. Wills, *J. Org. Chem.*, **66**, 3284–3290 (2001).
01EJOC553	G. J. T. Kuster, R. H. J. Steeghs, and H. W. Scheeren, *Eur. J. Org. Chem.*, (3), 553–560 (2001).
01OL135	F. Gallou, D. W. C. MacMillan, L. E. Overman, L. A. Paquette, L. D. Pennington, and J. Yang, *Org. Lett.*, **3**, 135–137 (2001).
01OL2907	S. E. Denmark and L. Gomez, *Org. Lett.*, **3**, 2907–2910 (2001).
01T3221	S. K. Bur and S. F. Martin, *Tetrahedron*, **57**, 3221–3242 (2001).
01T6855	P. L. Fuchs, *Tetrahedron*, **57**, 6855–6875 (2001).
01TL5109	J. M. Gardiner and J. Procter, *Tetrahedron Lett.*, **42**, 5109–5111 (2001).
02AG2753	M. C. Carreño, M. Ribagorda, and G. H. Posner, *Angew. Chem.*, **41**, 2753–2755 (2002).
02AG4693	B. M. Trost and M. R. Machacek, *Angew. Chem., Int. Ed.*, **41**, 4693–4697 (2002).
02COC1181	R. Schobert and G. J. Gordon, *Curr. Org. Chem.*, **6**, 1181–1196 (2002).
02H(56)613	S. Cacchi, G. Fabrizi, and A. Goggiomani, *Heterocycles*, **56**, 613 (2002).
02H(58)667	S. Cacchi, G. Fabrizi, and L. M. Parisi, *Heterocycles*, **58**, 667 (2002).
02JA13398	J. P. Marino, M. B. Rubio, G. Cao, and A. de Dios, *J. Am. Chem. Soc.*, **124**, 13398–13399 (2002).
02JA2560	P. Janvier, X. Sun, H. Bienaymé, and J. Zhu, *J. Am. Chem. Soc.*, **124**, 2560–2567 (2002).
02JA4521	M. M. -C. Lo and G. C. Fu, *J. Am. Chem. Soc.*, **124**, 4521–4573 (2002).
02JHC1049	R. A. Bunce, S. V. Kotturi, C. J. Peeples, and E. M. Holt, *J. Heterocycl. Chem.*, **39**, 1049 (2002).
02JOC1380	O. Arjona, A. G. Csáký, R. Medel, and J. Plumet, *J. Org. Chem.*, **67**, 1380–1383 (2002).
02EJOC221	P. Langer and M. Döring, *Eur. J. Org. Chem.*, (2), 221–234 (2002).

02JOC3412   A. Padwa, J. D. Ginn, S. K. Bur, C. K. Eidell, and S. M. Lynch, *J. Org. Chem.*, **64**, 3412–3424 (2002).

02EJOC3536 F. Denes, F. Chemla, and J. F. Normant, *Eur. J. Org. Chem.*, (21), 3536–3542 (2002).

02JOC4325   N. Bushmann, A. Rückert, and S. Blechert, *J. Org. Chem.*, **67**, 4325–4329 (2002).

02JOC5928   A. Padwa, T. M. Heidelbaugh, J. T. Kuethe, M. S. McClure, and Q. Wang, *J. Org. Chem.*, **67**, 5928–5937 (2002).

02JOC6456   C. Stapper and S. Blechert, *J. Org. Chem.*, **67**, 6456–6460 (2002).

02JOC8500   S. E. Vaillard, A. Postigo, and R. A. Rossi, *J. Org. Chem.*, **67**, 8500–8506 (2002).

02JOC929   A. Padwa, D. M. Danca, K. I. Hardcastle, and M. S. McClure, *J. Org. Chem.*, **68**, 929–941 (2002).

02OL2565   A. M. Bernard, E. Cadoni, A. Frongia, P. P. Piras, and F. Secci, *Org. Lett.*, **4**, 2565–2567 (2002).

02OL289   R. Roggenbuck, A. Schmidt, and P. Eilbracht, *Org. Lett.*, **4**, 289–291 (2002).

02OL3835   M. Ueda, H. Miyabe, A. Nishimura, O. Miyata, T. Yoshiji, and T. Naito, *Org. Lett.*, **5**, 3835–3838 (2002).

02OL4135   S. K. Bur and A. Padwa, *Org. Lett.*, **4**, 4135–4137 (2002).

02OL4677   C. Lu and X. Lu, *Org. Lett.*, **4**, 4677 (2002).

02OL715   A. Padwa and D. M. Danca, *Org. Lett.*, **4**, 715–717 (2002).

02OL909   E. J. Tisdale, C. Chowdhury, B. G. Vong, H. Li, and E. A. Theodorakis, *Org. Lett.*, **4**, 909–912 (2002).

02T10309   K. C. Majumdar, U. K. Kundu, and S. Ghosh, *Tetrahedron*, **58**, 10309–10313 (2002).

02T1611   J. Gonda, M. Martinková, and J. Imrich, *Tetrahedron*, **58**, 1611–1616 (2002).

02T531   E. Ceulemans, M. C. Voets, S. Emmers, K. Uytterhoeven, L. Van Meervelt, and W. Dehaen, *Tetrahedron*, **58**, 531–544 (2002).

02TL203   K. Paulvannan, R. Hale, R. Mesis, and T. Chen, *Tetrahedron Lett.*, **43**, 203–207 (2002).

02TL5499   A. Basak, S. C. Ghosh, T. Bhowmick, A. K. Das, and V. Bertolasi, *Tetrahedron Lett.*, **43**, 5499–5501 (2002).

02TL6739   J. Zaminer, C. Stapper, and S. Blechert, *Tetrahedron Lett.*, **43**, 6739–6741 (2002).

02TL7851   R. A. J. Wybrow, L. A. Johnson, B. Auffray, W. J. Moran, H. Adams, and J. P. A. Harrity, *Tetrahedron Lett.*, **43**, 7851–7854 (2002).

03AG2681   Y. Nishibayashi, M. Yoshikawa, Y. Inada, M. D. Milton, M. Hidai, and S. Uemura, *Angew. Chem., Int. Ed.*, **42**, 2681–2684 (2003).

03AG4082   R. Shintani and G. C. Fu, *Angew. Chem., Int. Ed.*, **42**, 4082–4085 (2003).

03CC2286   S. A. Raw, C. D. Wilfred, and R. J. K. Taylor, *Chem. Commun.*, (18), 2286–2287 (2003).

03EJOC611   O. Arjona, A. G. Csákÿ, and J. Plumet, *Eur. J. Med. Chem.*, (4), 611–622 (2003).

03JA7942   D. Crich, K. Ranganathan, S. Neelamkavil, and X. Huang, *J. Am. Chem. Soc.*, **125**, 7942–7947 (2003).

03JOC5618   H. Miyabe, M. Ueda, K. Fujii, A. Nishimura, and T. Naito, *J. Org. Chem.*, **68**, 5618–5626 (2003).

03OBC1450   M. Rosillo, L. Casarrubios, G. Domínguez, and J. Pérez-Castells, *Org. Biomol. Chem.*, **1**, 1450–1451 (2003).

03OL1757   V. K. Aggarwal, G. Y. Fang, J. P. H. Charmant, and G. Meek, *Org. Lett.*, **5**, 1757–1760 (2003).

03OL1765	W. Du and D. P. Curran, *Org. Lett.*, **5**, 1765–1768 (2003).
03OL259	P. Pinho, A. J. Minnaard, and B. L. Feringa, *Org. Lett.*, **5**, 259–261 (2003).
03OL3843	S. Cacchi, G. Fabrizi, and L. M. Parisi, *Org. Lett.*, **5**, 3843–3846 (2003).
03OL5095	F. Marion, C. Courillon, and M. Malacria, *Org. Lett.*, **5**, 5095–5097 (2003).
03SA681	G. W. Gribble, in *"Synthetic Applications of 1,3-Dipolar Cycloaddition Chemistry toward Heterocycles and Natural Products"* (A. Padwa and W. H. Pearson, eds.), pp. 681–753, Wiley, Hoboken, NJ (2003).
03SL1443	M. C. Bagley, D. D. Hughes, H. M. Sabo, P. H. Taylor, and X. Xiong, *Synlett*, (10), 1443–1446 (2003).
03SL1479	S. Kamila, C. Kukherjee, and A. De, *Synlett*, (10), 1479–1481 (2003).
03SL2305	S. Cicchi, J. Revuelta, A. Zanobini, M. Betti, and A. Brandi, *Synlett*, (15), 2305–2308 (2003).
03TL7143	X. Wei and R. J. K. Taylor, *Tetrahedron Lett.*, **44**, 7143–7146 (2003).
03TL7429	K. Takasu, N. Nishida, and M. Ihara, *Tetrahedron Lett.*, **44**, 7429–7432 (2003).
04CC508	S. A. Raw and R. J. K. Taylor, *Chem. Commun.*, (5), 508–509 (2004).
04CR2199	A. Deiters and S. F. Martin, *Chem. Rev.*, **104**, 2199–2238 (2004).
04CR2401	For a review of the Pummerer reaction in heterocyclic synthesis, see S. K. Bur and A. Padwa, *Chem. Rev.*, **104**, 2401–2432 (2004).
04EJOC1897	J. T. Anders, H. Görls, and P. Langer, *Eur. J. Org. Chem.*, (9), 1897–1910 (2004).
04EJOC3872	M. -W. Ding, Y. -F. Chen, and N. -Y. Huang, *Eur. J. Org. Chem.*, (18), 3872–3878 (2004).
04JA3529	A. Klapars, S. Parris, K. W. Anderson, and S. L. Buchwald, *J. Am. Chem. Soc.*, **126**, 3529–3533 (2004).
04JOC1207	K. Paulvannan, *J. Org. Chem.*, **69**, 1207–1214 (2004).
04OBC788	S. A. Raw, C. D. Wilfred, and R. J. K. Taylor, *Org. Biomol. Chem.*, **2**, 788–796 (2004).
04OL2641	J. A. González-Vera, M. T. García-López, and R. Herranz, *Org. Lett.*, **6**, 2641–2644 (2004).
04SL1	P. M. Fesneda and P. Molina, *Synlett*, (1), 1–17 (2004).
04SL1628	C. D. Wilfred and R. J. K. Taylor, *Synlett*, (9), 1628–1630 (2004).
04T275	D. V. Vilarelle, C. P. Veira, and J. M. Quintela López, *Tetrahedron*, **60**, 275–283 (2004).
04T8181	J. Flisinska-Luczak, S. Lesniak, and R. B. Nazarski, *Tetrahedron*, **60**, 8181–8188 (2004).
04TL3493	J. S. Yadav, B. V. S. Reddy, D. Narsimhaswamy, P. Naga Lakshmi, K. Narsimulu, G. Srinivasulu, and A. C. Kunwar, *Tetrahedron Lett.*, **45**, 3493–3497 (2004).
04TL35	S. Kamijo, Y. Sasaki, and Y. Yamamoto, *Tetrahedron Lett.*, **45**, 35–38 (2004).
04TL6349	S. Singh and H. Han, *Tetrahedron Lett.*, **45**, 6349–6352 (2004).
05ARK98	S. Eguchi, *Arkivoc*, (1), 98–119 (2005).
05CC271	B. Gabriele, R. Mancuso, G. Salerno, and L. Veltri, *Chem. Commun.*, (2), 271–273 (2005).
05JOC2957	T. Kawaskaki, A. Ogawa, R. Terashima, T. Saheki, N. Ban, H. Sekiguchi, K. -E. Sakaguchi, and M. Sakamoto, *J. Org. Chem.*, **70**, 2957–2966 (2005).
05JOC6833	A. Padwa, S. K. Bur, and H. -J. Zahng, *J. Org. Chem.*, **70**, 6833–6841 (2005).

05OL913   Z. D. Aron and L. E. Overman, *Org. Lett.*, **7**, 913–916 (2005).
05TA177   M. S. M. Timmer, M. D. P. Risseeuw, M. Verdoes, D. V. Filippov,
                 J. R. Plaisier, G. A. van der Marel, H. S. Overkleeft, and J. H. van
                 Boom, *Tetrahedron Asymm.*, **16**, 177–185 (2005).
06OL3275   J. Mejia-Oneto and A. Padwa, *Org. Lett.*, **8**, 3275–3278 (2006).

Organometallic Chemistry of Polypyridine Ligands II

ALEXANDER P. SADIMENKO

Department of Chemistry, University of Fort Hare, Alice 5701, Republic of South Africa

Abbreviations

Ac	acetyl
Alk	alkyl
AN	acetonitrile
bpy	2,2′-bipyridine
Bu	butyl
Cp	cyclopentadienyl
Cp*	pentamethylcyclopentadienyl
DMF	dimethylformamide
DMSO	dimethylsulfoxide
dpene	1,2-bis(diphenylphosphino)ethylene
dppb	1,4-bis(diphenylphosphino)benzene
dppe	diphenylphosphinoethane
dppm	diphenylphosphinomethane
Et	ethyl
Im	imidazolyl
Me	methyl
MLCT	metal-to-ligand charge transfer
naph	1,8-naphthyridine
OTf	triflate
Ph	phenyl
phen	1,10-phenanthroline
Pr	propyl

ADVANCES IN HETEROCYCLIC CHEMISTRY
VOLUME 94 ISSN: 0065-2725   DOI: 10.1016/S0065-2725(06)94002-1

PTZ   phenothiazine
py   pyridine
pz   pyrazolyl
THF   tetrahydrofuran
Tol   tolyl
trpy   2,2':6',2''-terpyridine

I. Introduction

This chapter constitutes part 12 in a series covering the organometallic chemistry of heterocyclic ligands of which the previous parts have all been published in different volumes of Advances in Heterocyclic Chemistry.

In the previous chapter (07AHC(93)185), complexes of polypyridine ligands with non-transition and early transition metals were considered. Most publications, however, are dedicated to the rhenium(I) and ruthenium(II) complexes, and the number of sources is so high that they deserve separate chapters. Moreover, studies of such complexes become more and more popular due to their unique photochemical and electrochemical properties and ability to form molecular assemblies and nanocrystallites. Herein we consider organomanganese and organorhenium complexes of polypyridine ligands. As always in this series of chapters, emphasis will be on the synthetic and coordination aspects, as well as reactivity. We have attempted to document all the publications on applied aspects, but without analyzing them since this could be the subject of a separate chapter.

II. Organomanganese and Organorhenium Complexes

A. [M(CO)₃(LL)X] Type Complexes

Re(I)–polypyridyl complexes are among most popular. Their studies have a certain history (80AJC2369, 82MI1, 87IC1449) and their first synthesis refers to the 1950s and 1960s (58JCS3149, 59JCS1501, 63CB3035, 67ICA(1)172). [Mn(CO)₃(b-py)Cl)] was prepared from [Mn(CO)₅Cl] and 2,2'-bipyridine (87AX(C)792, 95IC1588, 95RCPB565). The radical complex [Mn(CO)₃(bpy)]• has a weakly distorted square pyramidal geometry (98IC6244). 2,2'-Bipyridine or 1,10-phenanthroline (LL) with [Mn(CO)₅Br] give [Mn(CO)₃(LL)Br] (LL = bpy, phen) (88TMC126). 2,2'-Bipyridine with [Re(CO)₅Cl] yields [Re(CO)₃(bpy)Cl] (82AJC2445, 82JA7658, 83OM552, 84JCS(CC)403). 2,2'-Bipyridine, 4,4'-dimethyl-2,2'-bipyridine, or 4,7-diphenyl-1,10-phenanthroline with [Re(CO)₅X] (X = Cl, Br, I) afford the ligand substitution products [(LL)Re(CO₃X] (LL = bpy, 4,4'-Me₂-bpy, 4,7-Ph₂phen; X = Cl, Br, I) (59JCS1501, 67ICA(1)172, 72GCI587, 75JA2073, 83JA1067, 99IC4181,

00JCS(CC)201, 00JCS(CC)1687, 01JCS(CC)103). There are two ways of transformation of this complex to [Re(CO)$_3$(bpy)(OTf)] – by reacting with triflic acid in methylene chloride at room temperature or by heating with silver triflate. Interaction of [Re(CO)$_5$Cl] with silver triflate in methylene chloride and, further, with 2,2′-bipyridine, 1,10-phenthroline, or 4,4′-dimethyl-2,2′-bipyridine (LL) gives the tetracarbonyl complexes [Re(CO)$_4$(LL)](OTf) (92IC4101). [Re(CO)$_4$(LL)(bpy)](OTf) (LL = 4,4′-(MeOOC)$_2$bpy, 4,4′-(MeO)$_2$bpy) can be prepared in a similar fashion (98JCS(D)2893). For the 1,10-phenanthroline derivative, the photochemical substitution with the triflate anion leading to [Re(CO)$_3$(phen)(OTf)] occurs. [Re(CO)$_3$(LL)Cl] (LL = bpy, 4,4′-Me$_2$bpy, phen, 2,9-Me$_2$phen) (94CPL426); (LL = phen, 4,7-Ph$_2$phen) (78JA5790, 84PNA1961, 85JA5518, 97JCS(CC)2375), [Re(CO)$_3$(3,3′-(OH)$_2$bpy)Cl] (03ICA(343)357) follow from the similar synthetic route.

Electrochemical properties of [Re(CO)$_3$(bpy)Cl] (92JCS(D)1455) and properties of its radical anion (77JOM(125)71, 94JCS(D)2977) were studied. Photochemical properties (82CCR(46)159, 83IC2444, 83IC3825, 83JCE834, 83JCE877, 83MI2, 84CPL332, 85CCR(64)41, 85CCR(65)65, 86IC256, 86IC3212, 86JCS(F2)2401, 86JPC5010, 87CPL365, 87OM553, 88CCR(84)85, 88JA6243, 89ACR163, 89JPC3885, 90IC2285, 90IC2792, 90IC4169, 90ICA(167)149, 90TCC75, 91JA1991, 91JCS(D)849, 91MI1, 91MI2, 91MI3, 92CIC359, 92IC1072, 92JPC257, 92MI1, 94JCS(D)2977, 95HCA619, 95JA7119, 95JCS(CC)1191, 95OM4034, 98CCR(171)221, 98CCR(177)127, 98JCS(D)185, 00IC1817, 01CCR(211)163, 01MI1) are characterized by luminescence ascribed to MLCT predominantly having a triplet character (74JA988, 78JA2257, 87CRV711, 89JPC24, 93JA8230). Behavior of *cis*-[Re(CO)$_4$(phen)]$^+$ (73IC1067) and photooxidative features of [Re(CO)$_4$(LL)]$^+$ (LL = phen, bpy, 4,4′-dimethyl-2,2′-bipyridine, 2,9-diphenyl-1,10-phenanthroline, 2,5,9-triphenyl-1,10-phenanthroline, 2,4,7,9-tetraphenyl-1,10-phenanthroline, 2,9-dimethyl-4,7-diphenyl-1,10-phenantroline) (82JCP3337, 86JA5344, 89IC1022, 90IC4335, 91IC3722, 93IC5629) are of interest. Uniqueness of the photochemical properties allowed application of these complexes in chemiluminescence (74CRV4801, 80IC860, 81AGE469, 81ICA(53)L35, 81JCS(D)1124, 84IC1440, 86ICA(115)193, 90IC1574, 90IC2327, 90JA9490, 95IC2033), electroluminescence (83JA7241, 87ACS155), exciplex emission (80ICA(45)L265), emission-based microenvironmental studies (85JPC1095, 89JPC2129, 92CPL299), light harvesting (01CCR(219)545), intermolecular electron and energy transfer, especially in the case of covalently linked donor–acceptor pairs (98MI1, 01CRV2655, 02CSR168), nonlinear optics (99AGE366, 99CEJ2464), photovoltaics (91N737, 95CRV49), luminescent probes in curing epoxy resins and acrylate-based polymers (91CM25, 91IC4871, 91PP124, 92CM675) self-assembly (macrocyclic polymetallic assemblies or molecular sequences containing the rhenium complexes as corners) (97IC5422, 99CRV2777, 00JCS(CC)1211, 02EJI357), amide-type anion binding (03IC3445), as DNA intercalators (94CSR327, 94MI1), photocatalysis (88JCS(CC)16, 91AGE844), conversion and storage of solar energy (79MI1, 80JA1383, 83MI1, 84IC2098, 85IC2755, 86PAC1193, 88CCR(84)47, 95JPP(A)61, 96MI1), building blocks of molecular dyads (00IC3590, 00JCS(D)2599), supramolecules (89JA7791, 98CCR(178)1299, 98CEJ406), molecular wires and switches (99AGE2722), catenanes and rotaxanes, light-harvesting molecular materials (99IC4382), thermal, photochemical, and electrochemical catalysts (88CCR(84)5, 96IC2242, 96IC6194), agents for biological labeling (95IC1629, 95ICA(240)169), sensor development (95IC6235, 97JCE685). [Re(CO)$_3$(bpy)Cl] is an active homogeneous catalyst of electro- and photoreduction of carbon dioxide (83JCS(CC)536, 84JCS(CC)328, 85IC3640, 85JCS(CC)1414, 85OM2161, 86ICA(114)L43, 86JEAC(207)315, 86JEAC(209)101, 86OM1100, 87NATO(C)113, 88ACS52, 88IC4326, 89CCR(93)245, 89CL765, 89JEAC(259)217, 89MI1, 91JCS(CC)787, 93JCS(CC)631, 95OM1115, 96CRV2063, 96JPP(A)171, 97OM4675). There is a discussion in the literature on the nature of the catalytic cycle of the carbon dioxide reduction. Thus, the hypothesis that one of the stages is the loss of the axial carbon monoxide ligand in [Re(CO)$_3$(bpy)Cl] (86HCA1990) did not find confirmation by the ESR studies of [Re(CO)$_3$(bpy)X]$^{\bullet-}$, and the data are interpreted as the preferential loss of the axial halide ligand (90IC2909, 90NJC831, 93MI1, 96OM236). The presence of [Re(CO)$_3$(4,4'-R$_2$bpy)(AN)]$^+$ in the process has been proven electrochemically (86JEAC(201)347). Besides, [Re^I(CO)$_3$(bpy)Cl] under the electrochemical conditions forms a dimer [Re^0(CO)$_3$(bpy)]$_2$ with features indicating the presence of the rhenium–rhenium bonds. Active species of the electrolytic reduction of carbon dioxide can also be [Re^I(CO)$_2$(bpy)Cl]$^-$.

2,2'-Bipyridine with [Re(CO)$_5$X] (X = Cl, Br, I) gives [Re(CO)$_3$(bpy)X] (X = Cl, Br, I) (96IC2902). Various 4,4'-disubstituted 2,2'-bipyridines react with [Re(CO)$_5$Cl] to yield complexes **1** (R = H, Me, Br, Ph, PhC$_6$H$_4$C≡C, H$_2$NC$_6$H$_4$C≡C, and O$_2$NC$_6$H$_4$C≡C) (02IC2909); (R = NMe$_2$, NO$_2$, COOH, COOPr-i, CH = CHC$_6$H$_4$NEt$_2$-p, CH$_2$CH$_2$C$_6$H$_4$NEt$_2$-p, and others) (88IC4007). 1,2-Bis(4'-methyl-2,2'-bipyrid-4-yl)ethane with [Re(CO)$_5$Cl] forms the mononuclear complex **2** (90IC1761). Various 2,2'-bipyridine ligands react with [Re(CO)$_5$Cl] in toluene/methylene chloride to produce complexes of the type **3** and **4** (02IC1662). Interaction of **2** and **3** with, first, silver triflate in methylene chloride/THF and second, with pyridine in ethanol gives cationic complexes with the same set of polypyridine ligands LL having the general formula [Re(CO)$_3$(LL)(py)](OTf). These were tested as sensors for glucose. The range of 2,2'-bipyridine ligands containing aryl- and ethynyl-substituents in positions 4 and 4' may be continued, **5–8** (04IC5961). Complexes with fullerene-substituted 2,2'-bipyridine ligands should be mentioned as well (02CEJ2314). Complexes [Re(CO)$_3$(LL)Cl] contain LL = 5,5'-dibromo-2,2'-bipyridine, 5,5'-bis(trimethylsilylethynyl)-2,2'-bipyridine, and 3,8-dibromo-1,10-phenanthroline (00JOM(598)55).

1   2

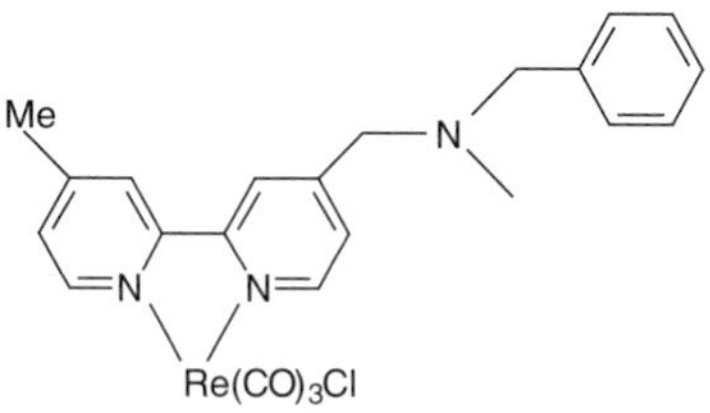
Me
N
N
N
Re(CO)$_3$Cl

3

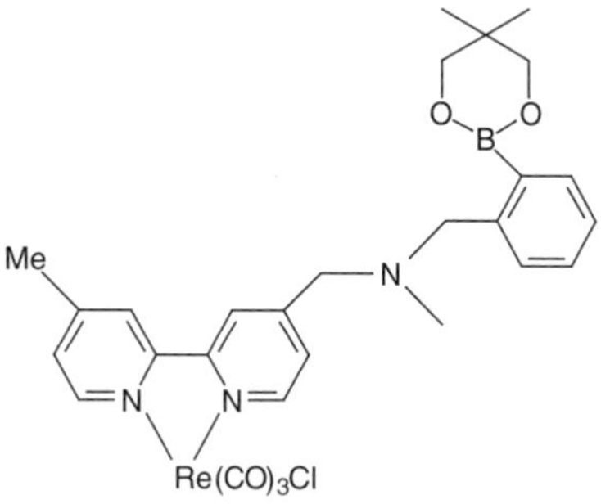
O
B
O
Me
N
N
N
Re(CO)$_3$Cl

4

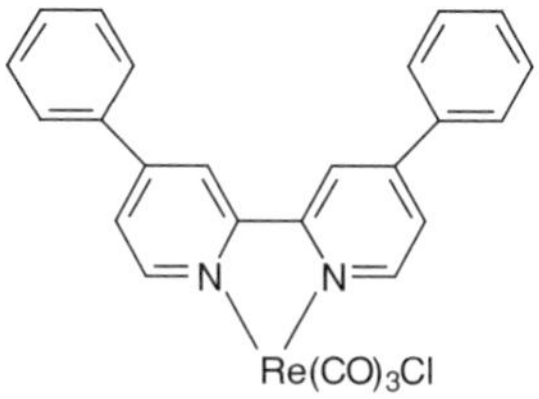
N
N
Re(CO)$_3$Cl

5

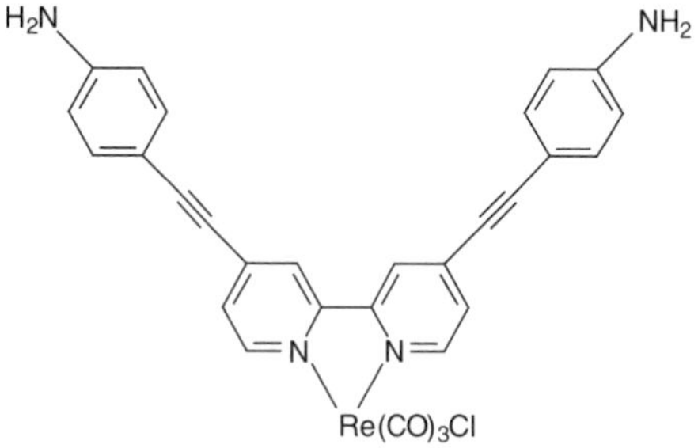
H$_2$N
NH$_2$
N
N
Re(CO)$_3$Cl

6

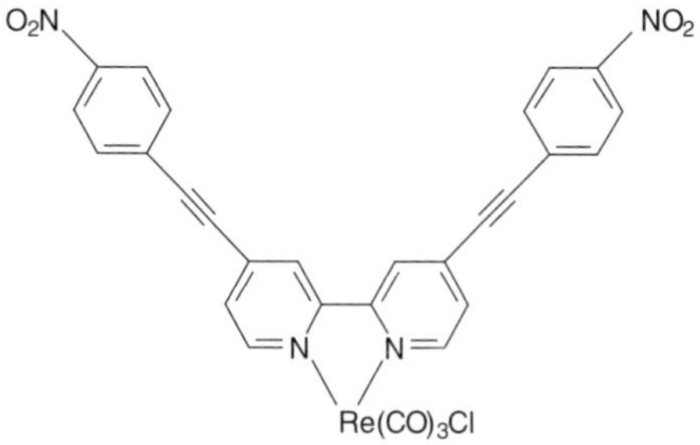
O$_2$N
NO$_2$
N
N
Re(CO)$_3$Cl

7

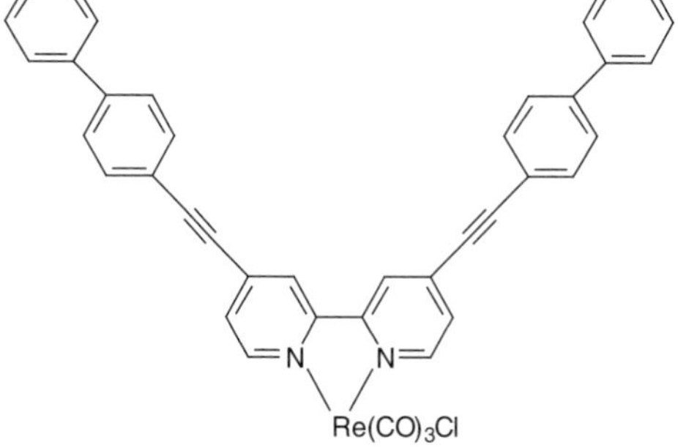
N
N
Re(CO)$_3$Cl

8

[Mn(CO)$_5$Br] with 1,10-phenanthroline gives [Mn(CO)$_3$(phen)Br], which with silver triflate transforms into [Mn(CO)$_3$(phen)(OTf)]. [Mn(CO)$_3$(bpy)X] (X = Cl, Br) can be prepared similarly (94JOM(482)15, 94OM2641, 97JPP(A)231, 98JA10871). The product with imidazole gives the cationic complex [Mn(CO)$_3$(phen)(Im)](OTf) (00ICA(299)231). A rhenium analogue can be prepared in a similar fashion (99ICA(288)150). Reaction of [Re(CO)$_5$Cl] with excess 4,7-diphenyl-1,10-phenanthroline (LL) gives [Re(CO)$_3$(LL)] (93IC2570). The 2,5-dimethylthienyl derivative of 1,10-phenanthroline reacts with [Re(CO)$_5$Cl] in benzene under reflux to yield **9** (04JA12734). N-Nitropyrido[3,2-a:2′,3′-c]phenazine, 11,12-dinitrodipyrido[3,2-a:2′,3′-c]phenazine, 11,12-dinitrodipyrido[3,2-a:2′,3′-c]phenazine, 11,12-diaminodipyrido[3,2-a:2′,3′-c]phenazine, and 11,12-bis(bromomethyl)dipyrido[3,2-a:2′,3′-c]phenazine react with [Re(CO)$_5$Cl] in toluene under reflux to yield **10** (R^1 = H, R^2 = NO$_2$; R^1 = R^2 = NO$_2$, NH$_2$, CH$_2$Br; R^1 = H, R^2 = NH$_2$) (06JOM1834). Complexes with R^1 = H, R^2 = NO$_2$, and R^1 = R^2 = NO$_2$ were described earlier (98JCS(D)609, 01JCC323). Compounds [Re(CO)$_3$(LL)X] (LL = 6-(pyrzol-1-yl)-2,2′-bipyridine, 6-(3,5-dimethylpyrazol-1-yl)-2,2′-bipyridine, and 6-(4-methylpyrazol-1-yl)-2,2′-bipyridine; X = Cl, Br, I) are coordinated in a normal chelating fashion (96JCS(D)203).

9   **10**

3-Ethynyl-1,10-phenanthroline or 3,8-bis(ethynyl)-1,10-phenanthroline react with [Re(CO)$_5$Cl] in benzene to yield complexes **11** and **12** (04JOM2905). The products enter the reaction with [Au(PPh$_3$)Cl] in THF in the presence of amine as a base to yield heteronuclear species **13** and **14**. Among the other products prepared by a traditional route, the following should be noted: [Re(CO)$_3$(1,6-bis(4-(2,2′-bipyridyl)pyrene)Cl] (02IC359), compound **15** containing a fluorene-based bipyridine ligand (01OM557), polymers based on [Re(CO)$_3$(phen)] (01CEJ4358). 4,4′-Distyryl-2,2′-bipyridine with [Re(CO)$_5$Cl] in dichloromethane forms **16** (87IC882). Complex **17** can be prepared similarly (90JA1333). Such species on electroreduction undergo polymerization (81JA56, 82IC2153, 83IC2151, 85JCS(CC)1416, 87IC2145, 88CCR(86)135, 89IC3923). Thus, on electroreduction in acetonitrile, complex **17** forms the stable and electrochemically active polymeric film (LL)(OC)$_3$Re-Re(CO)$_3$(LLH-LLH)(OC)$_3$Re(CO)$_3$(LL)] (LL = 4-methyl-4′-vinyl-2,2′-bipyridine), capable of electro- and photoelectrocatalytic reduction of CO$_2$ to CO on metallic and semiconducting electrodes (90JA1333). [Re(CO)$_3$(bpy)(OTf)] is reduced by sodium amalgam in THF to [Re(CO)$_3$(bpy)]$_2$ (04IC7636). A similar process for [Re(CO)$_3$(4,4′-Me$_2$bpy)(OTf)] is known (89JA5699, 91ACR325, 03JA11976).

2,2′-Biquinoline, 3,3′-dimethylene-2,2′-biquinoline, 3,3′-dimethyl-2,2′-biquinoline, and 3,3′-trimethylene-2,2′-biquinoline (LL) with [Re(CO)$_5$Br] in toluene give [Re(CO)$_3$(LL)Br] (92BSCQ43, 92BSCQ311, 94IC2341, 96BSCQ251). In the complexes [Re(CO)$_3$(3,3′-R,R′-2,2′-biquinoline)Br] (R = R′ = H, Me, R-R′ = dimethylene and trimethylene), the extent of conjugation over the 2,2′-biquinoline ligand is more extensive than that over the analogously substituted 2,2′-bipyridine ligand (92POL1665, 95JPC4929). Substitution of the bromide ligand with the p-substituted pyridine ligands in the presence of silver triflate leads, in particular, to [Re(CO)$_3$(3,3′-triemethylene-2,2′-biquinoline)L](OTf)   (L = py,   p-R-py,   R = OH,   Ph,   NC) (98POL2289). Manganese analogues **18** (R = H, Me) and **19** (n = 2, 3) were prepared from [Mn(CO)$_5$Br] at a later stage (02POL439).

2,2′:6′,2″-Terpyridine (97CCR(160)1) reacts with [Re(CO)₅X] (X = Cl, Br, I) to yield complexes **20** (X = Cl, Br, I) characterized by the bidentate coordination mode embracing only two N-heteroatoms out of three (92JCS(CC)303, 93IC237, 93JCS(D)597, 94JCS(D)3441, 96OM5442). However, in solutions, the products possess fluxional behavior with the re-switching of the pairs of pyridine heteroatoms participating in coordination. This reaction proceeds through the stage of the ter-coordinated complex. 4′-Phenyl-2,2′:6′,2″-terpyridine (LL) with [M(CO)₅Br] gives [M(CO)₃(LL)Br] (M = Mn, Re) (95JBCS29, 99BSCQ423, 01ICA(312)7). The polypyridine ligand is coordinated bidentately and reveals coordination dynamics in solution. Complexes [Re(CO)₃(LL)Br] (LL = 4-methyl-4′-(4-chlorophenyl)-2,2′:6′,2″-terpyridine, 4-methyl-4′-(4-chlorophenyl)-4″-methyl-2,2′:6′,2″-terpyridine, 4-methyl-4′methylthio-2,2′:6′,2″-terpyridine, and 4-*tert*-butyl-4′-methylthio-2,2′:6′,2″-terpyridine) prepared in a similar fashion also contain polydentate polypyridine ligands and possess fluxional behavior described by an associative mechanism (96JCS(CC)2329, 98JCS(D)937). Generally, the terpyridine complexes of rhenium are attractive in terms of the selectivity of the carbon dioxide catalytic reduction (91CCR(111)193, 91JPC7641). 2,2′:6′,2″:6″,2‴-quaterpyridine reacts with [Re(CO)₅Br in benzene to yield the bidentately coordinated complex **21**, where metallotropy is absent for steric reasons, and solution flexibility is related to the orientation of the uncomplexed part of the molecule (99POL1285). Under the altered ratio of the reagents, dinuclear complex formation is possible as in complexes **22** (R = H, 4-*t*-BuC₆H₄). 2,2′:6′,2″:6″,2‴:6‴,2⁗-Quinquepyridine with [Re(CO)₅Cl] in methanol gives complex **23** (96JCS(D)1411). The product reacts with silver perchlorate in pyridine to yield the dicationic compound **24**. Both complexes are photoluminescent. In both cases, the coordination mode is of the 2,2′-bipyridine type when one of the pyridine rings does not participate in coordination. 4′-Phenyl-2,2′:6′,2″-terpyridine and 2,2′:4,4⁗:6,2‴-quaterpyridine (LL) with [Re(CO)₅Cl] on refluxing in toluene and further on refluxing with silver triflate in acetonitrile form the η²(N)-coordinated complexes **25** and **26** (02JCS(D)4732).

20

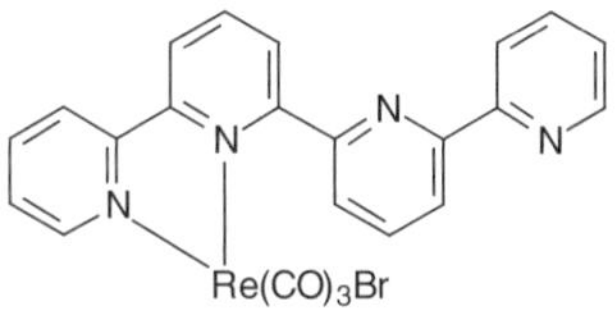

21

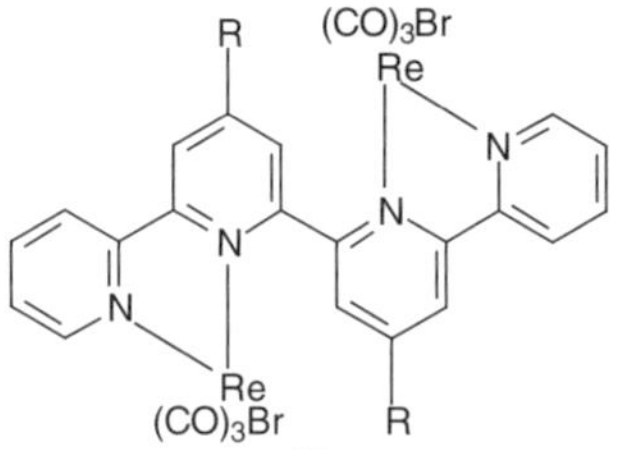

22

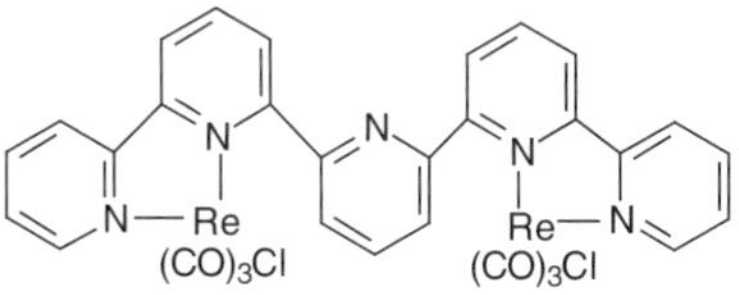

23

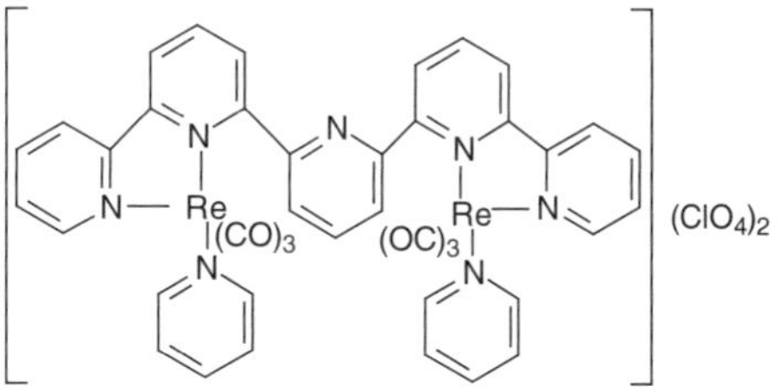

24

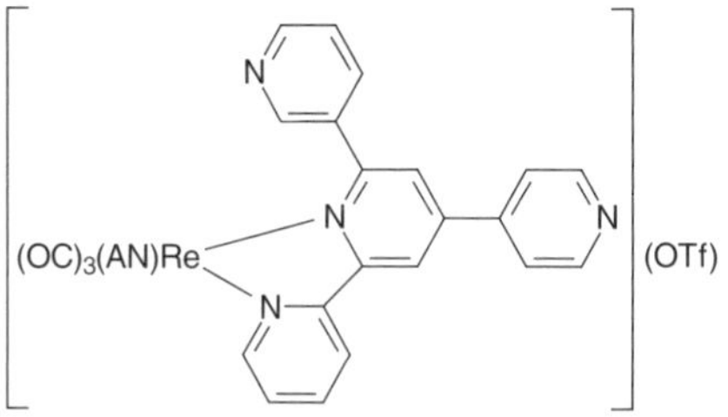

25

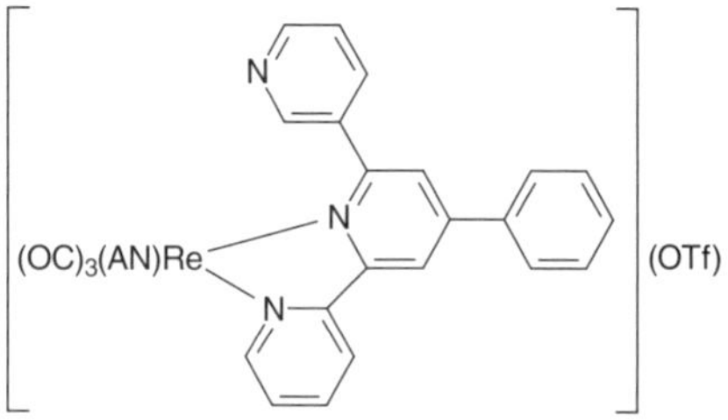

26

3,3'-Trimethylene-2,2'-bi-1,8-naphthyridine with [Re(CO)$_5$Br] in heptane gives the bidentately coordinated complex **27** (96OM3463). Compounds of the same composition, [Re(CO)$_3$(LL)Cl] (LL = 3,3'-dimethylene-2,2'-biquinoline) can be prepared similarly (94IC2341). p-(2,6-di-2-pyridyl-4-pyridyl)phenol reacts with [Re(CO)$_5$Cl] and forms complex **28** with bidentate coordination of the ligand (92JPP(A)259). The product is active in the electro- and photochemical reduction of carbon dioxide. Species [Re(CO)$_3$(2,7-bis(2'-pyridyl)-1,8-naphthyridine)Br] is known (83ICA(76)L29).

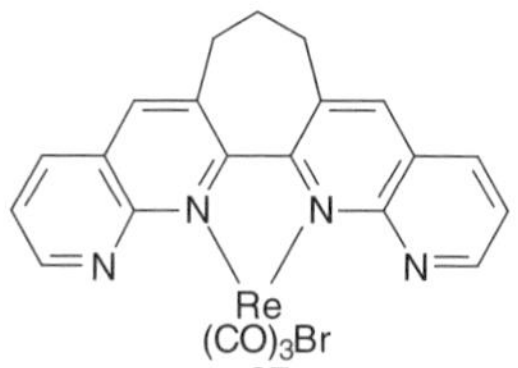

27

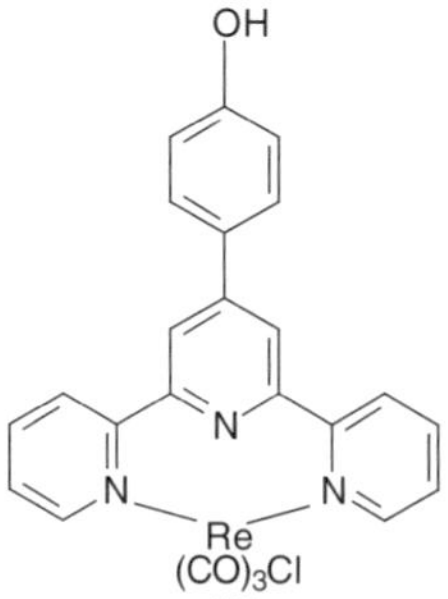

28

An alternative synthetic approach to the synthesis of the polypyridine complexes involves the interaction of bis(7'-azaindolyl)benzene ligands with $[Re_2(CO)_{10}]$ in diglyme under reflux leading to complexes **29** (R = H, F, Cl, Br, CF$_3$, MeO, N-azaindolyl) (04JOM1665).

29

Azine-containing ligands may give rise to dinuclear complexes. Thus, 4',7'-phenanthrolino-5,6':5,6-pyrazine (LL) reacts with $[Re(CO)_5Cl]$ in methanol to yield complex **30** (92IC5). The latter reacts with silver triflate in acetonitrile to give $[Re(CO)_3(LL)(AN)]_2(OTf)_2$. 2,3-Bis(2-pyridyl)pyrazine (LL) forms the mononuclear complex $[Re(CO)_3(LL)Cl]$ (95JCS(D)3677). 2,3-Bis(2-pyridyl)pyrazine and 2,3-bis(2'-pyridyl)quinoxalines (LL) react with $[Re(CO)_5Cl]$ in toluene under reflux to give $[Re(CO)_3(LL)Cl]$ (90JCS(D)1657). Subsequent reaction of the products with $[Re(CO)_5Cl]$ leads to the dinuclear complexes **31** ($ML_n = Re(CO)_3Cl$, $n = 0$) and **32** ($ML_n = Re(CO)_3Cl$; $n = 0$). Heteronuclear complexes **31** ($ML_n = Ru(bpy)_2$, $Os(bpy)_2$; $n = 2$) and **32** (M = Ru(bpy)$_2$, Os(bpy)$_2$; $n = 2$) follow from $[M(bpy)_2(LL)](PF_6)_2$ and excess $[Re(CO)_5Cl]$.

30   **31**   **32**

$[Re(CO)_5Cl]$ when reacted first with silver triflate and then 2,2'-bipyridine, 4,4'-dimethyl-2,2'-bipyridine, or 1,10-phenanthroline (LL) gives $[Re(CO)_4(LL)](OTf)$ (90JCS(CC)179, 95IC5578, 95ICA(240)453, 97JPC(A)9531, 00IC3107). This is the way of synthesis of $[M(bpy)(CO)_4](OTf)$ (M = Mn, Re) (78JOM(159)201). Compound $[Mn(CO)_4(bpy)](OTf)$ follows from $[Mn(CO)_5Cl]$, silver triflate, and 2,2'-bipyridine in dichloromethane (98JOM(598)136). Complex $[Re(CO)_3(phen)(H_2O)](OTf) \cdot H_2O$ was structurally characterized (99AX(C)913).

B. REACTIVITY OF $[M(CO)_3(LL)X]$ COMPLEXES

Photochemical irradiation of the solution of $[Re(CO)_3(bpy)Br]$ with triethylamine in DMF gives complex **33** (87JCS(CC)1153) containing the 5-ethyl-2,2'-bipyridine ligand.

33

Although the ligand-substitution reactions in [Re(CO)$_3$(phen)X] (X = Cl, Br, I) occur at the X ligand (79JA7415, 81JA5238, 83JPC952, 84JCS(CC)1244, 89JA5185), the loss of CO predominates under mass-spectral conditions (83AC1954, 85IC397). The X-substituted species include [Re(CO)$_3$(LL)Y]$^+$ (LL = bpy, phen; Y = MeCN, PhCN, pyridine, piperidine) and [Re(CO)$_3$(LL)Y] (LL = bpy, phen; Y = SnPh$_3$, GePh$_3$) (79JA7415, 80JA7892). Photochemical properties of the complexes [Re(L)(CO)$_3$(LL)] and [Re(L)(CO)$_3$(LL)]$^+$ (L = Cl, I, N-, or P-donor) (90JPC2229, 93CCR(125)101, 96JCS(CC)1587, 97CCR(165)239), [Re(CO)$_3$(4,4′-Me$_2$bpy)Br]$^-$ (96ICA(247)247), [Re(L)(CO)$_3$(LL)] (L = Mn(CO)$_5$, Re(CO)$_5$, Co(CO)$_4$, (η^5-Cp)Fe(CO)$_2$, SnPh$_3$) (76JA3931, 79JA1597, 85IC2934, 89IC318, 90CCR(104)39, 90ICA(178)185, 91IC599, 91ICA(187)133, 94IC2865, 95IC5183, 96ICA(250)5), [Re(L)(CO)$_3$(LL)] (L = Alk, CH$_2$Ph) (95JA11582, 95JOM(493)153, 96JPC18607) have been studied. [Re(CO)$_3$(LL)(EPh$_3$)] (E = Ge, Sn; LL = bpy, phen) are prepared by the reaction of [Re(CO)$_3$(LL)X] with Ph$_3$GeBr or Ph$_3$SnCl in THF (79JA1597). Substitution of the Br-ligand in [Re(CO)$_3$(LL)Br] (LL = bpy, phen) by the other functions allows creation of new materials with interesting photochemical properties, among them [Re(CO)$_3$(LL)(4-nitrobenzoate)] (98IC-2806).

[Re(CO)$_3$(4-styryl-4′-methyl-2,2′-bipyridine)Cl] gives rise to [Re(CO)$_3$(4-styryl-4′-methyl-2,2′-bipyridine)(AN)](PF$_6$) in the presence of AgBF$_4$ and then NH$_4$PF$_6$ (91JA389). 1,4-Bis{2′-(4′-methyl-2,2′-bipyrid-4-yl)ethenyl}benzene (LL) and [Re(CO)$_5$Cl] in the molar ratio 1:2 give [(OC)$_3$ClRe(μ-LL)Re(CO)$_3$Cl]. With silver tetrafluoroborate, dehalogenation takes place. Metathesis of the product with ammonium hexafluorophosphate yields [(OC)$_3$(AN)Re(μ-LL)Re(CO)$_3$(AN)](PF$_6$)$_2$. These complexes are illustrated as **34** and **35**. From [Re(CO)$_3$(H$_2$O)]$^+$, the following species were prepared: [Re(CO)$_3$(bpy)(H$_2$O)](OTf) and [Re(CO)$_3$(phen)(H$_2$O)] (NO$_3$)$_{0.5}$(OTf)$_{0.5}$ (03IC3516).

34

35

[Re(CO)$_3$(bpy)(PPh$_3$)]I on irradiation in the presence of triethanolamine and carbon dioxide gives the formate complex [Re(CO)$_3$(bpy){OC(O)H}], which represents the photochemical fixation of carbon dioxide (97JCS(D)1019). Photocatalytic reduction of carbon dioxide by [Re(CO)$_3$(LL)Cl] complexes was postulated to proceed through the stage of metallocarboxylic acids (M-COOH) (97CIC67). [Re(4,4'-Me$_2$bpy)(CO)$_4$(OTf)] with aqueous potassium hydroxide yields [Re(4,4'-Me$_2$bpy)(CO)$_3$(COOH)] (97OM4421, 98JA11200). The product after standing as the acetone solution undergoes photochemical transformation to the dinuclear complex **36** containing the μ_2-η^2-CO$_2$ bridge. The intermediate step of transformation to **36** involves partial decarboxylation of [Re(4,4'-Me$_2$bpy)(CO)$_3$(COOH)] to [Re(CO)$_3$(4,4'-Me$_2$bpy)(CO)$_3$H] and further reaction of these mononuclear complexes gives **36** and molecular hydrogen is evolved. [Re(CO)$_3$(4,4'-Me$_2$bpy)(OTf)] reacts with potassium hydroxide in acetone to yield the hydroxo-complex [Re(CO)$_3$(4,4'-Me$_2$bpy)(OH)], which eventually also leads to **36**. Complex [Re(CO)$_3$(4,4'-Me$_2$bpy)(OH)] in methanol affords the ester [Re(CO)$_3$(4,4'-Me$_2$bpy)(COOMe)]. On photolysis of [Re(CO)$_3$(4,4'-Me$_2$bpy)(COOH)] in deuterochloroform, the product is [Re(CO)$_3$(4,4'-Me$_2$bpy)Cl]. The same compound is formed from [Re(CO)$_3$(4,4'-Me$_2$bpy)H] and [Re(CO)$_3$(4,4'-Me$_2$bpy)(COOMe)] under identical conditions. [Re(CO)$_3$(4,4'-Me$_2$bpy)H] with excess CO$_2$ forms the formate complex (88JA7098, 98JA11200), probably as a result of the insertion of carbon dioxide into the Re–H bond. [Re(bpy)(CO)$_3$(COMe)] was also discovered (97JCS(P2)2569). Dissolution of this compound in acetone and prolonged photochemical treatment gives the bridged complex [(LL)(OC)$_3$Re(μ_2-η^2-CO$_2$)Re(CO)$_3$(LL)] through the stage of [Re(LL)(CO)$_3$H].

36

Reaction of [Re(LL)(CO)$_4$(OTf)] (LL = 4,4'-Me$_2$bpy) with aqueous potassium hydroxide yields complex **37** (99JCS(CC)1411, 01JCS(CC)2082). This species when allowed to stand after subsequent removal of the solvent (DMF, DMSO) forms the dinuclear species **38** along with the bicarbonato complex **39** (03OM337). In DMSO, the by-product is **40** containing the carbonato-bridge. Complex **37** reacts with CO$_2$ in

DMSO to yield the bicarbonato species **39**. The same product is formed on reaction of the dinuclear CO_2-bridged complex **38** with carbon dioxide. Complexes **37** and **39** react in DMSO to yield **40**. The latter reacts with $[Re(CO)_3(4,4'-Me_2bpy)(OTf)]$ to yield **41**. $[Re(CO)_3(bpy)(OTf)]$ interacts with aqueous potassium hydroxide to give $[Re(CO)_3(bpy)(OH)]$. Complex $[Re(CO)_3(4,4'-Me_2bpy)(CO)_4](OTf)$ reacts with the latter hydroxo-complex to yield $[(4,4'-Me_2bpy)(OC)_3Re(\mu-CO_2)Re(CO)_3(bpy)]$.

Complexes $[Re(CO)_3(LL)(CH_2R)]$ experience rhenium–carbon bond cleavage under photochemical conditions (93ICA(208)103, 95JOM(492)165, 96CEJ228, 96CEJ1556, 96ICA(247)215, 98OM2689). Thus, $[Re(CO)_3(bpy)(CH_2OH)]$ with methanol gives $[Re(CO)_3(bpy)(OMe)]$ (98OM2689). Attempts to protonate the methoxy group using $HBF_4 \cdot Et_2O$ gives $[Re(bpy)(CO)_3(OEt_2)](BF_4)$ and using triflic acid gives $[Re(CO)_3(bpy)(OTf)]$ (02IC4673). $[Re(CO)_3(4,4'-Me_2bpy)Br]$ (79JOM(170)235) with the relevant Grignard reagent in THF yields $[Re(CO)_3(4,4'-Me_2bpy)R]$ (R = Me, Et, *i*-Pr, $PhCH_2$) (99ICA(284)61).

[Re(CO)$_3$(bpy)Cl] with silver hexafluoroantimonate refluxed in acetonitrile affords [Re(CO)$_3$(bpy)(AN)](SbF$_6$) (97JOM(530)169). The product was reacted with a variety of phosphines or phosphites in THF to yield a series of substitution products [Re(CO)$_3$(bpy)(L)](SbF$_6$) (L = PBu-n_3, PEt$_3$, PPh$_3$, P(OMe)Ph$_2$, POPr-i_3, P(OEt)$_3$, P(OMe)$_3$, P(OPh)$_3$). [Re(CO)$_3$(bpy)Cl]reacts with P(OEt)$_3$ in the presence of silver triflate in THF and then ammonium hexafluorophosphate in methanol to afford [Re(CO)$_3$(bpy){P(OEt)$_3$}](PF$_6$) (00IC2777, 02JA11448). In a similar fashion, [Re(CO)$_3$(bpy)(PPh$_3$](PF$_6$), [Re(CO)$_3$(4,4'-Me$_2$bpy){P(OEt)$_3$}](PF$_6$), and [Re(CO)$_3$ {4,4'(CF$_3$)$_2$bpy}{P(OEt)$_3$}(PF$_6$) were prepared. They can be illustrated as **42** (R^1 = H, Me, CF$_3$; R = OEt; R^1 = H, R = Ph). [Re(CO)$_3$(bpy){P(OEt)$_3$}](PF$_6$) under photolysis in acetonitrile gives the CO-substitution product **43** (R^1 = H, R = OEt, L = AN). The same procedure carried out in THF in the presence of pyridine leads to **43** (R^1 = H, R = OEt, L = py) and in acetonitrile in the presence of triethylphosphite to **43** (R^1 = H, R = OEt, L = P(OEt)$_3$). Complex **42** (R = Ph, R^1 = H) under photolysis in acetonitrile gives **43** (R = Ph, R^1 = H, L = AN), while the same procedure but in the presence of triphenylphosphine gives **43** (R = Ph, R^1 = H, L = PPh$_3$). Complexes **42** (R = OEt, R^1 = Me, CF$_3$) as their triflate salts react with triethylphosphite in acetonitrile and further with ammonium hexafluorophosphate in methanol to give complexes **43** (R = OEt, R^1 = Me, CF$_3$, L = P(OEt)$_3$). Starting from **43** (R = OEt, R^1 = H, L = AN) and triphenylphosphine in acetone, species **43** (R = OEt, R^1 = H, L = PPh$_3$) resulted. Complex **43** (R = OEt, R^1 = H, L = AN) and triphenylphosphine in acetone form **43** (R = Ph, R^1 = H, L = PPh$_3$). Complexes [Re(CO)$_3$(LL)X] (X = Cl, CN; LL = bpy, phen) and [Re(CO)$_3$(phen)(AN)]$^+$ as well as [Re(CO)$_3$(4,4'-(CF$_3$)$_2$bpy)(AN)](OTf) are known (98JPC(B)4759). [Re(CO)$_3$(phen)(AN)]$^+$ efficiently catalyzes the electron transfer processes (87ACR214). [Re(CO)$_3$(phen)(AN)]$^+$ with triphenylphosphine under photochemical conditions gives [Re(CO)$_3$(phen)(PPh$_3$)]$^+$ (88CRV1189).

42   **43**

Complexes [Re(CO)$_3$(LL)Cl] (LL = 4,4'dimethyl-2,2'-bipyridine, 4,4'-di-trifluoromethyl-2,2'-bipyridine) on reaction with silver triflate and triethylphosphite in THF, and further with ammonium hexafluorophosphate in methanol give [Re(CO)$_3$(LL){P(OEt)$_3$}(PF$_6$) (97OM5724). They photocatalyze the reduction of carbon dioxide as well as [Re(CO)$_3$(bpy){P(OEt)$_3$}](PF$_6$) (95ECM621). [Re(CO)$_3$(bpy)Cl] on photochemical reaction with triphenylphosphine yields [Re(CO)$_2$(bpy)(PPh$_3$)$_2$]Cl (84IC2104, 98IC2618). [Re(CO)$_3$(bpy)Br], triethylamine, and triethylphosphite in THF under photolysis give rise to [Re(CO)$_2$(bpy){P(OEt)$_3$}$_2$]Br (87JCS(CC)1153, 94IC4712). From [Re(CO)$_3$(bpy)Cl], [Re(CO)$_2$(bpy)[P(OEt)$_3$}$_2$]Cl resulted. [Re(CO)$_2$ (bpy){P(OEt)$_3$}$_2$]Cl, [Re(CO)$_2$(bpy){P(OEt)$_3$}$_2$]Br, and sodium tetraphenylborate in methanol yield [Re(CO)$_2$(bpy){P(OEt)$_3$}$_2$](BPh$_4$). These complexes are efficient photochemical reductants of carbon dioxide. 2,2′-Bipyridine reacts with [Re(CO)$_3$(dppm)Cl] and thallium hexafluorophosphate in o-dichlorobenzene to yield complex **44** (X = PF$_6$) (98IC2618). The same product follows from [Re(CO)$_3$(bpy)Cl], diphenylphosphinomethane, and thallium hexafluorophosphate in o-dichlorobenzene. Reaction of [Re(CO)$_3$(bpy)(OTf)] with diphenylphosphinomethane in o-dichlorobenzene affords complex **44** (X = OTf). Reaction of [Re(CO)$_3$(bpy)(OTf)] with PPh$_3$ or Ph$_2$PCH = CHPPh$_2$ gives **45** and **46**, respectively.

          44   **45**   **46**

The reaction between [Mn(CO)$_3$(LL)(ClO$_4$)] and phosphite ligands, P(OR)$_3$ and P(OR)$_2$Ph (R = Me, Et) (L) in boiling ethanol gives [Mn(CO)$_2$(LL)L$_2$](ClO$_4$) (LL = bpy, phen) (81JOM(219)61. In the presence of excess phosphite in acetone under UV irradiation, the cationic products yield [Mn(CO)(LL)L$_3$](ClO$_4$) (81JOM(219)61, 87JOM(326)C71, 90JOM(394)275, 90JOM(397)309, 94JCS(D)3745, 00JPP(A)67). Species [Mn(CN)(CO)$_3$(LL)] (LL = bpy, 4,4′-Me$_2$bpy, phen) (86ICA(121)191) enter photochemical substitution with P(OR)$_3$ (R = Me, Et, Ph) (L) to afford [Mn(CO)(CN)(LL)L$_2$] (93JOM(452)91). Complexes [Mn(CO)$_3$(LL)(CNR)](ClO$_4$) (LL = bpy, phen; R = t-Bu, Ph) are known (87JOM(326)C71). Dinuclear complexes [(phen)(OC)$_3$Mn(µ-CN)Mn(CO)$_3$(phen)](PF$_6$) were isolated as well (86ICA(121)191).

[Re(CO)$_3$(bpy)Br] with silver triflate in methylene chloride gives [Re(CO)$_3$(bpy)(OTf)] (02IC4673). Further addition of NaB(3,5-(CF$_3$)$_2$C$_6$H$_3$)$_4$ in combination with a ligand yields the series [Re(CO)$_3$(bpy)L][B(3,5-(CF$_3$)$_2$C$_6$H$_3$)$_4$ (L = PPh$_3$, AN, PhCN, MeCOMe, PhCOMe, N(H)CPh$_2$, N(Me)CHPh, MeOH, Et$_2$O, THF, MeI) (02IC4673, 02OM1966). [Re(CO)$_3$(bpy)(THF)][B(3,5-(CF$_3$)$_2$C$_6$H$_3$)$_4$ reacts with lithium acetylides to generate the alkynyls [Re(CO)$_3$(bpy)(C≡CR)] (R = Ph, SiMe$_3$) (02IC4673), although some difficulties of synthesis of similar alkynyls were noted (95OM2749, 96OM1740, 01JCS(CC)789). Related complexes are [Re(CO)$_3$(bpy)(η^1-O = C(Me)R)][B(3,5-(CF$_3$)$_2$C$_6$H$_3$)$_4$ (01JCS(CC)2682). Complexes [Re(CO)$_3$(4,4′-R$_2$bpy)(C≡C-C≡CH)] (R = Me, t-Bu) are among the substitution products (98JCS(CC)2121). [Re(CO)$_3$(4,4′-t-Bu$_2$bpy)(C≡CC≡CH)] (99JCS(CC)1013) reacts with BrC≡CPh, piperidine, and copper(I) iodide in THF to yield [Re(CO)$_3$(4,4′-t-Bu$_2$bpy)(C≡CC≡CC≡CPh)] (00OM5092). If IC≡CSiMe$_3$ is used in this type of reaction, the product is [Re(CO)$_3$(4,4′-t-Bu$_2$bpy)(C≡CC≡CC≡CSiMe$_3$)]. The

4,4′-dimethyl-2,2′-bipyridine analogues can be prepared similarly. [Re(CO)$_3$(4,4′-t-Bu$_2$bpy)(C≡CH)] can be converted to the dinuclear complex [(t-Bu$_2$bpy)(O-C)$_3$Re(C≡CC≡C)Re(CO)$_3$(t-Bu$_2$bpy)] (96OM1740).

Complexes [Re(CO)$_3$(LL)Cl] (L = bpy, 4,4′-Me$_2$bpy, 4,4′-t-Bu$_2$bpy, phen) transform to the ethynyl derivatives using two routes (04JOM1393). The first route is the interaction of rhenium(I) complexes listed above with terminal acetylenes HC≡CR^1 and n-butyl lithium in ether. The products are **47** (R = H, Me, t-Bu; LL may be phen; n = 1, R^1 = t-Bu, SiMe$_3$, Ph, C$_6$H$_4$OMe-4, C$_6$H$_4$Et$_4$, C$_6$H$_4$Ph-4, C$_5$H$_4$N-4, C$_6$H$_4$-C≡CH-4, n-C$_6$H$_{13}$, n-C$_8$H$_{17}$, n-C$_{10}$H$_{21}$, H, Ph). In the second approach, the same starting complexes react with H(C≡C)$_n$R^1 (n = 1, 2) in the presence of silver triflate and triethylamine in THF. The products are not only those listed above but also **47** (n = 2, R^1 = Ph, H, SiMe$_3$). Complexes **47** (R = H, Me, t-Bu; LL may be phen; n = 2, R^1 = H) react with IC≡CSiMe$_3$ or BrC≡CPh in the presence of copper(I) iodide and piperidine in THF to yield **48** (R = H, Me, t-Bu; LL may be phen; n = 3, R^1 = SiMe$_3$, Ph). Species **47** (R = H, Me, t-Bu; LL may be phen; n = 1, 2, R^1 = H) in the presence of copper(II) acetate in pyridine form the dinuclear products **48** (R = H, Me, t-Bu; LL may be phen; n = 1, 2).

[Re(CO)$_3$(4,4′-t-Bu$_2$bpy)Cl] reacts with R^{1}C≡CLi (R^1 = C$_6$H$_4$OMe-4, C$_6$H$_4$Ph-4, C$_8$H$_{17}$, C$_4$H$_3$S, C$_4$H$_2$S-C$_4$H$_3$S, C$_5$H$_4$N) in diethyl ether to yield the series of complexes **49** (R = t-Bu, R^1 = C$_6$H$_4$OMe-4, C$_6$H$_4$Ph-4, C$_8$H$_{17}$, C$_4$H$_3$S, C$_4$H$_2$S-C$_4$H$_3$S, C$_5$H$_4$N) (99JCS(CC)1013, 03JOM(670)205). Refluxing [Re(CO)$_3$(bpy)Cl], silver triflate, HC≡CR^1 (R^1 = Ph, C$_6$H$_4$Cl-4, C$_6$H$_4$C$_8$H$_{17}$-4, C$_4$H$_3$S, C$_4$H$_2$S-C$_4$H$_3$S), and triethylamine in THF affords **49** (R = H; R^1 = Ph, C$_6$H$_4$Cl-4, C$_6$H$_4$C$_8$H$_{17}$-4, C$_4$H$_3$S, C$_4$H$_2$S-C$_4$H$_3$S). Complex **49** (R = t-Bu, R^1 = C$_5$H$_4$N) further reacts with [Re(CO)$_3$(4,4′-(CF$_3$)$_2$bpy)(AN)](ClO$_4$) or [Re(CO)$_3$(5-NO$_2$phen)](ClO$_4$) in THF under reflux to yield the dinuclear species **50** (R = t-Bu; ML$_n$ = [Re(CO)$_3$(4,4′-(CF$_3$)$_2$bpy)(CO)$_3$](ClO$_4$), [Re(CO)$_3$(5-NO$_2$phen)](ClO$_4$). Reaction of [Re(CO)$_3$(4,4′-R$_2$bpy)Cl] (R = H, t-Bu) with [W(CO)$_5$(C$_5$H$_4$N-C≡CH)] gives the heterodinuclear complex **50** (R = H, t-Bu; ML$_n$ = M(CO)$_5$).

[Mn(CO)$_3$(LL)Br] (LL = bpy, phen) with silver perchlorate give the Br-substitution products [Mn(CO)$_3$(LL)(ClO$_4$)] (LL = bpy, phen) (88IC4385). On further interaction with 1,8-naphthyridine complexes **51** result where the 1,8-naphthyridine ligand is η^1-coordinated (86JOM(304)207, 88IC4385). Complexes are fluxional, and the manganese atom shuffles between the two nitrogen heteroatoms of the ligand (88IC4385). Complex [Mn(CO)$_3$(η^2-naph)Br] (75JINC1375) reacts with silver perchlorate to give [Mn(CO)$_3$(η^2-naph)(OClO$_3$)], which in excess 1,8-naphthyridine gives [Mn(CO)$_3$(η^1-naph)(η^2-naph)](ClO$_4$) (93JOM(463)143). [Mn(CO)$_5$Br] reacts with silver perchlorate and then 1,8-anphthyridine to afford [Mn(CO)$_4$(η^2-naph)](ClO$_4$) (77TMC123). At low temperatures, [Mn(CO)$_5$(η^1-naph)](ClO$_4$) results.

51

Complex [Re(CO)$_3$(bpy)Cl] with sodium tetrahydroborate gives the dinuclear species **52** (86OM1500), with carbon dioxide and triethanolamine under photochemical conditions **53**, and with sodium trihydroborate cyanide **54** (89JCS(D)1449). [Re(CO)$_3$(bpy)X] (X = Cl, Br) reacts with silver carboxylates (naphthalene-2-carboxylate, anthracene-9-carboxylate, pyrene-1-carboxylate) in methylene chloride to yield [Re(CO)$_3$(bpy)(OOCR)] (02JCS(D)2194). Thermal substitution of the triflate anion in [Re(CO)$_3$(bpy)(OTf)] gives [Re(CO)$_3$(bpy)(OOCH)] (92IC5243). [Re(CO)$_3$(bpy)Cl] reacts with silver triflate in acetone to produce [Re(CO)$_3$(bpy)(OTf)] and then with CO$_2$ in THF, [Re(CO)$_3$(bpy)(OOCH)] results (96OM3374). This product is a catalyst for carbon dioxide electroreduction giving CO, CO$_3^{2-}$, and COOH$^-$ among the products.

52   **53**   **54**

Complexes [M(CO)$_3$(bpy)(OTf)] (M = Mn, Re) react with sodium methoxide to yield [M(CO)$_3$(bpy)(OMe)] (M = Mn, Re) (02OM1750). The products with dimethylacetylene dicarboxylate give the new products of insertion of the incoming acetylene to the M-OMe bond, **55** (M = Mn, Re). [Re(CO)$_3$(4,4'-Me$_2$bpy)(OEt)] also inserts dimethylacetylene dicarboxylate.

55

[M(CO)$_3$(bpy)Br] (M = Mn, Re) can be chemically reduced to Na[M(CO)$_3$(bpy)] by using sodium amalgam (92OM2826, 95OM1115). These complex anions generated *in situ* on reaction with chloromethyl sulfides ClCH$_2$SR (R = Me, Ph) produce neutral complexes **56** (M = Mn, Re; R = Me, Ph) (02OM5312). With triflic acid, the products give [M(CO)$_3$(bpy)(OTf)]. Cationic complex **57** was prepared from [Re(CO)$_3$(bpy)(Me)], thioanisole, and further with (Ph$_3$C)(PF$_6$). Another route to the cationic complex lies through methyl triflate, which also results in methylation at the sulfur atom, **58** (M = Mn, Re; R = Me, Ph). The products (R = Ph) react with anionic nucleophiles such as potassium diphenylphosphide, potassium ethanethiolate, or tetraethylammonium iodide to restore complexes **56** (M = Mn, Re; R = Ph). Complexes **58** (R = Ph; M = Mn, Re) react with styrene in toluene to yield [M(CO)$_3$(bpy)(OTf)] (M = Mn, Re), MeSPh, and phenylcyclopropane. Complex [Re(CO)$_3$(bpy)(OH) on reaction with CS$_2$ gives [Re(CO)$_3$(bpy)(SH)] (03JCS(CC)328).

56   **57**   **58**

Cationic complexes [Mn(CO)$_3$(LL)(Me$_2$CO)](ClO$_4$) (LL = bpy, phen) readily react with isocyanide ligands to substitute the coordinated acetone ligand and yield [Mn(CO)$_3$(CNR)(LL)](ClO$_4$) (R = Ph, t-Bu; LL = bpy, phen) (84JOM(276)39). If the same reaction is run with boiling in the presence of the decarbonylating reagent Me$_3$NO, successive CO/CNR substitution occurs, and the following products are formed: [Mn(CO)$_2$(LL)(CNR)$_2$](ClO$_4$) (R = t-Bu, Ph; LL = bpy, phen), [Mn(CO)(t-BuNC)$_3$(LL)](ClO$_4$) (L = bpy, phen), and [Mn(CNPh)$_4$(LL)](ClO$_4$) (LL = bpy, phen). Similar products are [Mn(CO)$_2$(phen)(CNBu^t)(CNPh)](ClO$_4$) and [Mn(CO)$_3$(phen)(CNBu^t)](ClO$_4$) (85AX(C)1312, 86AX(C)417). Complex [Re(CO)$_3$(bpy)(NH(p-Tol))] reacts with isocyanates (02JCS(CC)1814), and [Re(OMe)(CO)$_3$(bpy)] also reacts with isocyanates and to a lesser extent with isothiocyanates (02CEJ4510). Thus, with ethylisothiocyanate, two products, [Re(CO)$_3$(bpy){SC(= NEt)(OMe)}] and [Re(CO)$_3$(bpy){SC(= S)(OMe)}] were isolated. Complex **59** when reacted with p-tolylisothiocyanate in THF gives the insertion product **60** (03OM257). With diphenylketene in THF, complex **59** gives rise to **61**, the product of insertion of the ketene $C=C$ bond into the amido N–H bond of species **59**. Complex **62** (02AGE3858) obtained from [Re(CO)$_3$(bpy)(OTf)] and KN=CPh$_2$ reacts with diphenylketene to afford **63** (R = R^1 = Ph) containing a lactam rings useful in the synthesis of β-lactams (03JA3706). On reaction of **62** with ethylphenylketene and cycloheptylketene, the products are **63** (R = Ph, R^1 = Et; R = R^1 = (CH$_2$)$_6$). With alkynes, RC≡CR^1 (R = R^1 = COOMe, COOEt; R = COOMe, R^1 = H), metallacycles **64** result (03OM257). Treatment of **64** (R = R^1 = COOMe) first with KN(SiMe$_3$)$_2$ and then with methyl triflate gives **65**. The alkylidene amido complex **62** with dimethylacetylene dicarboxylate produces the metallacyclic species **66**, where one of the carbonyl ligands is activated (05OM1772).

Similar synthetic procedures are involved in the preparation of [Re(CO)$_3$ (bpy)(NC(CH$_2$)$_n$Me)]$^+$ (n = 0–17) (86JA5344, 88JA5051). [Re(CO)$_3$(LL)(OTf) with isonitriles give [Re(CO)$_3$(LL)NCR)](OTf) (LL = 3,4,7,8-Me$_4$phen, R = t-Bu, n-Bu, n-octyl, n-C$_{12}$H$_{25}$; 4,7-Me$_2$phen, R = t-Bu, n-Bu, n-C$_{12}$H$_{25}$; R = 4-Mephen, R = t-Bu; 5-Mephen, R = t-Bu; phen, R = t-Bu; 5-Clphen, R = t-Bu; 4,7-Ph$_2$phen, R = t-Bu; bpy, R = t-Bu) (93JA8230). The complexes are long-lived and highly luminescent and are attractive as molecular probes.

Complex **67** reacts with dimethylacetylene dicarboxylate to give **68** (R = COOMe) where the 1,10-phenanthroline ligand has undergone activation (05OM1772). Similar reaction of **67** with MeOOC-C≡CH also yields the 1,10-phenanthroline activated product **68** (R = H) (05JCS(CC)116). Complex **68** (R = COOMe) differs from **68** (R = H) by its instability and tendency to transform to **69** by 1,3-migration of the hydrogen atom attached to the dearomatized pyridine ring (05OM1772).

Reaction of [Re(CO)$_3$(LL)(OTf)] (LL = bpy, 4,4'-Me$_2$bpy) with potassium hydroxide in water/methylene chloride gives the hydroxo-complexes [Re(CO)$_3$(LL)(OH)] (04CEJ1765). The same reaction for the 2,2'-bipyridine complex but run in anhydrous methylene chloride leads to the formation of the dicationic dinuclear complex [(bpy)(OC)$_3$Re(μ-OH)Re(CO)$_3$(bpy)](OTf). Complex [Re(CO)$_3$(4,4'-Me$_2$bpy)(OH)] reacts with phenyl acetate or vinyl acetate to afford the acetato-complex [Re(CO)$_3$(4,4'-Me$_2$bpy)(OAc)]. [Re(CO)$_3$(bpy)(OH)] with phenyl isocyanate gives [Re(CO)$_3$(bpy)(OC(O)NHPh)], the product of insertion of PhNCO into the O–H bond. [Re(CO)$_3$(4,4'-Me$_2$bpy)(OH)] inserts ethyl- or p-tolylisothiocyanate into the Re–O bond to yield [Re(CO)$_3$(4,4'-Me$_2$bpy){SC(O)NHEt] or [Re(CO)$_3$(4,4'-Me$_2$bpy){SC(O)NH(p-Tol)}], respectively. The starting hydroxo-complexes react with dimethylacetylene dicarboxylate, which is accompanied by the activation of the CO and OH ligands and formation of the metallacycle **70** (R = H, Me).

C. COMPLEXES WITH CHROMOPHORE LIGANDS

Complexes of the type $[Re(CO)_3(bpy)(L)]X$ (L is a chromophore ligand) participate in photoinduced intramolecular energy transfer and electron transfer reactions (68JA3722, 83IC1617, 84JA2613, 84OM1242, 85IC2596, 85JCS(CC)796, 86JA7567, 89JA8305, 90ACS101, 90JA5378, 90JPC2740, 90JPC8745, 91CCR(111)221, 91JPC5850, 92JA1897, 93CCR(122)63, 93IC4994, 93JA5675).

Complex 71 ($R^1 = R^2 = H$; $X = PF_6$) follows from $[Re(CO)_3(bpy)(OTf)]$, pyridine, and ammonium hexafluorophosphate (94ICA(225)41, 95JPC1961, 96JPC5408). Compounds 71 ($R^1 = H$, $R^2 = COOMe$, $X = PF_6$; $R^1 = R^2 = Me$, $X = PF_6$) are prepared from the corresponding species, $[Re(CO)_3(LL)Cl]$, pyridine, and silver hexafluorophosphate (91JA7470). These products were converted to the chloride by metathesis using an anion-exchange resin. Further metathesis of the chloride complexes with $Na[Co(CO)_4]$ gives series 71 ($R^1 = R^2 = H$; $R^1 = H$, $R^2 = COOMe$; $R^1 = R^2 = Me$; $X = Co(CO)_4$) (97IC6224). 1,10-Phenanthroline, 4,7-dimethyl-, 2,9-dimethyl-, 2,9-dimethyl-4,7-diphenyl-1,10-phenenthroline, 3,4,7,8-tetramethyl-, 5,6-dimethyl-, 4,7-diphenyl-, and 5-phenyl-1,10-phenanthroline (LL) react with $[Re(CO)_5(OTf)]$ in methanol and further with pyridine to yield $[Re(CO)_3(LL)(py)](OTf)$ (93IC3836). Complex $[Re(CO)_3(4,7-Me_2phen)(py)]^+$ is known (95IC2875). Another synthetic approach involves the reaction of $[Re(CO)_5(OTf)]$ with LL (phen, $4,7-Me_2phen$, $2,9-Me_2phen$ $2,9-Me_2-4,7-Ph_2phen$, $3,4,7,8-Ph_4phen$, $5,6-Me_2phen$, $4,7-Ph_2phen$, 5-Phphen) and then pyridine, and the products are $[Re(LL)(CO)_3(py)](OTf)$ (95IC5210, 98JPC(A)45); LL = bpy (66IC2119, 77JOM(131)73).

71

$[Re(CO)_3(4,4'-R_2bpy)(OTf)]$ (R = H, Me, Ph, Et_2N, MeO, Cl, NO_2) as well as the analogue with $4,4',5,5'$-tetramethyl-2,2'-bipyridine react with 4-ethylpyridine in ethanol at elevated temperatures to give series 72 (R = H, $R^1 = H$, Me, Ph, Et_2N, MeO, Cl, NO_2; R = $R^1 = Me$) (82JCS(CC)915, 86JP5344). Such complexes are of use in solar energy storage schemes. Systems $[Re(CO)_3(phen)(4-Mepy)]^+$ and $[Re(CO)_3(bpy)(4-Etpy)]^+$ are also of photochemical interest (93IC2618, 96IC273, 98CIC165). Other known compounds are 73 (95IC6421), which are based on the $[Re(CO)_3(LL)Cl]$ precursor and series $[Re(CO)_3(LL)(4-Etpy)]^+$ (LL = phen, bpy, $4,4'-Me_2bpy$, $4,4'-(MeO)_2bpy$, $4,4'-(COOEt)_2bpy$) (02IC6071). $[Re(CO)_3(LL)(4-Etpy)](PF_6)$ (LL = phen, $5-NO_2phen$) were prepared from the corresponding triflate complex and ammonium hexafluorophosphate in methanol. $[Re(CO)_3(5-NO_2phen)(ImH)]_2(SO_4)$

results from [Re(CO)$_3$(5-NO$_2$phen)(L)] (L = Cl, OTf) and imidazole and further with aqueous ammonium sulfate (04IC4994). A series [Re(CO)$_3$(4,4'-R$_2$bpy)(4-Etpy)]$^+$ (R = H, Me, COOEt) was prepared as well (02JPC(A)4519, 04JPC(A)2363).

72

73

[Re(LL)(CO)$_3$(AN)](OTf) (LL = bpy, 4,4'-Me$_2$bpy, 4,4'-t-Bu$_2$bpy, phen, 2,9-Me$_2$phen, 4,7-Me$_2$phen, 3,4,7,8-Me$_4$phen, 5-Phphen, 5-Clphen, 4,7-Ph$_2$phen, 2,9-Me$_2$-4,7-Ph$_2$phen, 2,2'-biquinoline) with 3-aminopyridine (L) yield [Re(CO)$_3$(LL)(L)](OTf) (01JCS(D)2634). The products reacted with thiophosgene and calcium carbonate in acetone to generate [Re(CO)$_3$(LL)(L)](OTf) (L = 3-isothiocyanopyridine). Photochemical properties of [Re(L)(CO)$_3$(LL)] and [Re(L)(CO)$_3$(LL)]$^+$ (L = Cl, I, N- or P-donor) were studied (95IC1282, 95OM3303, 96CIC165, 98CCR(177)201, 00JPC(A)4291, 01CCR(216)127, 02JCS(CC)950).

[Mn(CO)$_5$Br] reacts with 1,10-phenanthroline in CH$_2$Cl$_2$ to yield [Mn(CO)$_3$ (phen)Br] (05ICA3735). The product with silver triflate gives [Mn(CO)$_3$(phen)(OTf)]. The latter on interaction with imidazole or isonicotinamide (L) in methylene chloride affords [Mn(CO)$_3$(phen)L](OTf), which is illustrated as **74** for L = isonicotinamide. Similar products are [Mn(CO)$_3$(phen)(Im-R)](OTf) (R = H, Me, Ph, 2-Me-5-NO$_2$) (04PP203). Complex [Re(CO)$_3$(phen)(Im)]$^+$ is known (95ICA(240)169). 1,10-Phenanthroline reacts with [Mn(CO)$_5$Br] in methylene chloride to yield [Mn(CO)$_3$ (phen)Br] (00ICA(299)231). Further reaction with silver triflate gives [Mn(CO)$_3$ (phen)(OTf)]. The product reacts with various imidazole ligands to produce complexes **75** (R = H, Me, Ph, R^1 = H; R = Me, R^1 = NO$_2$; instead of NH, CH$_2$OH). Neutral complexes [Re(CO)$_3$(LL)(3,5-(CF$_3$)$_2$pz)] (LL = bpy, phen) contain a pyrazolyl moiety (03IC1248). Complex **76** (04IC4523) can also be mentioned in this group.

74

75

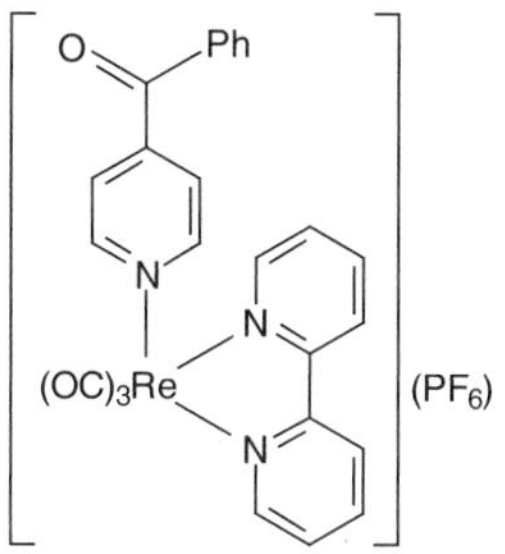

76

Diimine ligands, 2,2′-bipyridine, 4,4′-dimethyl-2,2′-bipyridine, 4,4′,5,5′-tetrame-thyl-2,2′-bipyridine, 4,4′-bis(diethylaminocarboxy)-2,2′-bipyridine, and 4,4′-bis (ethoxycarbonyl)-2,2′-bipyridine (LL) in toluene form [Re(CO)$_3$(LL)Cl] (91JA7470). They react with silver hexafluorophosphate and an appropriate pyridine ligand to yield the cationic complexes illustrated as **77** and **78** for the parent 2,2′-bipyridine ligand. Similar procedures lead to monomers **79** (R = R^1 = H; R = Me, R^1 = H; R = R^1 = Me; R = H, R^1 = Et$_2$N) (94IC1354). They react with [(η^5-Cp)Ru(AN)$_3$](PF$_6$) in 1,2-dichloroethane; [(η^5-Cp)Fe(η^6-4-benzylpyridine)] reacts with [Re(CO)$_3$(bpy)Cl] and silver hexafluorophosphate in DMF. The products are **80** (M = Ru, R = R^1 = H, R = Me, R^1 = H, R = R^1 = Me, R = H, R^1 = Et$_2$N; M = Fe, R = R^1 = H), the dinuclear species. A similar product is **81** (02IC132).

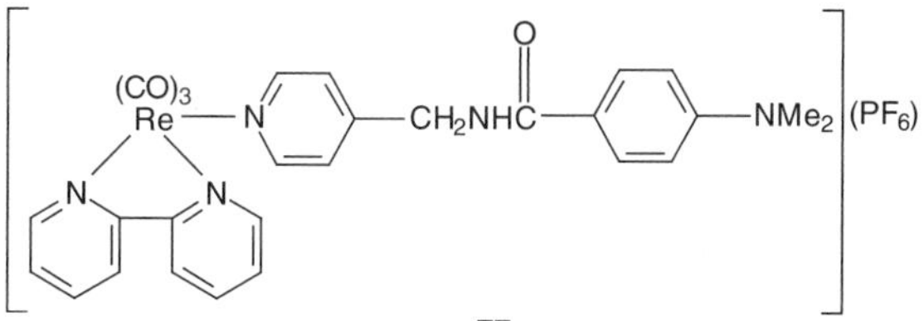

77

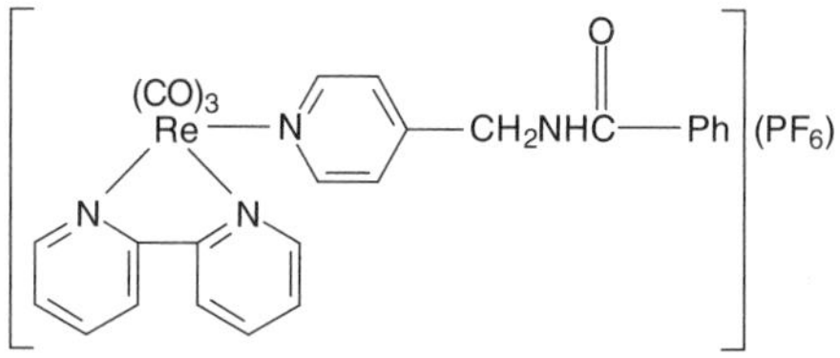

78

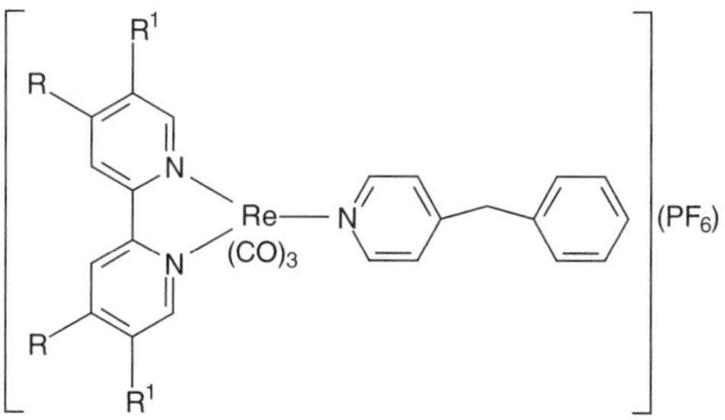

79

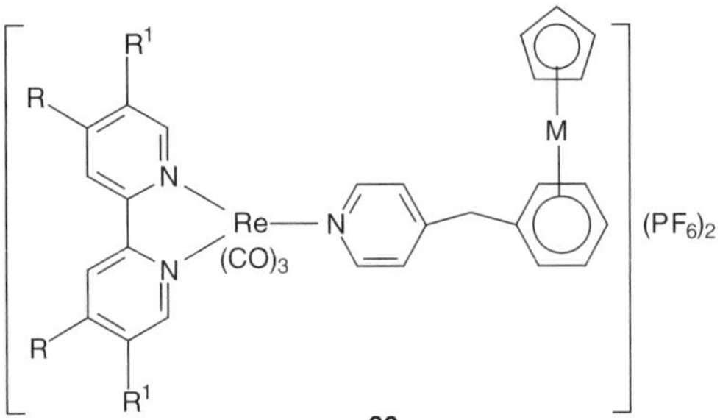

80

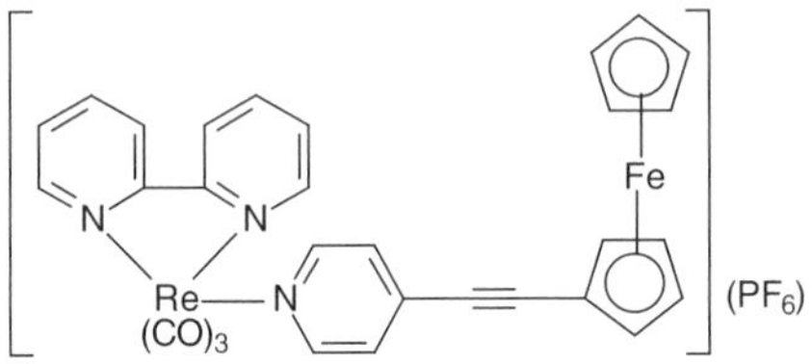

81

[Re(CO)$_3$(LL)(*N*-methyl-4,4′-bipyridinium)](PF$_6$)$_2$ (LL = bpy, 4,4′-Me$_2$bpy, 4,4′-(COOEt)$_2$bpy) are known (78CPL389, 82JOM(240)413, 83JA61, 83JA5952, 86JOM(300)139, 87IC1116, 87JA2519, 88JPC3708, 89ACR163, 89IC2271, 94IC793, 95JCS(CC)259, 96JPC15145, 98JPC(A)3042, 99IC2924, 00IC485, 02JCS(D)701, 02JPC(A)7795, 03IC7995, 04JPC(A)556). [Re(CO)$_3$(bpy)(*N*-methyl-4,4′-bipyridinium)](PF$_6$)$_2$ (89IC1596) was thoroughly studied using spectroelectrochemical methods (98IC5664). On reduction, electron transfer from the 2,2′-bipyridine to the 4,4′-bipyridinium ligand occurs. Refluxing [Re(CO)$_3$(LL)(OTf)] (LL = bpy, 4,4′-Me$_2$bpy) with *N*-methyl-4,4′-bipyridinium hexafluorophosphate or *N*-phenyl-4,4′-bipyridinium hexafluorophosphate (RL) leads to [Re(RL)(CO)$_3$(LL)](PF$_6$)$_2$ (R = Me, Ph; LL = bpy, 4,4′-Me$_2$bpy) (96JA9782, 04ICA167). [Re(CO)$_3$(bpy)(OTf)] and 10-(4-picolyl)phenothiazine (L) in ethanol and then ammonium hexafluorophosphate gives **82** (87IC1116, 94JPC8959) prepared from [Re(CO)$_3$(bpy-PTZ)Cl] (93JPC13126). The 3,3′-trimethylene-2,2′-biquinoline complex **83** reacts with silver triflate in methylene chloride and subsequently with pyridines and forms a series of cationic complexes **83** (R = H, CN, CH$_2$Ph, OH) (98POL2289). The following chromophore-quencher complexes should be mentioned: [Re(CO)$_3$(4,4′-R$_2$bpy)(L)]$^{n+}$ (R = Me, *t*-Bu; L = benz[*g*]isoquinoline-5,10-dione, 2-oxy-1,4-naphthoquinone, 2,6-dihydroxyantraquinone dianion, 1-methyl-6-oxyquinoline) (02JPC(A)7795). Interaction of [Re(CO)$_3$(bpy)(COMe)] with various π-acceptor ligands has been studied (00JOM(598)136).

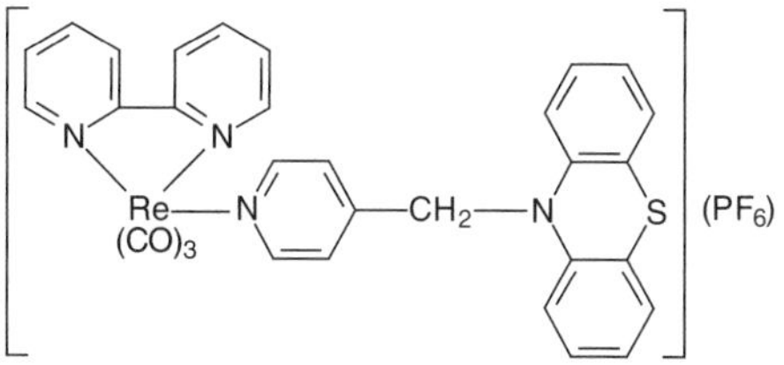

82 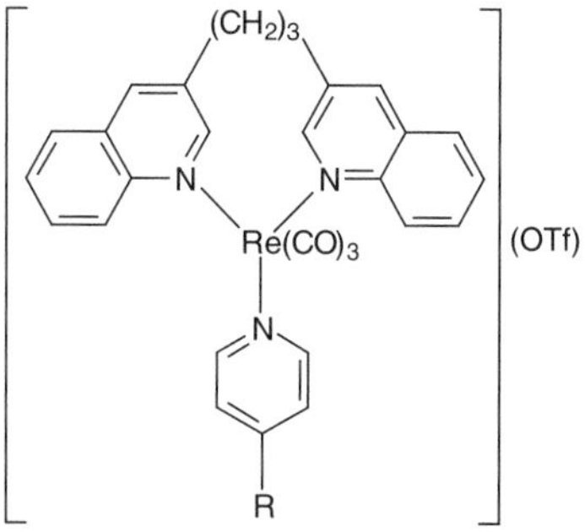   **83**

1,10-Phenanthroline with $[Re(CO)_5Cl]$ gives $[Re(CO)_3(phen)Cl]$ (01ICA(313)149). The product reacts with methyl triflate to yield $[Re(CO)_3(OTf)(phen)]$, which on interaction with *trans*-1,2-bis(4-pyridyl)ethylene (L) and ammonium hexafluorophosphate generates $[Re(CO)_3(phen)L](PF_6)$. Two complexes, $[Re(CO)_3(phen)(OTf)]$ and $[Re(CO)_3(phen)L](PF_6)$ afford in the presence of extra ammonium hexafluorophosphate in DMF the dinuclear species **84**. $[Re(CO)_3(bpy)Br]$ reacts with pyrazine-2-carboxylate (L) and silver triflate in acetonitrile to yield $[Re(CO)_3(LL)(L)](OTf)$ (01OM2842). $[Re(CO)_3(2,2'-biquinoline)Br]$ when treated with silver triflate in methylene chloride gives $[Re(CO)_3(2,2'-biquinoline)(OTf)]$. Further reaction with pyrazine gives $[Re(CO)_3(2,2'-biquinoline)(pyrazine)](OTf)$ and with 4,4'-bipyridine $[Re(CO)_3(2,2'-biquinoline)(4,4'-bipyridine)](OTf)$. $[Re(CO)_3(LL)Cl]$ (LL = bpy, 4,4'-Me$_2$bpy, phen, 3,4,7,8-Me$_4$phen) were first converted to $[Re(CO)_3(LL)(OTf)]$ and then reacted with 1,2-bis(4-pyridyl)ethylene or 1,2-bis(4-pyridyl)ethane (L) in methanol to give $[Re(LL)(CO)_3L](PF_6)$ (03JPC(A)4092, 03JPP(A)27, 04IC2043). Other related monomeric complexes were prepared from $[Re(CO)_3(bpy)(OTf)]$ and an appropriate ligand in the presence of ammonium hexafluorophosphate in THF (00JCS(CC)1865, 02JPC(A)12202, 03JCS(CC)2858, 04JCS(D)1376). They are presented as **85–88**. Complexes with the following substituents on the pyridine ring of the entering ligand $CH_2C_6H_4NHCH(Ph)CH(Ph)(C_5H_{10}N)$, $CH_2C_6H_4NMe_2$, and CH_2Ph were prepared from $[Re(CO)_3(bpy)(OTf)]$ (95JPC11801).

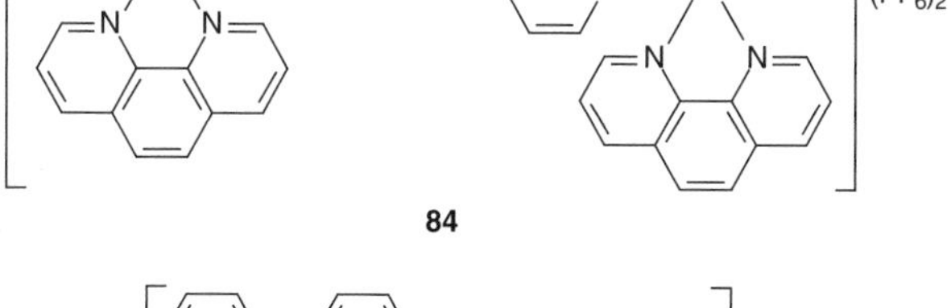

84

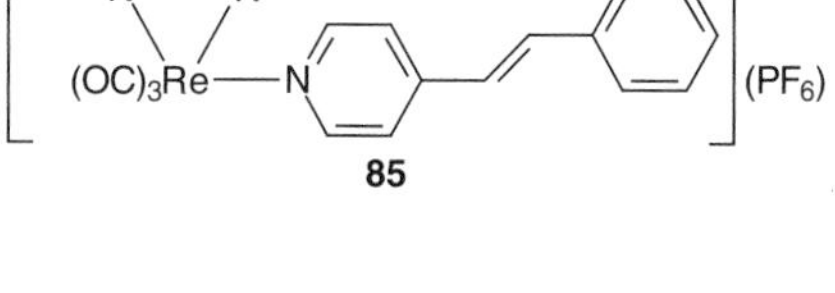

85

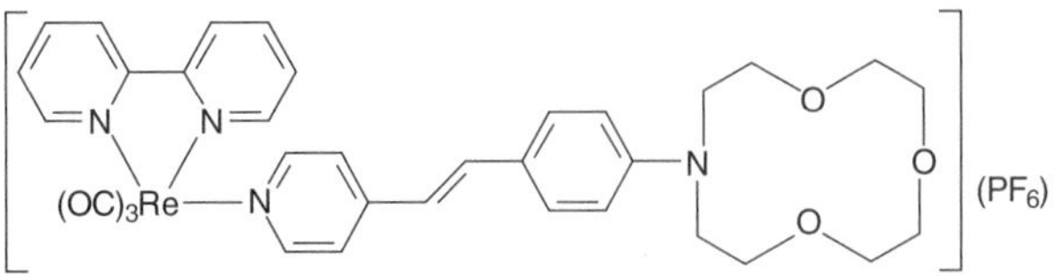

86

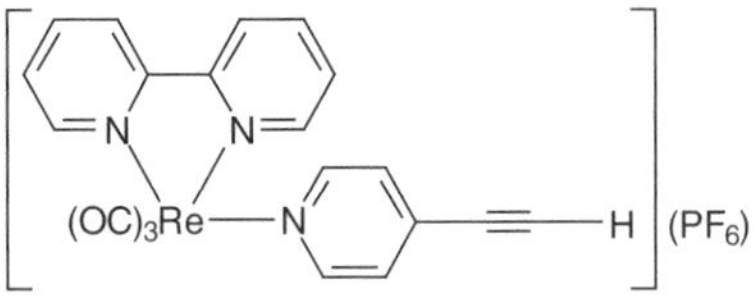

87

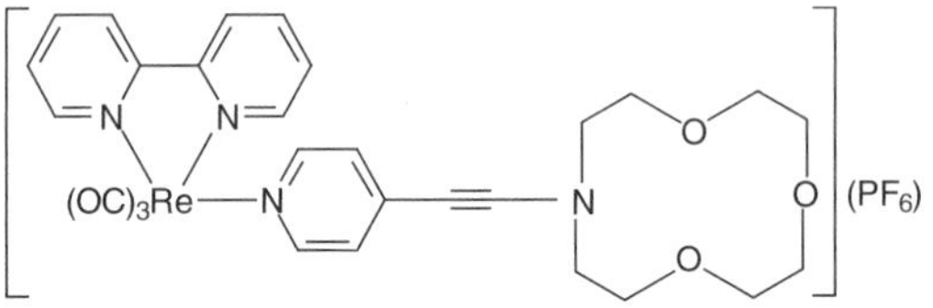

88

[Re(CO)$_3$(4,4′-Me$_2$bpy)(OTf)] with 4,4′-bipyridine in THF and further with ammonium hexafluorophosphate gives [Re(CO)$_3$(4,4′-Me$_2$bpy)(4,4′-bipyridine)](PF$_6$) (89JPC3885, 90IC2285, 91JPC1105). [Re(CO)$_3$(bpy)(OTf)] in a similar procedure gives [Re(CO)$_3$(bpy)(4,4′-bipyridine)](PF$_6$). Another representative in this series is [Re(CO)$_3$(4,4′-(COOEt)$_2$bpy)(4,4′-bipyridine)](PF$_6$). A number of dinuclear rhenium(I)–rhenium(I) complexes containing 4,4′-bipyridine as bridging ligands were prepared. [(4,4′-R$_2$bpy)(OC)$_3$Re(μ-4,4′-bipyridine)Re(CO)$_3$)4,4′-R$_2$bpy)](PF$_6$)$_2$ (R = H, Me, NH$_2$, COOEt) follow from 4,4′-bipyridine, [Re(CO)$_3$(4,4′-R$_2$bpy)(OTf)], and ammonium hexafluorophosphate. The list of representatives containing different 2,2′-bipyridine ligands on both sides includes [(4,4′-R$_2$bpy)(OC)$_3$Re(μ-4,4′-bipyridine)Re(CO)$_3$(4,4′-R$_2$bpy)](PF$_6$)$_2$ (R = NH$_2$, Me; R′ = H; R = NH$_2$, H, R′ = COOEt). The reactions leading to these complexes are conducted between [Re(CO)$_3$(4,4′-R$_2$bpy)(OTf)], [Re(CO)$_3$(4,4′-bpy)Re(4,4′-bipyridine)](PF$_6$) and ammonium hexafluorophosphate in THF. The reaction between [Re(CO)$_3$(bpy)(3,3′-Me$_2$-4,4′-bipyridine)](PF$_6$) and [Re(CO)$_3$(4,4′-(COOEt)$_2$bpy)(OTf)] in THF in the presence of ammonium hexafluorophosphate gives [(bpy)(OC)$_3$Re(3,3′-Me$_2$-4,4′-bipyridine)Re(CO)$_3$(4,4′-(COOEt)$_2$bpy)](PF$_6$)$_2$.

A vinylpyridine-based pendant chromophore [Re(CO)$_3$(phen)]$^+$ occurs in the polymers **89** and **90** (04IC1551). These and other rhenium polypyridine-based polymers have found application in supramolecular chemistry (87MI1, 91MI2, 99MI1, 99MI2, 00JPC(A)9281, 03SMC143).

89   **90**

The photochemical properties of [Re(CO)$_3$(LL)(L)] and [Re(CO)$_3$(LL)(L)]$^+$ (L = Cl, I, N- or P-donor) were thoroughly studied (91IC4754, 95CIC319, 97JCS(CC)1593, 98CCR(169)201, 98JMC89, 98JPC(A)5577, 98OM2440, 99OM5252, 00OM1820, 02OM39). [Re(LL)(CO)$_3$(AN)](OTf) (LL = phen, 2,9-Me$_2$phen, 3,4,7,8-Me$_4$phen, 4,7-Ph$_2$phen, 2,9-Me$_2$-4,7-Ph$_2$phen, 1,1′-biisoquinoline) react with N-(3-pyridyl)maleimide (L) to yield [Re(LL)(CO)$_3$(L)](OTf), for example **91** (02IC40). Such and similar rhenium(I) polypyridine complexes are used as nucleic acid or uracil-dimer photocleavage agents (97JCS(D)2067, 00JCS(CC)188), as DNA intercalators (96NJC791), to study the electron tunneling in metalloproteins (99PAC1753, 01JA3181, 01JA11623, 03JA14220), as anisotropy probes for protein hydrodynamics (97AB179, 98AC632, 00BC533), for recognition of anionic species (98JCS(CC)825, 99JCS(CC)1755, 00CCR(205)311, 01JCS(D)2188), as Re(II)/Re(I) redox couples (02HCA1261). The photochemical properties of [Re(CO)$_3$(phen)Cl] in the presence of 1,3-dimethyluracil cyclobutane dimer were studied (00ICC188). In a search for labeling reagents for biomolecules, the reaction of [Re(CO)$_3$(LL)(AN)](OTf) with N-{(4-pyridyl)methyl}biotinamide and potassium hexafluorophosphate in THF gives products **92** (R^1 = R^2 = R^3 = H; R^1 = H, R^2 = R^3 = Me; R^1 = Me, R^2 = H, R^3 = Ph) (02JA9344, 03JCS(CC)2704). The [Re(CO)$_3$(bpy)] moiety was linked by a styryl pyridine to an amine or an aza crown ether (00JCS(CC)1865). The crown-ether-linked assemblies **93** combine the [Re(CO)$_3$(bpy)] donor and nitrobenzene-containing acceptor counterparts (from dicesium 5-nitroisophthalate) (92IC3192, 93JA2048). If pyridine replaces nitrobenzene, the assembly is presented as **94** (96IC2032) and manifests photoinduced electron transfer reactivity (from dicesium 3,5-pyridine dicarboxylate). 1,10-Phenanthroline serves as a building block for macrocyclic ligands, thus the Re(CO)$_3$Cl complexes can be prepared (97IC5329, 98IC1121, 98JA5480). An acyclic ligand containing a 2,9-bis(p-phenyl)-1,10-phenanthroline unit with [Re(CO)$_5$Cl] forms **95** (97IC5329). A macrocycle carrying two (CH$_2$)$_5$Me substituents in the 4, 7 positions of the phenanthroline framework with [Re(CO)$_5$Cl] gives **96**.

91

O
HN
NH
S
H
N
(CO)3
Re
R1
R2
R3
N
N
(PF6)

92

N
N
Re
(CO)3
N
O
NO2
(PF6)

93

N
N
Re
(CO)3
N
O
(PF6)

94

OMe
N
Re(CO)5Cl
N
OMe

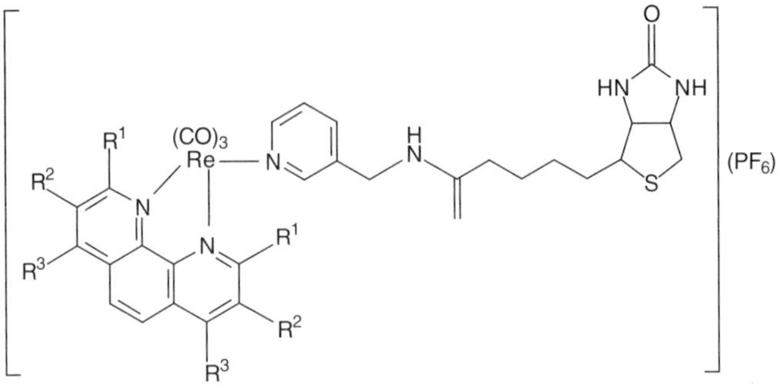

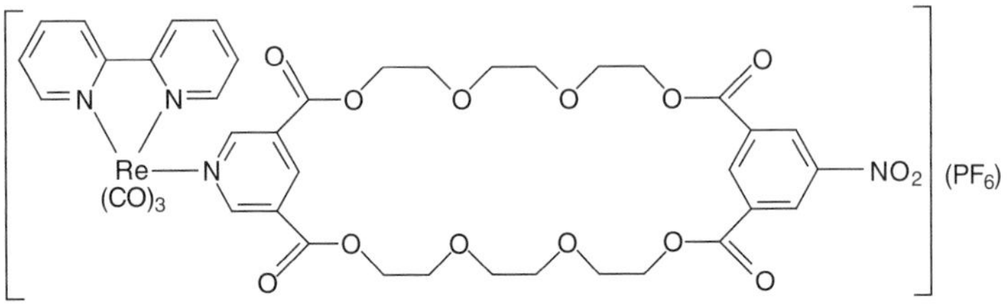

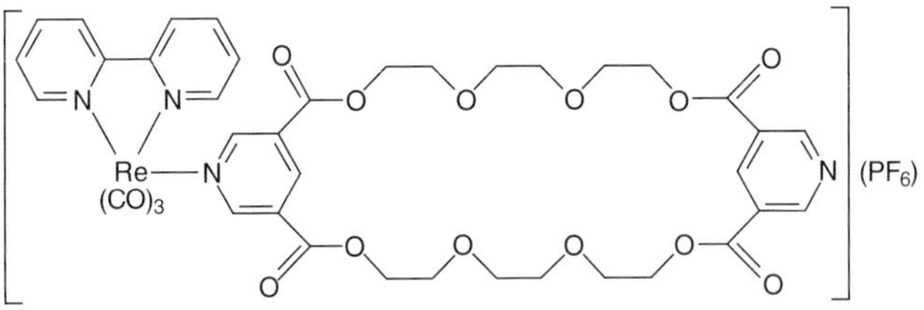

95

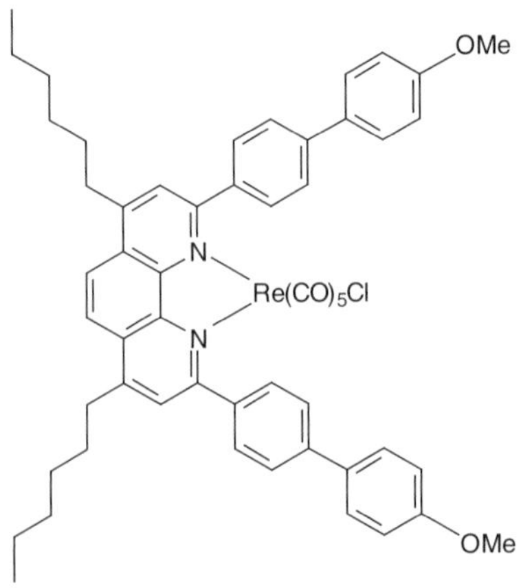

96

Reaction of [Re(CO)₃(3,4,7,8-Me₄phen)(AN)](OTf) with *N*-(1-antraquinolyl)-*N'*-(4-pyridinylmethyl)thiourea and *N*-(1-phenyl)-*N'*-(4-pyridinylmethyl)thiourea in refluxing THF gives **97** and **98** (04OM1098). [Re(CO)₅Cl] with 9-(4-octadecyloxy-4'-benzen-amino)-4,5-diazafluorene, 9-(4-octadecyloxy-4'-stilbenamino)-4,5-diazafluorene, and *trans*-4-methyl-4'-(2-(4-octadecyloxyphenyl)vinyl-2,2'-bipyridine (LL) give [Re(CO)₃(LL)Cl] depicted as **99** (X = N, CH) and **100** (01OM4911). The reaction of [Re(CO)₃(bpy)(AN)](OTf) with *trans*-4-(4'-octadecyloxyphenylazo)pyridine, *trans*-(4-(4'-octadecyloxystyryl)pyridine, 4-(4'-(*N*-octadecylamide)styryl)pyridine (L) leads to the formation of [Re(CO)₃(bpy)L]PF₆ shown as **101** (X = N, CH) and **102**. Stilbene-containing ligands coordinated to the rhenium(I) center were widely studied (90JA1117, 98JCS(D)1461). Complexes [Re(CO)₃(bpy)L](PF₆) (L = *trans*-4-do-decyloxy-4'stilbazole, *trans*-4-octadecyloxy-4'-stilbazole, 4-(4'-dodecyloxyphenylethy-nyl)pyridine should be mentioned (03EJI4035).

97   **98**

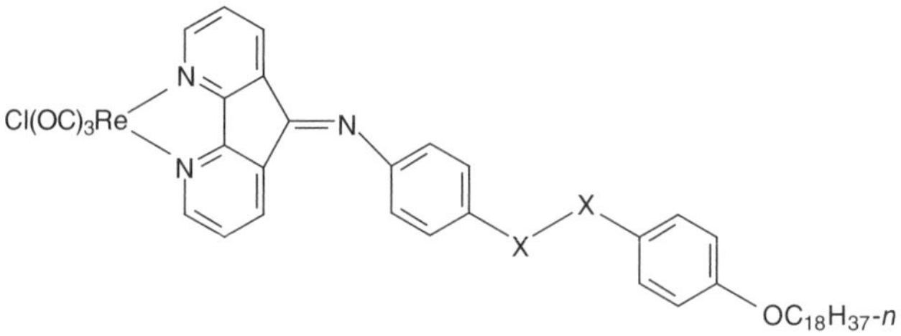

99

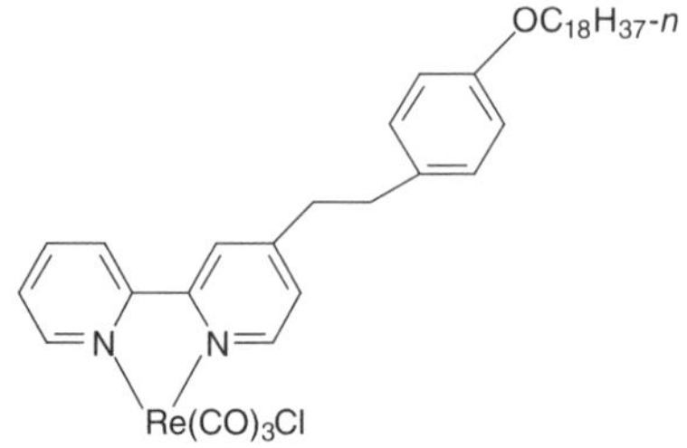

100

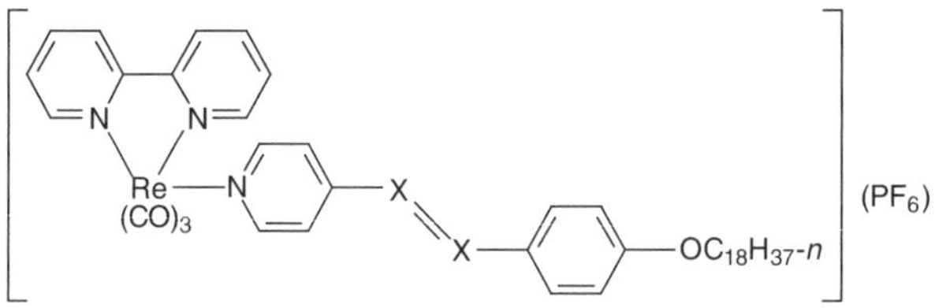

101

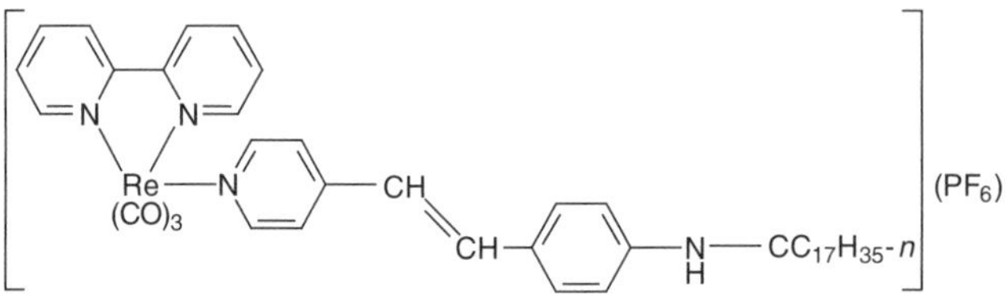

102

Dehalogenation of [Re(CO)$_3$(LL)Cl] (LL = bpy, 4,4′-bis{p-(diethylamino)-α-sty-ryl}-2,2′-bpy) in the presence of silver triflate in acetone leads to [Re(LL)(CO)$_3$](OTf) (98ACR26, 98IC5061). The products with 1,10-bis(4-pyridyl)-3,8-dimethyl-1,3,5,7,9-decaheptaene (L) in ethanol give [Re(CO)$_3$(LL)L](OTf). [Re(CO)$_3$(4,4′-bis{p-(di-ethylamino)-α-styryl}-2,2′-bipyridine)Cl] reacts with silver triflate in ethanol and then in dichloromethane with [Re(CO)$_3$(LL)L(CO)$_3$](OTf) described above to yield the dinuclear species [(LL)(OC)$_3$Re(μ-L)Re(CO)$_3$(LL)](OTf)$_2$, e.g. **103**.

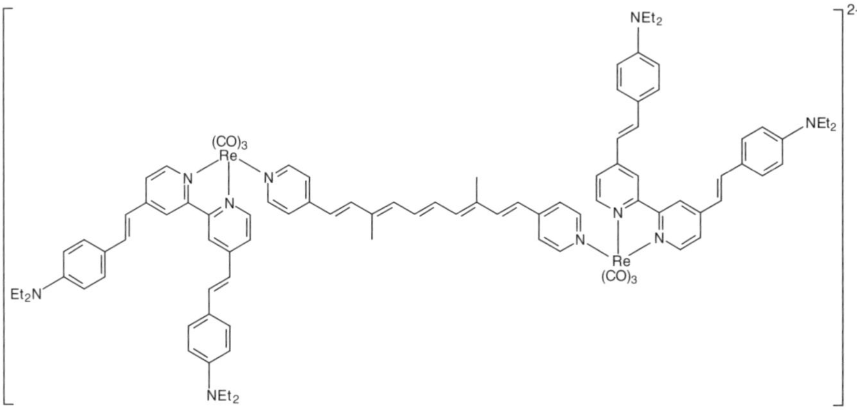

103

[Re(CO)$_3$(bpy)(OTf)] reacts with ligand **104** in ethanol to afford the mononuclear complex **105** (97JPC(B)3174). The product reacts further with [Re(CO)$_3$(bpy)(OTf)] to yield the dinuclear complex **106**. When anchored on the surface of silver or gold, **106** is rearranged to the surface mononuclear complex **107**. Another surface complex on silica formed from the rhenium(I) species with similar ligands may be formulated as –[Si](CH$_2$)$_4$(Mebpy)$_2$Re(CO)$_3$Cl (92JCS(CC)1615).

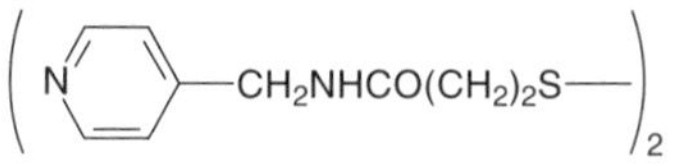

104

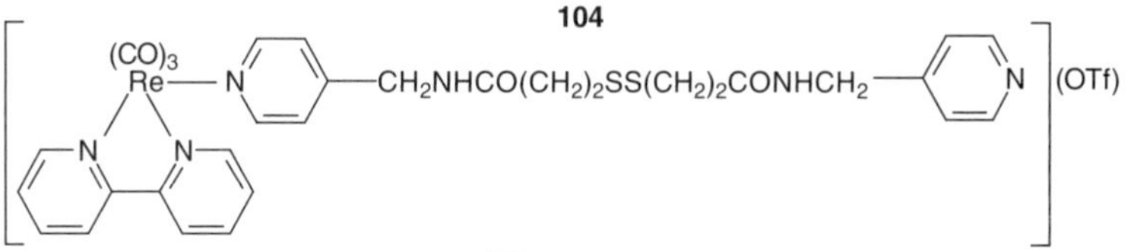

105

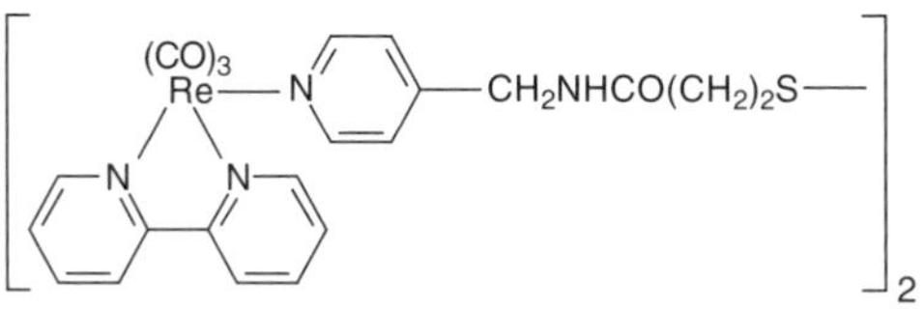

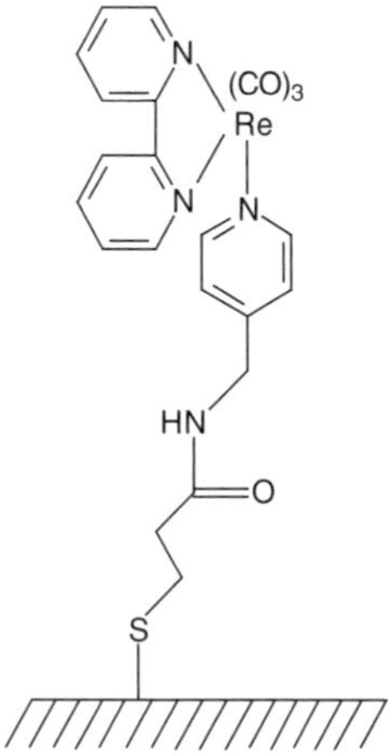

106

107

D. POLYNUCLEAR COMPLEXES

[Re(CO)$_3$(LL)Br] (LL = bpy, 4,4'-Me$_2$bpy, phen) and [Re(CO)$_5$]$^-$ form [(OC)$_5$Re-Re(CO)$_3$(LL)] (85CRV187, 90AOC523, 91IC42). [Re(L)(CO)$_3$(LL)] (L = Mn(CO)$_5$, Re(CO)$_5$, Co(CO)$_4$, (η^5-Cp)Fe(CO)$_2$, SnPh$_3$) (85IC4411, 88ICA(149)77, 88OM1100, 89IC75), [L$_n$M'M(CO)$_3$(LL)] (L$_n$M' = (OC)$_5$Mn, (OC)$_5$Re, (OC)$_4$Co, (η^5-Cp)(OC)$_2$Fe, Ph$_3$Sn; M = Mn, Re) (91OM3668), L$_n$M' = (OC)$_5$Mn, LL = 4,4'-Me$_2$bpy (79JOM(175)73), L$_n$M' = (OC)$_5$Mn, (OC)$_5$Re (94JCS(CC)63), (OC)$_5$MnMn(CO)$_3$(LL) (85JMS11, 85JOM(290)63, 85JOM(294)59) are illustrative examples of homo- and heterodinuclear complexes. In the binuclear rhenium(I) complexes, electronic coupling between the rhenium(I) sites is significantly less when the bridging ligands is 3,3'-dimethyl-4,4'-bipyridine compared with the unsubstituted 4,4'-bipyridine (81IC2348).

[Re(CO)$_3$(4,4'-Me$_2$bpy)Cl] with silver triflate and K$_3$[Fe(CN)$_6$] affords K$_2$[(4,4'-Me$_2$bpy)(OC)$_3$Re(μ-NC)Fe(CN)$_5$] (99IC606). The molar ratio of the reagents in this reaction is 1:1:1. When it is 2:2:1, the product is K[(4,4'-Me$_2$bpy)(OC)$_3$Re(μ-NC)]$_2$[Fe(CN)$_4$]. When the ratio is 4:4:1, **108**, the neutral heteronuclear complex is the product. In [(OC)$_3$(NC)Re{Ru^{II}(bpy)$_2$CN}$_n$Ru^{II}(bpy)$_2$(CN)]$^{(n+1)+}$ (n = 0–3) every Ru^{II} is bonded to one C and N atom of the bridging cyano ligand (94IC1652). In these complexes, there is a long-range energy transfer up the terminal ruthenium center (96CIC77, 98CRV1335). [(phen)Re^I(μ-CN)Ru^{II}(bpy)$_2$(CN)](PF$_6$) follows from sodium azide in methanol, [Ru(bpy)$_2$(NO(CN))](PF$_6$)$_2$, and [Re(CO)$_3$(phen)(CN)] (91IC1330), while [(phen)Re^I(μ-NC)Ru^{II}(bpy)$_2$(CN)(PF$_6$) is the result of the interaction of [Re(CO)$_3$(phen)Cl] with [Ru(bpy)$_2$(CN)$_2$] in methanol and then ammonium hexafluorophosphate (99JCS(D)3729). For the multinuclear species, [Re(CO)$_3$(-phen)(CN){Ru(bpy)$_2$(CN)•nRu(bpy)$_2$(CN)}] (n = 0–2), its mass-spectral pattern and photochemical properties were studied in detail (91CCR(111)297, 93IC1222). [(phen)(CO)$_3$Re(NC)Ru(bpy)$_2$(CN)]$^+$ was isolated (88IC651, 92JA8727, 93CP585, 93JA10996). 2,2':3',2'',6'',2'''-Quaterpyridine also forms dinuclear rhenium(I)–ruthenium(II) complexes **109** and **110** (93JCS(D)1321, 94JCS(D)3095, 98IC2598). Complex **111** was synthesized using the same preparative ideas (95IC2438). [Ru(LL)$_2$(bpy)](PF$_6$)$_2$ (LL = 2,2':3',2'':6'',2'''-quaterpyridine) and [Re(CO)$_5$Cl] in DMF yield **112** (97IC2601). [Ru(LL)$_3$](PF$_6$)$_2$ (LL = 2,2':3',2'':6'',2'''-quaterpyridine) and excess [Re(CO)$_5$Cl] in DMF give **113**. Another combination in the tetranuclear complexes is **114** (01JCS(CC)2540).

108

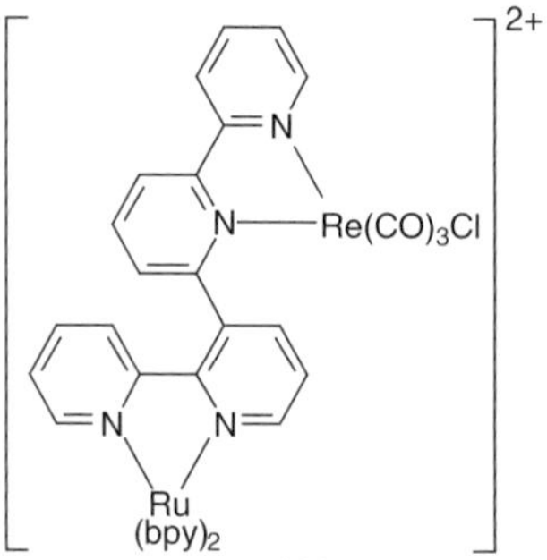

109

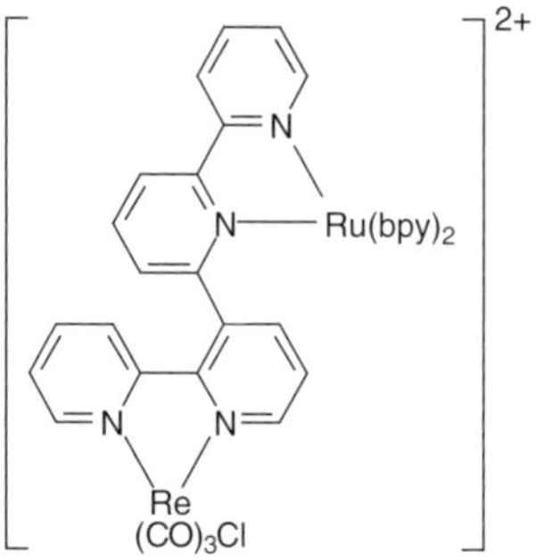

110

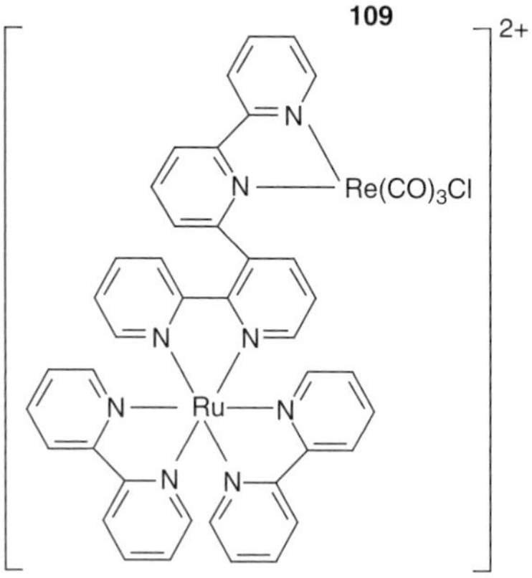

111

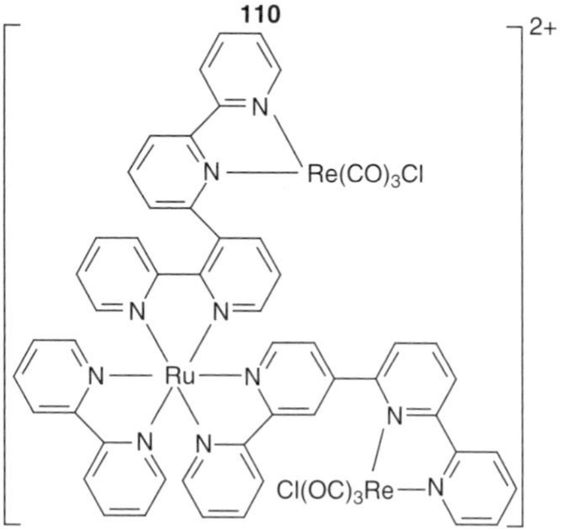

112

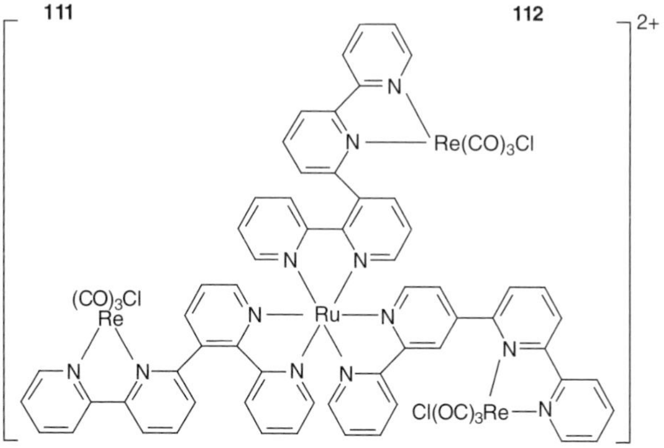

113

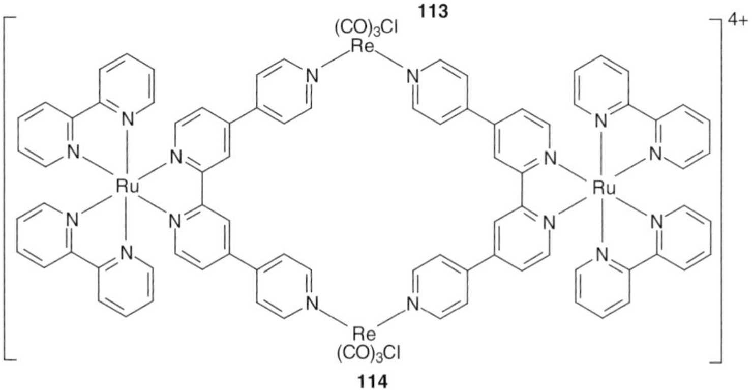

114

Reviews of Re(I)–Ru(II) and Re(I)–Os(II) systems exist (94CRV993, 96CR V759). The following di- and trinuclear chemistry should be mentioned: [Re(CO)₃(4,4′-t-Bu₂bpy)Ru(dppe)₂(C≡Cpy-4)H](OTf) (04NJC43), [Rh₂(N,N′-di-p-to-lylformamidinate)₄(C≡Cpy-4){Re(CO)₃(bpy)}](OTf), [Rh₂(N,N′-di-p-tolylformami-dinate)₄(C≡Cpy-4){Re(CO)₃(4,4′-t-Bu₂bpy)}](OTf) (03EJI449). [Re(CO)₃(bpy)(4,4′-bipyridine)](OTf) with M₂(OOCR)₄ (M = Mo, R = CF₃; M = Rh, R = Me) yields the tetranuclear species with general formula [{Re(CO)₃(bpy)(4,4-bipyridine)}₂ (LM₂(OOCR)₄](OTf) (01POL791). Reaction of [Re(CO)₃(4,4′-bipyridine)₂Br] and M₂(OOCR)₄ gives oligomers having stoichiometry M₂(OOCR)₄–Re(CO)₃(4,4′-bipyri-dine)₂Br and zigzag chain structure. Complexes where [Re(CO)₃(bpy)Cl] and [Ru(bpy)]²⁺ moieties are linked by a bis(bipyridyl ethynyl)benzene bridge in the dinuclear complex are of interest (01JCS(CC)277). Among the other products it is necessary to mention [(phen)Re(CO)₃(μ-L)Os(trpy)(bpy)]³⁺ (L = 4,4′-bpy, *trans*-1, 2-bis(4-pyridyl)ethylene, 1,2-bis(4-pyridyl)ethane (01IC6885), [Re(CO)₃(bpy)(C≡C-C₆H₄-C≡C)Fe(η⁵-Cp*)(dppe)], its oxidized species [Re(CO)₃(bpy)(C = -C₆H₄-C =)Fe(η⁵-Cp*)(dppe)](PF₆), and a related vinylidene complex [Re(CO)₃ (bpy)(C≡C-C₆H₄(H)C = C)Fe(η⁵-Cp*)(dppe)](PF₆) (03IC7086). Species **115** is adsorbed on the surface of titania, which brings about new photochemical proper-ties of the resultant solid material (97IC2). Similar structures based on coordination compounds of 3,5-bis(pyridine-2-yl)-1,2,4-triazole were studied (01IC5343).

115

Among the dinuclear complexes, one is based on [Re(CO)₃(bpy)(AN)]⁺ containing the N≡C-CH = C(Me)NH bridge (98JCS(CC)135). [Re(CO)₃(phen)(AN)](OTf) with Et₃NHSC₆H₄Me-*p* gives [(phen)(OC)₃Re(μ-SC₆H₄Me-*p*)Re(CO)₃(phen)](OTf) (97OM1729). In a similar manner, [(phen)(OC)₃Re(μ-SC₆H₄OMe-*p*)Re(CO)₃(phe-n)](OTf), [(phen)(OC)₃Re(μ-SC₆H₄Bu-*t-p*)Re(CO)₃(phen)](OTf), [(phen)(OC)₃Re(μ-SC₆H₄Cl-*p*)Re(CO)₃(phen)](OTf), [(bpy)(OC)₃Re(μ-SC₆H₄Me-*p*)Re(CO)₃(bpy)](OTf), [(4,4′-Me₂bpy)(OC)₃Re(μ-SC₆H₄Me-*p*)Re(CO)₃(4,4′-Me₂bpy)](OTf) were prepared. The structure of the representative complex is illustrated as **116**. Complexes possess long-lived luminescence with a ³MLCT state.

116

Photolysis of **117** (L = CO) in acetonitrile gives **117** (L = AN) (01IC5056). Photolysis of **117** (L = CO) in the presence of excess tetra-*n*-butylammonium chloride gives **118** (L = Cl, X = Cl) and in excess tetra-*n*-butylammonium triflate **118** (L = OTf, X = OTf). Reaction of **117** (L = AN) with triphenylphosphine gives **117** (L = PPh$_3$), and reactions of the same starting materials with 1,2-bis(diphenylphosphino)ethylene or 1,4-bis(diphenylphosphino)benzene gives **117** (L = η^1-dpene, η^1-dppb) with monodentate coordination of the diphosphine ligand. Complex **119** should also be mentioned (01JCS(CC)1514).

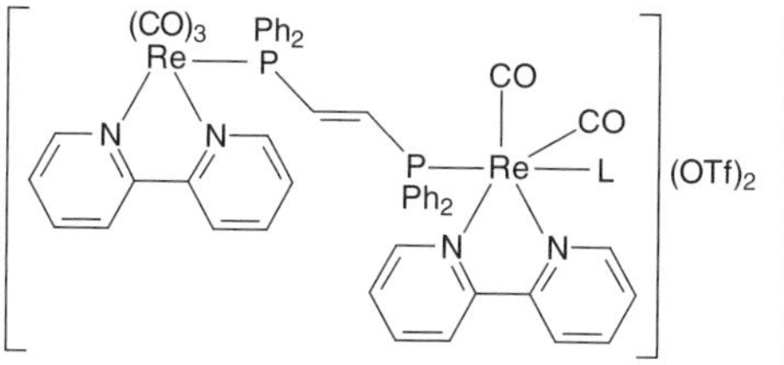

117 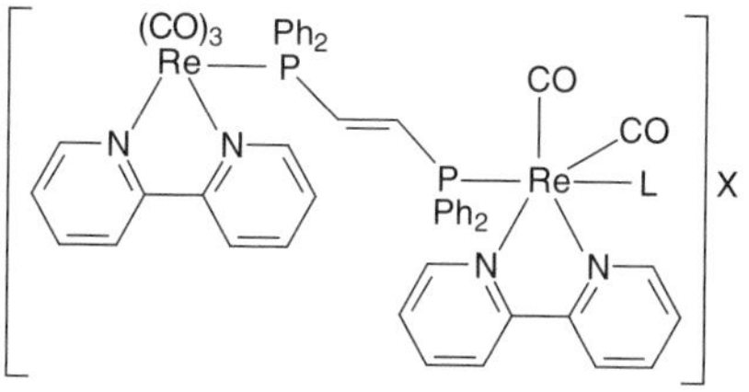   **118**

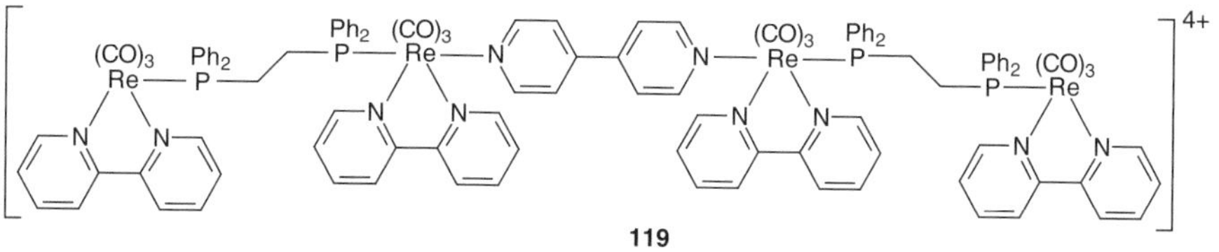

119

[Re(CO)$_3$(bpy)(OTf)] with (PPh$_4$)[Au(C≡Cpy)$_2$] in the molar ratio 2:1 in THF gives the transmetallation product **120** (04OM5096). In a similar way, [Re(CO)$_3$(bpy)(OTf)] with (PPh$_4$)[Au(C≡C-C$_6$H$_4$-C≡Cpy)$_2$] gives the dinuclear species [(OC)$_3$(bpy)Re(μ-C≡CD-C$_6$H$_4$-C≡Cpy)Re(CO)$_3$(bpy)](OTf); [Re(CO)$_3$(4,4'-*t*-Bu$_2$bpy)(OTf)] with (PPh$_4$)[Au(C≡Cpy)$_2$] gives [(4,4'-*t*-Bu$_2$bpy)(OC)$_3$Re(μ-C≡Cpy)Re(CO)$_3$(4,4'-*t*-Bu$_2$bpy)](OTf). When [Re(CO)$_3$(bpy)(OTf)] and (PPh$_4$)[Au(C≡Cpy)$_2$] are taken in a 1:1 molar ratio, the product is [Re(CO)$_3$(bpy)(C≡|Cpy)], a mononuclear species.

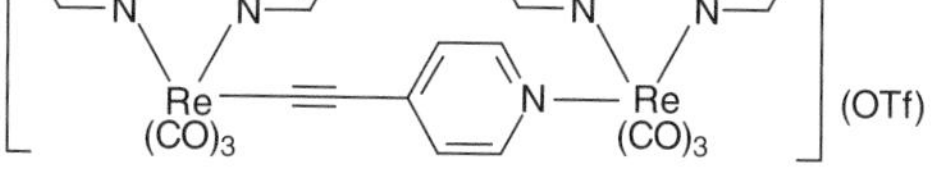

120

5-Bis{2-(3,5-dimethyl-1-pyrazolyl)ethyl}amino-1,10-phenanthroline and 5-dimethylamino-1,10-phenanthroline (LL) with [Re(CO)$_5$Br] in methylene chloride give

[Re(CO)$_3$Br(LL)] (01JCS(D)1813). The first of these two complexes when reacted with [Cu(AN)$_4$](PF$_6$) forms [Re(CO)$_3$Br(LL-Cu^I)](PF$_6$), **121**. [Cu$_3$ (μ-PP){μ$_3$-η-C≡CC$_6$H$_2$R$_2$-2,5-C≡C-*p*-Re(CO$_3$)(LL)}$_2$]$^+$ can be also mentioned in this regard (PP = dppm, *n*-PrPNPPr-*n*; LL = bpy, 4,4′-*t*-Bu$_2$bpy; R = H, Me) (98JCS(CC)777).

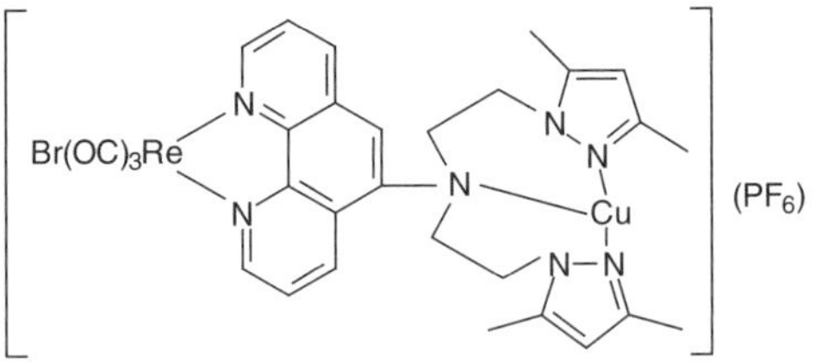

121

E. MOLECULAR ASSEMBLIES

Some ethynyl derivatives of the polypyridine complexes lead to the formation of supramolecules (03CCR(245)39, 03JCS(CC)2446, 04JOM1393). Thus complexes [Re(CO)$_3$(4,4′-R$_2$bpy)(C≡C(C$_6$H$_4$)$_n$C≡CH (R = H, Me, *t*-Bu; *n* = 0, 1) react with [(AN)M(μ-PP)$_2$M(AN)]$^{2+}$ (M = Cu, Ag; PP = PhPCH$_2$PPh$_2$, Ph$_2$PN(*n*-Pr)PPh$_2$) to yield the trinuclear **122** (R = H, Me, *t*-Bu; *n* = 0, 1; PP = Ph$_2$PCH$_2$PPh$_2$, Ph$_2$PN(*n*-Pr)PPh$_2$). When *n* = 0, another product, a polynuclear complex **123**, is formed. A mixed-metal dinuclear rhenium(I)–iron(II) complex [Re(CO)$_3$(bpy) (C≡CC$_6$H$_4$C≡C)Fe(η^5-Cp*)(dppe)] is the product of the reaction of [Re(CO)$_3$ (bpy)(C≡CC$_6$H$_4$C≡C-H)] and [(η^5-Cp*)Fe(dppe)Cl] in the presence of potassium hexafluorophosphate and potassium *t*-butylate in methanol. Oxidation of the product using [(η^5-Cp)Fe](PF$_6$) gives **124**.

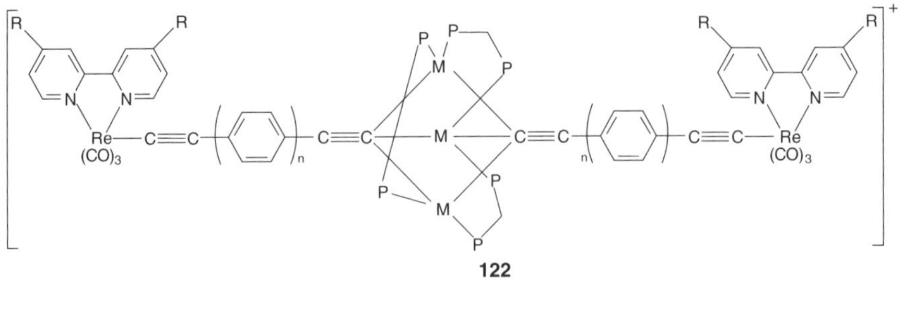

122

123

124

1,2-Bis(2-pyridyl)ethylene reacts with $[Re_2(CO)_8(AN)_2]$ in THF or dichloromethane to yield **125** and **126** (04JOM2486). Under refluxing conditions in THF, other products are formed, **127–129**.

Supramolecular systems have acquired popularity in the organorhenium(I) chemistry of polypyridines (02CCR(230)170). Self-assembly of $[Re(CO)_5Br]$ with $[M(4-C_5H_4Ntrpy)_2](PF_6)_2$ (M = Fe, Ru, Os) gives heteropolynuclear complexes **130** (M = Fe, Ru, Os) (00IC1344). Molecular squares based on $Re(CO)_3Cl$ metallocorners and bipyridine bridging ligands are of interest (02JA4554). Fluorene- and carbazole-bridged bipyridine ligands are used for the construction of supramolecules based on $[Re(CO)_5X]$ (X = Cl, Br) (01OM2262). The pyridine-containing phenylacetylenic hexagonal macrocycles (L) were used to prepare $[(4,4'-t-Bu_2bpy)Re(CO)_3]_2(\mu-L)(PF_6)_2$ (01OM2353). On refluxing with $[(t-Bu_2bpy)Re(CO)_3(AN)](PF_6)$ in THF, dinuclear complexes **131** and **132** are formed. These complexes in turn offer the opportunity of ready substitution of the chloride ligands and modification of the polypyridine ligand (88IC3652, 89IC1228, 89JOC1731).

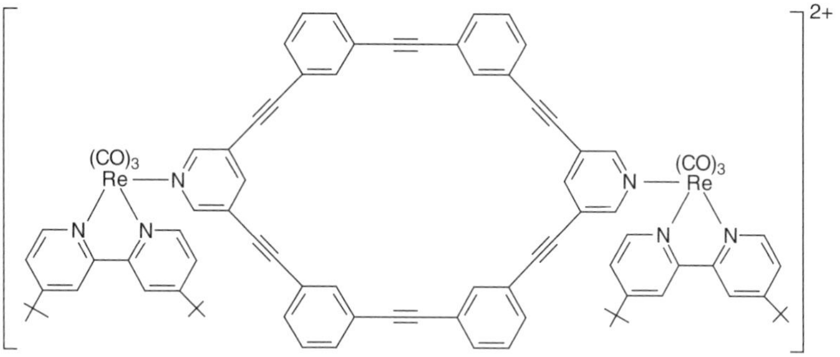

131

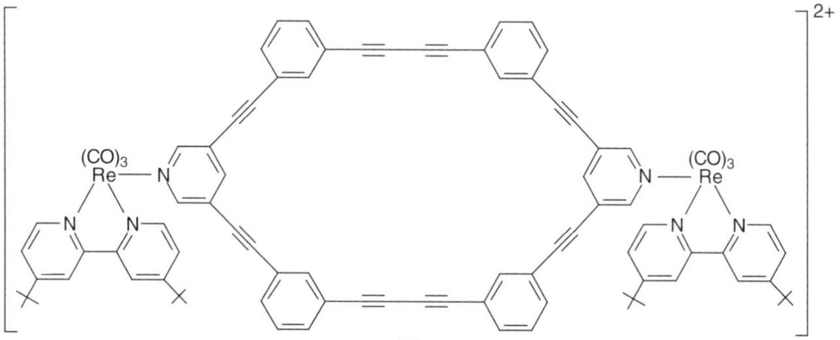

132

Species $[Re(CO)_3(bpy)(OTf)]$ enters the substitution reactions with diphosphines (PP) to yield $[Re(bpy)(CO)_3(PP)](OTf)$ but in excess diphosphine $[Re(bpy)(CO)_2(PP)_2]$ are formed (84JCS(CC)1244) with further oligomerization to various structures, e.g. $[(OC)_3(bpy)Re(\mu_2\text{-}PP)Re(bpy)(CO)_2(\mu_2\text{-}PP)Re(bpy)(CO)_3](OTf)_3$ (99IC4378) or $[Re(bpy)(PP)(CO)_2]_4(OTf)_4$ (99IC4380). Thus, $[Re(CO)_3(LL)(OTf)]$ (LL = bpy, phen) react with diphosphine ligands (1,2-bis(diphenylphosphino)ethylene, 1,2-bis(diphenylphosphino)acetylene, and 1,2-bis(diphenylphosphino)benzene (PP)) in ethanol on reflux to yield $[(LL)(OC)_3Re(\mu\text{-}PP)Re(CO)_3(LL)](OTf)_2$. In excess diphosphine in chlorobenzene on reflux $[(LL)(OC)_3Re(\mu\text{-}PP)Re(CO)_3(LL)(\eta^1\text{-}PP)](OTf)_2$ result. When the initial reaction is conducted in dichlorobenzene under reflux and an excess of a diphosphine, the product is $[Re(CO)_3(LL)(\eta^1\text{-}PP)_2](OTf)$. It reacts with $[Re(CO)_3(LL)(OTf)]$ to give the trinuclear complex $[(LL)(OC)_3Re(\mu\text{-}PP)Re(CO)_3(LL)(\mu\text{-}PP)Re(CO)_3(LL)](OTf)_3$. This complex and $[(LL)(OC)_3Re(\mu\text{-}PP)Re(CO)_3(LL)](OTf)_2$ give the tetranuclear **133**.

133

F. COMPLEXES WITH RHENIUM IN ITS HIGHER OXIDATION STATES

The reaction of *cis*-[Re(= O)Me$_2$(PMe$_3$)Cl] with 2,2'-bipyridine in the presence of pyridine-*N*-oxide in benzene gives the *cis*-rhenium(V) complex **134** (99JCS(D)4487). *Trans*-complex **135** follows from [ReOCl$_3$(bpy)] and AlMe$_3$ in dichloromethane. Isomerization of *cis*-**134** into *trans*-**135** occurs on heating **134** in benzene, the process being reversible. Complex **134** reacts with silver salts AgX (X = PF$_6$, BF$_4$, BPh$_4$) in acetonitrile to afford the cationic complexes **136** (X = PF$_6$, BF$_4$, BPh$_4$). *Trans*-**135** does not enter into this reaction. Products **136** (X = PF$_6$) react with various phosphines and phosphites to yield complexes **137** (R$_3$ = Me$_3$, (OMe)$_3$, (*n*-Bu)$_3$, Me$_2$Ph, MePh$_2$). Complex **134** prepared *in situ* reacts with silver hexafluorophosphate and then with an excess of MeOOCC≡CCOOMe to give **138** containing a metallacycle that is the product of insertion of the alkyne moiety into the rhenium–methyl bond (99JOM(599)112). With alkynes RC≡CR^1 (R = R^1 = H, Me, Et, Ph; R = H, R^1 = Me, Ph) in AN in the presence of silver hexafluorophosphate it yields different products, **139** (R = R^1 = H, Me, Et, Ph; R = H, R^1 = Me, Ph). Some of the alkyne complexes react with phosphines PR$_3^2$ (R^2 = Me, Ph) to generate the ylide complexes **140** (R = R' = H, R'' = Me, Ph; R = H, R' = Ph; R'' = Me).

134   135   136

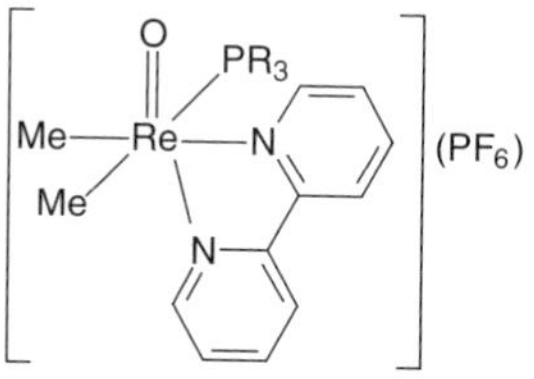

137 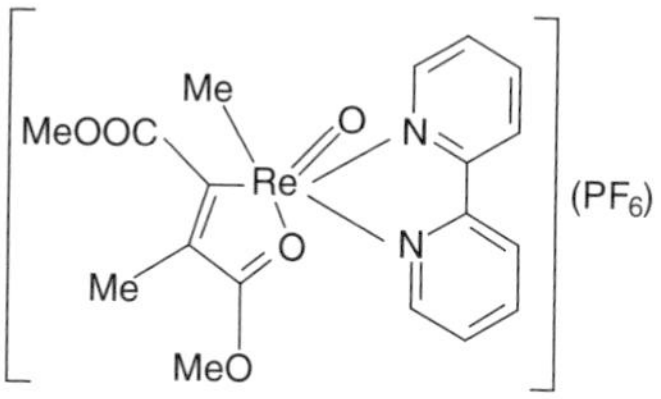   138

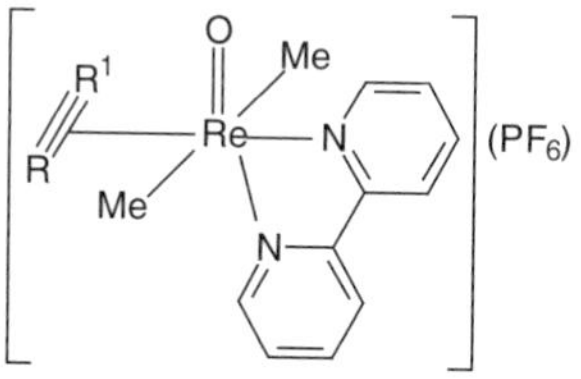

139 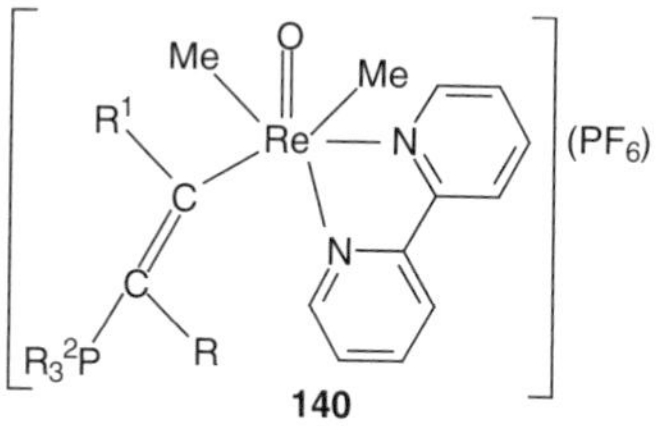   140

Methyltrioxorhenium(VII) forms distorted octahedral adducts with 4,4'-dimethyl-2,2'-bipyridine and 1,10-phenanthroline, which serve as epoxidation catalysts (01IC5834). Mesoporous silica, pristine, was first covalently grafted to (3-chloro-phenyl)trimethoxysilane, and then the pendant chloro groups were partially substituted by the anion 4-CH$_2$-4'-Me-2,2'-bipyridine (02EJI1100). Further reaction of the heterogenized ligand with the methoxyrhenium(VII) gives **141**. Reaction of 4,4'-dimethyl-2,2'-bipyridine or 4-methyl-4'-n-butyl-2,2'-bipyridine with MeReO$_3$ in THF gives **142** (R = Me, n-Bu).

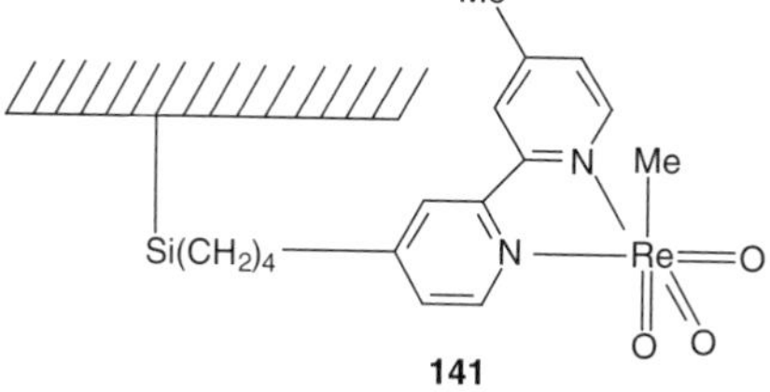

141 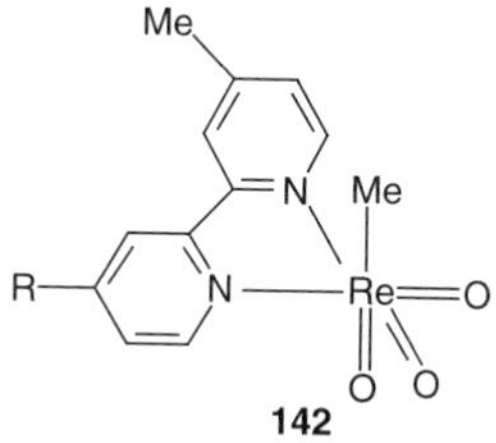   142

The diolate complexes of rhenium(VI), [Re(bpy)(= O)(OCH(R^1)CH(R^2)O)]$^+$ when treated with DCl produce carbene complexes **143** and **144** (R^1 = R^2 = COO-Me; R^1 = COOEt, R^2 = Ph; R^1 = Me, R^2 = H) (04OM3437).

143   144

G. Complexes of 4,4′-Bipyridine and Analogues

4,4′-Bipyridine with $[Re(CO)_5X]$ (X = Cl, Br, I) on heating in *iso*-octane gives $[Re(CO)_3(4,4′-pipyridine)X]$ (X = Cl, Br, I) (79JA2888, 88IC3325, 89IC85, 89IC2154, 89JA6602, 90IC3866). The luminescence of these complexes can be ascribed to MLCT or ligand-centered transfer as a function of X. The X-ray structure determination of the chloro-derivative, $[Re(CO)_3Cl(4,4′-bipyridine)]$ exists (98AX(C)1596) as well as spectroelectrochemical (94IC3246) and photochemical (96IC1421) studies. The luminescent properties of $[Re(CO)_3(LL)X]$ (LL = 4,4′-bipyridine; X = Cl, Br, I) are different (87JCS(CC)1752, 88JA3892). The photochemical properties of $[Re(CO)_3(LL)Cl]$ (LL = 4,4′-bipyridine) prepared from 4,4′-bipyridine and $[Re(CO)_5Cl]$ are also of interest (89JCS(CC)1655). 4,4′-Bipyridine (LL) with $[Re(CO)_5(OTf)]$ in polar media gives dinuclear complexes with luminescent properties, $[(RO)(OC)_3Re(\mu-LL)Re(CO)_3(OR)]$ (R = H, Me, Et, OCH_2CH_2OH) (98IC5406). Excess 4,4′-bipyridine with $[Re(CO)_5Cl]$ in *iso*-octane gives $[Re(CO)_3(4,4′-bipyridine)_2Cl]$ (94IC3246), which is a good precursor for such complexes (95JA11813, 96IC4096) characterized by their long-lived MLCT excited state (89JCS(CC)1655).

$[Re(CO)_5Cl]$ with 4,4′-dipyridylbutadiyne (LL) in methylene chloride gives $[Re(CO)_3(\eta^1-LL)_2Cl]$ (96JOM(517)217). 1,4-Bis(4′-pyridylethynyl)benzene (LL) reacts with $[Re(CO)_4(PPh_3)Cl]$ in benzene to yield $[Cl(Ph_3P)(OC)_3Re(\mu-LL)Re(CO)_3(PPh_3)Cl]$. In a similar fashion, two more dinuclear complexes with L = PPh_3 and $P(OMe)_3$ and LL = 4,4′-dipyridylbutadiyne were prepared. Another two dinuclear complexes follow from 4,4′-dipyridylbutadiyne or 1,4-bis(4′-pyridylethynyl)benzene (LL) and $[Re(CO)_3(bpy)(AN)](PF_6)$: $[(bpy)(OC)_3Re(\mu-LL)Re(CO)_3(bpy)](PF_6)_2$. The products are of interest as materials for non-linear optics.

4,4′-Bipyridine with $[Re_2(CO)_{10}]$ in the presence of 1-butanol, 1-octanol, and 1-dodecanol forms the alkoxy-bridged molecular rectangles **145** (R = $-(CH_2)_3Me$, $-(CH_2)_7Me$, $-(CH_2)_{11}Me$) (02IC5323). A supramolecular compound based on the 4,4′-bipyridine complex of rhenium, **146**, is known (98JA12982, 00IC4977). Another illustration is complex **147** (00AX(C)963). Molecular assembly **148** was prepared as nanocrystalline thin films consisting of molecular squares (99JA557, 04IC132, 04IC2013). Molecular rectangles is the subject of a publication (00AGE2891).

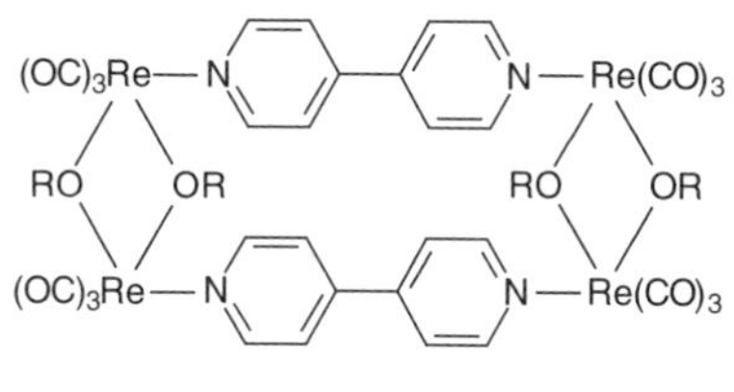

145

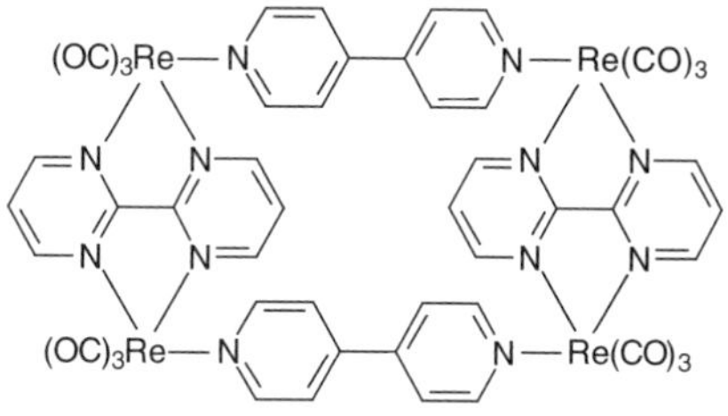

146

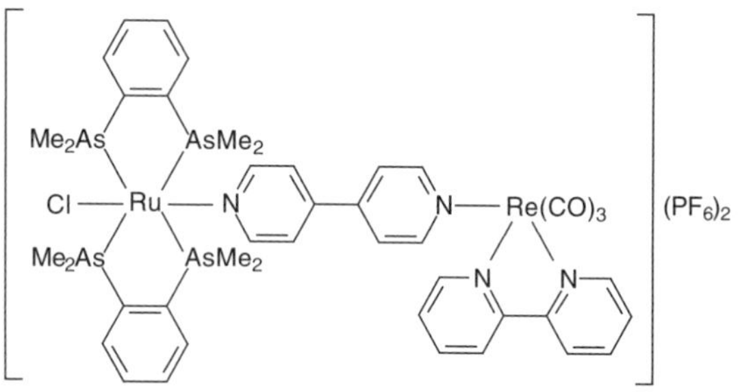

147

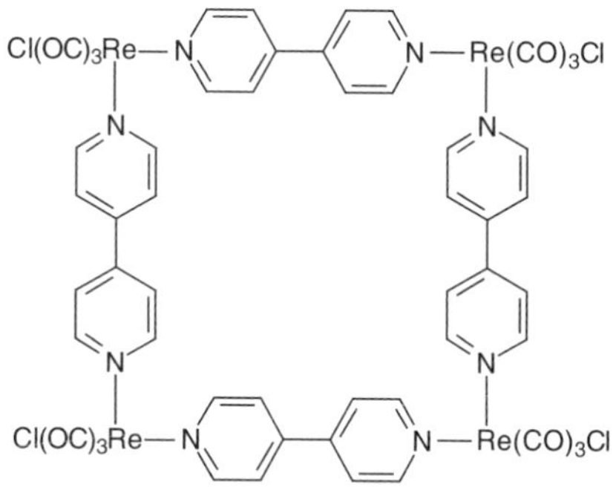

148

4,4'-Dipyridylbutadiyne or 4,4'-azopyridine react with [Re(CO)$_5$Cl in toluene/THF to give square complexes **149** and **150** (99IC4181, 00JA8956). 2,5-Bis (4-pyridylethynylthiophene) reacts with [Re(CO)$_5$Cl] in toluene/THF under prolonged reflux to give the dimer complex **151**. Triangular complexes **152** (R = n-C$_6$H$_{13}$, n-C$_{12}$H$_{25}$) follow from 1,4-bis(4'-pyridylethynyl)-2,5-di-n-hexyloxybenzene or 1,4-bis(4'-pyridylethynyl)-2,5-di-n-dodecyloxybenzene) in benzene. [Re(CO)$_5$Br] and 4,4'-bis(pyridyl)butadiyne or 4,4'-azobipyridine (LL) in *iso*-octane under reflux give corner complexes [Re(CO)$_3$(LL)$_2$Br].

149

150

151

152

III. Conclusion

Manganese(I) and rhenium(I) predominantly form polypyridine complexes of composition $[M(CO)_3(LL)X]$ with a typical chelating η^2-mode of coordination having attractive photo- and electrochemical properties and explaining their wide variety and applicability. The predominant reactivity path of these complexes is related to the substitution reactions of the X functionality (mainly Cl, Br, I, OTf) leading to neutral and ionic substitution products, among them those containing chromophoric ligands with added photochemical applications. Other possible reactions are related to the modification of the polypyridine ligands, insertion reactions, and molecular assembly processes leading to polynuclear complexes, molecular corners, triangles, squares, and rectangles. The latter aspect makes organorhenium complexes of polypyridines especially attractive in materials chemistry.

REFERENCES

58JCS3149   E. W. Abel, G. B. Hargreaves, and G. Wilkinson, *J. Chem. Soc.*, 3149 (1958).

59JCS1501   E. W. Abel and G. Wilkinson, *J. Chem. Soc.*, 1501 (1959).

63CB3035   R. Kruck and M. Hofler, *Chem. Ber.*, **96**, 3035 (1963).

66IC2119   L. W. Houk and G. R. Dobson, *Inorg. Chem.*, **12**, 2119 (1966).

67ICA(1)172   F. Zingales, M. Graziani, F. Foraone, and U. Belucco, *Inorg. Chim. Acta*, **1**, 172 (1967).

68JA3722   E. M. Kober, J. V. Caspar, R. S. Lumpkin, and T. J. Meyer, *J. Am. Chem. Soc.*, **90**, 3722 (1968).

72GCI587   D. Vitali and F. Calderazzo, *Gaz. Chim. Ital.*, **102**, 587 (1972).

73IC1067   R. J. Angelici and R. W. Brink, *Inorg. Chem.*, **12**, 1067 (1973).

74CRV4801   M. Wrighton, *Chem. Rev.*, **74**, 4801 (1974).

74JA988	M. Wrighton and D. L. Morse, *J. Am. Chem. Soc.*, **96**, 988 (1974).
75JA2073	M. S. Wrighton, D. L. Morse, and L. Pdungsap, *J. Am. Chem. Soc.*, **97**, 2073 (1975).
75JINC1375	J. R. Wagner and D. G. Hendricker, *J. Inorg. Nucl. Chem.*, **37**, 1375 (1975).
76JA3931	D. L. Morse and M. S. Wrighton, *J. Am. Chem. Soc.*, **98**, 3931 (1976).
77JOM(125)71	D. L. Morse and M. S. Wrighton, *J. Organomet. Chem.*, **125**, 71 (1977).
77JOM(131)73	D. A. Edwards and J. Marshalsea, *J. Organomet. Chem.*, **131**, 73 (1977).
77TMC123	R. Uson, V. Riera, J. Gimeno, and M. Laguna, *Transition Met. Chem.*, **2**, 123 (1977).
78CPL389	B. P. Sullivan, H. D. Abruna, H. O. Finklea, D. L. Salmon, J. K. Nagile, T. J. Meyer, and H. Sprintrchnik, *Chem. Phys. Lett.*, **58**, 389 (1978).
78JA2257	P. Giordano, S. M. Fredericks, M. S. Wrighton, and D. L. Morse, *J. Am. Chem. Soc.*, **100**, 2257 (1978).
78JA5790	J. C. Luong, L. Nadjo, and M. S. Wrighton, *J. Am. Chem. Soc.*, **100**, 5790 (1978).
78JOM(159)201	H. Behrens, R. J. Lampe, P. Merbach, and M. Moll, *J. Organomet. Chem.*, **159**, 201 (1978).
79JA1597	J. C. Luong, R. A. Faltynek, and M. S. Wrighton, *J. Am. Chem. Soc.*, **101**, 1597 (1979).
79JA2888	P. J. Giordano and M. Wrighton, *J. Am. Chem. Soc.*, **101**, 2888 (1979).
79JA7415	S. M. Fredericks, J. C. Luong, and M. S. Wrighton, *J. Am. Chem. Soc.*, **101**, 7415 (1979).
79JOM(170)235	L. H. Staal, A. Oskam, and K. Vrieze, *J. Organomet. Chem.*, **170**, 235 (1979).
79JOM(175)73	L. H. Staal, G. van Koten, and K. Vrieze, *J. Organomet. Chem.*, **175**, 73 (1979).
79MI1	G.L. Geoffroy and M.S. Wrighton, *"Organic Photochemistry"* Academic Press, New York (1979).
80AJC2369	E. Horn and M. R. Snow, *Aust. J. Chem.*, **33**, 2369 (1980).
80IC860	B. Durham, J. L. Walsh, C. L. Carter, and T. J. Meyer, *Inorg. Chem.*, **19**, 860 (1980).
80ICA(45)L265	A. Vogler and H. Kunkely, *Inorg. Chim. Acta*, **45**, L265 (1980).
80JA1383	E. M. Kober, B. P. Sullivan, W. J. Dressick, J. V. Caspar, and T. J. Meyer, *J. Am. Chem. Soc.*, **102**, 1383 (1980).
80JA7892	J. C. Luong, R. A. Faltynek, and M. S. Wrighton, *J. Am. Chem. Soc.*, **102**, 7892 (1980).
81AGE469	A. Vogler and H. Kunkely, *Angew. Chem., Int. Ed. Engl.*, **20**, 469 (1981).
81IC2348	M. Taker and A. Ludi, *Inorg. Chem.*, **20**, 2348 (1981).
81ICA(53)L35	A. Vogler, L. El-Sayed, R. G. Jones, J. Namnath, and A. W. Adamson, *Inorg. Chim. Acta*, **53**, L35 (1981).
81JA56	H. D. Abruna, P. Denisevich, M. Umana, T. J. Meyer, and R. W. Murray, *J. Am. Chem. Soc.*, **101**, 56 (1981).
81JA5238	D. P. Summers, L. S. Luong, and M. S. Wrighton, *J. Am. Chem. Soc.*, **103**, 5238 (1981).
81JCS(D)1124	R. W. Balk, D. J. Stufkens, and A. Oskam, *J. Chem. Soc., Dalton Trans.*, 1124 (1981).
81JOM(219)61	G. A. Carriedo, J. Gimeno, M. Laguna, and V. Riera, *J. Organomet. Chem.*, **219**, 61 (1981).

82AJC2445   D. S. Black, G. B. Deacon, and N. C. Thomas, *Aust. J. Chem.*, **35**, 2445 (1982).

82CCR(46)159   K. Kalyanasundaram, *Coord. Chem. Rev.*, **46**, 159 (1982).

82IC2153   P. Denisevich, H. D. Abruna, H. D. Leidner, T. J. Meyer, and R. W. Murray, *Inorg. Chem.*, **21**, 2153 (1982).

82JA7658   P. A. Lay, R. H. Magnuson, J. Sen, and H. Taube, *J. Am. Chem. Soc.*, **104**, 7658 (1982).

82JCP3337   O. A. Salman and H. G. Drickamer, *J. Chem. Phys.*, **77**, 3337 (1982).

82JCS(CC)915   G. A. Neyhart, J. L. Marshall, W. J. Dressick, B. P. Sullivan, P. A. Watkins, and T. J. Meyer, *J. Chem. Soc., Chem. Commun.*, 915 (1982).

82JOM(240)413   D. Micholova and A. Vlcek, *J. Organomet. Chem.*, **240**, 413 (1982).

82MI1   N. N. Boog and H. D. Goesz, in *"Comprehensive Organometallic Chemistry"* (E. W. Abel, F. G. A. Stone and G. Wilkinson, eds.), Vol. 4, p. 161, Pergamon, Oxford (1982).

83IC1617   D. P. Rillema, G. Allen, T. J. Meyer, and D. Conrad, *Inorg. Chem.*, **22**, 1617 (1983).

83AC1954   R. L. Cerny, B. P. Sullivan, M. M. Bursey, and T. J. Meyer, *Anal. Chem.*, **55**, 1954 (1983).

83IC2151   J. M. Calvert, R. H. Schmehl, B. P. Sullivan, J. S. Facci, T. J. Meyer, and R. W. Murray, *Inorg. Chem.*, **22**, 2151 (1983).

83IC2444   J. V. Caspar and T. J. Meyer, *Inorg. Chem.*, **22**, 2444 (1983).

83IC3825   D. M. Manuta and A. J. Lees, *Inorg. Chem.*, **22**, 3825 (1983).

83ICA(76)L29   W. Tikkanen, W. C. Kaska, S. Moya, T. Layman, R. Kane, and C. Kruger, *Inorg. Chim. Acta*, **76**, L29 (1983).

83JA61   J. W. Hershberger, R. J. Klinger, and J. J. Kochi, *J. Am. Chem. Soc.*, **105**, 61 (1983).

83JA1067   W. K. Smoothers and M. S. Wrighton, *J. Am. Chem. Soc.*, **105**, 1067 (1983).

83JA5952   T. D. Westmoreland, H. Le Bozec, R. W. Murray, and T. J. Meyer, *J. Am. Chem. Soc.*, **105**, 5952 (1983).

83JA7241   H. Kunkely, A. Merz, and A. Vogler, *J. Am. Chem. Soc.*, **105**, 7241 (1983).

83JCE834   R. J. Watts, *J. Chem. Educ.*, **60**, 834 (1983).

83JCE877   M. S. Wrighton, *J. Chem. Educ.*, **60**, 877 (1983).

83JCS(CC)536   J. Hawecker, J. M. Lehn, and R. Ziessel, *J. Chem. Soc., Chem. Commun.*, 536 (1983).

83JPC952   J. V. Caspar and T. J. Meyer, *J. Phys. Chem.*, **87**, 952 (1983).

83MI1   M. Gratzel, *"Energy Resources Through Photochemistry and Catalysis"* Academic Press, New York (1983).

83MI2   O. Horvath and R. L. Stevenson, *"Charge Transfer Photochemistry of Coordination Compounds"* VCH, New York (1983).

83OM552   J. V. Caspar, B. P. Sullivan, and T. J. Meyer, *Organometallics*, **2**, 552 (1983).

84CPL332   P. A. Mabrouk and M. S. Wrighton, *Chem. Phys. Lett.*, **103**, 332 (1984).

84IC1440   D. V. Pinnick and B. Durham, *Inorg. Chem.*, **23**, 1440 (1984).

84IC2098   J. V. Caspar, B. P. Sullivan, and T. J. Meyer, *Inorg. Chem.*, **23**, 2098 (1984).

84IC2104   J. V. Caspar, B. P. Sullivan, and T. J. Meyer, *Inorg. Chem.*, **23**, 2104 (1984).

84JA2613   G. A. Allen, R. P. White, D. P. Rillema, and T. J. Meyer, *J. Am. Chem. Soc.*, **106**, 2613 (1984).

84JCS(CC)328   J. Hawecker, J. M. Lehn, and R. Ziessel, *J. Chem. Soc., Chem. Commun.*, 328 (1984).

84JCS(CC)403   B. P. Sullivan and T. J. Meyer, *J. Chem. Soc., Chem. Commun.*, 403 (1984).

84JCS(CC)1244   B. P. Sullivan and T. J. Meyer, *J. Chem. Soc., Chem. Commun.*, 1244 (1984).

84JOM(276)39   F. J. G. Alonso, V. Riera, F. Villafane, and M. Vivanco, *J. Organomet. Chem.*, **276**, 39 (1984).

84OM1242   B. P. Sullivan, J. V. Caspar, S. R. Johnson, and T. J. Meyer, *Organometallics*, **3**, 1242 (1984).

84PNA1961   J. K. Barton, L. A. Basile, A. Danishefsky, and A. Alexandrescu, *Proc. Natl. Acad. Sci. USA*, **81**, 1961 (1984).

85AX(C)1312   M. L. Valin, D. Moreiras, X. Solans, M. Font-Altaba, F. J. Garcia-Alonso, V. Riera, and M. Vivanco, *Acta Crystallogr. C*, **41**, 1312 (1985).

85CCR(64)41   G. A. Crosby, K. A. Highland, and K. A. Truesdell, *Coord. Chem. Rev.*, **64**, 41 (1985).

85CCR(65)65   M. K. DeArmond, K. M. Hanck, and D. W. Wertz, *Coord. Chem. Rev.*, **65**, 65 (1985).

85CRV187   T. J. Meyer and J. V. Caspar, *Chem. Rev.*, **85**, 187 (1985).

85IC397   R. L. Cerny, B. P. Sullivan, M. M. Bursey, and T. J. Meyer, *Inorg. Chem.*, **24**, 397 (1985).

85IC2596   T. D. Westmoreland, K. S. Schanze, P. E. Neveaux, E. Danielson, B. P. Sullivan, P. Chen, and T. J. Meyer, *Inorg. Chem.*, **24**, 2596 (1985).

85IC2755   E. M. Kober, J. L. Marshall, W. J. Dressick, B. P. Sullivan, and T. J. Meyer, *Inorg. Chem.*, **24**, 2755 (1985).

85IC2934   M. W. Kokkes, D. J. Stufkens, and A. Oskam, *Inorg. Chem.*, **24**, 2934 (1985).

85IC3640   B. P. Sullivan, D. Conrad, and T. J. Meyer, *Inorg. Chem.*, **24**, 3640 (1985).

85IC4411   M. W. Kokkes, D. J. Stufkens, and A. Oskam, *Inorg. Chem.*, **24**, 4411 (1985).

85JA5518   C. V. Kumar, J. K. Barton, and N. Turro, *J. Am. Chem. Soc.*, **107**, 5518 (1985).

85JCS(CC)796   B. P. Sullivan, C. M. Bolinger, D. Conrad, W. J. Vining, and T. J. Meyer, *J. Chem. Soc., Chem. Commun.*, 796 (1985).

85JCS(CC)1414   B. P. Sullivan, C. M. Bolinger, D. Conrad, W. J. Vining, and T. J. Meyer, *J. Chem. Soc., Chem. Commun.*, 1414 (1985).

85JCS(CC)1416   T. R. O'Toole, L. D. Margerum, T. D. Westmoreland, W. J. Vining, R. W. Murray, and T. J. Meyer, *J. Chem. Soc., Chem. Commun.*, 1416 (1985).

85JMS11   M. W. Kokkes, T. J. Snoeck, D. J. Stufkens, and A. Oskam, *J. Mol. Struct.*, **131**, 11 (1985).

85JOM(290)63   R. R. Andrea, D. J. Stufkens, and A. Oskam, *J. Organomet. Chem.*, **290**, 63 (1985).

85JOM(294)59   M. W. Kokkes, W. G. J. De Lange, D. J. Stufkens, and A. Oskam, *J. Organomet. Chem.*, **294**, 59 (1985).

85JPC1095   W. J. Vining, J. V. Caspar, and T. J. Meyer, *J. Phys. Chem.*, **89**, 1095 (1985).

85OM2161   C. Kutal, M. A. Weber, G. Ferraudi, and D. Geiger, *Organometallics*, **4**, 2161 (1985).

86AX(C)417   M. L. Valin, D. Moreiras, X. Solans, M. Font-Altaba, and F. J. Garcia-Alonso, *Acta Crystallogr. C*, **42**, 417 (1986).

86HCA1990   J. Hawecker, J. M. Lehn, and R. Ziessel, *Helv. Chim. Acta*, **69**, 1990 (1986).

86IC256   A. Juris, R. Belser, F. Barigeletti, A. Zelewsky, and V. Blazani, *Inorg. Chem.*, **25**, 256 (1986).

86IC3212   D. M. Manuta and A. J. Lees, *Inorg. Chem.*, **25**, 3212 (1986).

86ICA(114)L43   H. Hukkanen and T. A. Pakkanen, *Inorg. Chim. Acta*, **114**, L43 (1986).

86ICA(115)193   A. Vogler and J. Kisslinger, *Inorg. Chim. Acta*, **115**, 193 (1986).

86ICA(121)191   G. A. Carriedo, M. C. Crespo, V. Riera, M. L. Valin, D. Moreiras, and X. Solans, *Inorg. Chim. Acta*, **121**, 191 (1986).

86JA5344   G. A. Reitz, W. J. Dressick, J. N. Demas, and B. A. DeGraff, *J. Am. Chem. Soc.*, **108**, 5344 (1986).

86JA7567   W. J. Dressick, J. T. Cline, J. N. Demas, and B. A. DeGraff, *J. Am. Chem. Soc.*, **108**, 7567 (1986).

86JCS(F2)2401   K. Kalyanasundaram, *J. Chem. Soc., Faraday Trans. 2*, **82**, 2401 (1986).

86JEAC(201)347   A. I. Breikiss and H. D. Abruna, *J. Electroanal. Chem.*, **201**, 347 (1986).

86JEAC(207)315   S. Cosnier, A. Deronzier, and J. C. Moutet, *J. Electroanal. Chem.*, **207**, 315 (1986).

86JEAC(209)101   C. R. Cabrera and H. D. Abruna, *J. Electroanal. Chem.*, **209**, 101 (1986).

86JOM(300)139   J. K. Kochi, *J. Organomet. Chem.*, **300**, 139 (1986).

86JOM(304)207   M. J. Bermejo, B. Martinez, and J. Vinaixa, *J. Organomet. Chem.*, **304**, 207 (1986).

86JP5344   J. P. Otruba, G. A. Neyhart, J. L. Marshall, W. J. Dressick, B. P. Sullivan, P. A. Watkins, and T. J. Meyer, *J. Photochem.*, **108**, 5344 (1986).

86JPC5010   W. Kaim and S. Ernst, *J. Phys. Chem.*, **90**, 5010 (1986).

86OM1100   R. R. Andrea, W. G. J. de Lange, T. van der Graaf, M. Rijkoff, D. J. Stufkens, and A. Oskam, *Organometallics*, **7**, 100 (1986).

86OM1500   B. P. Sullivan and T. J. Meyer, *Organometallics*, **5**, 1500 (1986).

86PAC1193   T. J. Meyer, *Pure Appl. Chem.*, **58**, 1193 (1986).

87ACR214   M. Chanon, *Acc. Chem. Res.*, **20**, 214 (1987).

87ACS155   A. Vogler and H. Kunkely, *ACS Symp. Ser. No. 333*, 155 (1987).

87AX(C)792   E. Horn, M. R. Snow, and E. R. T. Tiekink, *Acta Crystallogr. C*, **43**, 792 (1987).

87CPL365   W. Kaim and S. Kohlmann, *Chem. Phys. Lett.*, **139**, 365 (1987).

87CRV711   A. J. Lees, *Chem. Rev.*, **87**, 711 (1978).

87IC882   C. R. Leidner, B. P. Sullivan, R. A. Reed, B. A. White, M. T. Crimmins, R. W. Murray, and T. J. Meyer, *Inorg. Chem.*, **26**, 882 (1987).

87IC1116   P. Chen, T. D. Westmoreland, E. Danielson, K. S. Schanze, D. Anthon, P. Noveux, and T. J. Meyer, *Inorg. Chem.*, **26**, 1116 (1987).

87IC1449   M. Stebner, A. Gutierrez, A. Ludi, and H. B. Burgi, *Inorg. Chem.*, **26**, 1449 (1987).

87IC2145   B. P. Sullivan, T. J. Meyer, and J. V. Caspar, *Inorg. Chem.*, **26**, 2145 (1987).

87JA2519   E. Danielson, C. M. Elliot, C. M. Merkert, and T. J. Meyer, *J. Am. Chem. Soc.*, **109**, 2519 (1987).

87JCS(CC)1153   O. Ishitani, I. Namura, Y. Yanagida, and C. Pac, *J. Chem. Soc., Chem. Commun.*, 1153 (1987).

87JCS(CC)1752   M. M. Glezen and A. J. Lees, *J. Chem. Soc., Chem. Commun.*, 1752 (1987).

87JOM(326)C71	F. J. G. Alonso, V. Riera, M. L. Valin, D. Moreiras, M. Vivanco, and X. Solans, *J. Organomet. Chem.*, **326**, C71 (1987).
87NATO(C)113	R. Ziessel, *NATO ASI Ser. C*, **C206**, 113 (1987).
87MI1	J. D. Petersen, in *"Supramolecular Photochemistry"* (V. Balzani ed.), p. 135, Reidel, Dodrecht (1987).
87OM553	C. Kutal, A. J. Corbin, and G. Ferraudi, *Organometallics*, **6**, 553 (1987).
88ACS52	B. P. Sullivan, M. R. M. Bruce, T. R. O'Toole, C. M. Bolinger, E. Megehee, H. Thorp, and T. J. Meyer, *ACS Symp. Ser. No. 363,* 52 (1988).
88CCR(84)5	F. Barigeletti, S. Campagna, P. Belser, and A. Zelewsky, *Coord. Chem. Rev.*, **84**, 5 (1988).
88CCR(84)47	A. P. Zipp, *Coord. Chem. Rev.*, **84**, 47 (1988).
88CCR(84)85	A. Juris, V. Balzani, F. Barigelletti, S. Campagna, P. Belser, and A. Zelewsky, *Coord. Chem. Rev.*, **84**, 85 (1988).
88CCR(86)135	H. D. Abruna, *Coord. Chem. Rev.*, **86**, 135 (1988).
88CRV1189	D. Astruc, *Chem. Rev.*, **88**, 1189 (1988).
88IC651	N. Kitamura, M. Sato, H. B. Kim, R. Obata, and S. Tazuke, *Inorg. Chem.*, **27**, 651 (1988).
88IC3325	M. M. Zulu and A. J. Lees, *Inorg. Chem.*, **27**, 3325 (1988).
88IC3652	A. Juris, S. Campagna, V. Balzani, G. Gremaud, and A. Zelewski, *Inorg. Chem.*, **27**, 3652 (1988).
88IC4007	A. Juris, S. Campagna, I. Bidd, J. M. Lehn, and R. Ziessel, *Inorg. Chem.*, **27**, 4007 (1988).
88IC4326	C. F. Shu and M. S. Wrighton, *Inorg. Chem.*, **27**, 4326 (1988).
88IC4385	M. J. Bermejo, J. I. Ruiz, X. Solans, and J. Vinaixa, *Inorg. Chem.*, **27**, 4385 (1988).
88ICA(149)77	R. R. Andrea, W. G. J. de Lange, D. J. Stufkens, and A. Oskam, *Inorg. Chim. Acta*, **149**, 77 (1988).
88JA3892	M. M. Glezen and A. J. Lees, *J. Am. Chem. Soc.*, **110**, 3892 (1988).
88JA5051	G. A. Reitz, J. N. Demas, and B. A. DeGraff, *J. Am. Chem. Soc.*, **110**, 5051 (1988).
88JA6243	M. M. Glezen and A. J. Lees, *J. Am. Chem. Soc.*, **110**, 6243 (1988).
88JA7098	M. A. Bennett, G. B. Robinson, A. Rokicki, and W. A. Wickramasinghe, *J. Am. Chem. Soc.*, **110**, 7098 (1988).
88JCS(CC)16	R. Ziessel, *J. Chem. Soc., Chem. Commun.*, 16 (1988).
88JPC3708	P. Chen, E. Danielson, and T. J. Meyer, *J. Phys. Chem.*, **92**, 3708 (1988).
88OM1100	R. R. Andrea, W. G. J. de Lange, T. van der Graaf, M. Rijkoff, D. J. Stufkens, and A. Oskam, *Organometallics*, **7**, 1100 (1988).
88TMC126	R. Ali, J. Burgess, and P. Guardado, *Transition Met. Chem.*, **13**, 126 (1988).
89ACR163	T. J. Meyer, *Acc. Chem. Res.*, **22**, 163 (1989).
89CCR(93)245	J. P. Collin and J. P. Sauvage, *Coord. Chem. Rev.*, **93**, 245 (1989).
89CL765	C. Pac, K. Ishii, and S. Yanagida, *Chem. Lett.*, 765 (1989).
89IC75	H. K. van Dijk, J. van der Haar, D. J. Stufkens, and A. Oskam, *Inorg. Chem.*, **28**, 75 (1989).
89IC85	M. M. Zulu and A. J. Lees, *Inorg. Chem.*, **28**, 85 (1989).
89IC318	R. R. Andrea, W. G. J. de Lange, D. J. Stufkens, and A. Oskam, *Inorg. Chem.*, **28**, 318 (1989).
89IC1022	R. Sahai, D. P. Rillema, R. Shaver, S. Van Wallendael, D. C. Jackman, and M. Boldaji, *Inorg. Chem.*, **28**, 1022 (1989).
89IC1228	W. Hosek, W. A. Tysoe, H. D. Gafney, D. A. Baker, and T. C. Strekas, *Inorg. Chem.*, **28**, 1228 (1989).

89IC1596   L. W. Winslow, D. P. Rillema, J. H. Welch, and P. Singh, *Inorg. Chem.*, **28**, 1596 (1989).

89IC2154   K. A. Rawlins and A. J. Lees, *Inorg. Chem.*, **28**, 2154 (1989).

89IC2271   P. Chen, M. Curry, and T. J. Meyer, *Inorg. Chem.*, **28**, 2271 (1989).

89IC3923   T. R. O'Toole, J. N. Younathan, B. P. Sullivan, and T. J. Meyer, *Inorg. Chem.*, **28**, 3923 (1989).

89JA5185   M. R. Burke and T. L. Brown, *J. Am. Chem. Soc.*, **111**, 5185 (1989).

89JA5699   T. R. O'Toole, B. P. Sullivan, and T. J. Meyer, *J. Am. Chem. Soc.*, **111**, 5699 (1989).

89JA6602   M. M. Glezen and A. J. Lees, *J. Am. Chem. Soc.*, **111**, 6602 (1989).

89JA7791   C. O. Dietrich-Buchecker, J. P. Sauvage, and J. M. Kern, *J. Am. Chem. Soc.*, **111**, 7791 (1989).

89JA8305   P. Chen, R. Duesing, G. Tapolsky, and T. J. Meyer, *J. Am. Chem. Soc.*, **111**, 8305 (1989).

89JCS(CC)1655   P. Glyn, M. W. George, P. M. Hodges, and J. J. Turner, *J. Chem. Soc., Chem. Commun.*, 1655 (1989).

89JCS(D)1449   J. Guilhem, C. Pascard, J. M. Lehn, and R. Ziessel, *J. Chem. Soc., Dalton Trans.*, 1449 (1989).

89JEAC(259)217 T. R. O'Toole, B. P. Sullivan, M. R. M. Bruce, L. G. Margerum, R. W. Murray, and T. J. Meyer, *J. Electroanal. Chem.*, **259**, 217 (1989).

89JOC1731   L. Della Ciana, I. Hamachi, and T. J. Meyer, *J. Org. Chem.*, **54**, 1731 (1989).

89JPC24   B. P. Sullivan, *J. Phys. Chem.*, **93**, 24 (1989).

89JPC2129   J. McKiernan, J. C. Pouxviel, B. Dunn, and J. I. Zink, *J. Phys. Chem.*, **93**, 2129 (1989).

89JPC3885   G. Tapolsky, R. Duesing, and T. J. Meyer, *J. Phys. Chem.*, **93**, 3885 (1989).

89MI1   T. J. Meyer, in "*Photochemical Energy Conversion*" (J. R. Norris and D. Meisel, eds.), p. 75, Elsevier, Amsterdam (1989).

90ACS101   L. A. Cabana and K. S. Schanze, *Adv. Chem. Ser.*, **226**, 101 (1990).

90AOC523   P. A. Anderson, F. R. Keene, E. Horn, and E. R. T. Tiekink, *Appl. Organomet. Chem.*, **4**, 523 (1990).

90CCR(104)39   D. J. Stufkens, *Coord. Chem. Rev.*, **104**, 39 (1990).

90IC1574   R. Ruminski and R. T. Cambion, *Inorg. Chem.*, **29**, 1574 (1990).

90IC1761   S. van Wallendael, R. J. Shaver, D. P. Rillema, B. J. Yoblinski, M. Stathis, and T. F. Guarr, *Inorg. Chem.*, **29**, 1761 (1990).

90IC2285   G. Tapolsky, R. Duesing, and T. J. Meyer, *Inorg. Chem.*, **29**, 2285 (1990).

90IC2327   J. A. Baiano, D. L. Carlson, G. M. Wolosh, D. E. DeJesus, C. F. Knowles, E. G. Szabo, and W. R. Murphy, *Inorg. Chem.*, **29**, 2327 (1990).

90IC2792   L. Della Ciana, W. J. Dressick, D. Sandrini, M. Maestri, and M. Ciano, *Inorg. Chem.*, **29**, 2792 (1990).

90IC2909   W. Kaim and S. Kohlmann, *Inorg. Chem.*, **29**, 2909 (1990).

90IC3866   K. A. Rawlins, A. J. Lees, and A. W. Adamson, *Inorg. Chem.*, **29**, 3866 (1990).

90IC4169   R. Lin, T. F. Guarr, and R. Duesing, *Inorg. Chem.*, **29**, 4169 (1990).

90IC4335   L. Sacksteder, A. P. Zipp, E. A. Brown, J. Streich, J. N. Demas, and B. A. DeGraff, *Inorg. Chem.*, **29**, 4335 (1990).

90ICA(167)149   R. Lin and T. F. Guarr, *Inorg. Chim. Acta*, **167**, 149 (1990).

90ICA(178)185   P. C. Servaas, G. J. Stor, D. J. Stufkens, and A. Oskam, *Inorg. Chim. Acta*, **178**, 185 (1990).

90JA1117   J. R. Shaw, R. T. Webb, and R. H. Schmehl, *J. Am. Chem. Soc.*, **112**, 1117 (1990).
90JA1333   S. R. Snyder, H. S. White, S. Lopez, and H. D. Abruna, *J. Am. Chem. Soc.*, **112**, 1333 (1990).
90JA5378   R. Duesing, G. Tapolsky, and T. J. Meyer, *J. Am. Chem. Soc.*, **112**, 5378 (1990).
90JA9490   S. Gould, T. O'Toole, and T. J. Meyer, *J. Am. Chem. Soc.*, **112**, 9490 (1990).
90JCS(CC)179   R. J. Shaver, D. P. Rillema, and C. J. Woods, *J. Chem. Soc., Chem. Commun.*, 179 (1990).
90JCS(D)1657   K. Kalyanasundaram and M. K. Nazeeruddin, *J. Chem. Soc., Dalton Trans.*, 1657 (1990).
90JOM(394)275   G. A. Carriedo and V. Riera, *J. Organomet. Chem.*, **394**, 275 (1990).
90JOM(397)309   G. A. Carriedo, M. C. Crespo, C. Diaz, and V. Riera, *J. Organomet. Chem.*, **397**, 309 (1990).
90JPC2229   T. A. Perkins, W. Humer, T. L. Netzel, and K. S. Schanze, *J. Phys. Chem.*, **94**, 2229 (1990).
90JPC2740   K. S. Schanze and L. A. Cabana, *J. Phys. Chem.*, **94**, 2740 (1990).
90JPC8745   T. A. Perkins, B. T. Hauser, J. R. Eyler, and K. S. Schanze, *J. Phys. Chem.*, **94**, 8745 (1990).
90NJC831   S. Cosnier, A. Deronzier, and J. C. Moutet, *Nouv. J. Chim.*, **14**, 831 (1990).
90TCC75   F. Scandola, M. T. Indeli, C. Chiorboli, and C. A. Bignozzi, *Top. Curr. Chem.*, **158**, 75 (1990).
91ACR325   D. R. Tyler, *Acc. Chem. Res.*, **24**, 325 (1991).
91AGE844   R. Ziessel, *Angew. Chem., Int. Ed. Engl.*, **30**, 844 (1991).
91CCR(111)193   R. Beer, G. Calzaferri, J. Li, and B. Waldeck, *Coord. Chem. Rev.*, **111**, 193 (1991).
91CCR(111)221   M. Furue, M. Naiki, Y. Kanematsu, T. Kushida, and M. Kamachi, *Coord. Chem. Rev.*, **111**, 221 (1991).
91CCR(111)297   S. Van Wallendael and D. P. Rillema, *Coord. Chem. Rev.*, **111**, 297 (1991).
91CM25   T. G. Kotch, A. J. Lees, S. J. Fuerniss, and K. I. Papathomas, *Chem. Mater.*, **3**, 25 (1991).
91IC42   L. J. Larson, A. Oskam, and J. J. Zink, *Inorg. Chem.*, **30**, 42 (1991).
91IC599   T. van der Graaf, D. J. Stufkens, A. Oskam, and K. Goubitz, *Inorg. Chem.*, **30**, 599 (1991).
91IC1330   R. M. Leasure, L. A. Stacksteder, D. Nesselrodt, G. A. Reitz, J. N. Demas, and B. A. DeGraff, *Inorg. Chem.*, **31**, 1330 (1991).
91IC3722   R. M. Leasure, L. Sacksteder, D. Nesselbrodt, G. A. Reitz, J. N. Demas, and B. A. DeGraff, *Inorg. Chem.*, **30**, 3722 (1991).
91IC4754   R. N. Dominey, B. Hauser, J. Hubbard, and J. Dunham, *Inorg. Chem.*, **30**, 4754 (1991).
91IC4871   T. G. Kotch, A. J. Lees, S. J. Fuerniss, K. I. Papathomas, and R. Snyder, *Inorg. Chem.*, **30**, 4871 (1991).
91ICA(187)133   T. van der Graaf, A. van Roy, D. J. Stufkens, and A. Oskam, *Inorg. Chim. Acta*, **187**, 133 (1991).
91JA389   J. R. Shaw and R. H. Schmehl, *J. Am. Chem. Soc.*, **113**, 389 (1991).
91JA1991   J. R. Shaw and R. H. Schmehl, *J. Am. Chem. Soc.*, **113**, 1991 (1991).
91JA7470   D. B. MacQueen and K. S. Schanze, *J. Phys. Chem.*, **113**, 7470 (1991).
91JCS(CC)787   C. Pac, S. Kaneda, K. Ishii, and S. Yanagida, *J. Chem. Soc., Chem. Commun.*, 787 (1991).
91JCS(D)849   L. A. Worl, R. Duesing, P. Chen, L. Della Ciana, and T. J. Meyer, *J. Chem. Soc., Dalton Trans.*, 849 (1991).

91JPC1105   G. Tapolsky, R. Duesing, and T. J. Meyer, *J. Phys. Chem.*, **91**, 1105 (1991).

91JPC5850   P. Chen, R. Duesing, D. K. Graff, and T. J. Meyer, *J. Phys. Chem.*, **95**, 5850 (1991).

91JPC7641   E. Amouyal, M. Mouallem-Bahout, and G. Calzaferri, *J. Phys. Chem.*, **95**, 7641 (1991).

91MI1   C. Pac, S. Kaseda, K. Ishii, S. Yanagida, and O. Ishitani, in "*Photochemical Processes in Organized Molecular Systems*" (K. Honda ed.), p. 177, Elsevier, Amsterdam (1991).

91MI2   V. Balzani and F. Scandola, "*Supramolecular Photochemistry*" Ellis Horwood, New York (1991).

91MI3   T. J. Meyer, "*Photochemical Processes in Organized Molecular Systems*" p. 133, Elsevier, Yokohama (1991).

91N737   M. Gratzel and B. O'Regan, *Nature*, **353**, 737 (1991).

91OM3668   T. van der Graaf, R. M. J. Hofstra, P. G. Schilder, M. Rijkoff, D. J. Stufkens, and J. G. M. van der Linden, *Organometallics*, **10**, 3668 (1991).

91PP124   T. G. Kotch, A. J. Lees, S. J. Fuerniss, K. I. Papathomas, and R. Snyder, *Polym. Prepr.*, **32**, 124 (1991).

92BSCQ43   S. A. Moya, R. Pastene, R. Schmidt, J. Guerrero, J. Sariego, and R. Sartori, *Bol. Soc. Chil. Quim.*, **37**, 43 (1992).

92BSCQ311   R. Sartori, J. Guerrero, R. Pastene, R. Schmidt, R. Sariego, and S. A. Moya, *Bol. Soc. Chil. Quim.*, **37**, 311 (1992).

92CIC359   D. J. Stufkens, *Comments Inorg. Chem.*, **13**, 359 (1992).

92CM675   T. G. Kotch, A. J. Lees, S. J. Fuerniss, and K. I. Papathomas, *Chem. Mater.*, **4**, 675 (1992).

92CPL299   J. M. Lang, Z. A. Dreger, and H. G. Drickamer, *Chem. Phys. Lett.*, **192**, 299 (1992).

92IC5   B. J. Yoblinski, M. Stathis, and T. F. Guarr, *Inorg. Chem.*, **31**, 5 (1992).

92IC1072   J. K. Hino, L. Della Ciana, W. Dressick, and B. P. Sullivan, *Inorg. Chem.*, **31**, 1072 (1992).

92IC3192   D. I. Yoon, C. A. Berg-Brennan, H. Lu, and J. T. Hupp, *Inorg. Chem.*, **31**, 3192 (1992).

92IC4101   R. J. Shaver and D. P. Rillema, *Inorg. Chem.*, **31**, 4101 (1992).

92IC5243   K. Kalyanasundaram, M. Gratzel, and M. K. Nazeeruddin, *Inorg. Chem.*, **31**, 5243 (1992).

92JA1897   D. B. MacQueen, J. R. Eyler, and K. S. Schanze, *J. Am. Chem. Soc.*, **114**, 1897 (1992).

92JA8727   C. A. Bignozzi, R. Argazzi, C. G. Garcia, F. Scandola, J. R. Schoonover, and T. J. Meyer, *J. Am. Chem. Soc.*, **114**, 8727 (1992).

92JCS(CC)303   E. W. Abel, N. J. Long, K. G. Orrell, A. G. Osborne, H. M. Pain, and V. Sik, *J. Chem. Soc., Chem. Commun.*, 303 (1992).

92JCS(CC)1615   S. Paulson, K. Morris, and B. P. Sullivan, *J. Chem. Soc., Dalton Trans.*, 1615 (1992).

92JCS(D)1455   P. Christensen, A. Hamnett, A. V. G. Muir, and J. A. Timney, *J. Chem. Soc., Dalton Trans.*, 1455 (1992).

92JPC257   M. Feliz, G. Ferraudi, and H. Altmiller, *J. Phys. Chem.*, **96**, 257 (1992).

92JPP(A)259   G. Calzaferri, K. Hadener, and J. Li, *J. Photochem. Photobiol. A. Chem. A*, **64**, 259 (1992).

92MI1   K. Kalyanasundaram, "*Photochemistry of Polypyridine and Porphyrin Complexes*" Academic Press, New York (1992).

92OM2826   F. J. Garcia-Alonso, A. Llamazares, M. Vivanco, V. Riera, and S. Garcia-Granda, *Organometallics*, **11**, 2826 (1992).

92POL1665   S. A. Moya, R. Pastene, R. Schmidt, J. Guerrero, and R. Sartori, *Polyhedron*, **11**, 1665 (1992).

93CCR(122)63 K. S. Schanze, D. B. MacQueen, T. A. Perkins, and L. A. Cabana, *Coord. Chem. Rev.*, **122**, 63 (1993).

93CCR(125)101 J. J. Turner, M. W. George, F. P. A. Johnson, and J. R. Westwell, *Coord. Chem. Rev.*, **125**, 101 (1993).

93CP585   K. A. Peterson, R. B. Dyer, K. C. Gordon, W. H. Woodruff, J. R. Schoonover, T. J. Meyer, and C. R. Bignozzi, *Chem. Phys.*, **55**, 585 (1993).

93IC237   E. R. Civitello, P. S. Dragovich, P. S. Karpishin, S. G. Novick, G. Bierach, J. F. O'Connell, and T. D. Westmoreland, *Inorg. Chem.*, **32**, 237 (1993).

93IC1222   R. Argazzi, C. A. Bignozzi, and O. Bortollini, *Inorg. Chem.*, **32**, 1222 (1993).

93IC2570   T. G. Kotch, A. J. Lees, S. J. Fuerniss, K. I. Papathomas, and R. W. Snyder, *Inorg. Chem.*, **32**, 2570 (1993).

93IC2618   J. R. Schoonover, G. F. Strouse, P. Chen, W. D. Bates, and T. J. Meyer, *Inorg. Chem.*, **32**, 2618 (1993).

93IC3836   L. Wallace and D. P. Rillema, *Inorg. Chem.*, **32**, 3836 (1993).

93IC4994   N. B. Thornton and K. S. Schanze, *Inorg. Chem.*, **32**, 4994 (1993).

93IC5629   A. P. Zipp, L. A. Sacksteder, J. Streich, A. Cook, J. N. Demas, and B. A. DeGraff, *Inorg. Chem.*, **32**, 5629 (1993).

93ICA(208)103 L. A. Lucia, R. D. Burton, and K. S. Schanze, *Inorg. Chim. Acta*, **208**, 103 (1993).

93JA2048   C. A. Berg-Brennan, D. I. Yoon, and J. T. Hupp, *J. Am. Chem. Soc.*, **115**, 2048 (1993).

93JA5675   Y. Wang, R. D. Burton, B. T. Hauser, M. M. Rooney, and K. S. Schanze, *J. Am. Chem. Soc.*, **115**, 5675 (1993).

93JA8230   L. Sacksteder, M. Lee, J. N. Demas, and B. A. DeGraff, *J. Am. Chem. Soc.*, **115**, 8230 (1993).

93JA10996 J. R. Schoonover, K. C. Gordon, R. Argazzi, C. A. Bignozzi, R. B. Dyer, and T. J. Meyer, *J. Am. Chem. Soc.*, **115**, 10996 (1993).

93JCS(CC)631 T. Yoshida, K. Tsutsumida, S. Teratani, K. Yasufuku, and M. Kaneko, *J. Chem. Soc., Chem. Commun.*, 631 (1993).

93JCS(D)597 E. W. Abel, V. S. Dimitrov, N. J. Long, K. G. Orrell, A. G. Osborne, H. M. Pain, V. Sik, M. B. Hursthouse, and M. A. Mazid, *J. Chem. Soc., Dalton Trans.*, 597 (1993).

93JCS(D)1321 M. D. Ward, *J. Chem. Soc., Dalton Trans.*, 1321 (1993).

93JOM(452)91 G. A. Carriedo, C. Carriedo, M. C. Crespo, and P. Gomez, *J. Organomet. Chem.*, **452**, 91 (1993).

93JOM(463)143 M. J. Bermejo, J. I. Ruiz, X. Solans, and J. Vinaixa, *J. Organomet. Chem.*, **463**, 143 (1993).

93JPC13126 P. Chen, S. L. Mecklenburg, and T. J. Meyer, *J. Phys. Chem.*, **97**, 13126 (1993).

93MI1   R. Ziessel, in *"Photosensitization and Photocatalysis Using Inorganic and Organometallic Compounds"* (K. Kalyanasundaram and M. Gratzel, eds.), p. 217, Kluwer, Dodrecht (1993).

94CPL426   D. R. Striplin and G. A. Crosby, *Chem. Phys. Lett.*, **221**, 426 (1994).

94CRV993   J. P. Sauvage, J. C. Collin, S. Chambron, C. Gutierrez, V. Coudret, V. Balzani, F. Barigeletti, L. De Cola, and L. Flamingi, *Chem. Rev.*, **94**, 993 (1994).

94CSR327   P. G. Sammes and G. Yahigolu, *Chem. Soc. Rev.*, **23**, 327 (1994).

94IC793   J. R. Schoonover, P. Chen, W. D. Bates, R. B. Dyer, and T. J. Meyer, *Inorg. Chem.*, **33**, 793 (1994).

94IC1354   V. Wang and K. S. Schanze, *Inorg. Chem.*, **33**, 1354 (1994).

94IC1652   C. A. Bignozzi, R. Argazzi, C. Chiorboli, F. Scandola, R. B. Dyer, J. R. Schoonover, and T. J. Meyer, *Inorg. Chem.*, **36**, 1652 (1994).

94IC2341   S. A. Moya, J. Guerrero, R. Pastene, R. Schmidt, R. Sariego, R. Sartori, J. Sans-Aparicio, I. Fonseca, and M. Martinez-Ripoll, *Inorg. Chem.*, **33**, 2341 (1994).

94IC2865   B. D. Rossenaar, T. van der Graaf, R. van Eldik, C. H. Langford, D. J. Stufkens, and A. Vlcek, *Inorg. Chem.*, **33**, 2865 (1994).

94IC3246   D. R. Gamelin, M. W. George, P. Glyn, F. W. Grevels, F. P. A. Johnson, W. E. Klotzbucher, S. L. Morrison, G. Russel, K. Schaffner, and J. J. Turner, *Inorg. Chem.*, **33**, 3246 (1994).

94IC4712   O. Ishitani, M. W. George, T. Ibusuki, F. P. A. Johnson, K. Koike, K. Nozaki, C. Pak, J. J. Turner, and J. R. Westwell, *Inorg. Chem.*, **33**, 4712 (1994).

94ICA(225)41   L. A. Lucia and K. S. Schanze, *Inorg. Chim. Acta*, **225**, 41 (1994).

94JCS(CC)63   B. D. Rossenaar, C. J. Kleverlaan, D. J. Stufkens, and A. Oskam, *J. Chem. Soc., Chem. Commun.*, 63 (1994).

94JCS(D)2977   M. W. George, F. P. A. Johnson, J. R. Westwell, P. M. Hodges, and J. J. Turner, *J. Chem. Soc., Dalton Trans.*, 2977 (1994).

94JCS(D)3095   M. D. Ward, *J. Chem. Soc., Dalton Trans.*, 3095 (1994).

94JCS(D)3441   E. W. Abel, K. G. Orrell, A. G. Osborne, H. M. Pain, V. Sik, M. B. Hursthouse, and K. M. A. Malik, *J. Chem. Soc., Dalton Trans.*, 3441 (1994).

94JCS(D)3745   N. C. Brown, G. A. Carriedo, N. G. Connely, F. J. G. Alonso, I. C. Quarmby, A. L. Rieger, P. H. Rieger, V. Riera, and M. Vivanco, *J. Chem. Soc., Dalton Trans.*, 3745 (1994).

94JOM(482)15   G. J. Stor, M. van der Vis, D. J. Stufkens, A. Oskam, J. Fraanje, and K. Goubitz, *J. Organomet. Chem.*, **482**, 15 (1994).

94JPC8959   N. E. Katz, S. L. Mecklenburg, D. K. Graff, P. Chen, and T. J. Meyer, *J. Phys. Chem.*, **98**, 8959 (1994).

94MI1   J. K. Barton, in *"Bioinorganic Chemistry"* (I. Bertini, H. B. Gray, S. J. Lippard and J. S. Valentine, eds.), University Science Books, Mill Valley, CA (1994).

94OM2641   G. J. Stor, S. L. Morrison, D. L. Stufkens, and A. Oskam, *Organometallics*, **13**, 2641 (1994).

95CIC319   A. J. Lees, *Comments Inorg. Chem.*, **17**, 319 (1995).

95CRV49   A. Hagfeldt and M. Gratzel, *Chem. Rev.*, **95**, 49 (1995).

95ECM621   H. Hori, O. Ishitani, K. Koike, F. P. A. Johnson, and I. Ibusuki, *Energy Convers. Manage.*, **36**, 621 (1995).

95HCA619   R. Deschenaux, T. Ruch, P. F. Deschenaux, A. Juris, and R. Ziessel, *Helv. Chim. Acta*, **78**, 619 (1995).

95IC1282   N. E. Katz, S. L. Mecklenburg, and T. J. Meyer, *Inorg. Chem.*, **34**, 1282 (1995).

95IC1588   G. J. Stor, D. J. Stufkens, P. Vernooijs, E. J. Baerends, J. Fraanje, and K. Goubitz, *Inorg. Chem.*, **34**, 1588 (1995).

95IC1629   T. A. Ostrikovich, P. S. White, and H. H. Thorp, *Inorg. Chem.*, **34**, 1629 (1995).

95IC2033   L. E. Helberg, J. Barrera, M. Sabat, and W. Harman, *Inorg. Chem.*, **34**, 2033 (1995).

95IC2438   D. A. Bardwell, F. Barigeletti, R. L. Cleary, L. Flamingi, M. Guardigli, J. C. Jeffery, and M. D. Ward, *Inorg. Chem.*, **34**, 2438 (1995).

95IC2875   L. Wallace, C. Woods, and D. P. Rillema, *Inorg. Chem.*, **34**, 2875 (1995).

95IC5183   J. W. M. Outersterp, D. J. Stufkens, and A. Vlcek, *Inorg. Chem.*, **34**, 5183 (1995).

95IC5210   L. Wallace, D. C. Jackman, D. P. Rillema, and J. W. Merkert, *Inorg. Chem.*, **34**, 5210 (1995).

95IC5578   G. F. Strouse, H. U. Gudel, V. Bertolasi, and V. Feretti, *Inorg. Chem.*, **34**, 5578 (1995).

95IC6235   Y. Shen and B. P. Sullivan, *Inorg. Chem.*, **34**, 6235 (1995).

95IC6421   J. R. Schoonover, W. D. Bates, and T. J. Meyer, *Inorg. Chem.*, **34**, 6421 (1995).

95ICA(240)169   W. B. Connick, A. J. DiBilio, M. G. Hill, J. R. Winkler, and H. B. Gray, *Inorg. Chim. Acta*, **240**, 169 (1995).

95ICA(240)453   G. F. Strouse and H. U. Gudel, *Inorg. Chim. Acta*, **240**, 453 (1995).

95JA7119   H. D. Stoeffler, N. B. Thornton, S. L. Temkin, and K. S. Schanze, *J. Am. Chem. Soc.*, **117**, 7119 (1995).

95JA11582   B. D. Rossenaar, M. W. George, F. P. A. Johnson, D. J. Stufkens, J. J. Turner, and A. Vlcek, *J. Am. Chem. Soc.*, **117**, 11582 (1995).

95JA11813   R. V. Slone, D. I. Yoon, R. M. Callhoun, and J. T. Hupp, *J. Am. Chem. Soc.*, **117**, 11813 (1995).

95JBCS29   S. A. Moya, R. Pastene, R. Sartori, P. Dixneuf, and H. Le Bozec, *J. Braz. Chem. Soc.*, **6**, 29 (1995).

95JCS(CC)259   V. W. W. Yam, V. C. Y. Lau, and K. K. Cheung, *J. Chem. Soc., Chem. Commun.*, 259 (1995).

95JCS(CC)1191   V. W. W. Yam, K. K. W. Lo, K. K. Cheung, and R. Y. C. Kong, *J. Chem. Soc., Chem. Commun.*, 1191 (1995).

95JCS(D)3677   Y. F. Lee, J. R. Kirchhoff, R. M. Berger, and D. Gosztola, *J. Chem. Soc., Dalton Trans.*, 3677 (1995).

95JOM(492)165   B. D. Rossenaar, C. J. Kleverlaan, M. C. E. van de Ven, D. J. Stufkens, A. Oskam, J. Fraanje, and K. Goubitz, *J. Organomet. Chem.*, **492**, 165 (1995).

95JOM(493)153   B. D. Rossenaar, C. J. Kleverlaan, M. C. E. van de Ven, D. J. Stufkens, A. Oskam, J. Fraanje, and K. Goubitz, *J. Organomet. Chem.*, **493**, 153 (1995).

95JPC1961   Y. Wang, L. A. Lucia, and K. S. Schanze, *J. Phys. Chem.*, **99**, 1961 (1995).

95JPC4929   D. Ferraudi, M. Felix, E. Wolcan, T. Hsu, S. A. Moya, and J. Guerrera, *J. Phys. Chem.*, **99**, 4929 (1995).

95JPC11801   L. A. Lucia, Y. Wang, K. Nafisi, T. L. Netzel, and K. S. Schanze, *J. Phys. Chem.*, **99**, 11801 (1995).

95JPP(A)61   E. Ruiz, E. Wolcan, A. L. Caparelli, and M. R. Feliz, *J. Photochem. Photobiol. A: Chem.*, **A89**, 61 (1995).

95OM1115   G. J. Stor, F. Hartl, J. W. M. Van Outersterp, and D. J. Stufkens, *Organometallics*, **14**, 1115 (1995).

95OM2749   V. W. W. Yam, V. C. Y. Lau, and K. K. Cheung, *Organometallics*, **14**, 2749 (1995).

95OM3303   J. W. M. van Outersterp, F. Hartl, and D. J. Stufkens, *Organometallics*, **14**, 3303 (1995).

95OM4034   V. W. W. Yam, K. M. C. Cheung, V. W. M. Lee, K. K. W. Lo, and V. K. Cheung, *Organometallics*, **14**, 4034 (1995).

95RCPB565   F. Hartl, B. D. Rossenaar, G. J. Stor, and D. J. Stufkens, *Recl. Chim. Pays-Bays*, **114**, 565 (1995).

96BSCQ251   S. Moya, R. Pastene, A. J. Pardey, and P. Barialli, *Bol. Soc. Chil. Quim.*, **41**, 251 (1996).

96CEJ228   D. J. Stufkens and A. Vlcek, *Chem. Eur. J.*, **2**, 228 (1996).

96CEJ1556   M. P. Aarnts, D. J. Stufkens, M. P. Wilms, E. J. Baerends, A. Vlcek, I. P. Clark, M. W. George, and J. J. Turner, *Chem. Eur. J.*, **2**, 1556 (1996).

96CIC77   C. A. Bignozzi, J. R. Schoonover, and R. B. Dyer, *Comments Inorg. Chem.*, **18**, 77 (1996).

96CIC165   J. R. Schoonover, G. F. Strouse, K. M. Omberg, and R. B. Dyer, *Comments Inorg. Chem.*, **18**, 165 (1996).

96CRV759   V. Balzani, A. Juris, M. Venturi, S. Campagna, and S. Serroni, *Chem. Rev.*, **96**, 759 (1996).

96CRV2063   D. H. Gibson, *Chem. Rev.*, **96**, 2063 (1996).

96IC273   J. R. Schoonover, R. B. Dyer, G. F. Strouse, W. D. Bates, P. Chen, and T. J. Meyer, *Inorg. Chem.*, **35**, 273 (1996).

96IC1421   S. Trummell, P. A. Goodson, and B. P. Sullivan, *Inorg. Chem.*, **35**, 1421 (1996).

96IC2032   C. A. Berg-Brennan, D. I. Yoon, R. V. Slone, A. P. Kazala, and J. T. Hupp, *Inorg. Chem.*, **35**, 2032 (1996).

96IC2242   J. A. Treadway, B. Loeb, R. Lopez, P. Anderson, F. R. Keene, and T. J. Meyer, *Inorg. Chem.*, **35**, 2242 (1996).

96IC2902   B. D. Rossenaar, D. J. Stufkens, and A. Vlcek, *Inorg. Chem.*, **35**, 2902 (1996).

96IC4096   R. V. Slone, J. T. Hupp, C. L. Stern, and T. E. Albrecht-Schmitt, *Inorg. Chem.*, **35**, 4096 (1996).

96IC6194   B. D. Rossenaar, F. Hartl, and D. J. Stufkens, *Inorg. Chem.*, **35**, 6194 (1996).

96ICA(247)215   B. D. Rossenaar, D. J. Stufkens, A. Oskam, J. Fraanje, and K. Goubitz, *Inorg. Chim. Acta*, **247**, 215 (1996).

96ICA(247)247   B. D. Rossenaar, D. J. Stufkens, and A. Vlcek, *Inorg. Chim. Acta*, **247**, 247 (1996).

96ICA(250)5   B. D. Rossenaar, E. Lindsay, D. J. Stufkens, and A. Vlcek, *Inorg. Chim. Acta*, **250**, 5 (1996).

96JA9782   J. P. Claude, D. S. Williams, and T. J. Meyer, *J. Am. Chem. Soc.*, **118**, 9782 (1996).

96JCS(CC)1587   I. P. Clark, M. W. George, F. P. A. Johnson, and J. J. Turner, *J. Chem. Soc., Chem. Commun.*, 1587 (1996).

96JCS(CC)2329   E. W. Abel, A. Gelling, K. G. Orrell, A. G. Osborne, and V. Sik, *J. Chem. Soc., Chem. Commun.*, 2329 (1996).

96JCS(D)203   A. Gelling, K. G. Orrell, A. G. Osborne, V. Sik, M. B. Hursthouse, and S. J. Coles, *J. Chem. Soc., Dalton Trans.*, 203 (1996).

96JCS(D)1411   P. K. K. Ho, K. K. Cheung, S. M. Peng, and C. M. Che, *J. Chem. Soc., Dalton Trans.*, 1411 (1996).

96JOM(517)217   J. T. Lin, S. S. Sun, J. J. Wu, Y. C. Liaw, and K. J. Lin, *J. Organomet. Chem.*, **517**, 217 (1996).

96JPC5408   Y. Wang and K. S. Schanze, *J. Phys. Chem.*, **100**, 5408 (1996).

96JPC15145   S. L. Mecklenburg, K. A. Opperman, P. Chen, and T. J. Meyer, *J. Phys. Chem.*, **100**, 15145 (1996).

96JPC18607   C. J. Kleverlaan, D. M. Martino, H. van Willingen, D. J. Stufkens, and A. Oskam, *J. Phys. Chem.*, **100**, 18607 (1996).

96JPP(A)171   H. Hori, F. P. A. Johnson, K. Koike, O. Ishitani, and T. Ibusuki, *J. Photochem. Photobiol. A: Chem.*, **A96**, 17 (1996).

96MI1   Molecular Level Artificial Photosynthetic Materials (G. J. Meyer, ed.), *Progr. Inorg. Chem.*, **44**(1996).

96NJC791   N. B. Thornton and K. S. Schanze, *New J. Chem.*, **20**, 791 (1996).

96OM236   A. Klein, C. Vogler, and W. Kaim, *Organometallics*, **15**, 236 (1996).

96OM1740   V. W. W. Yam, V. C. Y. Lau, and K. K. Cheung, *Organometallics*, **15**, 1740 (1996).

96OM3374   F. P. A. Johnson, M. W. George, F. Hartl, and J. J. Turner, *Organometallics*, **15**, 3374 (1996).

96OM3463   S. A. Moya, R. Schmidt, R. Pastene, R. Sartori, U. Muller, and G. Frenzen, *Organometallics*, **15**, 3463 (1996).

96OM5442   S. Ramdeehul, L. Barloy, J. A. Osborn, A. de Cian, and J. Fischer, *Organometallics*, **15**, 5442 (1996).

97AB179   X. Q. Guo, F. N. Castellano, L. Li, H. Szmacinski, J. R. Lakowicz, and J. Sipior, *Anal. Biochem.*, **254**, 179 (1997).

97CCR(160)1   A. M. W. C. Thompson, *Coord. Chem. Rev.*, **160**, 1 (1997).

97CCR(165)239   J. R. Schoonover, C. A. Bignozzi, and T. J. Meyer, *Coord. Chem. Rev.*, **165**, 239 (1997).

97CIC67   V. Sutin, C. Creutz, and E. Fujita, *Comments Inorg. Chem.*, **19**, 67 (1997).

97IC2   R. Argazzi, C. A. Bignozzi, T. A. Heimer, and G. J. Meyer, *Inorg. Chem.*, **36**, 2 (1997).

97IC2601   R. L. Cleary, K. J. Byrom, D. A. Bardwell, J. C. Jeffery, M. D. Ward, G. Calogera, N. Armaroli, L. Flamingi, and F. Barigeletti, *Inorg. Chem.*, **36**, 2601 (1997).

97IC5329   J. M. Kern, J. P. Sauvage, J. L. Weidmann, N. Armaroli, L. Flamingi, P. Ceroni, and V. Balzani, *Inorg. Chem.*, **36**, 5329 (1997).

97IC5422   R. V. Slone and J. T. Hupp, *Inorg. Chem.*, **36**, 5422 (1997).

97IC6224   L. A. Lucia, K. Abboud, and K. S. Schanze, *Inorg. Chem.*, **36**, 6224 (1997).

97JCE685   Y. Shen and B. P. Sullivan, *J. Chem. Educ.*, **74**, 685 (1997).

97JCS(CC)1593   R. Ziessel, A. Juris, and M. Venturi, *J. Chem. Soc., Chem. Commun.*, 1593 (1997).

97JCS(CC)2375   H. K. Kim, P. Lincoln, B. Norden, and E. Tuite, *J. Chem. Soc., Chem. Commun.*, 2375 (1997).

97JCS(D)1019   H. Hori, F. P. A. Johnson, K. Koike, K. Takeuchi, T. Ibusuki, and O. Ishitani, *J. Chem. Soc., Dalton Trans.*, 1019 (1997).

97JCS(D)2067   V. W. W. Yam, K. K. W. Lo, K. K. Cheung, and R. Y. C. Kong, *J. Chem. Soc., Dalton Trans.*, 2067 (1997).

97JCS(P2)2569   T. Scheiring, A. Klein, and W. Kaim, *J. Chem. Soc., Perkin Trans. II*, 2569 (1997).

97JOM(530)169   H. Hori, K. Koike, M. Ishizuka, K. Takeuchi, T. Ibusuki, and O. Ishitani, *J. Organomet. Chem.*, **530**, 169 (1997).

97JPC(A)9531   K. M. Omberg, J. R. Schoonover, and T. J. Meyer, *J. Phys. Chem.*, **A101**, 9531 (1997).

97JPC(B)3174   T. T. Ehler, N. Malmberg, K. Carron, B. P. Sullivan, and L. J. Noe, *J. Phys. Chem.*, **B101**, 3174 (1997).

97JPP(A)231   C. J. Kleverlaan, F. Hartl, and D. J. Stufkens, *J. Photochem. Photobiol.*, **A102**, 231 (1997).

97OM1729   V. W. W. Yam, K. M. C. Wong, and K. K. Cheung, *Organometallics*, **16**, 1729 (1997).

97OM4421   D. H. Gibson, B. A. Sleadd, M. S. Mashuta, and J. F. Richardson, *Organometallics*, **16**, 4421 (1997).

97OM4675   B. D. Rossenaar, F. Hartl, D. J. Stufkens, C. Amatore, E. Maisonhaute, and J. N. Verpeaux, *Organometallics*, **16**, 4675 (1997).

97OM5724   K. Koike, H. Hori, M. Ishizuka, J. R. Westwell, K. Takeuchi, T. Ibusuki, K. Enjouji, H. Konno, K. Sakamoto, and O. Ishitani, *Organometallics*, **16**, 5724 (1997).

98AC632   X. Q. Guo, F. N. Castellano, L. Li, and J. R. Lakowicz, *Anal. Chem.*, **70**, 632 (1998).

98ACR26   V. Balzani, S. Campagna, G. Denti, A. Juris, S. Serroni, and M. Venturi, *Acc. Chem. Res.*, **31**, 26 (1998).

98AX(C)1596   S. Belanger, J. T. Hupp, and C. L. Stern, *Acta Crystallogr.*, **C54**, 1596 (1998).

98CCR(169)201 J. C. Vites and M. M. Lynam, *Coord. Chem. Rev.*, **169**, 201 (1998).

98CCR(171)221 R. V. Slone, K. D. Benkstein, S. Belanger, J. T. Hupp, I. A. Guzei, and A. L. Rheingold, *Coord. Chem. Rev.*, **171**, 221 (1998).

98CCR(177)127 D. J. Stufkens and A. Vlcek, *Coord. Chem. Rev.*, **177**, 127 (1998).

98CCR(177)201 M. W. George and J. J. Turner, *Coord. Chem. Rev.*, **177**, 201 (1998).

98CCR(178)1299 J. C. Chambron, J. P. Collin, J. O. Dalbavie, C. O. Dietrich-Buchecker, F. Heitz, F. O. Lobel, N. Solladie, and J. P. Sauvage, *Coord. Chem. Rev.*, **178–180**, 1299 (1998).

98CEJ406   N. Armaroli, F. E. Diederich, C. O. Dietrich-Buchecker, L. Flamingi, J. F. Nierengarten, and J. P. Sauvage, *Chem. Eur. J.*, **4**, 406 (1998).

98CIC165   J. R. Schoonover, G. F. Strouse, K. M. Omberg, and R. B. Dyer, *Comments Inorg. Chem.*, **18**, 165 (1998).

98CRV1335   J. R. Schoonover and G. F. Strouse, *Chem. Rev.*, **98**, 1335 (1998).

98IC1121   P. J. Connors, D. Tzalis, A. L. Dunnick, and Y. Tor, *Inorg. Chem.*, **37**, 1121 (1998).

98IC2598   J. R. Schoonover, A. P. Shreve, R. P. Dyer, R. L. Cleary, M. D. Ward, and C. A. Bignozzi, *Inorg. Chem.*, **37**, 2598 (1998).

98IC2618   E. Schutte, J. B. Helms, S. M. Woessner, J. Bowen, and B. P. Sullivan, *Inorg. Chem.*, **37**, 2618 (1998).

98IC2806   G. Ferraudi and M. R. Feliz, *Inorg. Chem.*, **37**, 2806 (1998).

98IC5061   R. Ziessel, A. Juris, and M. Venturi, *Inorg. Chem.*, **37**, 5061 (1998).

98IC5406   S. M. Woessner, J. B. Helms, Y. Shen, and B. P. Sullivan, *Inorg. Chem.*, **37**, 5406 (1998).

98IC5664   S. Berger, A. Klein, W. Kaim, and J. Fiedler, *Inorg. Chem.*, **37**, 5664 (1998).

98IC6244   A. Rosa, G. Riccardi, E. J. Baerends, and D. J. Stufkens, *Inorg. Chem.*, **37**, 6244 (1998).

98JA5480   P. Ceromi, F. Paolucci, C. Paradisi, A. Juris, S. Roffia, S. Serroni, S. Campagna, and A. J. Bard, *J. Am. Chem. Soc.*, **120**, 5480 (1998).

98JA10871   C. J. Kleverlaan, D. J. Stufkens, I. P. Clark, M. W. George, J. J. Turner, D. M. Martino, H. van Willingen, and A. Vlcek, *J. Am. Chem. Soc.*, **120**, 10871 (1998).

98JA11200   D. H. Gibson and X. Yin, *J. Am. Chem. Soc.*, **120**, 11200 (1998).

98JA12982   K. D. Benkstein, J. T. Hupp, and C. L. Stern, *J. Am. Chem. Soc.*, **120**, 12982 (1982).

98JCS(CC)135   V. W. W. Yam, K. M. C. Wong, and K. K. Cheung, *J. Chem. Soc., Chem. Commun.*, 135 (1998).

98JCS(CC)777   V. W. W. Yam and W. K. M. Fung, *J. Chem. Soc., Chem. Commun.*, 777 (1998).

98JCS(CC)825   P. D. Beer and S. W. Dent, *J. Chem. Soc., Chem. Commun.*, 825 (1998).

98JCS(CC)2121 V. W. W. Yam, S. H. F. Chong, and K. K. Cheung, *J. Chem. Soc., Chem. Commun.*, 2121 (1998).

98JCS(D)185   M. R. Waterland, T. J. Simpson, K. C. Gordon, and A. K. Burrell, *J. Chem. Soc., Dalton Trans.*, 185 (1998).

98JCS(D)609   M. R. Waterland, K. C. Gordon, J. J. McGarvey, and P. M. Jayaweera, *J. Chem. Soc., Dalton Trans.*, 609 (1998).

98JCS(D)937 A. Gelling, K. G. Orrell, A. G. Osborne, and V. Sik, *J. Chem. Soc., Dalton Trans.*, 937 (1998).

98JCS(D)1461 V. W. W. Yam, V. C. Y. Lau, and L. X. Wu, *J. Chem. Soc., Dalton Trans.*, 1461 (1998).

98JCS(D)2893 F. W. M. Wahlenmont, M. V. Rajasekharan, H. U. Gudel, S. C. Capelli, J. Hauser, and H. B. Burgi, *J. Chem. Soc., Dalton Trans.*, 2893 (1998).

98JMC89 V. W. W. Yam, V. C. Y. Lau, K. Z. Wang, K. K. Cheung, and C. H. Huang, *J. Mater. Chem.*, **8**, 89 (1998).

98JOM(598)136 T. Scheiring, W. Kaim, and J. Fiedler, *J. Organomet. Chem.*, **598**, 136 (1998).

98JPC(A)45 J. A. Brozik and G. A. Crosby, *J. Phys. Chem.*, **A102**, 45 (1998).

98JPC(A)3042 P. Chen, R. A. Palmer, and T. J. Meyer, *J. Phys. Chem.*, **A102**, 3042 (1998).

98JPC(A)5577 K. S. Schanze, L. A. Lucia, M. Cooper, K. A. Walters, H. F. Ji, and O. Sabina, *J. Phys. Chem.*, **A102**, 5577 (1998).

98JPC(B)4759 F. Paolucci, M. Marcaccio, C. Paradisi, S. Roffa, C. A. Bignozzi, and C. Amatore, *J. Phys. Chem.*, **B102**, 4759 (1998).

98MI1 K. S. Schanze and K. A. Walters, in *"Photoinduced Electron Transfer in Metal-Organic Dyads"* (V. Ramamurthy and K. S. Schanze, eds.), Vol. 2, p. 75, Marcel Dekker, New York (1998).

98OM2440 V. W. W. Yam, K. Z. Wang, C. R. Wang, Y. Yang, and K. K. Cheung, *Organometallics*, **17**, 2440 (1998).

98OM2689 D. H. Gibson, B. A. Sleadd, X. Yin, and A. Vij, *Organometallics*, **17**, 2689 (1998).

98POL2289 S. A. Moya, R. Guerrero, R. Pastene, A. J. Pardey, and P. Baricelli, *Polyhedron*, **17**, 2289 (1998).

99AGE2722 H. S. Joshi, P. Jamshidi, and Y. Tor, *Angew. Chem., Int. Ed. Engl.*, **38**, 2722 (1999).

99AGE366 B. J. Coe, S. Houbrechts, J. Asselberghs, and A. Persoons, *Angew. Chem., Int. Ed. Engl.*, **38**, 366 (1999).

99AX(C)913 W. B. Connick, A. J. de Bilio, W. P. Schaeffer, and H. B. Gray, *Acta Crystallogr.*, **C55**, 913 (1999).

99BSCQ423 S. A. Moya, R. Pastene, H. Le Bozec, P. Baricelli, and J. A. Pardey, *Bol. Soc. Chil. Quim.*, **44**, 423 (1999).

99CEJ2464 B. J. Coe, *Chem. Eur. J.*, **5**, 2464 (1999).

99CRV2777 S. S. Sun and A. J. Lees, *Chem. Rev.*, **99**, 2777 (1999).

99IC606 B. W. Pfennig, J. L. Cohen, I. Sosnowski, N. M. Novotny, and D. M. Ho, *Inorg. Chem.*, **38**, 606 (1999).

99IC2924 R. Lopez, A. M. Levia, F. Zuloaga, B. Loeb, E. Norambuena, K. M. Omberg, J. R. Schoonover, D. Striplin, M. Devenney, and T. J. Meyer, *Inorg. Chem.*, **38**, 2924 (1999).

99IC4181 S. S. Sun and A. J. Lees, *Inorg. Chem.*, **38**, 4181 (1999).

99IC4378 S. M. Woessner, J. B. Helms, K. M. Lantzky, and B. P. Sullivan, *Inorg. Chem.*, **38**, 4378 (1999).

99IC4380 S. M. Woessner, J. B. Helms, J. F. Houlis, and B. P. Sullivan, *Inorg. Chem.*, **38**, 4380 (1999).

99IC4382 D. S. Tyson and F. N. Castellano, *Inorg. Chem.*, **38**, 4382 (1999).

99ICA(284)61 C. J. Kleverlaan and D. J. Stufkens, *Inorg. Chim. Acta*, **284**, 61 (1999).

99ICA(288)150 M. L. Leirer, G. Knor, and A. Vogler, *Inorg. Chim. Acta*, **288**, 150 (1999).

99JA557 S. Belanger, J. T. Hupp, C. L. Stern, R. V. Slone, D. F. Watson, and T. G. Carrell, *J. Am. Chem. Soc.*, **121**, 557 (1999).

99JCS(CC)1013   V. W. W. Yam, S. H. F. Chong, K. M. C. Wong, and K. K. Cheung, *J. Chem. Soc., Chem. Commun.*, 1013 (1999).

99JCS(CC)1411   D. H. Gibson and X. Yin, *J. Chem. Soc., Chem. Commun.*, 1411 (1999).

99JCS(CC)1755   P. D. Beer, V. Timoshenko, M. Maestri, P. Passaniti, and V. Balzani, *J. Chem. Soc., Chem. Commun.*, 1755 (1999).

99JCS(D)3729   D. H. Thompson, J. R. Schoonover, T. J. Meyer, R. Argazzi, and C. A. Bignozzi, *J. Chem. Soc., Dalton Trans.*, 3729 (1999).

99JCS(D)4487   J. H. Jung, T. A. Albright, D. M. Hoffman, and T. R. Lee, *J. Chem. Soc., Dalton Trans.*, 4487 (1999).

99JOM(599)112   J. H. Jung, D. M. Hoffman, and T. R. Lee, *J. Organomet. Chem.*, **599**, 112 (1999).

99MI1   W. E. Jones, L. Hermans, and B. Jiang, in "*Multimetallic and Macromolecular Inorganic Photochemistry*" (V. Ramamurthy and K. S. Schanze, eds.), Vol. 4, Marcel Dekker, New York (1999) Chapter 1.

99MI2   M. Y. Ogawa, in "*Multimetallic and Macromolecular Inorganic Photochemistry*" (V. Ramamurthy and K. S. Schanze, eds.), Vol. 4, Marcel Dekker, New York (1999) Chapter 3.

99OM5252   V. W. W. Yam, Y. Yang, H. P. Yang, and K. K. Cheung, *Organometallics*, **19**, 5252 (1999).

99PAC1753   J. R. Winkler, A. J. Di Bilio, N. A. Farrow, J. H. Richards, and H. B. Gray, *Pure Appl. Chem.*, **71**, 1753 (1999).

99POL1285   A. Gelling, K. G. Orrell, A. G. Osborne, and V. Sik, *Polyhedron*, **18**, 1285 (1999).

00AGE2891   K. D. Benkstein, J. T. Hupp, and C. L. Stern, *Angew. Chem., Int. Ed. Engl.*, **39**, 2891 (2000).

00AX(C)963   B. O. Coe, C. I. McDonald, S. J. Coles, and M. B. Hursthouse, *Acta Crystallogr.*, **C56**, 963 (2000).

00BC533   J. D. Dattelbaum, O. O. Abugo, and J. R. Lakowicz, *Bioconjug. Chem.*, **11**, 533 (2000).

00CCR(205)311   P. D. Beer and J. Cadman, *Coord. Chem. Rev.*, **205**, 311 (2000).

00IC485   D. J. Liard and A. Vlcek, *Inorg. Chem.*, **39**, 485 (2000).

00IC1344   S. S. Sun, A. S. Silva, I. M. Brinn, and A. J. Lees, *Inorg. Chem.*, **39**, 1344 (2000).

00IC1817   F. W. M. Vanhelmont and J. T. Hupp, *Inorg. Chem.*, **39**, 1817 (2000).

00IC2777   K. Koike, J. Tanabe, S. Toyama, K. Sakamoto, J. R. Westwell, F. P. A. Johnson, H. Hori, H. Saitoh, and O. Ishitani, *Inorg. Chem.*, **39**, 2777 (2000).

00IC3107   S. Bernhard, K. M. Omberg, G. F. Strouse, and J. R. Schoonover, *Inorg. Chem.*, **39**, 3107 (2000).

00IC3590   A. Juris, L. Prodi, A. Harriman, R. Ziessel, M. Hissler, A. El-Ghayoury, F. Wu, E. C. Riesgo, and R. P. Thummel, *Inorg. Chem.*, **39**, 3590 (2000).

00IC4977   H. Hartmann, S. Berger, R. Winter, J. Fiedler, and W. Kaim, *Inorg. Chem.*, **49**, 3977 (2000).

00ICA(299)231   R. M. Carlos, I. A. Carlos, B. S. L. Neto, and M. G. Neumann, *Inorg. Chim. Acta*, **299**, 231 (2000).

00ICC188   H. Kunkely and A. Vogler, *Inorg. Chem. Commun.*, **3**, 188 (2000).

00JA8956   S. S. Sun and A. J. Lees, *J. Am. Chem. Soc.*, **122**, 8956 (2000).

00JCS(CC)188   H. Kunkely and A. Vogler, *J. Chem. Soc., Chem. Commun.*, 188 (2000).

00JCS(CC)201   S. S. Sun, E. Robson, N. Dunwoody, A. S. Silva, and I. M. Brinn, *J. Chem. Soc., Chem. Commun.*, 201 (2000).

00JCS(CC)1211   K. Chichak and N. R. Branda, *J. Chem. Soc., Chem. Commun.*, 1211 (2000).

00JCS(CC)1687   S. S. Sun and A. J. Lees, *J. Chem. Soc., Chem. Commun.*, 1687 (2000).

00JCS(CC)1865   J. D. Lewis, R. N. Perutz, and J. N. Moore, *J. Chem. Soc., Chem. Commun.*, 1865 (2000).

00JCS(D)2599   L. S. Kelso, T. A. Smith, A. C. Schutz, P. C. Junk, R. N. Warrener, K. P. Chiggino, and F. R. Keene, *J. Chem. Soc., Dalton Trans.*, 2599 (2000).

00JOM(598)55   Y. B. Dong, L. Yang, K. K. Cheung, and A. Mayr, *J. Organomet. Chem.*, **598**, 55 (2000).

00JOM(598)136   T. Scheiring, W. Kaim, and J. Fiedler, *J. Organomet. Chem.*, **598**, 136 (2000).

00JPC(A)4291   Y. Wang, J. B. Asbury, and T. Lian, *J. Phys. Chem.*, **A104**, 4291 (2000).

00JPC(A)9281   E. Wolcan and G. Ferraudi, *J. Phys. Chem.*, **A104**, 9281 (2000).

00JPP(A)67   R. M. Carlos and M. G. Neumann, *J. Photochem. Photobiol.*, **A131**, 67 (2000).

00OM1820   V. W. W. Yam, C. C. Ko, L. X. Wu, K. M. C. Wong, and K. K. Cheung, *Organometallics*, **19**, 1820 (2000).

00OM5092   V. W. W. Yam, S. H. F. Chong, C. C. Ko, and K. K. Cheung, *Organometallics*, **19**, 5092 (2000).

01CCR(211)163   D. R. Striplin and G. A. Cosby, *Coord. Chem. Rev.*, **211**, 163 (2001).

01CCR(216)127   M. D. Ward and F. Barigiletti, *Coord. Chem. Rev.*, **216**, 127 (2001).

01CCR(219)545   V. Balzani, P. Ceroni, A. Juris, M. Venturi, F. Puntariero, S. Campagna, and S. Seroni, *Coord. Chem. Rev.*, **219–221**, 545 (2001).

01CEJ4358   P. K. Ng, X. Gong, S. H. Chan, L. S. M. Lam, and W. K. Chan, *Chem. Eur. J.*, **7**, 4358 (2001).

01CRV2655   K. D. Demadis, C. M. Hartshorn, and T. J. Meyer, *Chem. Rev.*, **101**, 2655 (2001).

01IC5056   A. S. DelNegro, S. M. Woessner, B. P. Sullivan, D. M. Dattelbaum, and J. R. Schoonover, *Inorg. Chem.*, **40**, 5056 (2001).

01IC5343   A. C. Lees, C. J. Kleverlaan, C. A. Bignozzi, and J. G. Vos, *Inorg. Chem.*, **40**, 5343 (2001).

01IC5834   P. Ferreira, W. M. Xue, E. Bencze, E. Herdtweck, and F. E. Kuhn, *Inorg. Chem.*, **40**, 5834 (2001).

01IC6885   R. Argazzi, E. Bertolasi, C. Chiorboli, C. A. Bignozzi, M. K. Itokazu, and N. Y. M. Iha, *Inorg. Chem.*, **40**, 6885 (2001).

01ICA(312)7   S. Moya, R. Pastene, H. Le Bozec, P. J. Baricelli, A. J. Pardey, and J. Gimeno, *Inorg. Chim. Acta*, **312**, 7 (2001).

01ICA(313)149   M. K. Itokazu, A. S. Polo, D. L. A. de Faria, C. A. Bignozzi, and N. Y. M. Iha, *Inorg. Chim. Acta*, **313**, 149 (2001).

01JA3181   A. J. Di Bilio, B. R. Crane, W. A. Wehbi, C. N. Kiser, M. M. Abu-Omar, R. M. Carlos, J. H. Richards, J. R. Winkler, and H. B. Gray, *J. Am. Chem. Soc.*, **123**, 3181 (2001).

01JA11623   B. R. Crane, A. J. de Bilio, J. R. Winkler, and H. B. Gray, *J. Am. Chem. Soc.*, **123**, 11623 (2001).

01JCC323   A. Aranacabia, J. Concepcion, N. Daire, G. Leiva, A. M. Leiva, B. Loeb, R. del Rio, R. Dias, A. Francois, and M. Saldivia, *J. Coord. Chem.*, **54**, 323 (2001).

01JCS(CC)103   S. S. Sun and A. J. Lees, *J. Chem. Soc., Chem. Commun.*, 103 (2001).

01JCS(CC)277   S. Encinas, F. Barigelletti, A. M. Barthram, M. D. Ward, and S. Campagna, *J. Chem. Soc., Chem. Commun.*, 277 (2001).

01JCS(CC)789   V. W. W. Yam, *J. Chem. Soc., Chem. Commun.*, 789 (2001).

01JCS(CC)1514 O. Ishitani, K. Kanai, Y. Yamada, and K. Sakamoto, *J. Chem. Soc., Chem. Commun.*, 1514 (2001).

01JCS(CC)2082 D. H. Gibson and D. H. He, *J. Chem. Soc., Chem. Commun.*, 2082 (2001).

01JCS(CC)2540 P. de Wolf, S. L. Heath, and J. A. Thomas, *J. Chem. Soc., Chem. Commun.*, 2540 (2001).

01JCS(CC)2682 K. H. Anderson and A. G. Orpen, *J. Chem. Soc., Chem. Commun.*, 2682 (2001).

01JCS(D)1813 M. Riklin, D. Tran, X. Bu, L. E. Lovermann, and P. C. Ford, *J. Chem. Soc., Dalton Trans.*, 1813 (2001).

01JCS(D)2188 L. H. Uppadine, F. R. Keene, and P. D. Beer, *J. Chem. Soc., Dalton Trans.*, 2188 (2001).

01JCS(D)2634 K. K. W. Lo, D. C. M. Ng, W. K. Hui, and K. K. Cheung, *J. Chem. Soc., Dalton Trans.*, 2634 (2001).

01MI1 A. Vlcek, in *"Electron Transfer in Chemistry"* (D. Astruc ed.), Vol. 2, p. 804, Wiley-VCH, Weinheim (2001).

01OM557 K. J. Thomas, J. T. Lin, H. M. Lin, C. P. Chang, and C. H. Chuen, *Organometallics*, **20**, 557 (2001).

01OM2262 K. R. J. Thomas, J. T. Lin, Y. Y. Lin, C. Tsai, and S. S. Sun, *Organometallics*, **20**, 2262 (2001).

01OM2353 S. S. Sun and A. J. Lees, *Organometallics*, **20**, 2353 (2001).

01OM2842 J. Guerrero, O. E. Piro, E. Wolcan, M. R. Feliz, G. Ferraudi, and S. A. Moya, *Organometallics*, **20**, 2842 (2001).

01OM4911 V. W. W. Yam, Y. Yang, J. Zhang, B. W. K. Chu, and N. Zhu, *Organometallics*, **20**, 4911 (2001).

01POL791 W. M. Xue, F. E. Kuhn, and E. Herdtweck, *Polyhedron*, **20**, 791 (2001).

02AGE3858 E. Hevia, J. Perez, V. Riera, and D. Miguel, *Angew. Chem., Int. Ed. Engl.*, **41**, 3858 (2002).

02CCR(230)170 S. S. Sun and A. J. Lees, *Coord. Chem. Rev.*, **230**, 170 (2002).

02CEJ2314 N. Armaroli, G. Accorsi, D. Felder, and J. F. Nierengarten, *Chem. Eur. J.*, **8**, 2314 (2002).

02CEJ4510 E. Hevia, J. Perez, L. Riera, V. Riera, D. Miguel, L. del Rio, and S. Garcia-Granda, *Chem. Eur. J.*, **8**, 4510 (2002).

02CSR168 B. S. Brunschwig, C. Creutz, and N. Sutin, *Chem. Soc. Rev.*, **31**, 168 (2002).

02EJI357 K. Chichak, U. Jacquemard, and N. R. Branda, *Eur. J. Inorg. Chem.*, 357 (2002).

02EJI1100 C. D. Nunes, M. Pillinger, A. A. Valente, I. S. Goncalves, J. Rocha, P. Ferreira, and F. E. Kuhn, *Eur. J. Inorg. Chem.*, 1100 (2002).

02HCA1261 G. T. Ruiz, M. P. Jiliarena, P. O. Lezna, E. Wolcan, M. R. Feliz, and G. Ferraudi, *Helv. Chim. Acta*, **85**, 1261 (2001).

02IC40 K. K. W. Lo, W. K. Hui, D. C. M. Ng, and K. K. Cheung, *Inorg. Chem.*, **41**, 40 (2002).

02IC132 S. S. Sun, D. T. Tran, O. S. Odongo, and A. J. Lees, *Inorg. Chem.*, **41**, 132 (2002).

02IC359 A. Del Guerzo, S. Leroy, F. Fages, and R. H. Schmehl, *Inorg. Chem.*, **41**, 359 (2002).

02IC1662 D. R. Cary, N. P. Zaitseva, K. Gray, K. E. O'Day, C. B. Darrow, S. M. Lani, T. A. Peyser, J. H. Satcher, W. P. Van Antwerp, A. J. Nelson, and J. G. Reynolds, *Inorg. Chem.*, **41**, 1662 (2002).

02IC2909 K. A. Walters, Y. J. Kim, and J. T. Hupp, *Inorg. Chem.*, **41**, 2909 (2002).

02IC4673   E. Hevia, J. Perez, V. Riera, D. Miguel, S. Kassel, and A. Rheingold, *Inorg. Chem.*, **41**, 4673 (2002).

02IC5323   B. Manimaran, P. Thanasekaran, T. Rajendran, R. J. Lin, I. J. Chang, G. H. Lee, S. M. Peng, S. Rajagopal, and K. L. Lu, *Inorg. Chem.*, **41**, 5323 (2002).

02IC6071   D. M. Dattelbaum, K. M. Omberg, J. R. Schoonover, R. L. Martin, and T. J. Meyer, *Inorg. Chem.*, **41**, 6071 (2002).

02JA4554   S. J. Lee and W. Lin, *J. Am. Chem. Soc.*, **124**, 4554 (2002).

02JA9344   K. K. W. Lo, W. K. Hui, and D. C. M. Ng, *J. Am. Chem. Soc.*, **124**, 9344 (2002).

02JA11448   K. Koike, N. Okoshi, H. Hori, K. Takeuchi, O. Ishitani, H. Tsubaki, I. P. Clark, M. W. George, F. P. A. Johnson, and J. J. Turner, *J. Am. Chem. Soc.*, **124**, 11448 (2002).

02JCS(CC)950   A. Gabrielsson, F. Hartl, J. R. L. Smith, and R. N. Perutz, *J. Chem. Soc., Chem. Commun.*, 950 (2002).

02JCS(CC)1814   E. Hevia, J. Perez, V. Riera, and D. Miguel, *J. Chem. Soc., Chem. Commun.*, 1814 (2002).

02JCS(D)701   A. Vlcek, I. R. Farrell, D. J. Liard, P. Matousek, M. Towrie, A. W. Parker, D. C. Grills, and M. W. George, *J. Chem. Soc., Dalton Trans.*, 701 (2002).

02JCS(D)2194   E. Wolcan, G. Torchia, J. Tocho, O. E. Piro, P. Juliarena, G. Ruiz, and M. R. Feliz, *J. Chem. Soc., Dalton Trans.*, 2194 (2002).

02JCS(D)4732   C. Metcalfe, C. Spey, H. Adams, and J. A. Thomas, *J. Chem. Soc., Dalton Trans.*, 4732 (2002).

02JPC(A)4519   D. M. Dattelbaum and T. J. Meyer, *J. Phys. Chem.*, **A106**, 4519 (2002).

02JPC(A)7795   J. P. Claude, K. M. Ombert, D. S. Williams, and T. J. Meyer, *J. Phys. Chem.*, **A106**, 7795 (2002).

02JPC(A)12202   J. D. Lewis, R. N. Perutz, and J. N. Moore, *J. Phys. Chem.*, **A106**, 12202 (2002).

02OM39   S. S. Sun and A. J. Lees, *Organometallics*, **21**, 39 (2002).

02OM1750   E. Hevia, J. Perez, L. Riera, V. Riera, and D. Miguel, *Organometallics*, **21**, 1750 (2002).

02OM1966   E. Hevia, J. Perez, V. Riera, and D. Miguel, *Organometallics*, **21**, 1966 (2002).

02OM5312   E. Hevia, J. Perez, V. Riera, and D. Miguel, *Organometallics*, **21**, 5312 (2002).

02POL439   S. Moya, J. Guerrero, R. Pastene, I. Azocar-Guzman, and A. J. Pardey, *Polyhedron*, **21**, 439 (2002).

03CCR(245)39   V. W. W. Yam, W. Y. Lo, C. H. Lam, W. K. M. Fung, K. M. C. Wong, V. C. Y. Lau, and N. Zhu, *Coord. Chem. Rev.*, **245**, 39 (2003).

03EJI449   J. L. Zuo, F. F. de Biani, A. M. Santos, K. Kohler, and F. E. Kuhn, *Eur. J. Inorg. Chem.*, 449 (2003).

03EJI4035   V. W. W. Yam, B. Li, Y. Yang, B. W. K. Chu, K. M. C. Wong, and K. K. Cheung, *Eur. J. Inorg. Chem.*, 4035 (2003).

03IC1248   S. Ranjan, S. Y. Lin, K. C. Hwang, Y. Chi, W. L. Ching, and C. S. Liu, *Inorg. Chem.*, **42**, 1248 (2003).

03IC3445   S. S. Sun, A. J. Lees, and P. Y. Zavalij, *Inorg. Chem.*, **42**, 3445 (2003).

03IC3516   B. Salignac, P. V. Grundler, S. Cayemittes, U. Frey, R. Scopelliti, A. Merbach, R. Hedinger, K. Hegetschweiler, R. Alberto, U. Prinz, G. Raabe, U. Kolle, and S. Hall, *Inorg. Chem.*, **42**, 3516 (2003).

03IC7086   K. M. C. Wong, S. C. F. Lam, C. C. Ko, N. Zhu, V. W. W. Yam, S. Rove, C. Lapinte, S. Fathallah, K. Costuas, S. Kahlal, and J. F. Halet, *Inorg. Chem.*, **42**, 7086 (2003).

03IC7995   D. J. Liard, C. J. Kleverlaan, and A. Vlcek, *Inorg. Chem.*, **42**, 7995 (2003).

03ICA(343)357   H. Kunkely and A. Vogler, *Inorg. Chim. Acta*, **343**, 357 (2003).

03JA3706   E. Hevia, J. Perez, V. Riera, D. Miguel, P. Campomanes, M. I. Menendez, T. L. Sordo, and S. Garcia-Granda, *J. Am. Chem. Soc.*, **125**, 3706 (2003).

03JA11976   Y. Hayashi, S. Kita, B. S. Braunschweig, and E. Fujita, *J. Am. Chem. Soc.*, **125**, 11976 (2003).

03JA14220   J. E. Miller, C. Gradinaru, B. R. Crane, A. J. Di Bilio, W. A. Wehbi, S. Un, J. R. Winkler, and H. B. Gray, *J. Am. Chem. Soc.*, **125**, 14220 (2003).

03JCS(CC)328   D. C. Gerbino, E. Hevia, D. Morales, M. E. N. Clemente, J. Perez, L. Riera, V. Riera, and D. Miguel, *J. Chem. Soc., Chem. Commun.*, 328 (2003).

03JCS(CC)2446   V. W. W. Yam, W. Y. Lo, and N. Zhu, *J. Chem. Soc., Chem. Commun.*, 2446 (2003).

03JCS(CC)2704   K. K. W. Lo, K. H. K. Tsang, W. K. Hui, and N. Zhu, *J. Chem. Soc., Chem. Commun.*, 2704 (1983).

03JCS(CC)2858   J. D. Lewis and J. N. Moore, *J. Chem. Soc., Chem. Commun.*, 2858 (2003).

03JOM(670)205   V. W. W. Yam, K. M. C. Wong, S. H. F. Chong, V. C. Y. Lau, S. C. F. Lam, L. Zhang, and K. K. Cheung, *J. Organomet. Chem.*, **670**, 205 (2003).

03JPC(A)4092   D. M. Dattelbaum, M. T. Itokazu, N. Y. M. Iha, and T. J. Meyer, *J. Phys. Chem.*, **A107**, 4092 (2003).

03JPP(A)27   M. K. Itokazu, A. S. Polo, and N. Y. M. Iha, *J. Photochem. Photobiol.*, **A160**, 27 (2003).

03OM257   E. Hevia, J. Perez, V. Riera, and D. Miguel, *Organometallics*, **22**, 257 (2003).

03OM337   D. H. Gibson, X. Yin, H. He, and M. S. Mashuta, *Organometallics*, **22**, 337 (2003).

03SMC143   E. Wolcan, G. Ferraudi, M. R. Veliz, R. V. Gomez, and L. Mikelsons, *Supramolec. Chem.*, **15**, 143 (2003).

04CEJ1765   L. Cuesta, D. C. Gerbino, E. Hevia, D. Morales, M. E. N. Clemente, J. Perez, L. Riera, V. Riera, D. Miguel, I. del Rio, and S. Garcia-Granda, *Chem. Eur. J.*, **10**, 1765 (2004).

04IC132   T. P. Ortiz, J. A. Marshall, L. A. Emmert, J. Yang, W. Choi, A. L. Costello, and J. A. Brozik, *Inorg. Chem.*, **43**, 132 (2004).

04IC1551   M. R. Feliz and G. Ferraudi, *Inorg. Chem.*, **43**, 1551 (2004).

04IC2013   C. R. Graves, M. L. Merlau, G. A. Morris, S. S. Sun, S. B. T. Nguyen, and J. T. Hupp, *Inorg. Chem.*, **43**, 2013 (2004).

04IC2043   O. S. Wenger, L. M. Henling, M. W. Day, J. R. Winkler, and H. B. Gray, *Inorg. Chem.*, **43**, 2043 (2004).

04IC4523   M. Busby, P. Matousek, M. Towrie, I. P. Clark, M. Motevalli, F. Hartl, and A. Vlcek, *Inorg. Chem.*, **43**, 4523 (2004).

04IC4994   M. Busby, A. Gabrielson, P. Matousek, M. Towrie, A. D. Di Bilio, H. B. Gray, and A. Vlcek, *Inorg. Chem.*, **43**, 4994 (2004).

04IC5961   L. Yang, A. M. Ren, J. K. Feng, X. D. Liu, Y. G. Ma, and H. X. Zhang, *Inorg. Chem.*, **43**, 5961 (2004).

04IC7636   E. Fujita and J. T. Muckerman, *Inorg. Chem.*, **43**, 7366 (2004).

04ICA167   M. Busby, D. J. Liard, M. Motevalli, H. Toms, and A. Vlcek, *Inorg. Chim. Acta*, **357**, 167 (2004).

04JA12734   V. W. W. Yam, C. C. Ko, and N. Zhu, *J. Am. Chem. Soc.*, **126**, 12734 (2004).

04JCS(D)1376 J. D. Lewis and J. N. Moore, *J. Chem. Soc., Dalton Trans.*, 1376 (2004).

04JOM1393   V. W. W. Yam, *J. Organomet. Chem.*, **689**, 1393 (2004).

04JOM1665   K. Tani, H. Sakurai, H. Fujii, and T. Hirao, *J. Organomet. Chem.*, **689**, 1665 (2004).

04JOM2486   R. A. Machado, D. Rivillo, A. J. Price, L. D'Ornelas, Y. De Sanctis, R. Atencio, T. Gonzalez, and E. Galarza, *J. Organomet. Chem.*, **689**, 2486 (2004).

04JOM2905   Y. Yamamoto, M. Shiotsuka, and S. Onaka, *J. Organomet. Chem.*, **689**, 2905 (2004).

04JPC(A)556   D. J. Liard, M. Busby, I. R. Farrell, P. Matousek, M. Towrie, and A. Vlcek, *J. Phys. Chem.*, **A108**, 556 (2004).

04JPC(A)2363 D. J. Liard, M. Busby, P. Matousek, M. Towrie, and A. Vlcek, *J. Phys. Chem.*, **A108**, 2363 (2004).

04NJC43   F. E. Kuhn, J. L. Zuo, F. F. de Biani, A. M. Santos, Y. Zhang, J. Zhao, A. Sandulache, and E. Herdtweck, *New J. Chem.*, **28**, 43 (2004).

04OM1098   K. K. W. Lo, J. S. Y. Lau, V. W. Y. Fong, and N. Zhu, *Organometallics*, **23**, 1098 (2004).

04OM3437   X. Zhang, X. Chen, and P. Chen, *Organometallics*, **23**, 3437 (2004).

04OM5096   M. Ferrer, L. Rodriguez, O. Rossell, J. C. Lima, P. Gomez-Sal, and A. Martin, *Organometallics*, **23**, 5096 (2004).

04PP203   R. M. Carlos, B. S. Lima-Neto, and M. G. Neumann, *Photochem. Photobiol.*, **80**, 203 (2004).

05ICA3735   C. Pereira, H. G. Ferreira, M. S. Schultz, J. Milanez, M. Izidoro, P. C. Leme, R. H. A. Santos, M. T. P. Gambardella, E. E. Castellano, B. S. Lima-Neto, and R. M. Carlos, *Inorg. Chim. Acta*, **358**, 3735 (2005).

05JCS(CC)116 L. Cuesta, E. Hevia, D. Morales, J. Perez, V. Riera, E. Rodriguez, and D. Miguel, *J. Chem. Soc., Chem. Commun.*, 116 (2005).

05OM1772   L. Cuesta, E. Hevia, D. Morales, J. Perez, V. Riera, M. Seitz, and D. Miguel, *Organometallics*, **24**, 1772 (2005).

06JOM1834   A. Kleineweischede and J. Mattay, *J. Organomet. Chem.*, **691**, 1834 (2006).

07AHC(93)185 A. P. Sadimenko, *Adv. Heterocycl. Chem.*, **93**, 185 (2007).

Aminoisoxazoles: Preparations and Utility in the Synthesis of Condensed Systems

V.P. Kislyi, E.B. Danilova, and V.V. Semenov

N.D. Zelinsky Institute of Organic Chemistry, Russian Academy of Sciences, 47 Leninsky prosp., 117913, Moscow, Russian Federation

I. Introduction

The chemistry of aminoisoxazoles (AI) is a subsection of isoxazole chemistry; nevertheless, they have special properties in many respects. The available amino group changes distinctly the properties and reactivity of the isoxazole ring, and some reactions and rearrangements that are unusual in isoxazole chemistry become typical. An amino group is convenient in combinatorial chemistry and facilitates the use of these compounds in modern drug discovery. The studies on AI have been mainly concerned with biological activities and the preparation of fused heterocyclic systems, such as isoxazolopyridines, isoxazolopyrimidines, and isoxazolodiazepines. Often, AI-based condensed heterocyclic systems are more biologically promising than AI themselves. Indeed, AI derivatives possess cytostatic (03PCT013517, 03MI2, 03PJC1001, 03BMCL1487, 03JMC1706), antibacterial (01MI2, 04PCT048385, 04GEP10256186), fungicidal (88FA753, 99FA90, 01MI1, 02MI1), herbicidal (91H1153, 02JAP(K)363171), immunological (03MI1), hypocholesterolemic (96JMC4382), and anticonvulsant activities (02MI2) among other activities (82JMC12, 87JHC1291, 94JMC2721, 95JMC1344, 00JMC3111). Thus the general reactivity and differences in the reactivity of AI isomers are of

ADVANCES IN HETEROCYCLIC CHEMISTRY
VOLUME 94 ISSN: 0065-2725 DOI: 10.1016/S0065-2725(06)94003-3

practical interest. There exist three isomers of AI, namely, 3-amino-, 4-amino-, and 5-aminoisoxazoles.

3-AI   **4-AI**   **5-AI**

Syntheses and some reactions of AI were briefly discussed in a few reviews on the general chemistry of isoxazoles (62HC1, 63AHC365, 79AHC147, 84CHC1, 91HC1) and on the ring openings of isoxazoles (75S20, 81KGS1155), but the subject has not been specifically covered in the literature. The last comprehensive review on isoxazole chemistry (62HC1, 91HC1) lacks a chapter regarding the methods for AI preparation or their reactions. The preparation of condensed systems was discussed in reviews on the chemistry of several related heterocyclic systems. For example, the synthesis of isoxazolopyrimidines was considered in the review dealing with the chemistry of condensed pyrimidines (82AHC1). Hence, a new review is needed. This survey includes references to publications up to the beginning of 2005.

II. Structure

Physical properties of simple AI have been surveyed (62HC1, 79AHC147, 91HC1). All AI are potentially tautomeric, and Beilstein categorizes 5-AI as the imino derivatives of the corresponding isoxazolones. Tautomerization was examined by all spectral methods to reach the conclusion that all AI exist solely as an amino form (61T51, 84CHC1). X-ray diffraction studies of 3-amino- (85CJC3012, 88JHC607, 02MI2, 05JA5512), 4-amino- (88BCJ2881), and 5-aminoisoxazoles (82AX(B)1845, 91H1913, 98JCX373) and AI complexes with metals (01JCX369, 04AX(E)330) were carried out. Molecular dimensions of AI and isoxazoles without amino groups (79AHC147, 84CHC1) are closely related. During some reactions, AI may rearrange (Section III.D). Specifically, in the case of trisubstituted isoxazoles, the identification of product structures by ^{1}H-NMR spectroscopy may be misleading (91H1765, 98FA513). X-ray determinations are necessary in such cases, but it is often difficult to grow the monocrystal. Hence, ^{13}C-NMR spectroscopy seems to be an indispensable tool (76JHC825, 80OMR235, 83JOC575, 84JOC3423, 85JCS(P1)1871, 85JHC1663, 87H2419, 87BCJ4480, 89JHC1335, 90GA725, 90JHC1481, 90JHC1617, 91H1765, 90H1823, 95JCS(P1)1153, 96TL3339, 02MC99, 02JOC8558, 05IZV461).

The differences in the chemical behavior of AI isomers are closely related to some physicochemical parameters, such as basicity and protonation site. Basicity of 4-AI is higher than that of 3-AI or 5-AI (Table 1), and there is a significant difference in the site of protonation. In 75% H_2SO_4, 4-amino-3,5-dimethylisoxazole is ca. 50% diprotonated. The diprotonation of 3-amino-5-methyl isoxazole starts in 75% H_2SO_4, while the diprotonation of 5-amino-3-methylisoxazole does not begin (89JCS(P2)1941). All the AI isomers are only monoprotonated in trifluoroacetic

Table 1.   Ionization constants of isoxazoles and aminoisoxazoles at 20 °C

Compound	pK_a	References
Isoxazole (unsubstituted)	−2.28	61T41
3,5-Dimethyl	−1.27	61T41
3-Amino-5-methyl	0.47	61T51
4-Amino-3,5-dimethyl	3.8	61T51
5-Amino-3-methyl	0.64	61T51
5-Amino-3,4-dimethyl	1.16	61T51
	1.28	83PJC1161

Table 2. ^{15}N-NMR spectra of aminoisoxazoles (89JCS(P2)1941)

	Solvent			
	DMSO		TFA[a]	
Compound	NH$_2$	N$_{ring}$	NH$_2$	N$_{ring}$
Isoxazole				
3-Amino-5-methyl	−357.1	−78.6	−341.0	−189.7
4-Amino-3,5-dimethyl	−378.2	−37.4	−372.7	−57.7
4-Amino-3,5-dimethyl, hydrochloride	−363.4	−30.9		
5-Amino-3-methyl	−344.7	−54.3	−333.4	−188.5

[a]Trifluoroacetic acid.

acid (H$_0$ trifluoroacetic acid, 3.03; H$_0$ 45% H$_2$SO$_4$, 2.85). From the study of their UV spectra (61T51) an assumption was made that 3-AI and 5-AI are protonated at the ring nitrogen atom, while 4-AI is protonated at the amino group. This assumption was confirmed 28 years ago (89JCS(P2)1941) by ^{15}N-NMR spectroscopy (Table 2).

As established by X-ray analysis, metals coordinate with the ring nitrogen of 3-AI similarly to protonation (01JCX369, 04ACE330). X-ray diffraction data is lacking for the complexes of 4-AI or 5-AI with metals. However, for 4-AI, the coordination of metal ions was ascribed to the amino group on the basis of their IR-spectra (77JNC45, 78SA(A)967). This great difference in the coordination sites of both protons and metal cations is likely to originate from dissimilar electron distributions on the isoxazole ring atoms for different AI isomers.

As noted earlier (79AHC147, 84CHC1), with isoxazoles the results of molecular orbital calculations have not provided agreement with experimental findings. Recently, some new calculations of the HOMO and LUMO orbitals and geometrical parameters for 5-AI have been performed by the AM1 method (03MI3, 03PJC1461, 04MI1). The simulated ^{13}C-NMR spectra were in good agreement with the observed spectra for AI containing no strong acceptor group adjacent to the amino group (Table 3). When such a substituent is available, the calculations deviate from the observed values in the range of 15–30 ppm, indicating that electron densities on the

Table 3. Selected ^{13}C-NMR spectra of aminoisoxazoles

R1	R2	R3	C-3	C-4	C-5	References
tBuOOCNH	H	Me	159.0	95.0	170.0	96TL3339, 91HET1765
			165	97.0	158.5	Calculated[a]
NH$_2$	Ph	NH$_2$	167.0	87.0	168.0	83JOC575
Ph	NH$_2$	COOH	155.5	134.0	140.0	90GA725
			149	118.7	144	Calculated[a]
Ph	NHCOCH$_2$CH$_2$CH$_2$		158.0	115.0	156.0	90JHC1617
Me	NHCOCH$_2$CHArCH$_2$		153.0	116.0	156.0	90HET1823
PhNHCO	NH$_2$	COPh	147.8	135.0	146.5	05IZV461
			151.3	117.3	135.3	Calculated[a]
Me	Me	NH$_2$	161.0	85.0	167.0	80OMR235, 84JOC3423
Ph	H	NH$_2$	164.0	78.0	169.0	02MC99
			162	85	172	Calculated[a]

[a]Calculations with ACD-Labs ^{13}C-NMR version 4.56.

isoxazole ring carbons differ markedly from normal values due to conjugation between the substituents.

III. Reactivity and Annulation of Another Ring to an Isoxazole Ring

A. REACTIVITY AT THE RING ATOMS

Numerous examples of nucleophilic substitution in isoxazoles have been reported; these reactions have been used for the preparation of AI (Section IV.B.2). A strong electron-donating effect of the amino group hampers the nucleophilic substitution, which is rarely observed in AI. Treatment of 5-chloro-4-AI **1** with the lithium salt of 2-aminoethanethiol gave 4-AI **2** (89MI1) (Scheme 1).

The conversion of 5-AI into the corresponding isoxazol-5-ones **3** occurs in 10% H$_2$SO$_4$ in aqueous alcohol (10 h, 100 °C, 80%) (54CB1175). The conversion of 3-AI to the corresponding isoxazol-3-ones proceeds more slowly (32G457). Under conditions where the complete conversion of 5-AI was attained, transformation of 3-AI did not begin (91H1153). The conversion of 4-AI to 4-hydroxyisoxazoles is possible only through diazotization followed by thermal decomposition of the corresponding diazonium salts (Section III.B) (Scheme 2).

Meanwhile, electrophilic substitution in AI is noticeably more feasible than that in isoxazoles without amino groups. However, examples of electrophilic substitution were reported only for 3-AI and 5-AI. There is no data for 4-AI on electrophilic substitution in positions 3 or/and 5, despite the strong electron-donating effect of the

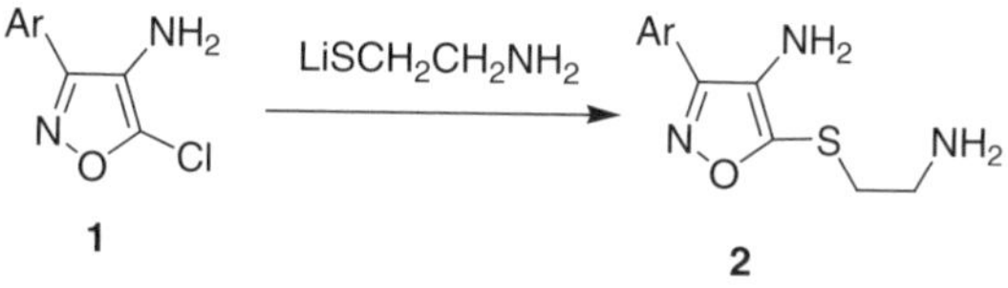
Ar
NH2
N
O
Cl
1
LiSCH2CH2NH2
Ar
NH2
N
O
S
NH2
2

Scheme 1

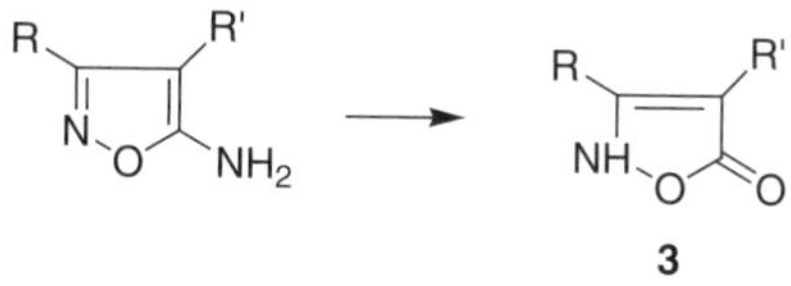
R
R'
N
O
NH2
R
R'
NH
O
O
3

Scheme 2

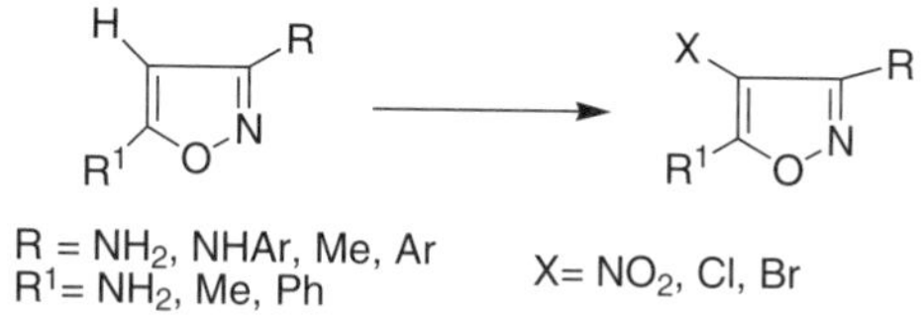
H
R
R1
O
N
X
R
R1
O
N
R = NH2, NHAr, Me, Ar
R1= NH2, Me, Ph
X= NO2, Cl, Br

Scheme 3

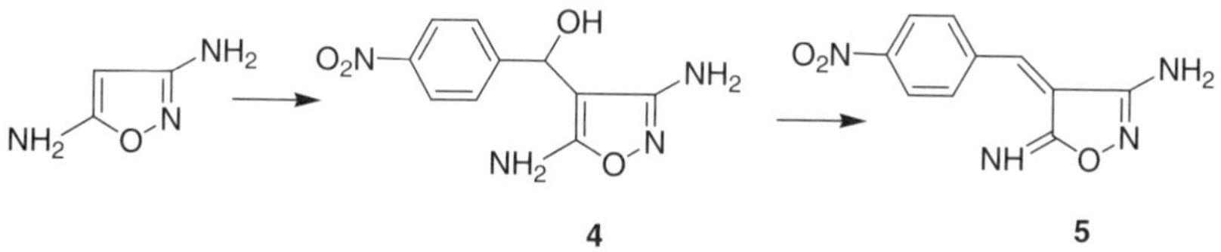
NH2
NH2
O
N
O2N
OH
NH2
NH2
O
N
4
O2N
NH2
NH
O
N
5

Scheme 4

amino group. For both 3-AI and 5-AI, nitration (37JA1486, 38JA1198, 68T4907), chlorination with SO_2Cl_2 (77JMC934), and bromination (63AHC365, 37JA1486, 38JA1198, 85JOC5723) were described (62HC1) (Scheme 3).

A few instances of substitution with C-electrophiles are also documented for 5-AI and 3,5-diaminoisoxazoles. In the case of highly activated 3,5-diaminoisoxazole, the reaction with 4-nitrobenzaldehyde gave benzyl alcohol **4** (65JOC2862) and then 5-iminoisoxazole **5** (53G192, 84H618), instead of the expected Schiff base (Scheme 4).

In a similar way, 5-AI reacts with ketones in acetic acid (84S1050), pyridinium salts (03TL391), or quinoline N-oxides in the presence of benzoyl chloride (78CPB2759), and also under the Vilsmeier conditions (77H51, 80CPB1832) to give the products **6**, **7**, **8**, and **9**, respectively (Scheme 5).

Unlike isoxazoles for which alkylation of the ring nitrogen is a well-known and widely used process, alkylation of AI has not been adequately studied. Common alkylating agents can attack either a ring nitrogen atom or an exocyclic amino group

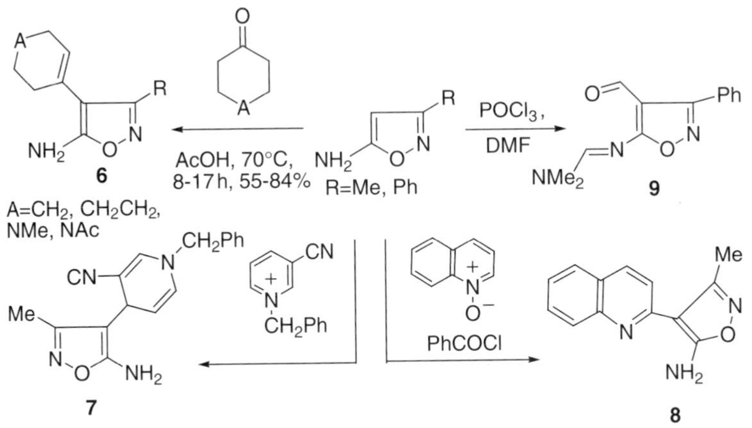

Scheme 5

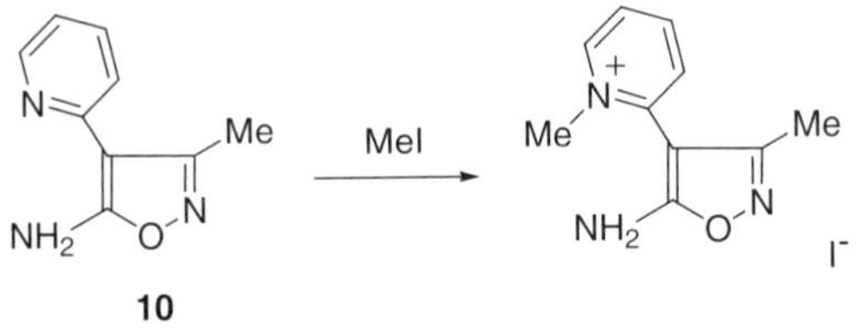

Scheme 6

(or both). On treatment with alkylating agents, alkylation of the ring nitrogen is more probable for 3-AI and 5-AI than for 4-AI, just as protonation of 3-AI and 5-AI involves the ring nitrogen, while protonation of 4-AI is directed to the amino group (Section II). However, 3(5)-amino-2-alkylisoxazolium salts were not formed in this way. 3-Amino-2-alkylisoxazolium salts were synthesized by replacing the chlorine atom in 3-chloro-2-isoxazolium salts (Section IV.B.2). On boiling in dichlorobenzene, these compounds lose methyl iodide to give 3-AI (84CPB530). One example of alkylation of 5-amino-3-methyl-4-pyrid-2-ylisoxazole **10** with methyl iodide involving only the pyridine nitrogen atom has been reported (98BMC1623) (Scheme 6).

These two examples, though indirect, show the significant stability of the ring nitrogen in AI against alkylation.

B. Reactivity of the Substituents

In the chemistry of AI, the amino group participates in the overwhelming majority of reactions. Few transformations of other substituents can take place when the amino group has not been protected. Numerous mono- and diacylated AI have been prepared for drug discovery by acylation of AI with carboxylic (34CB1062, 46G87,

Scheme 7

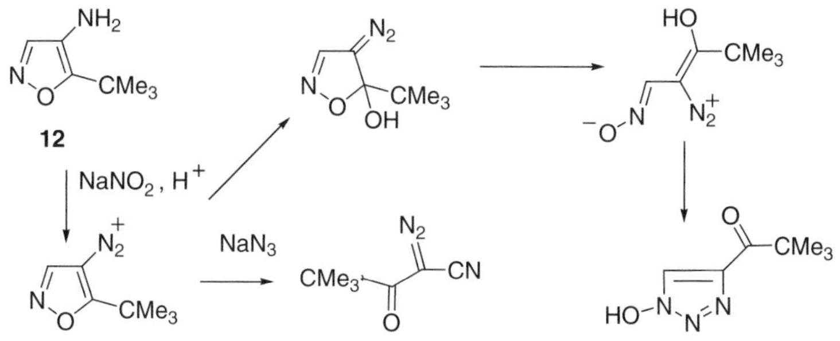

Scheme 8

53G192, 65JOC2862, 66CPB1277, 67T687, 68T4907, 77JMC934, 77JCS(P1)1616, 82JMC12 88FA753), sulfonic (66CPB1277, 82FA701, 95JMC1344, 00JMC3111), and phosphinic acids (90ZOB1778). In most cases, acylation of AI stops when simple acylated AI have been formed, but in the case of 3-AI, many rearrangements of 3-acetamidoisoxazoles are possible (Section III.D.1). There are many examples of the preparation of Schiff bases of AI (85JHC1663, 46G87, 56JPJ1311, 79M1387, 88IJC(B)686, 90BCJ1851), although aldehydes and ketones can also attack the carbon atom in position 4 of the isoxazole ring (Section III.A). N-Acylated AI and Schiff bases of AI are important intermediates in the syntheses of fused systems. The reduction of some acylated AI by $BH_3 \times SMe_2$ (90GA725) or LiAlH4 (78CPB2759) is a convenient route to the corresponding N-alkylated AI **11** (Scheme 7).

Diazotization of 4-AI and 3-AI yields stable diazoniaisoxazoles, which enter into typical reactions (62HC1), in particular, substitution of the iodo (21JCS1546, 77JMC934), azido (77JHC1299), or hydroxy group (in the presence of urea) (46GA87), and coupling with acetylacetone (21JCS1546). Only in the case of 3-unsubstituted 4-AI **12**, a mixture of diazo ketone **13** and N-hydroxytriazole **14** was formed by opening of 4-diazonium-isoxazole (83TL3149). The opening of 3-diazonium isoxazoles proceeds more easily then that of AI (Section III.D.1) (Scheme 8).

Diazotization of 5-AI proceeds in a different manner depending on the presence or the absence of an electron-withdrawing function in position 4 of the isoxazole ring. In the case of 5-AI containing no group of this type, the isoxazol-5-yl radical is formed giving rise to 5-iodoisoxazoles, isoxazol-5-ones, and 5-aryl-substituted isoxazoles (76HCA1705, 79AHC147). However, in the presence of an electron-withdrawing

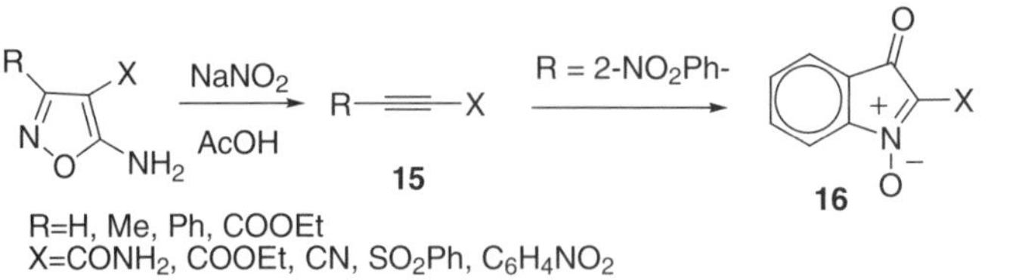

R=H, Me, Ph, COOEt
X=CONH$_2$, COOEt, CN, SO$_2$Ph, C$_6$H$_4$NO$_2$

Scheme 9

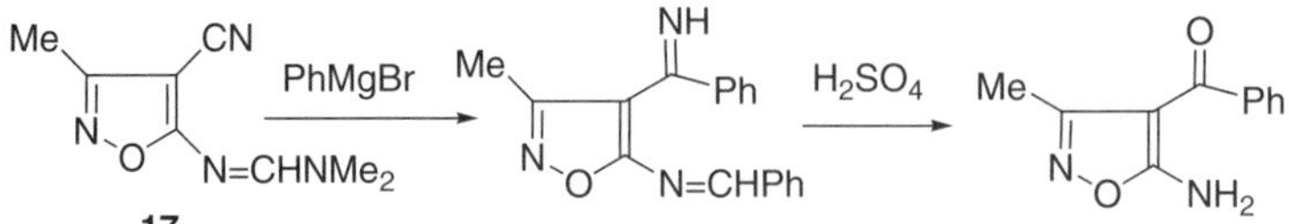

Scheme 10

group, the isoxazole ring loses NOC-R to yield acetylenes **15** (85JOC2372). A putative multistep mechanism of this conversion involving radical formation does not seem persuasive. Acetylenes **15** with the 2-nitrophenyl group are unstable under the reaction conditions, being cyclized immediately to give ^{3}H-indole N-oxides **16** (Scheme 9).

There are a few examples of transformations of substituents other then the amino group. The methyl 4-amino-5-phenylisoxazole-5-carboxylate (or the *N*-hydroxysuccinimide derivative of the same acid) reacts with amines to produce amides (87H2419, 90G725). A preliminary protection of the amino group is necessary for the oxidation of 4-acetamido-5-styrylisoxazole to 4-acetamidoisoxazole-5-carboxylic acid (67G25) and for the addition of phenylmagnesium bromide to nitrile **17** (74HCA1934). 4-Benzoyl-5-aminoisoxazole **19** formed on hydrolysis of Schiff base **18** in H$_2$SO$_4$ cannot be prepared by other methods (Section IV.B.2) (Scheme 10).

C. Syntheses of Condensed Heterocyclic Systems from Aminoisoxazoles

1. From Aminoisoxazoles with an Amino Group Adjacent to a Replaceable Hydrogen Atom

The electrophilic substitution in position 4 of the isoxazole ring is widespread for 3-AI and 5-AI; however, the electrophilic substitution in position 3 or 5 of 4-AI has not been observed (Section III.A). Theoretically, one should expect the formation of isoxazolo[3,4-*b*]pyridines **20** and isoxazolo[5,4-*b*]pyridines **21** from 3-AI and 5-AI, correspondingly. A synthesis of isoxazolo[4,3-*b*]pyridines **22** or isoxazolo[4,5-*b*]pyridines **23** from 4-AI, with the substitution of hydrogen, seems impracticable (Scheme 11).

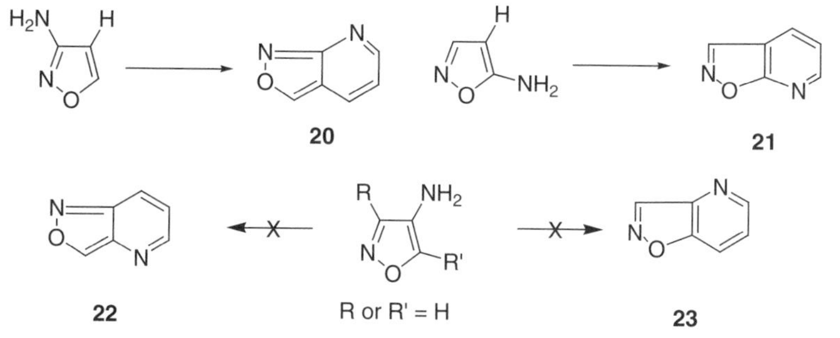

Scheme 11

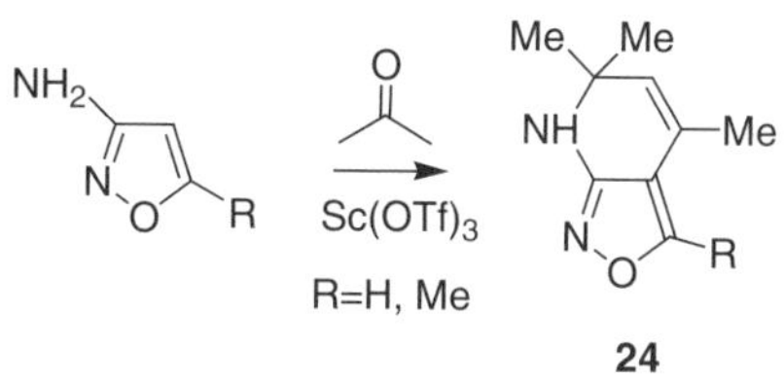

Scheme 12

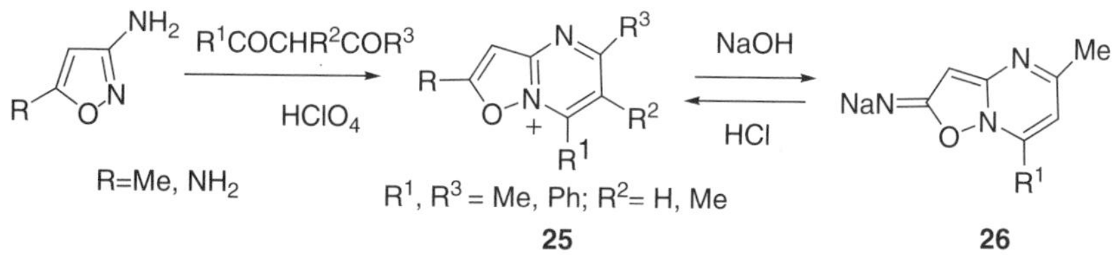

Scheme 13

However, there are numerous communications considering the preparation of isoxazolo[5,4-b]pyridines **21** from 5-AI (03BMCL1487, 72MI1, 75JCS(P1)693, 79S449, 81JHC607, 81JHC619, 87CPB3676, 88JHC231, 97JMC2674, 92CB2259, 01TL8931, 01T9123, 03S1531) and only one work devoted to the synthesis of isoxazolo[3,4-b]pyridines **20**. The reaction of 3-AI with acetone and Sc(OTf)$_3$ yielded dihydroisoxazolo[3,4-b]pyridines **24** (02TL3907) (Scheme 12).

Actually, the reaction between 3-AI and dielectrophiles resulted in annulation of a new ring at the b-bond of the isoxazole ring, position 4 remaining intact. Instead of the expected isoxazolo[3,4-b]pyridines **20**, isoxazolo[2,3-a]pyrimidinium salts **25** were formed by the reaction of 3-AI with diketones (76MI1, 83JOC575). 2-Amino-isoxazolo[2,3-a]pyrimidinium salts **25** prepared from 3,5-diaminoisoxazole are highly stable; they may be recovered unchanged after deprotonation with NaOH and acidification sodium salts **26** (83JOC575) (Scheme 13).

The reactions of 3-amino-5-methylisoxazole with 1,3-bis-(dimethylamino)-2-azapropene **27** (94JHC535), isocyanate (87JHC501), and trichloromethylsulfenyl chloride (75JOC2600) gave isoxazolo[2,3-*a*]1.3.5-triazines **28, 29,** and isoxazolo[3,2-*c*]thiadiazoles **30**, respectively (Scheme 14).

This reaction with N-substituted 3-AI resulted in mesoionic diketoisoxazolo[2,3-*a*]pyrimidine **31** (87JHC1291) (Scheme 15).

Numerous examples of the formation of isoxazolo[5,4-*b*]pyridines from 5-AI are known. The condensation of 5-AI with β-dicarbonyl compounds (75JCS(P1)693, 77H51, 03S1531) or with 3-ethoxypropenal (79S449) yields isoxazolo[5,4-*b*]pyridines **32**. The reaction of 5-AI with 2-ethoxymethylene-substituted and 2-arylidene-substituted 1,3-dicarbonyl compounds yields 7-hydroxyisoxazolo[5,4-*b*]pyridines **33** (72MI1, 88JHC231, 97JMC2674, 03BMCL1487) and 4,7-dihydro-isoxazolo[5,4-*b*]pyridines **34** (85T913), correspondingly (Scheme 16).

Isoxazolo[5,4-*b*]pyridines **35, 36** (81JHC607, 81JHC619), tricyclic system **37**, and tetracyclic system **38** (01T9123) have been prepared from 5-AI. These condensations might start from the electrophilic attack on the carbon atom in position 4 followed by involvement of the amino group in the process (Section III.A) (Scheme 17).

An alternative mechanism with initial formation of a Schiff base takes place for the stereoselective intramolecular cyclization into tetracyclic chromeno[4,3-*b*]isoxazolo[4,5-*e*]pyridine **41** (92CB2259, 01TL8931) (Scheme 18).

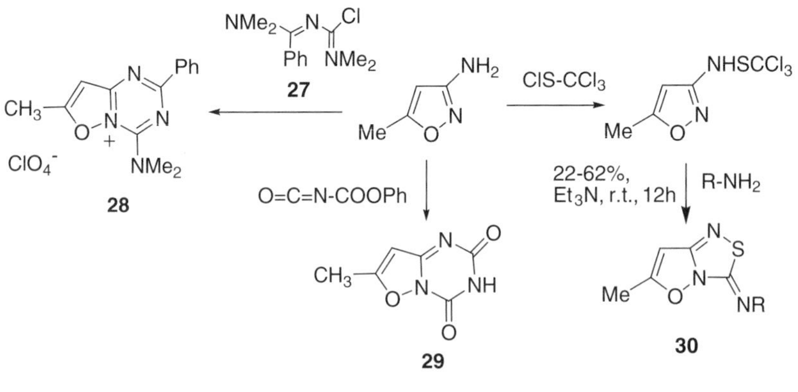

Scheme 14

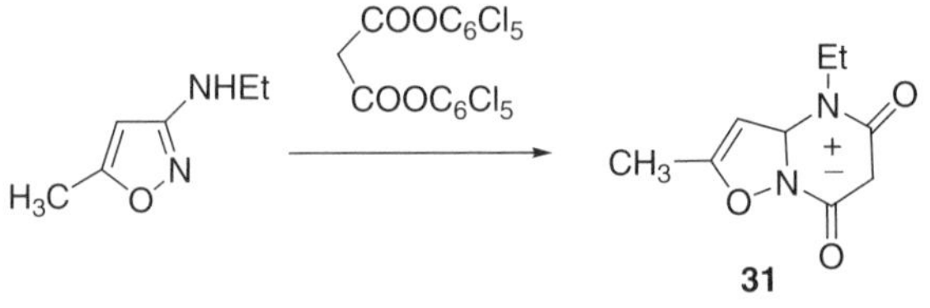

Scheme 15

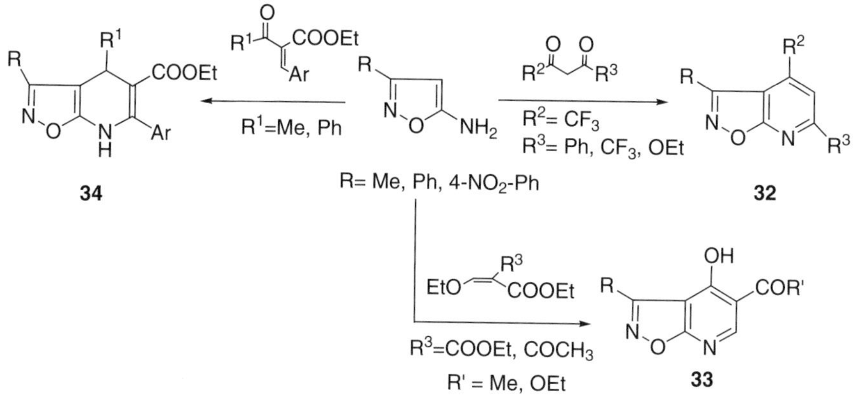

Scheme 16

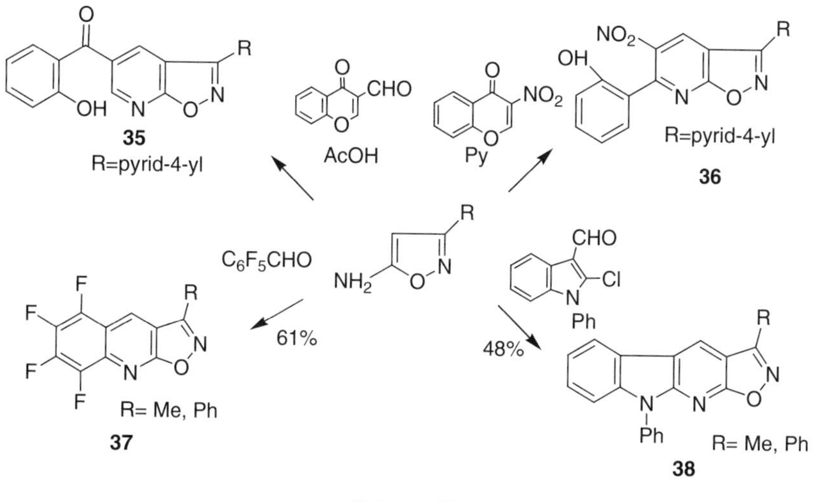

Scheme 17

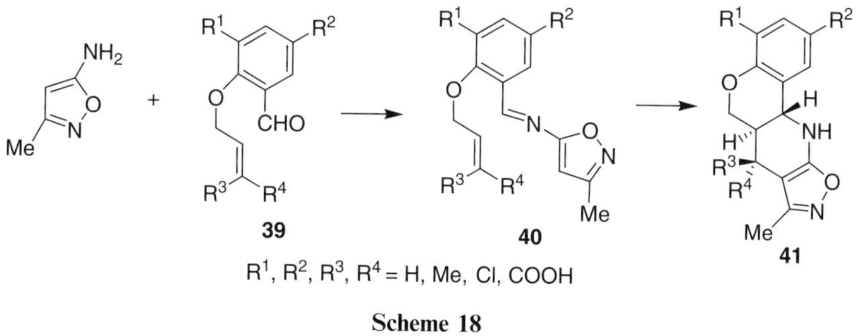

R^1, R^2, R^3, R^4 = H, Me, Cl, COOH

Scheme 18

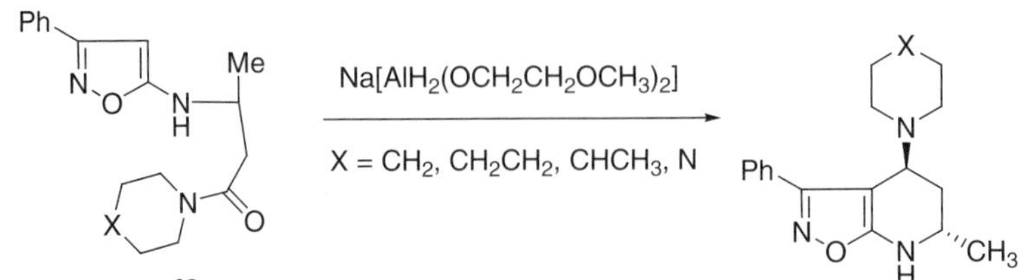

Scheme 19

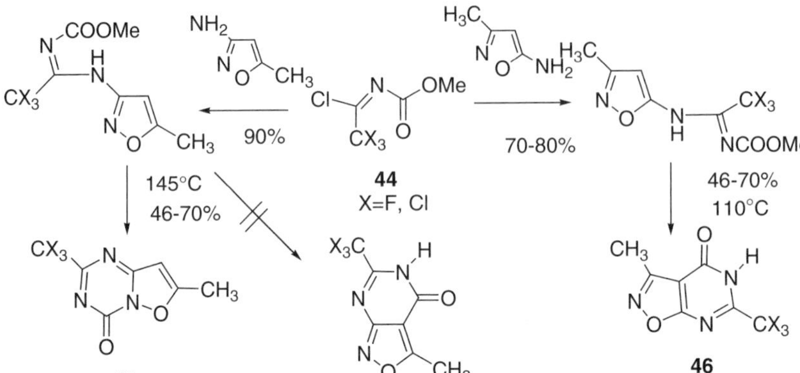

Scheme 20

An unusual type of cyclization was induced by SMEAH. Derivatives **42** were converted into tetrahydroisoxazolopyridines **43** (87CPB3676) (Scheme 19).

One synthesis of isoxazolo[5,4-*b*]pyrimidines from 4-unsubtituted 5-AI has been reported (Section III.C.2). As discussed above for isoxazolo[3,4-*b*]pyridines, the reaction of 3-amino-5-methylisoxazole with carbamate **44** does not produce isoxazolo [3,4-*b*]pyrimidines, isoxazolo[2,3-*a*]1.3.5-triazines **45** being formed as the only products. Meanwhile, the condensation of 5-amino-3-methylisoxazole with the same carbamate yielded isoxazolo[5,4-*b*]pyrimidine **46** (04KGS496) (Scheme 20).

2. *From Aminoisoxazoles with an Amino Group Adjacent to an Amide, Nitrile, or Ketone Group*

Diverse fused heterocyclic systems, such as isoxazolopyridines, isoxazolopyrimidines, isoxazolodiazepines, and isoxazolopyrazines were prepared from 4-AI and 5-AI with functional substituents in the *ortho*-position to an amino group.

Isoxazolo[5,4-*b*]pyridines **47** and their N-oxides **48** were prepared by condensation of 3-phenyl -5-aminoisoxazole-4-carbaldehyde with amidines or its oxime **49** with triethyl orthoacetate (77H51, 80CPB1832). These and methods mentioned below, that were used in the syntheses of isoxazolo[5,4-*d*]pyrimidines, are very similar to the methods described previously in detail for other amino heterocycles (82AHC1) (Scheme 21).

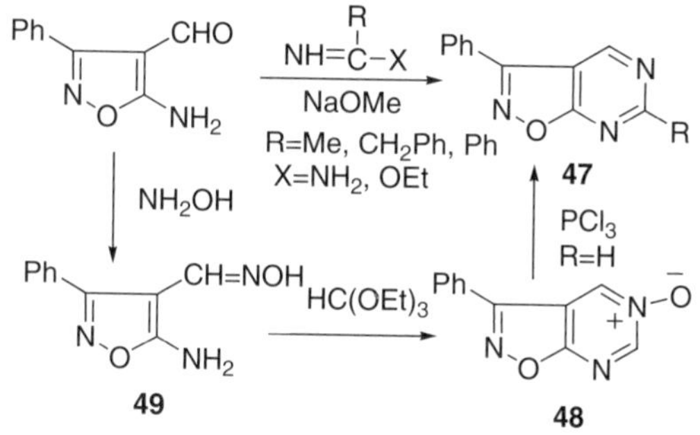

Scheme 21

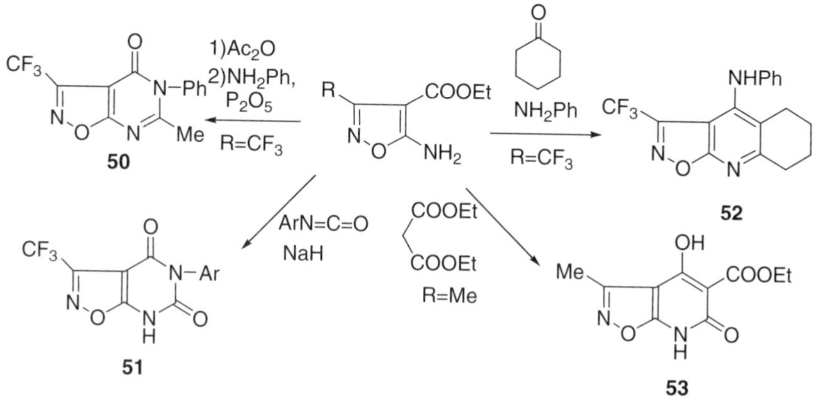

Scheme 22

The reactions of the esters of 5-aminoisoxazole-4-carboxylic acids with acetic anhydride and aryl isocyanates yield substituted isoxazolo[5,4-*d*]pyrimidines **50** and **51** (86ACS(B)760, 86JHC1535) (Scheme 22).

The same amino esters were condensed with cyclohexanone and malonic ester to give isoxazolo[5,4-*b*]quinoline **52** (86ACS(B)760) and isoxazolo[5,4-*b*]pyridine **53** (77JHC435), respectively. To prepare isomeric isoxazolo[4,5-*b*]pyridine **55**, the isolated intermediate **54** was cyclized on treatment with EtONa (82JCS(P1)2391) (Scheme 23).

The addition of 4-AI **56** to dimethyl acetylenedicarboxylate on refluxing in ethanol afforded a Michael-type adduct. Ring closure of this product was accomplished by refluxing in xylene for several hours, and 4,5-dihydroisoxazolo[4,5-*e*][1,4]diazepin-6,8-dione **57** was isolated (87H2419). A convenient route to isoxazolo [4,5-*e*][1,4]diazepin-8-one **58** was developed, starting from *N*-methylated amino ester **56** (90G725) (Scheme 24).

The standard procedure including treatment with bromoacetyl bromide and then with hexamethylenetetramine was used to prepare isoxazolo[5,4-*e*]1.4-diazepin-7-one **59** from 5-amino-4-benzoyl-isoxazole (74HCA1934) (Scheme 25).

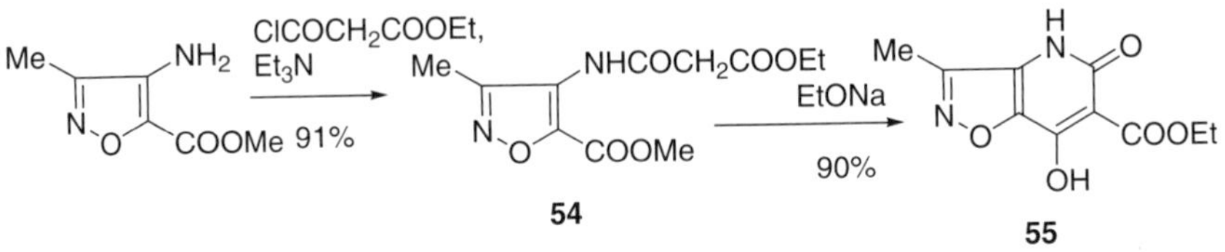

Scheme 23

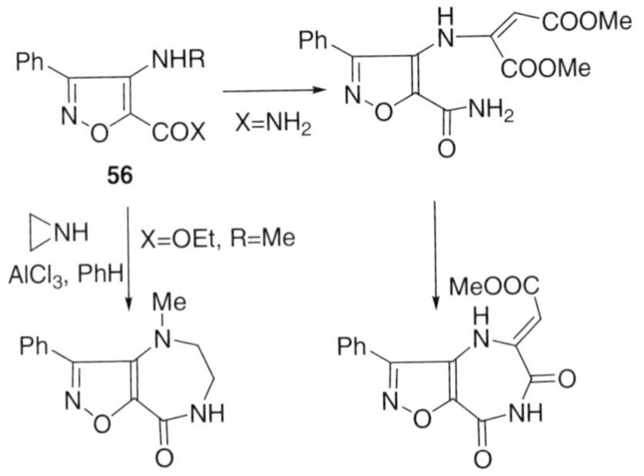

Scheme 24

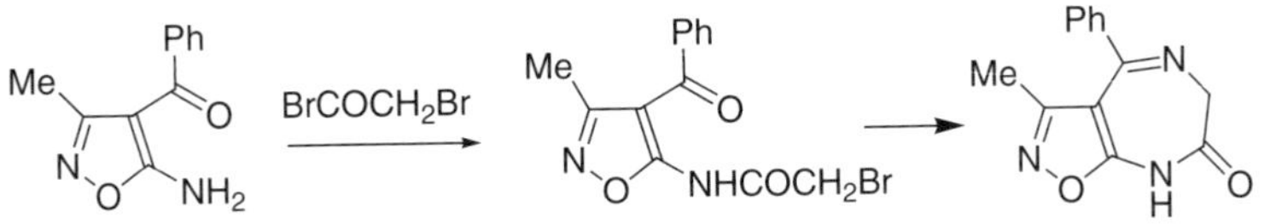

Scheme 25

Amides of 4-amino-5-benzoylisoxazole-3-carboxylic acid could give both iso-xazolo[4,5-b]pyridines and isoxazolo[4,3-b]pyridines, but only isoxazolo[4,3-b] pyrimidines **60** were obtained (80LA1623, 94FA529). The reaction of the same amides with malononitrile affords isoxazolo[4,5-b]pyridines **61** (80LA1623) (Scheme 26).

Orthoesters (64JOC2116, 91IJC(B)946, 02JOC8558) and esters of carboxylic acids condense with 5-aminoisoxazole-4-carboxamides and thioamides in the presence of EtONa (96H691) to give isoxazolo[5,4-d]pyrimidines **62**. The reaction of the same isoxazole carbox(thio)amides with carbonyl chlorides and trifluoroacetic anhydride gave mixtures of compounds **62** with acylated nitriles **63** (67T687, 81IJC(B)654). The oxidation of 5-amino-isoxazole-4-thioamides with hydrogen peroxide or iodine

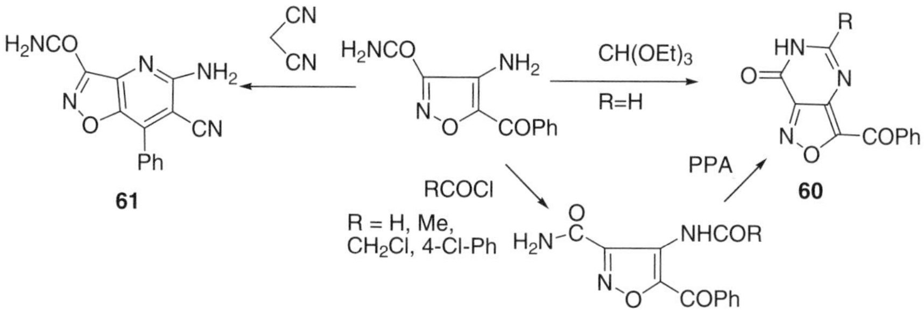

Scheme 26

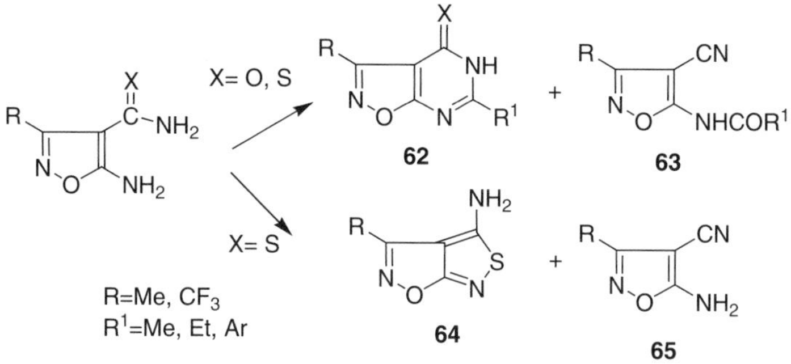

Scheme 27

affords a mixture of 4-aminoisothiazolo[4,3-*d*]isoxazole **64** and 4-cyano-5-amino-isoxazole **65** (90JHC1481). 4-Cyano-5-aminoisoxazoles were condensed with triethyl orthoformate and then treated with amines to give isoxazolo[5,4-*d*]pyrimidines **62** (64JOC2116, 74JMC451) (Scheme 27).

3-Phenyl-4,5-diaminoisoxazole was made to react with 1,2-diketones to give isoxazolo[4,5-*b*]pyrazines **66** (68T4907, 73JHC181, 91MI1), but condensation with acetylacetone yielded only enamine **67**, instead of the expected isoxazolodiazepine **68**. An attempt to prepare isoxazolotriazole from 3-phenyl-4,5-diaminoisoxazole was unsuccessful. Under all of the examined conditions, including the action of PPh_3 and NaOH, 4-diazo-5-iminoisoxazole **69** was recovered unchanged (70T1393) (Scheme 28).

The possibility of acetylene formation (Section III.B) may be responsible for the unusually low yields (4–9% after the chromatographic separation) obtained in the synthesis of isoxazolo[5.4-*d*][1,2,3]triazin-4-ones **70** (97FA105, 03MI1) (Scheme 29).

4-(3,4-Dimethoxyphenyl)-5AI (90KGS914) and 4-(cyclo(aza)alkenyl)-5-AI (84S1052) were condensed with acylating agents to give isoxazolo[5,4-*b*]quinolines **71** and **72** (Scheme 30).

On boiling in Ph_2O, Schiff bases **73** undergo thermally induced ring closure giving rise to 6,7-dihydroisoxazolo[4,5-*b*]pyridines **74** (83S839, 85JIC135, 90BCJ1851) (Scheme 31).

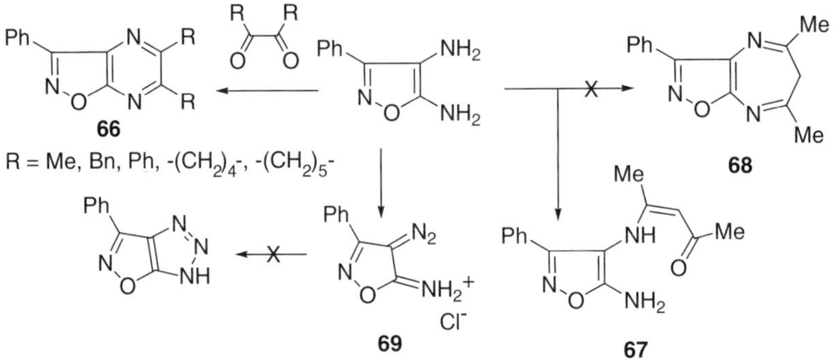

Scheme 28

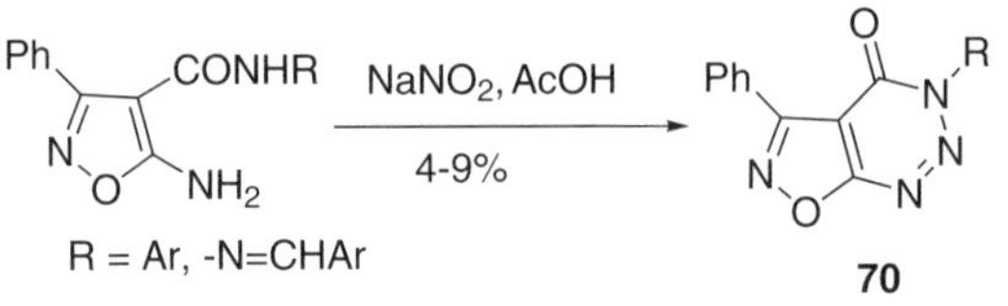

Scheme 29

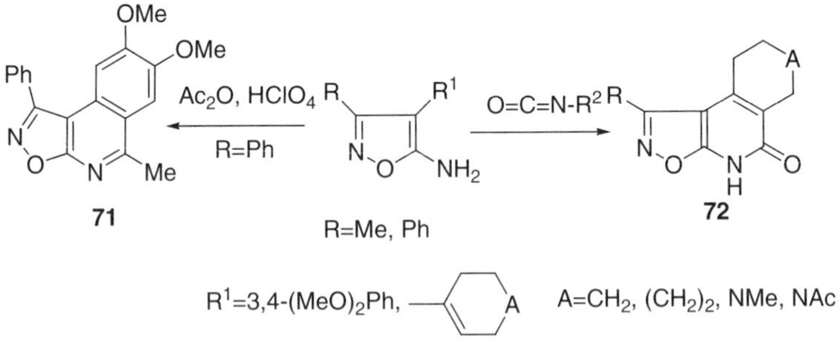

Scheme 30

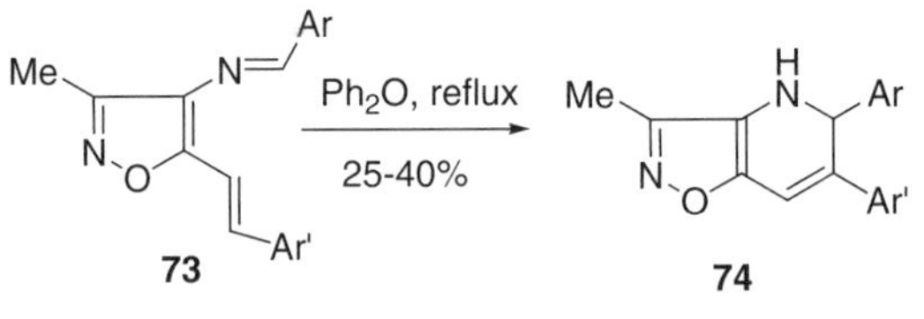

Scheme 31

D. RING OPENINGS AND REARRANGEMENTS

The reactions of AI that start with isoxazole ring opening often proceed with cyclization of intermediates to give diverse heterocyclic systems. In this review, the reactions are classified accordingly to the type of reagent used to cleave the ring. A few rearrangements of AI in acidic media have been recently discovered (Section III.D.4), in addition to the well-known reactions induced by bases (Section III.D.1), heating or irradiation (Section III.D.2), and reducing agents (Section III.D.3).

1. By Bases or Nucleophiles

Base-catalyzed opening of isoxazoles unsubstituted in position 3 is a well-published process (63AHC365, 75S20, 81KGS1155, 84CHC1). In a similar manner, 5-AI unsubstituted in position 3 (58CPB105, 73LA898) or esters of isoxazole-3-carboxylic acids (39G523) also readily yield tricarbonyl compounds **75** (Scheme 32).

The derivatives of 4-AI **76** and **77** unsubstituted in position 3 rearrange into 2-aminooxazoles **78** via typical ring opening (89HCA556) (Scheme 33).

Surprisingly, 5-(2-aminophenylamino)-4-cyano-isoxazole reacts with triethylamine to give benzoimidazole **80**, instead of expected amide **79**. This transformation may be reversed under acidic conditions (80CPB567, 89JHC277) (Scheme 34).

A few base-catalyzed rearrangements of AI with alkyl or aryl groups in position 3 are known. 3-Aryl-substituted 5-AI **81** rearrange to 4-cyanoisoxazol-5-one **82** under the action of EtONa (60CB1103) (Scheme 35).

The reaction of AI with hydrazines follows a complex pathway. Thus 5-aminoisoxazoles react with hydrazine to furnish a mixture of pyrazolone **83**, 4-aminopyrazol-5-one **84**, and 1-aminopyrazol-5-one **85** (72JHC1219). Bipyrazole **86** and isoxazole ring destruction product **87** were obtained from the reactions of 5-AI with phenylhydrazine (53JPJ387) and semicarbazide (54JPJ138), respectively. This issue seems to require reinvestigation; however, these reactions have a complex nature and

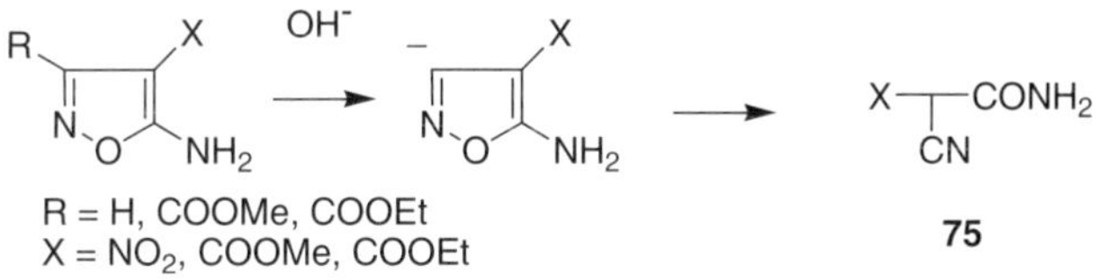

R = H, COOMe, COOEt
X = NO₂, COOMe, COOEt

Scheme 32

R = H, Me
R¹ = i-Pr, t-Bu, Ar

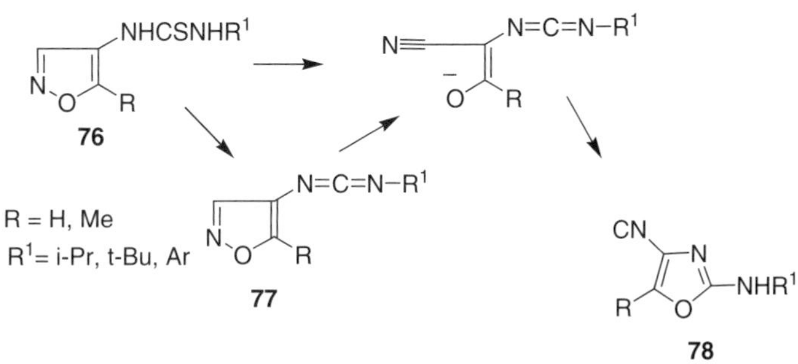

Scheme 33

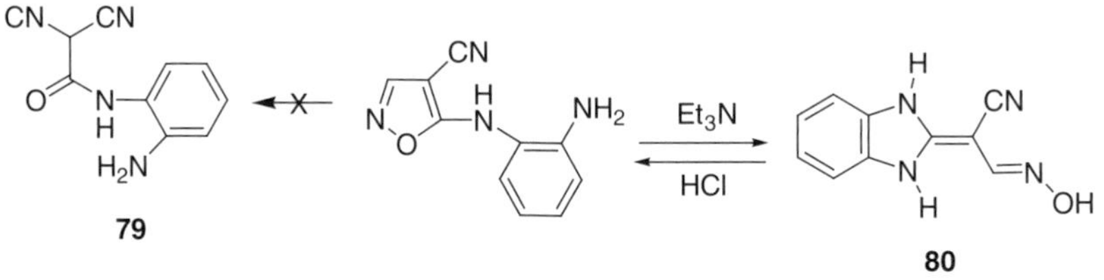

Scheme 34

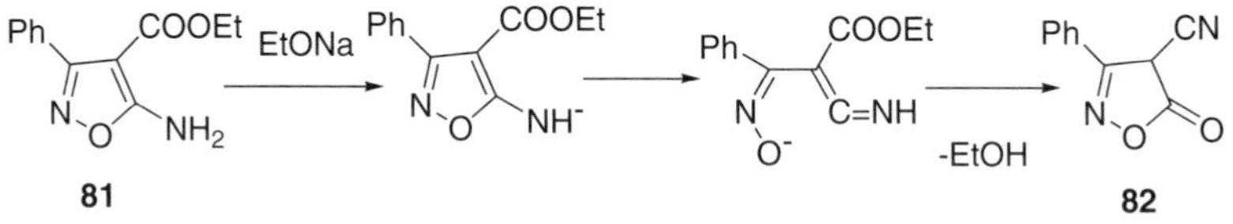

Scheme 35

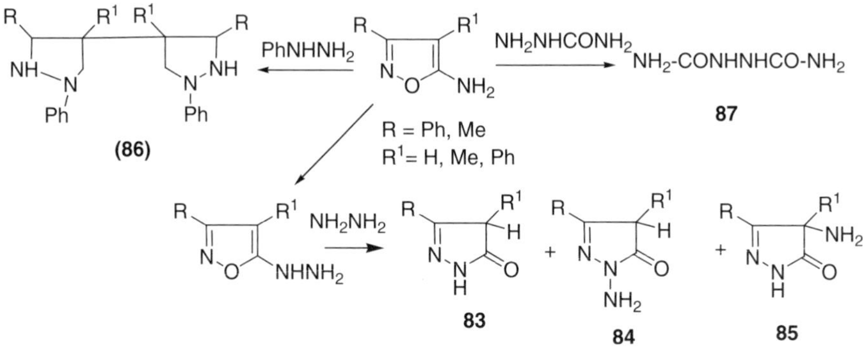

Scheme 36

low synthetic utility because the corresponding aminopyrazoles may be prepared more easily from acyclic precursors (Scheme 36).

Many rearrangements of acylated AI are documented. Thus 3-thioureidoisoxazoles rearrange into 1,2,4-thiadiazoles **88** (77JCS(P1)1616, 86H3433, 93TL6423, 94BCJ1701, 03BMC591). When 3-AI react with acyl isothiocyanate **89**, 1,2,4-thiadiazoles are the only isolable products (02ZOR599). The corresponding oxygen analogs, 3-ureidoisoxazoles, do not rearrange under any of the conditions studied (67JCS(C)2005, 88JHC607) (Scheme 37).

3-Acylaminoisoxazoles **90**, which are stable in acid media (99FA90), rearrange under basic conditions. Relying on NMR spectroscopic data, the structure of the enol form of 3-acetonyl-1.2.4-oxadiazole **92** was assigned to the product of rearrangement of 3-acylaminoisoxazoles (98FA513). Actually, this product has the structure of 2-aroylaminooxazole **91**, as established by ^{13}C-NMR spectroscopy (91H1765). The prepared 3-acetonyl-1,2,4-oxadiazoles **92** rearrange on treatment with MeONa to give 3-acylaminoisoxazoles **90** in high yields (83M373) (Scheme 38).

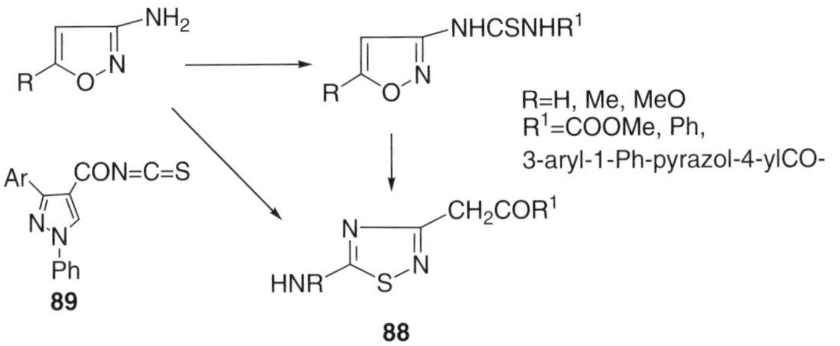

R=H, Me, MeO
R¹=COOMe, Ph,
3-aryl-1-Ph-pyrazol-4-ylCO-

Scheme 37

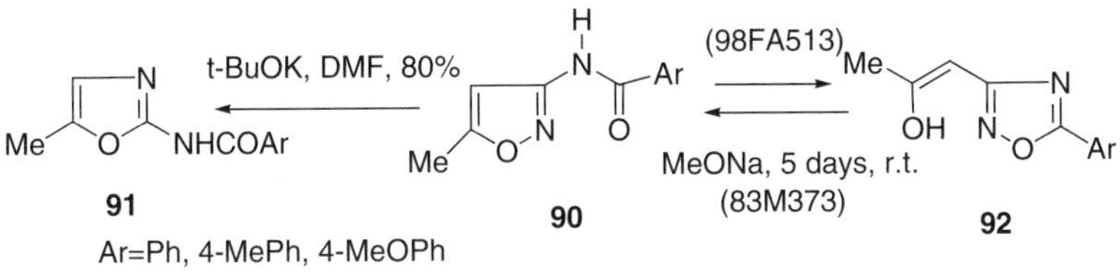

Ar=Ph, 4-MePh, 4-MeOPh

Scheme 38

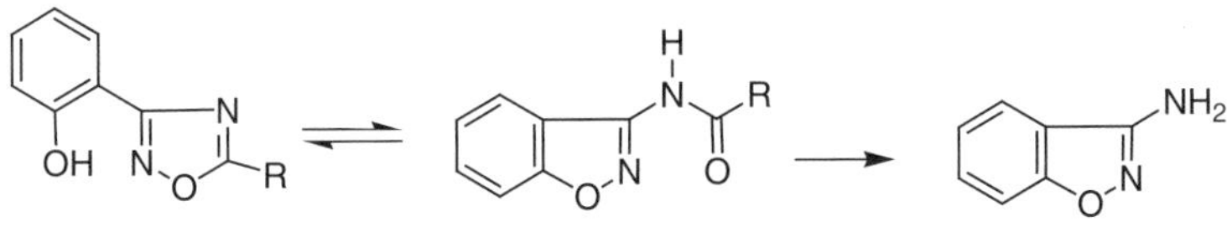

Scheme 39

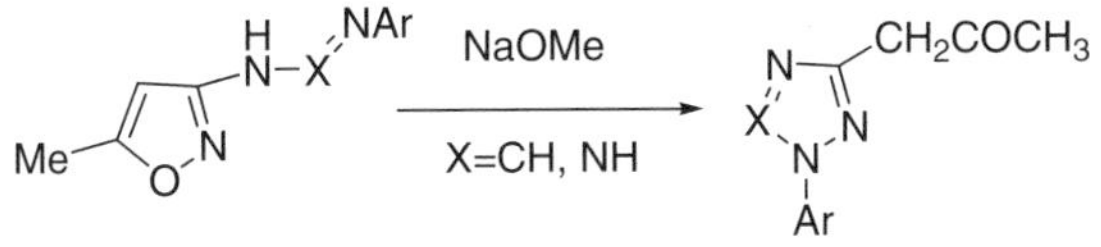

Scheme 40

3-Acetylaminobenzoisoxazoles **93** are in equilibrium with 3-(2-hydroxyphenyl)-
1.2.4-oxadiazoles **94**. Acylaminobenzoisoxazoles **93** predominate in the presence of
weak bases, while 3-(2-hydroxyphenyl)-1.2.4-oxadiazoles **94** predominate in the
presence of EtONa, although in this case, EtONa induces slow deacylation yielding
3-aminobenzoisoxazole **95** (73JHC957) (Scheme 39).

1,2,4-Triazoles (64T159) and tetrazoles **97** (64T461) were obtained from the corresponding 3-AI derivatives **96** (Scheme 40).

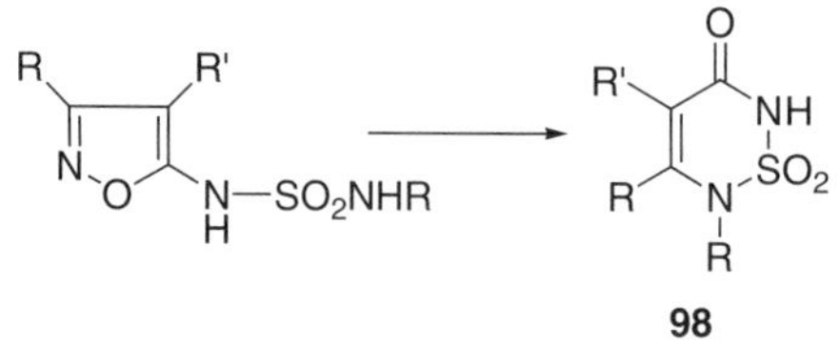

Scheme 41

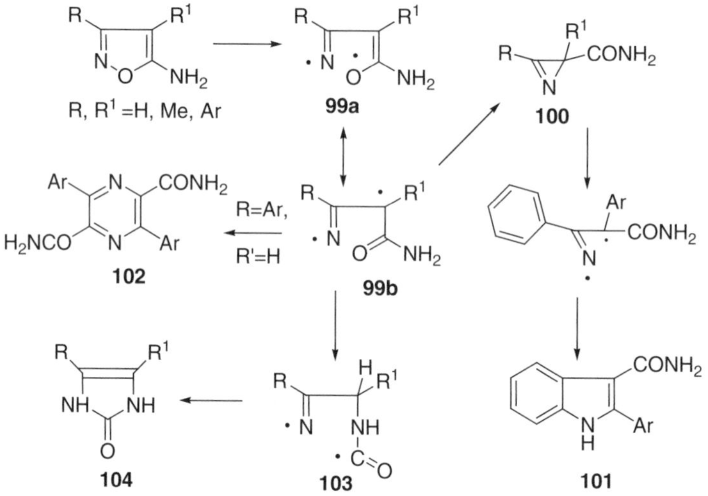

Scheme 42

Only one example of a rearrangement was reported for *N*-acylated 5-AI (see also Scheme 35). Cleavage of 5-sulfamidoisoxazoles gave 1.2.6-thiadiazine 1,1-dioxides **98**. Cleavage of 3-sulfamidoisoxazoles follows a similar pathway (79JOC4191) (Scheme 41).

2. *By Photolysis and Thermolysis*

Several works devoted to AI rearrangements induced by irradiation or heating were published in the 1970s, but interest in these reactions is limited because they often gave complex product mixtures. Homolytic rupture of the N–O bond in 5-AI induced by irradiation or heating furnishes biradical **99**. Depending on the reaction conditions (temperature, solvent, concentration), the biradical can produce a variety of heterocyclic systems, the reaction route being not always predictable. In the case of 3,4-diaryl-substituted 5-AI, biradical **99b** forms 2*H*-azirines **100** (71JCS(C)2644, 71JCS(C)2648) or indoles **101**, when the thermal reaction was performed in HMFTA (74JCS(P1)1867). In the case of 4-unsubstituted 5-AI, dimerization of the biradical **99** leads to pyrazines **102** (70JCS(C)1825, 71JCS(C)2644, 71JCS(C)2648) (Scheme 42).

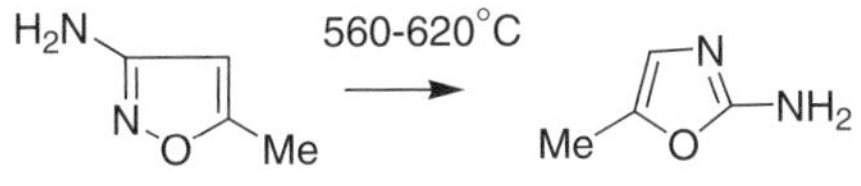

Scheme 43

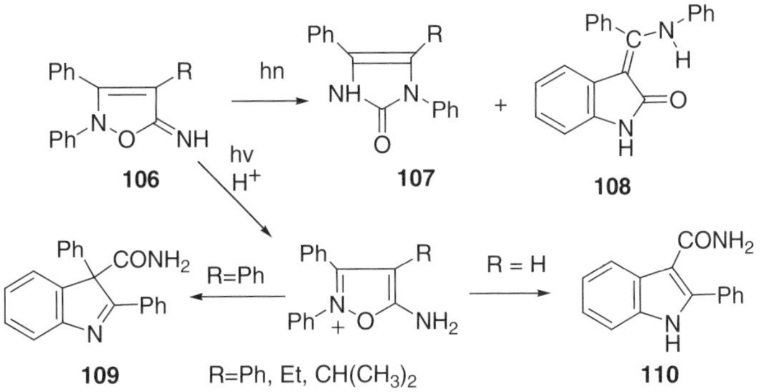

Scheme 44

In the condensed phase, thermal opening of 5-AI at 180–220 °C resulted in imidazol-2-ones **104** through the rearrangement of biradical **99** to biradical **103** (52JPJ148, 52JPJ1118). However, when the thermal rearrangement of 5-amino-3-methylisoxazole was carried out in a nitrogen flow at 500 °C, 2H-azirine **100** was the single isolable product (81JOC3505). Thermal rearrangement of 3-amino-5-methylisoxazole in nitrogen at 420–500 °C yielded only 2-aminooxazole **105**, while no 2H-azirine was isolated (81JOC3505) (Scheme 43).

On irradiation in neutral media, 5-imino-2-phenyl-2,5-dihydro-isoxazoles **106** were rearranged into imidazol-2-ones **107** and indol-2-ones **108** (70LA195, 74CB13), whereas irradiation in acidic media afforded 1H-indoles **110** or 3H-indoles **109** (74CB13) (Scheme 44).

3. By Reducing Agents

The reductive opening of monocyclic AI will be discussed in relation to hydrogenation of nitroisoxazoles (Section IV.B.1). Whereas cleavage of monocyclic AI is achievable by catalytic hydrogenation with Pd/C or, better, with Raney Ni and unachievable with the use of chemical hydrogenating agents, opening of bi- and tricyclic AI can be performed using many reducing agents. For example, benzoannulated 3-AI is cleaved under the action of $H_2 + Pd/C$, $NaBH_4/NiCl_2$, and even Zn/AcOH (02TL8777). Moreover, tricyclic system **111** was opened by a very mild hydrogenating agent, sodium dithionite (90HCA2147). Probably, the ring strain of condensed AI is a crucial factor for isoxazole ring opening (Scheme 45).

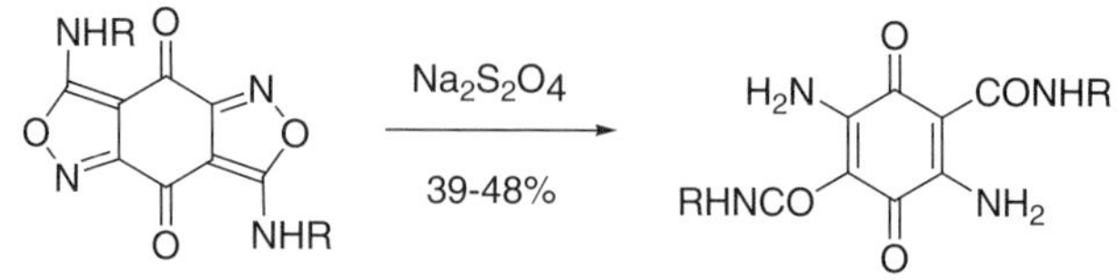

111 **112**

Scheme 45

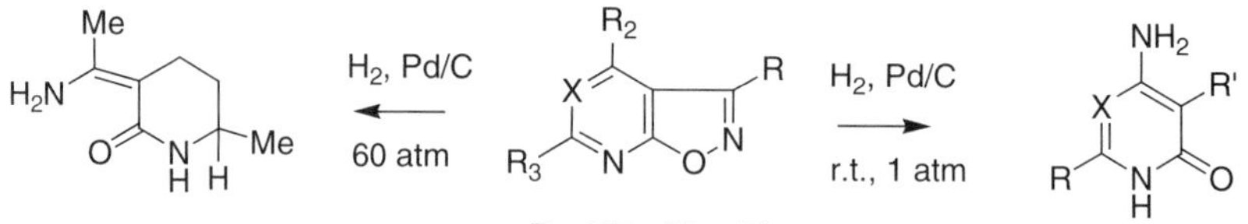

114

R = NH$_2$, Me, Ph
R^1, R^2 = H, NH$_2$,
X = CH, N;

113

R'=COMe, COPh, C(NH)NH$_2$

Scheme 46

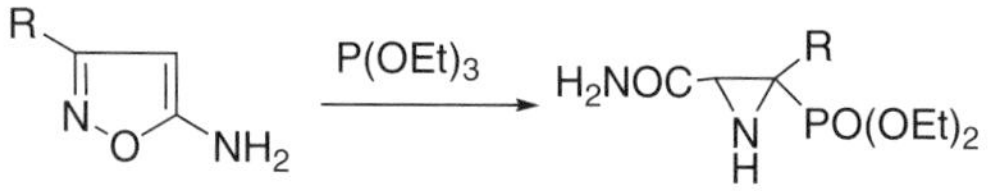

R = C$_6$H$_5$, 4-Cl-C$_6$H$_4$, 4-Me-C$_6$H$_4$ **115**

Scheme 47

Hydrogenation of isoxazolopyridines and isoxazolopyrimidines prepared from AI in the presence of Pd/C under atmospheric pressure and at room temperature led only to opening of the isoxazole ring to give monocyclic amino products **113** (64JOC2116, 99JOC8479). Hydrogenation under more drastic conditions involved both the isoxazole and pyridine rings, resulting in piperidones **114** (79S449) (Scheme 46).

4. *Miscellaneous*

Three rearrangements of AI that are difficult to classify. An unusual type of ring opening was found in the reaction of 5-AI with triethyl phosphite (71JCS(C)3021). 2-Phosphonoaziridines **115** prepared by this reaction were similar to the products formed on irradiation of AI, though phosphite acted as a reducing agent (Scheme 47).

4-Aminoisoxazole derivatives **116** rearranged into a mixture of 2-methylthioimidazole **117** and 2-aminooxazole **118** under the catalytic action of molybdenum hexacarbonyl and SnCl$_4$ (91HCA531) (Scheme 48).

3-Acylamino-1.2.5-thiadiazoles **119** were obtained from the reaction of 3-acylaminoisoxazoles and the S$_4$N$_4$ × SbCl$_5$ complex. A multistage mechanism of this interesting transformation has been proposed (98JCS(P1)2175) (Scheme 49).

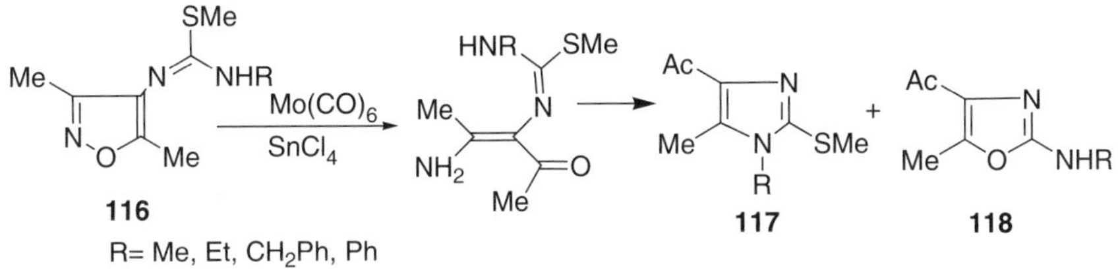

Scheme 48

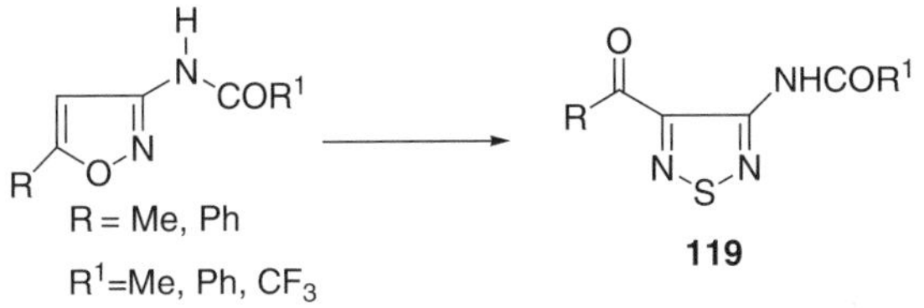

Scheme 49

IV. Synthesis of Aminoisoxazoles

A. PREPARATION FROM ACYCLIC PRECURSORS

Two approaches are commonly used to prepare isoxazoles (63AHC365, 79AHC147, 84CHC1, 91HC1): formation of bonds 1–5 and 2–3 (path A) and formation of bonds 1–5 and 3–4 (path B). Indeed, there are a lot of examples of AI syntheses by path A (Section IV.A.1). The addition of nitrile N-oxides to alkynes or alkenes is widely used for the preparation of isoxazoles without amino groups. However, to our knowledge, the reaction of 1-aminoacetylenes and nitrile N-oxides (path B) was not used to prepare 3(5)-AI. Instead, 3(5)-AI were produced by the reaction of substituted acetonitriles and chloro(bromo)oximes. The latter is not a concerted process, ring closure taking place when bond 2–3 has already been formed (path C). Actually, path B is similar to path C, so in this review paths B and C will be considered together (Section IV.A.2). The methods for preparing 4-AI often differ from the methods used to synthesize 3-AI and 5-AI. Thus, paths A and B were used to prepare 3-AI and 5-AI, but not 4-AI, which were obtained by path D (Section IV.A.3). However, hydrogenation of nitroisoxazoles is a good method for preparing 4-AI, but 3-AI and 5-AI cannot be obtained by this procedure (Scheme 50).

1. (CCC+NO) Reactions

To prepare 5-AI or 3-AI, hydroxylamine or N-hydroxyurea may be used as synthetic equivalents of the NO-synthon and malonodinitriles, β-ketonitriles and cyanoacetylenes, as equivalents of the C$_3$-synthon. The condensation of

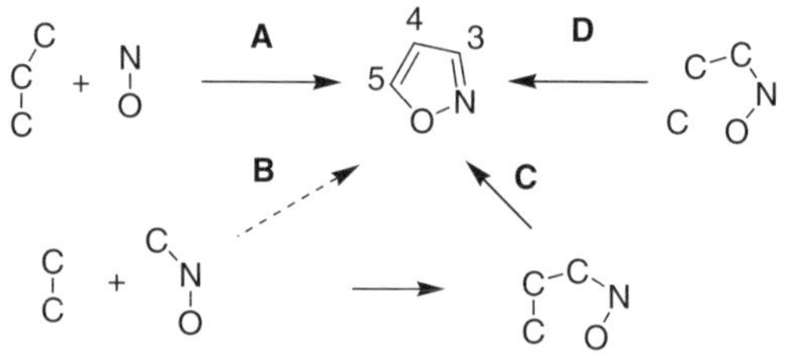

Scheme 50

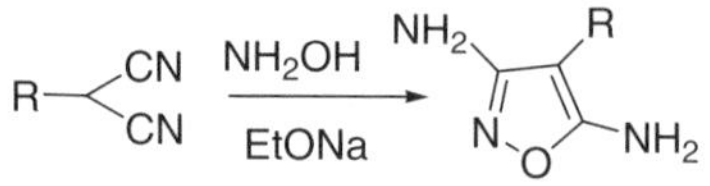

Scheme 51

R= H, sec-C_4H_9, CH_2Ph, Ar, -N=NAr

120

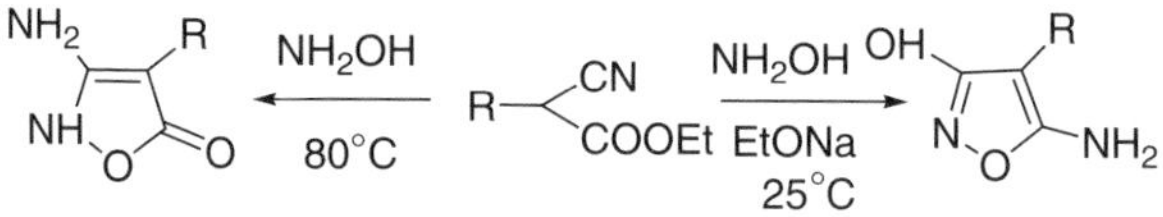

Scheme 52

R=H, Alk, CH_2Ph, Ar

121   **122**

hydroxylamine with 2-alkyl(aryl)-substituted malononitriles gave 3,5-di-aminoisoxazoles **120** (65JOC2862, 83JOC575, 87S33, 84ZC256, 94IJC(B)1048, 97IJC(B)394) (Scheme 51).

In the reaction of 2-substituted cyanoacetic esters, it is possible to generate both 3-amino-isoxazol-5-ones **121** and isomeric 5-aminoisoxazol-3-ones **122**. The product ratio is mainly dependent on the reaction conditions rather than on the structure of the starting compounds. When the reaction is carried out at 80 °C in neutral media, 3-amino-isoxazol-5-ones **2** are produced. 5-Aminoisoxazol-3-ones **3** are the major products in the presence of strong bases (61JOC4917, 67T4395) (Scheme 52).

Spiro-condensed 3-aminoisoxazol-5-ones **123** were formed regioselectively under the standard conditions (80 °C, hydroxylamine), no isomeric 5-aminoisoxazol-3-ones being detected (02JHC649, 02IJC(B)1670, 03HAC513) (Scheme 53).

Unlike cyanoacetic esters in which electrophilicities of the ethoxycabonyl group and nitrile group are comparable, in the case of β-ketonitriles, the electrophilicity of the ketone group is noticeably higher. There are many examples where 5-AI **124** was obtained as the only product on the reaction of β-ketonitriles with hydroxylamine (34CB1062, 32CB1857, 52G98, 54CB1175, 74CB2563, 90KGS914). However, the addition of hydroxylamine to the ketone group is reversible and cyclization of intermediate **125** at pH 8 proceeds at a low rate. The irreversible addition of hydroxylamine to the nitrile group is dramatically accelerated by heating, and

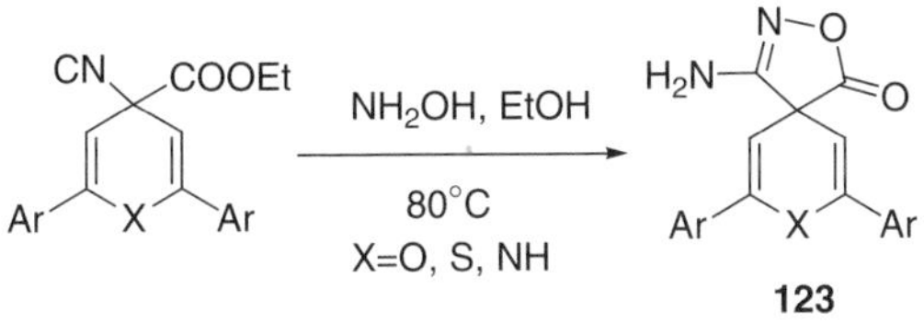

Scheme 53

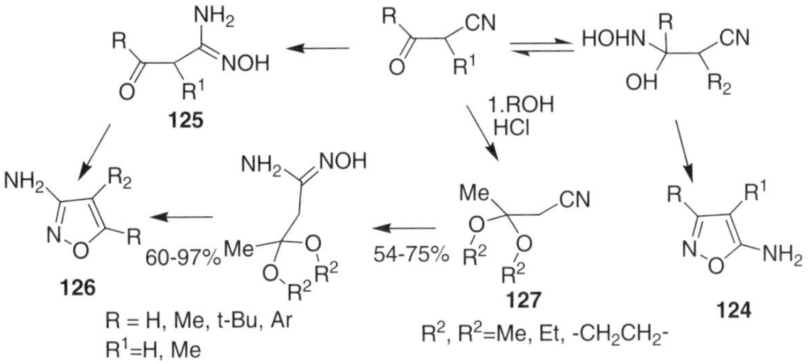

R = H, Me, t-Bu, Ar
R¹=H, Me

R^2, R^2=Me, Et, -CH₂CH₂-

Scheme 54

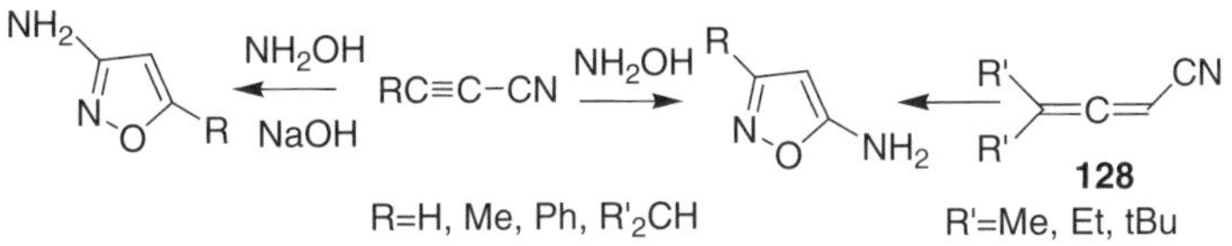

R=H, Me, Ph, R'₂CH

R'=Me, Et, tBu

Scheme 55

3-amino-5-*tert*-butylisoxazole is thus prepared in a more than 95% yield at 80–100 °C and pH 8 (91H1153). A substantial drawback of this synthesis is its inability to produce 3-AI **126** with lower alkyl or functional substituents. The ketal protection of the ketone group is an obvious way for preparing 3-AI from β-ketonitriles (78USP4152336); however this procedure has not been widely used (Scheme 54).

Since 3-AI show promise for drug discovery, new methods for the selective preparation of 3-AI were investigated. The hydroxylamine addition to substituted propynoic nitriles gave 3-AI in the presence of strong bases (66CPB1277) and on heating (66CR557, 84JCS(P1)1079) or 5-AI at low temperatures and in neutral media. The reaction with cyanoallenes **128** yields selectively 5-AI (84JCS(P1)1079) (Scheme 55).

The hydroxylamine addition to amidines of phenylpropynoic acid yields 5-AI derivatives **130** (72BCJ1846), although the addition to thioanilides of the same acid (17JA697, 37JA933, 37JA1486), to 1-aryl-3-methoxy-3-(alkylamino)propenones and their thio analogs gave 3-arylaminoisoxazoles **129** (63CB1088, 73LA256, 84S247) (Scheme 56).

N-Hydroxyurea reacts regioselectively with 2,3-dibromopropionitrile (70M1109, 74CB2563) and 2-bromoacrylonitrile (75HCA1735, 79HCA833, 84ZC93) to produce

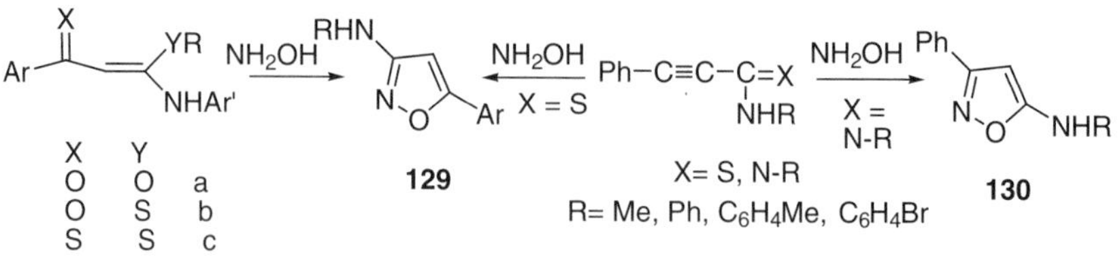

Scheme 56

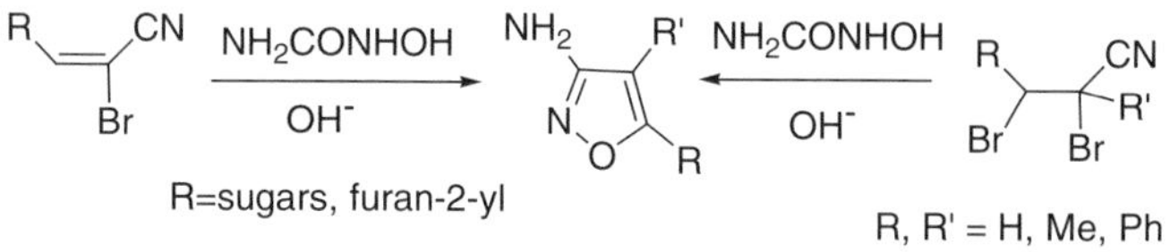

R=sugars, furan-2-yl

R, R' = H, Me, Ph

Scheme 57

X= COOEt, CN, SO₂Ph, CSNH₂, SEt
R = H, Alk, Ar
Y= OMe, OEt, Br

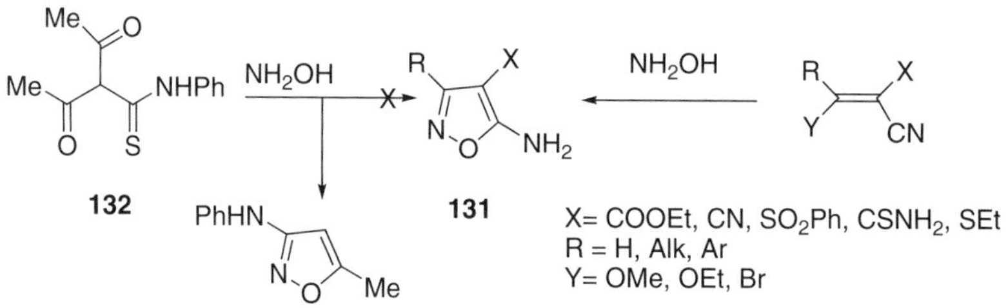

Scheme 58

3-AI. The utility of this available reagent has not been adequately investigated in the chemistry of AI. (Scheme 57)

The preparation of AI with different functional substituents (COOEt, CN, SO$_2$Ph) in the *ortho*-position to an amino group deserves attention because such AI **131** are the key intermediates in the synthesis of numerous fused systems (Section III.C.2). The attempted reaction of tricarbonyl compounds **132** with hydroxylamine was unsuccessful, giving only 4-unsubstituted 3-AI **129** (20JA1055). The action of hydroxylamine on 3-methoxy (ethoxy, bromo)-acrylonitriles provides a regioselective route to 5-AI with functional groups in position 4 (58CPB105, 60CB1103, 64JOC2116, 78KGS969, 80TL3755, 90JHC1481) (Scheme 58).

In a few examples, 3-AI were also identified after the above-mentioned reactions. In the case of 2-cyano- (58JA2829), 2-COOMe- (89JHC1335), and 1.2.4-oxadiazol-5-yl-arylonitriles **133** (96JHC1943) containing two donor groups at position 3 of the acrylic system, only 5-AI were isolated from the reaction mixtures, but in the case of 2-(*tert*-butoxycarbonyl)-3,3-dimethoxyacrylonitrile, a mixture of 5-AI **134** and 3-AI **135** was formed in the a 33:13 ratio (89JHC1335). 2-Unsubstituted 3,3-di-methoxyacrylonitrile **133** (X = H) condenses with hydroxylamine to produce 3-AI **135** as the only product (94BCJ1701) (Scheme 59).

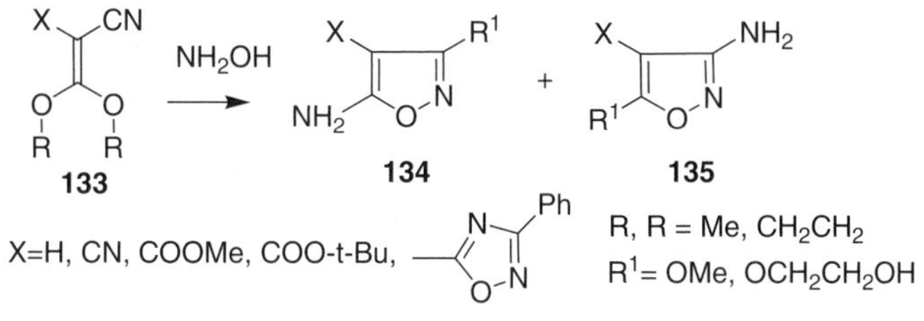

X=H, CN, COOMe, COO-t-Bu,

R, R = Me, CH$_2$CH$_2$
R^1= OMe, OCH$_2$CH$_2$OH

Scheme 59

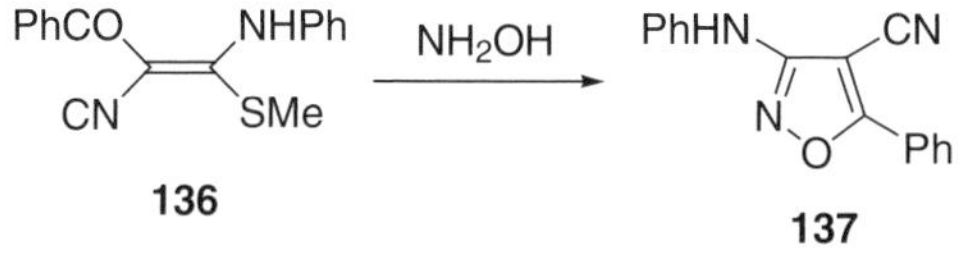

Scheme 60

An example of regioselective preparation of 3-AI with a functional substituent in position 4 was published (78JPR585). The reaction of 3,3-disubstituted acrylonitriles **136** with hydroxylamine produced 3-amino-4-cyanoisoxazoles **137**. Although the authors do not present convincing evidence for the formation of 3-amino-4-cyano-, and not isomeric 5-amino-4-cyano-isoxazoles (only elemental analysis data are given), this inference appears probable. Usually hydroxylamine adds initially to the β position of the acrylonitrile system and then the attack can be directed onto the keto group due to higher electrophilicity of this group compared to the nitrile group (Scheme 60).

2. (CNO+CC) and (CCCNO) Reactions

In 1936, Quilico (36MI1) was the first to propose a general synthesis of 5-AI **131** by the condensation of substituted acetonitriles with chlorooximes or, in some cases, bromooximes (87BCJ4480). The electron-withdrawing function (X) present in position 4 of the isoxazole ring could be CN, COOEt, (46G148, 67T3541, 81IJC(B)654, 85JOC2372, 87BCJ4480, 91IJC(B)946), C(NH$_2$) = C(CN)$_2$ (80CB1195, 87BCJ4480), NO$_2$ (75S664), or SO$_2$Ar (84CHC1, p. 70). The synthesis of 5-AI in this way is a useful alternative to the methods discussed in the previous section (Scheme 58). This base-catalyzed reaction is not concerted and proceeds via the same intermediate as the reaction with hydroxylamine (62HC1) (Scheme 61).

Some other syntheses involve closure of the acyclic intermediate prepared by addition to diverse alkenes. 5-AI without an electron-withdrawing group in position 4 can be prepared from alkenes by a two-step procedure including the action of NOCl and then cyanide. Using this method, 3,4-disubstituted 5-AI was synthesized in a high yield (69TL4817) (Scheme 62).

The addition of the cyanide ion to 2-aryl-1-aroyl-1-nitroarylalkenes yields a mixture of 2-aminofurans **138** and 5-AI **139** (85JHC1663). It is not clear from the publication (85JHC1663) how nitronitrile **140** was hydrogenated to oxime **141** under

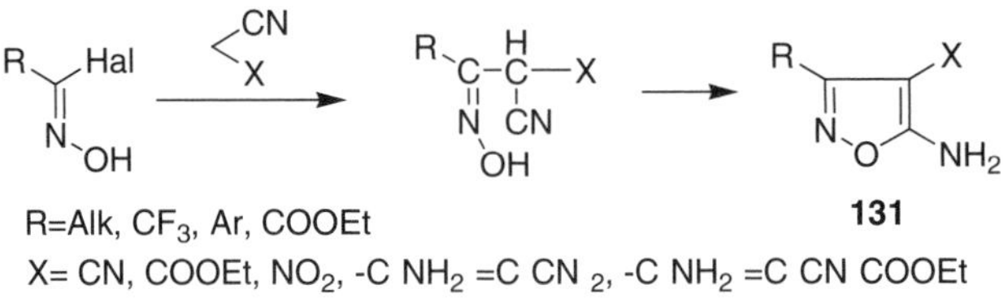
R=Alk, CF3, Ar, COOEt
X= CN, COOEt, NO2, -C NH2 =C CN 2, -C NH2 =C CN COOEt
131

Scheme 61

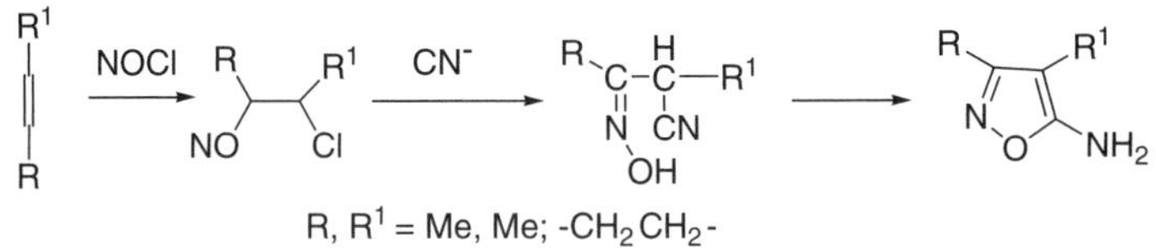
NOCl
CN-
R, R1 = Me, Me; -CH2CH2-

Scheme 62

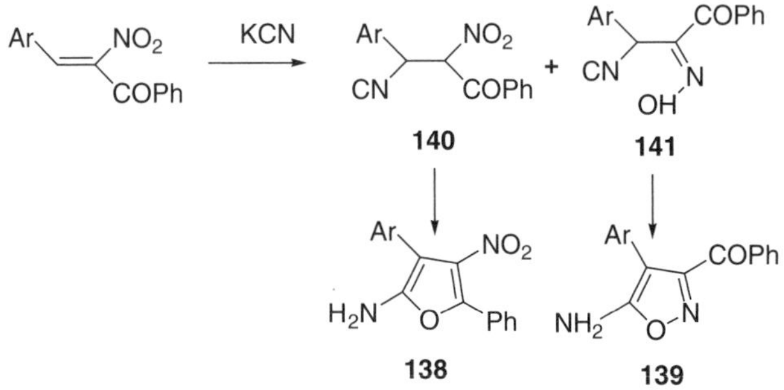
KCN
140
141
138
139

Scheme 63

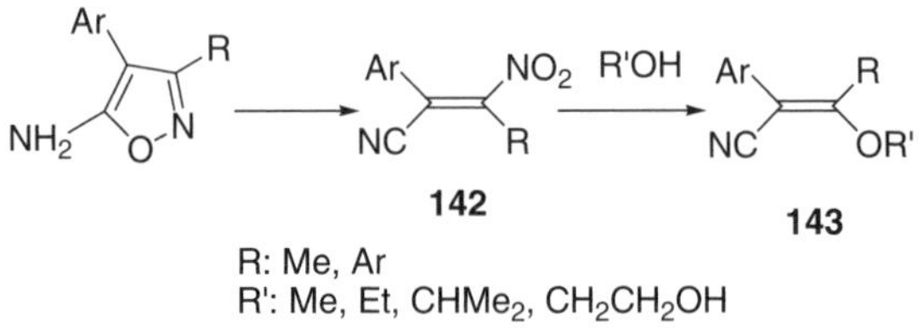
R'OH
142
143
R: Me, Ar
R': Me, Et, CHMe2, CH2CH2OH

Scheme 64

these conditions. Another molecule of nitronitrile or the cyanide ion could act as a reducing agent, but this issue has not been investigated (Scheme 63).

3-Nitroacrylonitriles **142** were electrochemically hydrogenated to 5-AI (79JHC1611). The same products were obtained by the reaction with BuSH; however the action of alcoholates gave normal products of nucleophilic substitution of the nitro group **143** (80BSF163) (Scheme 64).

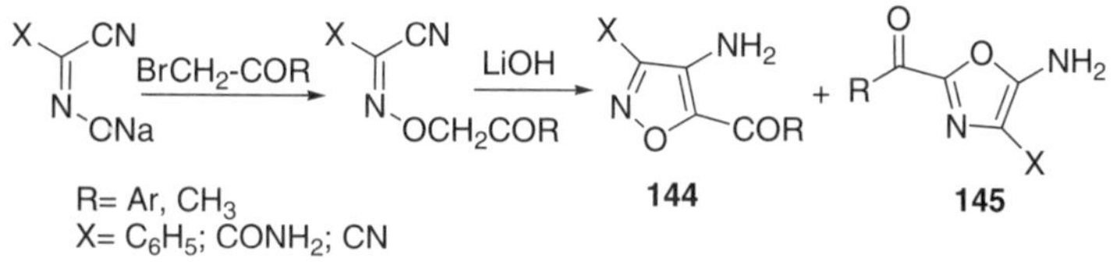

Scheme 65

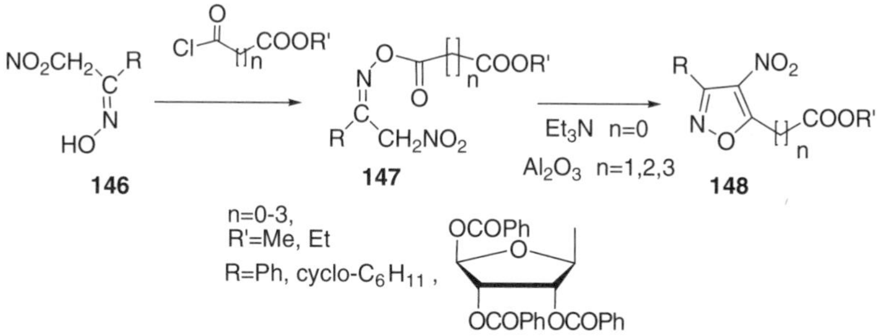

Scheme 66

3. (CCNO+C) Reactions

In the early 1980s, two new approaches to the synthesis of 4-AI with electron-withdrawing functions were proposed (80LA1623, 80TL3613). Using the first method, 5-acyl-4-aminoisoxazoles **144** were synthesized through a two-step procedure including alkylation of cyanooximes followed by cyclization under the action of LiOH (80LA1623). Although this procedure is two-stage, 4-AI were formed as mixtures with 5-aminooxazoles **145,** the target compounds being difficult to isolate (06IZV1773). A modified procedure including the interaction of O-alkylated oximes with lithium perchlorate prior to the action of LiOH furnishes 4-AI **144** as the only reaction products (05IZV1159) (Scheme 65).

In the other approach, nitromethyloximes **146** were acylated with diacid monochlorides and then the isolated O-acylated oximes **147** were cyclized by treatment with triethylamine (80TL3613, 85H1465) or Al_2O_3 (85H1465). 4-Nitroisoxazoles **148** were hydrogenated to the corresponding 4-AI by aluminum amalgam (high yield) or by powdered iron in the presence of $CaCl_2$ (78%) (87H2419) (Section IV.B.1). As noted by the authors, this method is suitable for a large-scale preparation of 4-AI, as no complex mixtures were formed and the yields were excellent (85H1465) (Scheme 66).

B. TRANSFORMATION OF OTHER SUBSTITUENTS TO AN AMINO GROUP

1. Hydrogenation of Nitroisoxazoles and Reductive Openings of Monocyclic Aminoisoxazoles

The choice of a reducing agent for the preparation of aminoisoxazoles is not always obvious because mild hydrogenating agents, which do not affect the isoxazole ring, can reduce the nitro group to azo or azoxy groups, but not to an amino group. On the other hand, many conventional agents used for hydrogenation of a nitro group to an amino group (Pd/C, Raney Ni) are capable of opening the isoxazole ring (75S20, 81KGS1155). Reductive cleavage of the N–O bond in isoxazoles is a well-known process, often employed in the synthesis of other heterocycles and natural products (54JCS665, 62HC1, 63AHC365, 79AHC147, 84CHC1, 91HC1). Hence, as a result of the action of strong reducing agents on nitroisoxazoles, ring-opening products could be obtained instead of AI. The following agents were used for reducing nitroisoxazoles: zinc in acetic acid (62HC1, p. 73, 83H1291), SnCl$_2$ in hydrochloric acid (31G970, 35G1203, 41G327, 46G87, 61T51), aluminum amalgam (21JCS697, 41G327, 61T51, 80TL3613, 77JINC45), and powdered iron in the presence of CaCl$_2$ (87H2419). As a rule, the yields were above 80%.

Hydrogenation of 4-nitroisoxazoles into 4-AI in the presence of Pd/C was used successfully in many works (67G25, 68T4907, 87JHC1003, 87JHC1309), but hydrogenation of 3-nitroisoxazole gave no 3-AI under any of the examined conditions (85JOC5723). Thus opening of isoxazolylisoxazole **149** is of considerable interest (71T379) (Scheme 67).

Hydrogenation over palladium produces isoxazole **150** and then 3-methylisoxazol-5-ylacetic acid **151**. The action of more powerful Raney nickel provides opening of both isoxazole rings, which results in γ-pyridone **152** after closure of the acyclic intermediate.

Since the amino group increases the electron density in the isoxazole ring, we may assume that hydrogenation of aminoisoxazole will occur more slowly then hydrogenation of the isoxazole ring without amino groups. However, in the presence of palladium, the aminoisoxazole ring was hydrogenated first, probably due to preferred adsorption of the aminoisoxazole ring on the palladium surface. Another question is why the isoxazole ring does not open during hydrogenation of 4-nitroisoxazoles in the presence of palladium. This dramatic difference in the behavior of AI isomers may be caused by the above-mentioned distinction in the sites of metal ion coordination to isoxazole rings of AI isomers. The protonation of 4-AI is directed to the amino group and that of 3-AI and 5-AI occurs at the ring nitrogen. Moreover, some data indicate that metal cations coordinate in the same way (Section

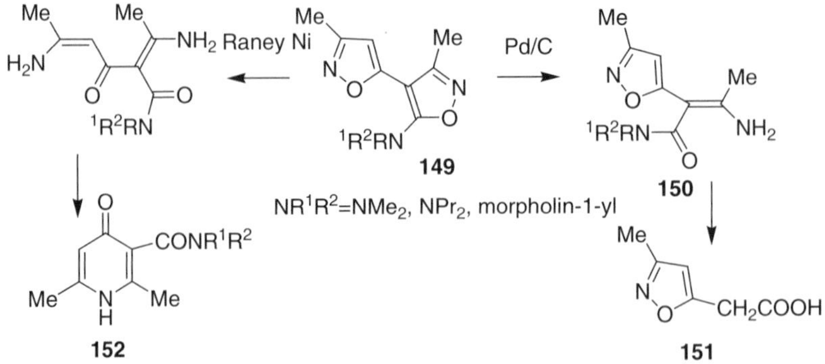

Scheme 67

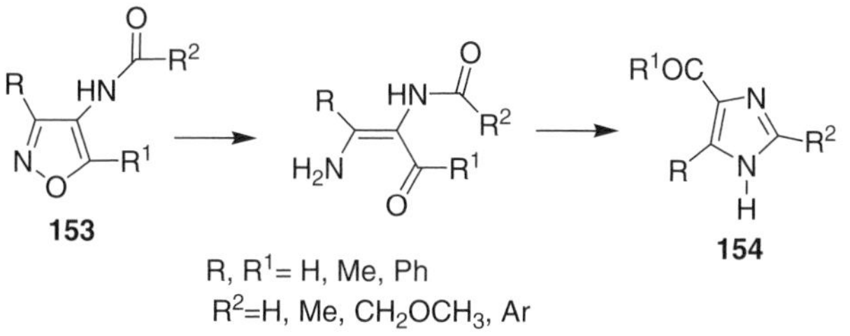

R, R^1= H, Me, Ph

R^2=H, Me, CH$_2$OCH$_3$, Ar

Scheme 68

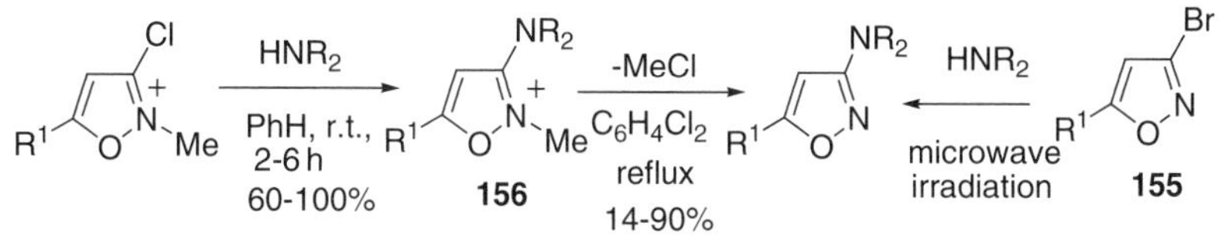

Scheme 69

II). Hence, the assumption that 3-AI (and perhaps 5-AI) are coordinated by the ring nitrogen to the palladium surface, while 4-AI is coordinated by the amino-group nitrogen does not seem unlikely.

Actually, opening of 4-AI can take place when the amino group is acylated. Hydrogenation of 4-acetamidoisoxazole **153** in the presence of palladium on carbon and recyclization of the intermediate yields 4-acylimidazole **154** (87JOC2714) (Scheme 68).

2. Nucleophilic Substitution

The replacement of an halogen atom by amines in 3-chloro(bromo)isoxazoles **155** does not occur on heating; however it can be induced by microwave irradiation (04TL3189). The halogen atom may be easily replaced in 2-methyl-3-chloroiso-xazolium salts **156** (84CPB530) (Scheme 69).

When an acceptor group is missing, the substitution of halogens in position 5 is noticeably hindered (53G192); however using lithium salts of amines, one can achieve high rates and high yields (89S275, 90MI1, 90MI2). All appropriate leaving groups including a methoxy group can be easily substituted if an acceptor group is present in position 4 of the isoxazole ring (60CB1103) (Scheme 70).

Structure **159** has been erroneously assigned to the product of the reaction bet-ween 3-benzoyl-3-methyl-4H-isoxazol-5-one **157** and SOCl$_2$ (53GA192). In reality, the compound has the structure of isoxazolone **158** (74HCA1934). The reactions of chloro compound **158** with ammonia, phenylenediamine, and phenylhydrazine pro-duce enamine **160**, isoxazolo[5,4-e]1.4-diazepine **161**, and isoxazolo[5,4-c]pyrazole **162**, respectively (59ZOB3446). The formation of **161** and **162** represents examples of substitution at the keto group in position 5 (Scheme 71).

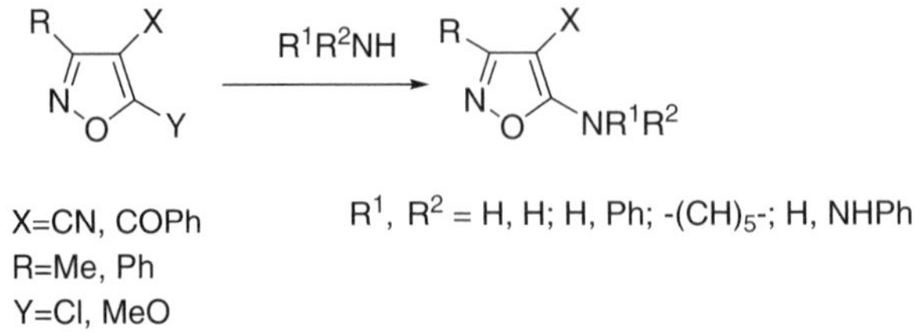

X=CN, COPh
R=Me, Ph
Y=Cl, MeO

R^1, R^2 = H, H; H, Ph; -(CH)$_5$-; H, NHPh

Scheme 70

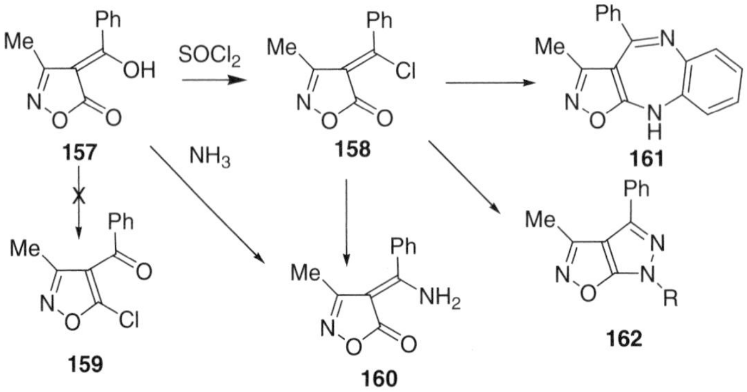

Scheme 71

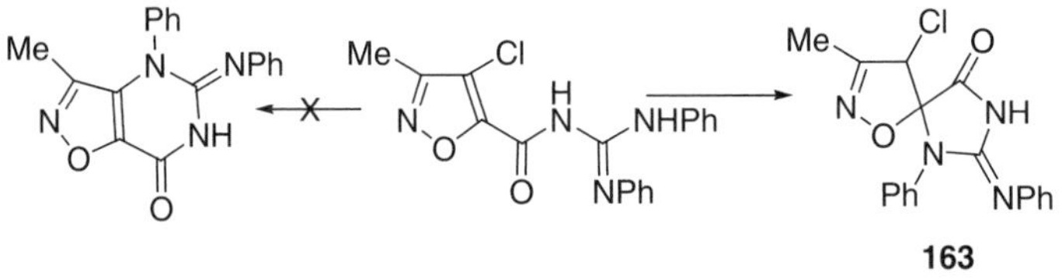

Scheme 72

The substitution of halogens in position 4 of the isoxazole ring is so disadvantageous that it does not take place even if an additional acceptor is present in position 5. The spiro-product **163** was only obtained after long refluxing in THF (85JCS(P1)1871) (Scheme 72).

3. Reactions Involving Nitrene Rearrangements

When hydrogen, alkyl or carboxyl is present in the *ortho*-position to the amide (carbonyl azide) group, the Hofmann and the Curtius reactions give the corresponding AI in good yields (57MI1, 82JCS(P1)2391, 32G457, 91JCS(P1)765). However, if the *ortho*-position is occupied by an amide group or a hydroxyalkyl group, the condensed systems **164** and **165** are formed in these reactions (67G25, 67T675, 81S315) instead of the desired AI. The Beckmann rearrangement yields acylated

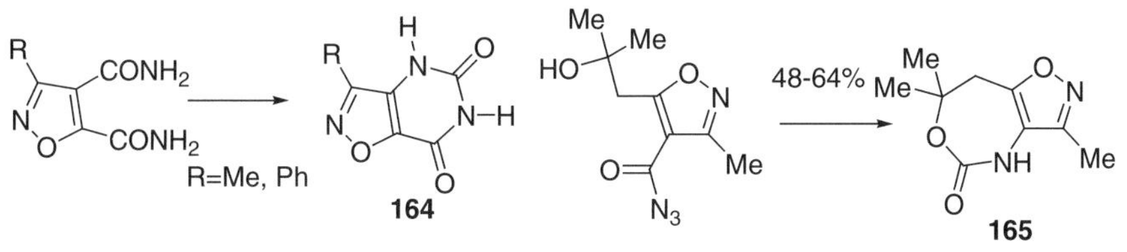

Scheme 73

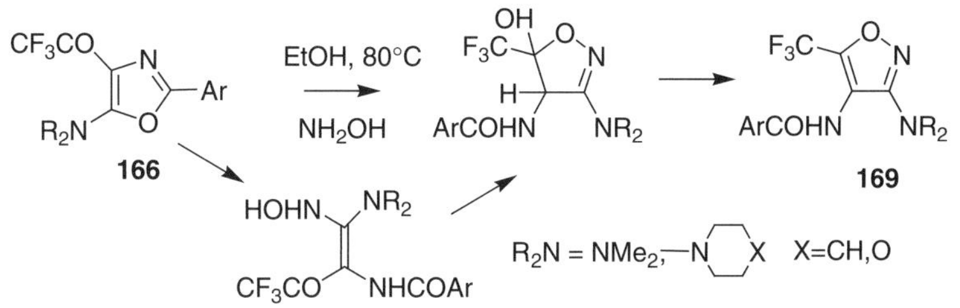

Scheme 74

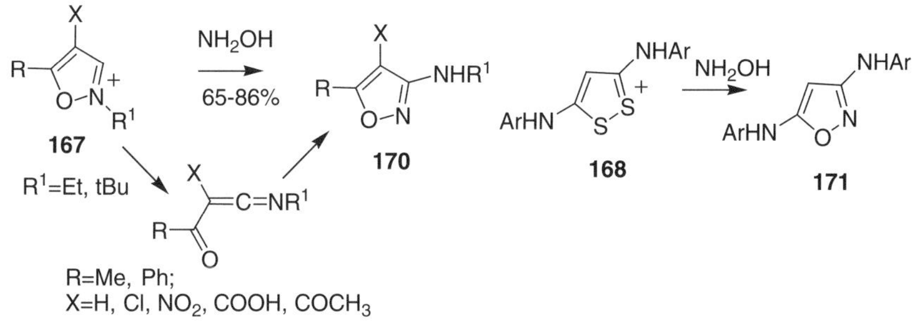

Scheme 75

5-AI mixed with the amides of isoxazole-5-carboxylic acids, the mixture being diffi-cult to separate (75ACS(B)65). Such problems are typical in the preparation of many other amino-substituted heterocycles by these nitrene rearrangements (Scheme 73).

C. Ring Transformations Leading to Aminoisoxazoles

A few rearrangements leading to AI were discovered, however these reactions give AI of unusual types. Thus the action of hydroxylamine on 5-dialkylaminooxazoles **166**, 2-alkylisoxazolium salts unsubstituted in position 3 **167**, and the 3,5-dianilino-[1,2]dithiolylium salts **168** results in acylated 3,4-diaminoisoxazoles **169** (76JHC825), 3-alkylaminoisoxazoles **170** with the functional group in position 4 (88S203), and N,N′-diaryl-3,5-diaminoisoxazoles **171** (67CB1389), respectively (Schemes 74 and 75).

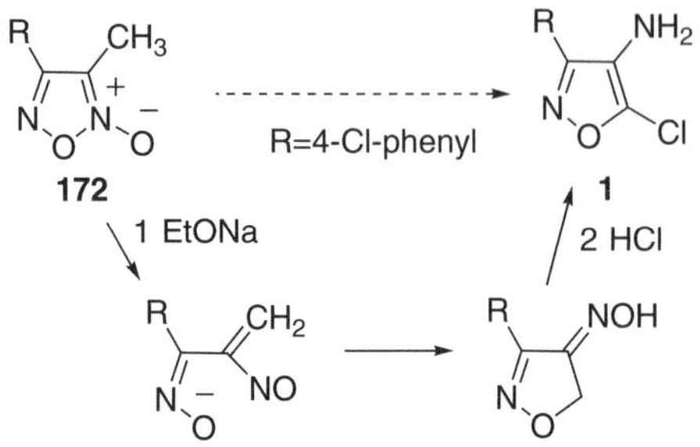

Scheme 76

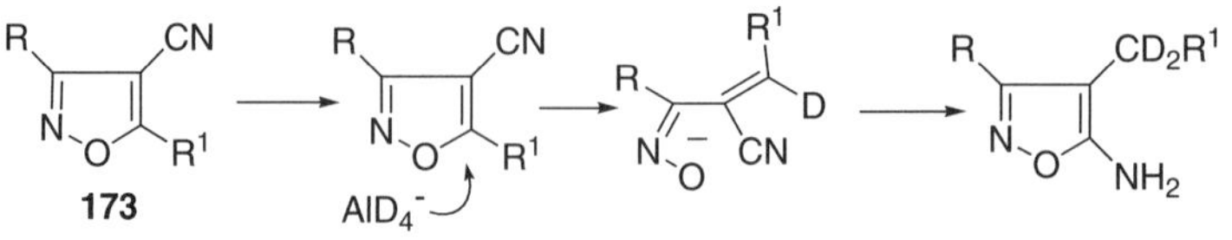

R, R^1= Alk, Ar

Scheme 77

A two-stage synthesis including opening of furoxan **172** on treatment with NaOEt and then HCl gives 5-chloro-4-aminoisoxazoles **1**, which cannot be obtained by other methods (89MI1) (Scheme 76).

Reducing 4-cyanoisoxazoles **173** with LiAlH$_4$ or LiAlD$_4$, but not with NaBH$_4$, gives 5-AI or selectively deuterated 5-AI (84JOC3423). With the reaction using NaBH$_4$, isoxazoles with electron-withdrawing groups at C-4 are reduced regiospecifically to 2-isoxazolines (Scheme 77).

These results demonstrate that the chemistry of AI can be similar to isoxazole chemistry, or to aminopyrazole chemistry, or dissimilar to both. The special properties of AI discussed in this survey originate from interaction of an amino group and the isoxazole ring and such understanding may facilitate a identification of biodegradation products of some drugs based on AI and biochemical pathways of AI in living cell.

REFERENCES

17JA697	D. E. Worrall, *J. Am. Chem. Soc.*, **39**, 697 (1917).
20JA1055	D. E. Worrall, *J. Am. Chem. Soc.*, **42**, 1055 (1920).
21JCS697	T. Morgan and H. Burges, *J. Chem. Soc.*, **119**, 697 (1921).
21JCS1546	T. Morgan and H. Burges, *J. Chem. Soc.*, **119**, 1546 (1921).
31G970	A. Quilico, *Gazz. Chim. Ital.*, **61**, 970 (1931).
32CB1857	E. Wahlberg, *Chem. Ber.*, **65**, 1857 (1932).
32G457	M. Freri, *Gazz. Chim. Ital.*, **62**, 457 (1932).
34CB1062	K. V. Auwers and H. Wunderling, *Chem. Ber.*, **67**, 1062 (1934).
35G1203	A. Quilico, *Gazz. Chim. Ital.*, **65**, 1203 (1935).

36MI1   A. Quilico and R. Fusco, *Rend. Ist. Lombargo Sci. [2]*, **69**, 436 (1936)
                 [*Chem. Abstr.*, **32**, 7454 (1936)].
37JA933   D. E. Worrall, *J. Am. Chem. Soc.*, **59**, 933 (1937).
37JA1486   D. E. Worrall, *J. Am. Chem. Soc.*, **59**, 1486 (1937).
38JA1198   D. E. Worrall, *J. Am. Chem. Soc.*, **60**, 1198 (1938).
39G523   C. Musante, *Gazz. Chim. Ital.*, **69**, 523 (1939).
41G327   A. Quilico and C. Musante, *Gazz. Chim. Ital.*, **71**, 327 (1941).
46G87   A. Quilico, *Gazz. Chim. Ital.*, **76**, 87 (1946).
46G148   A. Quilico and G. Speroni, *Gazz. Chim. Ital.*, **76**, 148 (1946).
52G98   S. Cusmano, V. Sprio, and F. Trapani, *Gazz. Chim. Ital.*, **82**, 98
                 (1952).
52JPJ148   H. Kano, *J. Pharm. Soc. Jpn.*, **72**, 148 (1952).
52JPJ1118   H. Kano, *J. Pharm. Soc. Jpn.*, **72**, 1118 (1952).
53G192   G. Speroni and E. Giachetti, *Gazz. Chim. Ital.*, **83**, 192 (1953).
53JPJ387   H. Kano, *J. Pharm. Soc. Jpn.*, **73**, 387 (1953).
54CB1175   W. Logemann, L. Almirante, and L. Caprio, *Chem. Ber.*, **8**, 1175
                 (1954).
54JCS665   G. Shaw and G. Sugowdz, *J. Chem. Soc.*, **665**(1954).
54JPJ138   H. Kano, *J. Pharm. Soc. Jpn.*, **74**, 138 (1954).
56JPJ1311   Y. Makizumi, *J. Pharm. Soc. Jpn.*, **74**, 138 (1954).
57MI1   H. Kano and K. Ogata, *Ann. Rep. Shionogi Res. Lab.*, **7**, 1 (1957)
                 [*Chem. Abstr.*, **51**, 17889 (1957)].
58JA2829   W. J. Middleton and V. A. Engelhardt, *J. Am. Chem. Soc.*, **80**, 2829
                 (1958).
58CPB105   H. Kano, Y. Makisumi, and K. Ogata, *Chem. Pharm. Bull.*, **6**, 105
                 (1958).
59ZOB3446   I. Ya. Postovskii and S. L. Sokolow, *Zh. Obshch. Khim.*, **29**, 3446
                 (1959).
60CB1103   A. Dornow and H. Teckenburg, *Chem. Ber.*, **98**, 1103 (1960).
61JOC4917   L. Bauer and C. V. N. Nambyry, *J. Org. Chem.*, **26**, 4917 (1961).
61T41   A. J. Boulton and A. R. Katritzky, *Tetrahedron*, **12**, 41 (1961).
61T51   A. J. Boulton and A. R. Katritzky, *Tetrahedron*, **12**, 51 (1961).
62HC1   A. Quilico, in *"The Chemistry of Heterocyclic Compounds"*
                 (A. Weissberger, ed.), Vol. 17, p. 1, Interscience Publishers.
                 A Division of John Wiley and Sons, NY (1962).
63AHC365   M. K. Kochetkov and S. D. Sokolov, *Adv. Heterocycl. Chem.*, **2**, 365
                 (1963).
63CB1088   H. -D. Stachel, *Chem. Ber.*, **96**, 1088 (1963).
64JOC2116   E. C. Taylor and E. E. Garcia, *J. Org. Chem.*, **29**, 2116 (1964).
64T159   H. Kano and E. Yamazaki, *Tetrahedron*, **20**, 159 (1964).
64T461   H. Kano and E. Yamazaki, *Tetrahedron*, **20**, 461 (1964).
65JOC2862   W. J. Fanshawe, V. J. Bauer, and S. R. Safir, *J. Org. Chem.*, **30**, 2862
                 (1965).
66CPB1277   I. Iwan and N. Nakamura, *Chem. Pharm. Bull.*, **14**, 1277 (1966).
66CR557   L. Lopez and J. Barrans, *C. R. Hebd. Seances Acad. Sci. Ser. C*, **263**,
                 557 (1966).
67CB1389   G. Barnikow, *Chem. Ber.*, **100**, 1389 (1967).
67G25   G. Desimoni and P. Grunanger, *Gazz. Chim. Ital.*, **97**, 25 (1967).
67JCS(C)2005   A. J. Boulton, A. R. Katritzky, and A. Majid Hamid, *J. Chem. Soc.
                 (C)*, 2005 (1967).
67T675   G. Desimoni, P. Grunanger, and V. Finzi, *Tetrahedron*, **23**, 675
                 (1967).
67T687   G. Desimoni, P. Grunanger, and V. Finzi, *Tetrahedron*, **23**, 687
                 (1967).

67T3541	R. Rajagopalan and C. N. Talaty, *Tetrahedron*, **23**, 3541 (1967).
67T4395	W. Barbieri, L. Bernardi, S. Coda, V. Colo, and G. Palamidessi, *Tetrahedron*, **23**, 4395 (1967).
68T4907	G. Dezimoni and G. Minoli, *Tetrahedron*, **24**, 4907 (1968).
69TL4817	M. Dines and M. L. Scheinbaum, *Tetrahedron Lett.*, **54**, 4817 (1969).
70JCS(C)1825	T. Nishivaki, A. Nakano, and H. Matsuoka, *J. Chem. Soc. (C)*, 1825 (1970).
70LA195	H. G. Aurich, *Liebigs Ann. Chem.*, **732**, 195 (1970).
70M1109	W. Kloetzer, H. Bretschneider, and E. Fritz, *Monash. Chem.*, **101**, 1109 (1970).
70T1393	G. Desimoni and G. Minoli, *Tetrahedron*, **26**, 1393 (1970).
71JCS(C)2644	T. Nishivaki, T. Sato, S. Onomura, and K. Kondo, *J. Chem. Soc. (C)*, 2644 (1971).
71JCS(C)2648	T. Nishivaki and T. Sato, *J. Chem. Soc. (C)*, 2648 (1971).
71JCS(C)3021	T. Nishivaki and T. Sato, *J. Chem. Soc. (C)*, 3021 (1971).
71T379	P. Caramella, R. Metelli, and P. Grunanger, *Tetrahedron*, **27**, 379 (1971).
72BCJ1846	H. Fujita and R. Endo, *Bull. Chem. Soc. Jpn.*, **45**, 1846 (1972).
72JHC1219	G. Adembri, F. Ponticelli, and P. Tadeshi, *J. Heterocycl. Chem.*, **9**, 1219 (1972).
72MI1	T. Denzel and H. Hoehn, *Arch. Pharm. (Weinheim Ger.)*, **305**, 833 (1972).
73JHC181	E. Abushanab, D. Y. Lee, and L. Goodman, *J. Heterocycl. Chem.*, **10**, 181 (1973).
73JHC957	H. Harsanyi, *J. Heterocycl. Chem.*, **10**, 957 (1973).
73LA256	F. Boberg and W. von Gentzkow, *Liebigs Ann. Chem.*, 256 (1973).
73LA898	C. Grundmann, R. K. Bansal, and P. S. Osmanski, *Liebigs Ann. Chem.*, 898 (1973).
74CB13	H. G. Aurich and G. Blinne, *Chem. Ber.*, **107**, 13 (1974).
74CB2563	K. Harsanyi, K. Takacs, and K. Horvath, *Chem. Ber.*, **107**, 2563 (1974).
74HCA1934	R. Jaunin, *Helv. Chim. Acta*, **57**, 1934 (1974).
74JCS(P1)1867	T. Nishiwaki, K. Azechi, and F. Fujiyama, *J. Chem. Soc., Perkin Trans. 1*, 1867 (1974).
74JMC451	H. A. Burch, L. E. Bengamin, and H. E. Russell, *J. Med. Chem.*, **17**, 451 (1974).
75ACS(B)65	S. B. Christensen and P. Krogsgaard-Larsen, *Acta Chem. Scand., Ser. B*, **29**, 65 (1975).
75HCA1735	J. Tronchet, O. Martin, J. B. Zumwald, N. Le Hong, and F. Perret, *Helv. Chim. Acta*, **58**, 1735 (1975).
75JCS(P1)693	M. A. Khan and F. K. Rafla, *J. Chem. Soc., Perkin Trans. 1*, 693 (1975).
75JOC2600	K. T. Potts and J. Kane, *J. Org. Chem.*, **40**, 2600 (1975).
75S20	T. Nishiwaki, *Synthesis*, 20 (1975).
75S664	V. D. Piaz and S. Pinzauti, *Synthesis*, 664 (1975).
76HCA1705	G. Vernin, G. Siv, S. Treppendahh, and J. Metzger, *Helv. Chim. Acta*, **59**, 1705 (1976).
76JHC825	D. Clerin, J. P. Fleury, and H. Fritz, *J. Heterocycl. Chem.*, **13**, 825 (1976).
76MI1	V. A. Chuiguk and N. A. Parkhomenko, *Sov. Prog. Chem.*, **42**, 37 (1976).
77H51	H. Yamanaka, T. Sakamoto, and A. Shiozawa, *Heterocycles*, **7**, 51 (1977).

77JCS(P1)1616   N. Vivona, G. Cusmano, and G. Macaluso, *J. Chem. Soc., Perkin Trans. 1*, 1616 (1977).

77JHC435   A. Camparini, F. Pouticelli, and P. Tedeschi, *J. Heterocycl. Chem.*, **14**, 435 (1977).

77JHC1299   R. Faure, J. -P. Galy, E. -J. Vincent, and J. Elguero, *J. Heterocycl. Chem.*, **14**, 1299 (1977).

77JINC45   G. Ponticelli, *J. Inorg. Nucl. Chem.*, **39**, 45 (1977).

77JMC934   J. B. Carr, H. G. Durhann, and D. K. Hass, *J. Med. Chem.*, **20**, 934 (1977).

78CPB2759   H. Yamanaka, N. Egawa, and T. Sakamoto, *Chem. Pharm. Bull.*, **26**, 2759 (1978).

78JPR585   W. D. Rudolf and M. Augustin, *J. Prakt. Chem.*, **320**, 585 (1978).

78KGS969   V. M. Neplyev, T. A. Sinenko, and P. S. Pelkis, *Khim. Geterotsikl. Soed.*, 969 (1978).

78SA(A)967   M. Biddau, M. Massacesi, R. Pinna, and G. Ponticelli, *Spectrochim. Acta*, **34A**, 967 (1978).

78USP4152336   M. Kuroki, S. Kono, and K. Shioka, US Pat. 4,152,336, Sogo Pharm Co., Ltd. [*Chem. Abstr.*, **91**, 56990 (1978)].

79AHC147   B. J. Wakefield and D. J. Wright, *Adv. Heterocycl. Chem.*, **25**, 147 (1979).

79HCA833   J. Tronchet, O. Martin, A. Groniller, and N. Sarda, *Helv. Chim. Acta*, **62**, 833 (1979).

79JHC1611   C. Bellec, D. Bertin, and R. Colau, *J. Heterocycl. Chem.*, **16**, 1611 (1979).

79JOC4191   H. A. Albrecht, J. F. Blount, F. M. Konzemann, and J. T. Plati, *J. Org. Chem.*, **44**, 4191 (1979).

79M1387   O. S. Wolfbeis and H. Junek, *Monash. Chem.*, **110**, 1387 (1979).

79S449   C. Skoetsch, I. Kohlmeyer, and E. Breitmaier, *Synthesis*, 449 (1979).

80CB1195   H. Janek, B. Thierricher, and G. Lukas, *Chem. Ber.*, **113**, 1195 (1980).

80CPB567   Y. Okamoto, K. Takagi, and T. Ueda, *Chem. Pharm. Bull.*, **28**, 567 (1980).

80CPB1832   T. Sakamoto, H. Yamanaka, and A. Shiozawa, *Chem. Pharm. Bull.*, **28**, 1832 (1980).

80BSF163   R. Colau and C. Viel, *Bull. Soc. Chim. Fr., Part 2*, 163 (1980).

80LA1623   K. Gewald, P. Bellmann, and H.-J. Jaensch, *Liebigs Ann. Chem.*, 1623 (1980).

80OMR235   G. Vernin, C. Siv, and L. Bouscasse, *Org. Magn. Reson.*, **14**, 235 (1980).

80TL3613   J. A. Deceuninck, D. K. Buffel, and G. J. Hoornaert, *Tetrahedron Lett.*, **21**, 3613 (1980).

80TL3755   F. Pochat, *Tetrahedron Lett.*, **21**, 3755 (1980).

81IJC(B)654   J. M. Bellary and V. V. Badiger, *Indian J. Chem., Sect. B.*, **20**, 654 (1981).

81JHC607   G. Hass, J. L. Stanton, A. von Sprecher, and P. Wenk, *J. Heterocycl. Chem.*, **18**, 607 (1981).

81JHC619   G. Haas, J. Stanton, and T. Winkler, *J. Heterocycl. Chem.*, **18**, 619 (1981).

81JOC3505   J. D. Perez, R. G. de Diaz, and G. I. Yranzo, *J. Org. Chem.*, **46**, 3505 (1981).

81KGS1155   A. A. Akhrem, F. A. Lakhvich, and V. A. Khripach, *Khim. Geterotsikl. Soed.*, 1155 (1981).

81S315   B. Chantegrel and S. Gelin, *Synthesis*, 315 (1981).

82AHC1   A. Albert, *Adv. Heterocycl. Chem.*, **32**, 1 (1982).

82AX(B)1845   C. Chatterjee, J. K. Dattagupta, and M. N. Saha, *Acta Crystallogr.,
                 Sect. B*, **38**, 1845 (1982).

82FA701   G. Vampa, P. Pecorari, A. Albasini, M. Rinaldi, and M. Melegari,
                 Farmaco Ed. Sci., **37**, 701 (1982).

82JCS(P1)2391   A. Camparini, F. Pontichelli, and P. Tedeschi, *J. Chem. Soc., Perkin
                 Trans. 1*, 2391 (1982).

82JMC12   H. Zinnes, J. Sircar, N. Lindo, M. Schwartz, and A. Fabian, *J. Med.
                 Chem.*, **25**, 12 (1982).

83H1291   J. F. Keana and G. M. Little, *Heterocycles*, **20**, 1291 (1983).

83JOC575   G. Zvilichovsky and M. David, *J. Org. Chem.*, **48**, 575 (1983).

83M373   B. Kuebel, *Monash. Chem.*, **114**, 373 (1983).

83PJC1161   M. Gabryszewski and M. Wisniewski, *Pol. J. Chem.*, **57**, 1161 (1983).

83S839   A. K. Murthy, S. Sailaja, E. Rajanarendar, and C. J. Rao, *Synthesis*,
                 839 (1983).

83TL3149   G. L'abbe, F. Godts, and S. Toppet, *Tetrahedron Lett.*, **24**, 3149
                 (1983).

84CHC1   S. A. Lang and Y.-I. Lin, in *"Compounds of Heterocyclic Chemistry"*
                 (A.R. Katritsky, ed.), Vol. 6, p. 1, Pergamon Press, New York
                 (1984).

84CPB530   S. Sugai, K. Sato, K. Kataoka, Y. Iwasaki, and K. Tomita, *Chem.
                 Pharm. Bull.*, **32**, 530 (1984).

84H618   T. Yamamori, Y. Hiramatsu, K. Sakai, and I. Adachi, *Heterocycles*,
                 21, 618 (1984).

84JCS(P1)1079   Z. T. Fomum, P. F. Asobo, S. R. Landor, and P. D. Landor, *J.
                 Chem. Soc., Perkin Trans. 1*, 1079 (1984).

84JOC3423   A. Alberola, A. M. Gonzalez, M. A. Laguna, and F. J. Pulido, *J. Org.
                 Chem.*, **49**, 3423 (1984).

84S247   A. Rahman, J. N. Vishwakarma, R. D. Yadav, H. Ila, and
                 H. Junjappa, *Synthesis*, 247 (1984).

84S1050   G. Winters, A. Sala, A. de Paoli, and M. Conti, *Synthesis*, 1050
                 (1984).

84S1052   G. Winters, A. Sala, A. de Paoli, and V. Ferri, *Synthesis*, 1052 (1984).

84ZC93   K. Jaehnisch, H. Seeboth, and N. Catasus, *Z. Chem.*, **24**, 93 (1984).

84ZC256   J. Wrubel and R. Mayer, *Z. Chem.*, **24**, 256 (1984).

85CJC3012   L. Maury, J. Rambaud, B. Pauvert, Y. Lasserre, G. Berge, and
                 M. Audran, *Can. J. Chem.*, **63**, 3012 (1985).

85H1465   R. Nesi, S. Chimichi, P. Sarti-Fantoni, A. Buzzi, and D. Giomi,
                 Heterocycles, **23**, 1465 (1985).

85JCS(P1)1871   R. Nesi, S. Chimichi, P. Sarti-Fantoni, P. Tedeschi, and D. Giomi, *J.
                 Chem. Soc., Perkin Trans. 1*, 1871 (1985).

85JHC1663   J. A. Ciller, C. Seoane, and J. L. Soto, *J. Heterocycl. Chem.*, **22**,
                 1663–1665 (1985).

85JIC135   S. Sailaja, E. Rajanarendar, C. J. Rao, and A. K. Murthy, *J. Indian
                 Chem. Soc.*, **62**, 135 (1985).

85JOC2372   E. M. Beccalli, A. Manfredi, and A. Marchesini, *J. Org. Chem.*, **50**,
                 2372 (1985).

85JOC5723   J. C. Sircar, T. Capiris, T. P. Bobovski, and C. F. Schwender, *J. Org.
                 Chem.*, **50**, 5723 (1985).

85T913   T. Yamamori, Y. Hiramatu, K. Sakai, and I. Adachi, *Tetrahedron*,
                 41, 913 (1985).

86ACS(B)760   K. Tanaka and E. B. Pedersen, *Acta Chem. Scand., Ser. B*, **40**, 760
                 (1986).

86JHC1535   K. Tanaka, T. Suzuki, S. Maeno, and K. Mitsuhashi, *J. Heterocycl.
                 Chem.*, **23**, 1535 (1986).

86H3433	G. Macaluso, G. Cusmano, S. Buscemi, V. Frenna, N. Vivona, and M. Ruccia, *Heterocycles*, **24**, 3433 (1986).
87BCJ4480	K. Tanaka, T. Suzuki, S. Maeno, and K. Mitsuhashi, *Bull. Chem. Soc. Jpn.*, **60**, 4480 (1987).
87CPB3676	T. Tatee, K. Narita, S. Kurashige, S. Ito, H. Miyazaki, H. Yamanaka, M. Mizugaki, T. Sakamoto, and H. Fukuda, *Chem. Pharm. Bull.*, **35**, 3676 (1987).
87H2419	R. Nesi, D. Giomi, L. Quartara, S. Papaleo, and P. Tedeschi, *Heterocycles*, **26**, 2419 (1987).
87JHC501	M. M. El-Kerdawy, S. M. Bayomi, I. A. Shehata, and R. A. Glennon, *J. Heterocycl. Chem.*, **24**, 501 (1987).
87JHC1003	K. Takagi, M. Tanaka, Y. Murakami, K. Ogura, and K. Ishii, *J. Heterocycl. Chem.*, **24**, 1003 (1987).
87JHC1291	I. A. Shehata and R. A. Glennon, *J. Heterocycl. Chem.*, **24**, 1291 (1987).
87JHC1309	A. A. Almerico, G. Dattolo, G. Cirrincione, G. Presti, and E. Aiello, *J. Heterocycl. Chem.*, **24**, 1309–1311 (1987).
87JOC2714	L. A. Reiter, *J. Org. Chem.*, **52**, 2714 (1987).
87S33	J. J. Vaquero, L. C. Fuentes, C. Del Juan, M. I. Perez, J. L. Garcia, and J. L. Soto, *Synthesis*, 33 (1987).
88BCJ2881	S. Zen, K. Harada, H. Nakamura, and Y. Iitaka, *Bull. Chem. Soc. Jpn.*, **61**, 2881 (1988).
88FA753	G. Daidone, D. Raffa, M. L. Bajardi, S. Plescia, and M. Milici, *Farmaco Ed. Sci.*, **43**, 753 (1988).
88IJC(B)686	E. T. Rao, E. Rajanarendar, and A. Krishnamurthy, *Indian J. Chem., Sect. B.*, **27**, 686 (1988).
88JHC231	D. Chiarino, M. Napoletano, and A. Sala, *J. Heterocycl. Chem.*, **25**, 231 (1988).
88JHC607	K. H. Pilgram and L. H. Gale, *J. Heterocycl. Chem.*, **25**, 607 (1988).
88S203	A. Alberola, L. F. Antolin, P. Cuadrado, A. M. Gonzalez, M. A. Laguna, and F. J. Pulido, *Synthesis*, 203 (1988).
89HCA556	A. Pascual, *Helv. Chim. Acta*, **72**, 556 (1989).
89JCS(P2)1941	A. Garrone, R. Fruttero, C. Tironi, and A. Gasco, *J. Chem. Soc., Perkin Trans. 2*, 1941 (1989).
89JHC277	Y. Okamoto and K. Takagi, *J. Heterocycl. Chem.*, **26**, 277 (1989).
89JHC1335	R. Neidlein, D. Kikelj, and W. Kramer, *J. Heterocycl. Chem.*, **26**, 1335 (1989).
89MI1	G. Dannhardt and I. Obergrusberger, *Arch. Pharm. (Weinheim Ger.)*, **322**, 513 (1989).
89S275	G. Dannhardt, S. Laufer, and I. Obergrusberger, *Synthesis*, 275 (1989).
90BCJ1851	E. T. Rao, E. Rajanarendar, and A. Krishnamurthy, *Bull. Chem. Soc. Jpn.*, **63**, 1851 (1990).
90G725	R. Nesi, D. Giomi, S. Papaleo, P. Tedeschi, and F. Ponticelli, *Gazz. Chim. Ital.*, **120**, 725 (1990).
90H1823	A. Baracchi, S. Chimichi, F. de Sio, D. Donati, C. Polo, and P. Sarti-Fantoni, *Heterocycles*, **31**, 1823 (1990).
90HCA2147	R. Neidlein, A. Bischer, and W. Kramer, *Helv. Chim. Acta*, **73**, 2147 (1990).
90JHC1481	C. B. Vicentini, A. C. Veronese, T. Poli, M. Guarneri, and P. Giori, *J. Heterocycl. Chem.*, **27**, 1481 (1990).
90JHC1617	G. M. Shutske, *J. Heterocycl. Chem.*, **27**, 1617 (1990).
90KGS914	Yu. A. Nikolyukin, L. V. Dulenko, and V. I. Dulenko, *Chem. Heterocycl. Compd. (Engl. Transl.)*, **26**, 914 (1990).

212   V.P. KISLYI ET AL.   [Refs.

90MI1   G. Dannhardt, P. Dominiak, and S. Laufer, *Arch. Pharm. (Weinheim Ger.)*, **323**, 517 (1990).
90MI2   G. Dannhardt, P. Dominiak, and S. Laufer, *Arch. Pharm. (Weinheim Ger.)*, **323**, 571 (1990).
90ZOB1778   S. N. Kuz'min, L. V. Razvodovskaya, V. V. Negrebetskii, and A. F. Grapov, *J. Gen. Chem. USSR (Engl. Transl.)*, **60**, 1778 (1990).
91H1153   A. Takase, A. Murabayashi, S. Sumimoto, S. Ueda, and Y. Makisumi, *Heterocycles*, **32**, 1153 (1991).
91H1765   S. Buscemi, V. Frenna, and N. Vivona, *Heterocycles*, **32**, 1765 (1991).
91H1913   R. Nesi, D. Giomi, S. Papaleo, and S. Turchi, *Heterocycles*, **32**, 1913 (1991).
91HC1   P. Grunanger and P. Vita-Finzi, in "*The Chemistry of Heterocyclic Compounds*" Vol. 49, p. 1, Interscience Publishers, NY (1991).
91HCA531   A. Pascual, *Helv. Chim. Acta*, **74**, 531 (1991).
91IJC(B)946   V. A. Adhikari and V. V. Badiger, *Indian J. Chem., Sect. B.*, **30**, 946 (1991).
91JCS(P1)765   M. M. Campbell, N. D. Cosford, D. R. Rae, and M. Sainsbury, *J. Chem. Soc., Perkin Trans. 1*, **14**, 765 (1991).
91MI1   G. Dannhardt, P. Dominiak, and S. Laufer, *Arch. Pharm. (Weinheim Ger.)*, **324**(3), 141 (1991).
92CB2259   L. F. Tietze and J. Utecht, *Chem. Ber.*, **125**, 2259 (1992).
93TL6423   K. Tatsuta, S. Miura, H. Gunji, T. Tamai, R. Yoshida, and T. Inagaki, *Tetrahedron Lett.*, **34**(40), 6423 (1993).
94BCJ1701   K. Tatsuta, S. Miura, H. Gunji, T. Tamai, R. Yoshida, T. Inagaki, and Y. Kurita, *Bull. Chem. Soc. Jpn.*, **67**, 1701 (1994).
94FA529   E. Wagner and K. Poreba, *Farmaco*, **50**, 183 (1995).
94JHC535   G. B. Okide, *J. Heterocycl. Chem.*, **31**, 535 (1994).
94JMC2721   A. Villalobos, J. F. Blake, C. K. Biggers, T. W. Butler, D. S. Chapin, Y. L. Chen, J. L. Ives, S. B. Jones, D. R. Liston, A. A. Nagel, D. M. Nason, J. A. Nielsen, I. A. Shalaby, and W. F. White, *J. Med. Chem.*, **37**, 2721 (1994).
94IJC(B)1048   V. J. Ram, M. Nath, and S. Chandra, *Indian J. Chem., Sect. B.*, **33**, 1048 (1994).
95JCS(P1)1153 W. Cocker, D. H. Grayson, and P. V. R. Shannon, *J. Chem. Soc., Perkin Trans. 1*, 1153 (1995).
95JMC1344   P. Stein, D. M. Floyd, S. Bisaha, J. Dickey, R. Girotra, J. Z. Gougoutas, M. Kozlowski, V. G. Lee, E. C. -K. Liu, M. F. Malley, D. McMullen, C. Mitchell, S. Moreland, N. Murugesan, R. Serafino, M. L. Webb, R. Zhang, and J. T. Hunt, *J. Med. Chem.*, **38**, 1344 (1995).
96H691   A. Miyashita, K. Fujimoto, T. Okada, and T. Higashino, *Heterocycles*, **42**, 691 (1996).
96JHC1943   R. Neidlein and S. Li, *J. Heterocycl. Chem.*, **33**, 1943 (1996).
96JMC4382   A. D. White, M. W. Creswell, A. W. Chucholowski, C. J. Blankley, M. W. Wilson, R. F. Bousley, A. D. Essenburg, K. L. Hamelehle, B. R. Krause, R. L. Stanfield, M. A. Dominick, and M. Neub, *J. Med. Chem.*, **39**, 4382 (1996).
96TL3339   T. Konoike, Y. Kanda, and Y. Araki, *Tetrahedron Lett.*, **37**, 3339 (1996).
97FA105   S. Ring, W. Malinka, and D. Dus, *Farmaco*, **52**, 105 (1997).
97IJC(B)394   J. Vishnu and M. Nath, *Indian J. Chem., Sect. B.*, **36**, 394 (1997).
97JMC2674   T. R. Elworthy, A. P. Ford, G. W. Bantle, D. J. Morgans, and R. S. Ozer, *J. Med. Chem.*, **40**, 2674 (1997).

98BMC1623   P. H. Olesen, M. D. Swedberg, and K. Rimvall, *Bioorg. Med. Chem.*, **6**, 1623 (1998).

98FA513   S. Adolphe-Pierre, S. Menager, F. Tombret, P. Verite, F. Lepage, and O. Lafont, *Farmaco*, **53**, 513 (1998).

98JCS(P1)2175 K. -J. Kim and K. Kim, *J. Chem. Soc., Perkin Trans. 1*, **14**, 2175 (1998).

98JCX373   S. Ryng and T. Glowiak, *J. Chem. Crystallogr.*, **28**, 373 (1998).

99FA90   D. Raffa, G. Daidone, B. Maggio, D. Schillaci, F. Plescia, and L. Torta, *Farmaco*, **54**, 90 (1999).

99JOC8479   M. Mascal, N. Hext, R. Warmuth, J. Arnall-Culliford, M. Moore, and J. P. Turkenburg, *J. Org. Chem.*, **64**, 8479 (1999).

00JMC3111   N. Murugesan, Z. Gu, P. D. Stein, S. Spergel, A. Mathur, and L. Leith, *J. Med. Chem.*, **43**, 3111 (2000).

01JCX369   Y. Baran and W. Linert, *J. Chem. Crystallogr.*, **31**, 369 (2001).

01MI1   J. Matysiak, E. Krajewska-Kulak, J. Karczewski, and A. Niewiadomy, *J. Agric. Food Chem.*, **49**, 5251 (2001).

01MI2   P. M. Bogdanov, C. S. Ortiz, A. J. Eraso, and I. Albesa, *Med. Chem. Res.*, **10**, 577 (2001).

01T9123   M. A. Abramov, E. J. Ceulemans, A. M. van der Carine, and W. Dehaen, *Tetrahedron*, **57**, 9123 (2001).

01TL8931   H. J. Mason, X. Wu, R. Schmitt, J. E. Macor, and G. Yu, *Tetrahedron Lett.*, **42**, 8931 (2001).

02IJC(B)1670   V. Padmavathi, A. Balaiah, K. V. Reddy, A. Padmaja, and D. B. Reddy, *Indian J. Chem., Sect. B.*, **41**, 1670 (2002).

02JAP(K)363171 A. Matsushita, K. Yoshii, M. Ogami, T. Nakamura, and S. Yamada, *Jpn. Kokai Tokkyo Koho* JP 2002363171 (2002).

02JHC649   V. Padmavathi, A. Balaiah, and D. B. Reddy, *J. Heterocycl. Chem.*, **39**, 649 (2002).

02JOC8558   M. Shtaiwi and C. Wentrup, *J. Org. Chem.*, **67**, 8558 (2002).

02MC99   A. V. Lesiv, S. L. Ioffe, Y. A. Strelenko, I. V. Bliznets, and V. A. Tartakovsky, *Mendeleev Commun.*, **3**, 99 (2002).

02MI1   D. Mares, C. Romagnoli, B. Tosi, R. Benvegn, A. Bruni, and C. Vicentini, *Fung. Gen. Biol.*, **36**, 47 (2002).

02MI2   N. Eddington, D. S. Cox, R. R. Roberts, R. J. Butcher, I. O. Edafiogho, J. P. Stables, N. Cooke, A. M. Goodwin, C. A. Smith, and K. R. Scott, *Eur. J. Med. Chem.*, **37**, 635 (2002).

02TL3907   M. -E. Theoclitou and L. A. Robinson, *Tetrahedron Lett.*, **43**, 3907 (2002).

02TL8777   E. Alberola, *Tetrahedron Lett.*, **43**, 8777 (2002).

02ZOR599   M. V. Vovk, N. V. Mel'nichenko, V. A. Chornous, and M. K.Bratenko, *Zh. Org. Khim.*, **38**, 599 (2002).

03BMC591   T. Glinka, K. Huie, A. Cho, M. Ludwikow, J. Blais, D. Griffith, S. Hecker, and M. Dudley, *Bioorg. Med. Chem.*, **11**(4), 591 (2003).

03BMCL1487   A. Zask, Y. Gu, J. D. Albright, X. Du, M. Hogan, J. I. Levin, J. M. Chen, L. M. Killar, A. Sung, J. F. DiJoseph, M. A. Sharr, C. E. Roth, S. Skala, G. Jin, R. Cowling, K. M. Mohler, D. Barone, R. Black, C. March, and J. S. Skotnicki, *Bioorg. Med. Chem. Lett.*, **13**, 1487 (2003).

03HAC513   V. Padmavathi, A. Balaiah, T. V. Reddy, B. J. Reddy, and D. B. Reddy, *Heteroatom. Chem.*, **14**, 513 (2003).

03JMC1706   W. -T. Li, D. -R. Hwang, Ch. -P. Chen, Ch. -W. Shen, C. -L. Huang, T. -W. Chen, C. -H. Lin, Y. -L. Chang, Y. -Y. Chang, Y. -K. Lo, H. -Y. Tseng, C. -C. Lin, J. -S. Song, H. -C. Chen, S. -J. Chen, S. -H. Wu, and C. -T. Chen, *J. Med. Chem.*, **46**, 1706 (2003).

03MI1   M. Maczynski, A. Jezierska, M. Zimecki, and S. Ryng, *Acta Pol.*
               Pharm., **60**, 147 (2003).

03MI2   K. Poreba, J. Wietrzyk, and A. Opolski, *Acta Pol. Pharm.*, **60**, 293
               (2003).

03MI3   A. Jezierska, J. Panek, S. Ryng, T. Glowiak, and A. Koll, *J. Mol.*
               Model., **9**, 159 (2003).

03PCT013517   M. Cavicchioli, P. Pevarello, B. Salom, and A. Vulpetti, PCT Int.
               Appl. WO 2003013517 (2003) [*Chem Abstr.*, **138**, 187773 (2003)].

03PJC1001   E. Wagner, A. Opolski, and J. Wietrzyk, *Pol. J. Chem.*, **77**(8), 1001
               (2003).

03PJC1461   A. Jezierska, J. Zygmunt, T. Glowiak, A. Koll, and S. Ryng, *Pol. J.*
               Chem., **77**, 1461 (2003).

03S1531   D. M. Volochnyuk, A. O. Pushechnikov, D. G. Krotko,
               D. A. Sibgatulin, S. A. Kovalyova, and A. A. Tolmachev,
               Synthesis, 1531 (2003).

03TL391   D. M. Volochnyuk, A. N. Kostyuk, A. M. Pinchuk, and A. A.
               Tolmachev, *Tetrahedron Lett.*, **44**, 391 (2003).

04AX(E)330   L. Shen, M. C. Li, Z. M. Jin, M. L. Hu, and R. C. Xuan, *Acta*
               Crystallogr., Sect. E: Struct. Rep. Online, **E60**, m330 (2004).

04GEP10256186 R. Bruns, G. Eberz, W. Kreiss, O. Kretschik, M. Kugler, H. Uhr, and
               P. Wachtler, Ger. Offen. DE 10256186, 9 June 2004 [*Chem. Abstr.*,
               141, 23520 (2004)].

04KGS496   M. V. Vovk, A. V. Bolbut, and V. I. Dorokhov, *Chem. Heterocycl.*
               Comp. (Engl. Transl.), **40**, 496 (2004).

04MI1   A. Jezierska, M. Maczynski, A. Koll, and S. Ryng, *Arch. Pharm.*,
               337, 81 (2004).

04PCT048385   N. Kujundzic, K. Bukvic, and K. Brajsa, PCT Int. Appl. WO
               2004043985, 27 May 2004, 21pp [*Chem. Abstr.*, **140**, 407072 (2004)].

04TL3189   J. E. Moore, D. Spinks, and J. Harrity, *Tetrahedron Lett.*, **45**(16),
               3189 (2004).

05JA5512   C. P. Price, A. L. Grzesiak, and A. J. Matzger, *J. Am. Chem. Soc.*,
               127, 5512 (2005).

05IZV1159   V. P. Kislyi, E. B. Danilova, and V. V. Semenov, *Izv. Akad. Nauk*
               SSSR, Ser. Khim., 1159 (2005).

06IZV1773   V. P. Kislyi, E. B. Danilova, V. V. Semenov, A. A. Jakovenko, and
               F. M. Dongushin, *Izv. Akad. Nauk SSSR, Ser. Khim.*, 1773 (2006).

Isothiazolium Salts and Their Use as Components for the Synthesis of S,N-Heterocycles

J. Wolf, and B. Schulze

Department of Organic Chemistry, University of Leipzig, Johannisallee 29, D-04103 Leipzig, Germany

Dedicated to Professor Dr. S. Hauptmann on the occasion of his 75th birthday.

I. Introduction

Isothiazole and its derivatives have a broad range of biological activities (81PMC117, 97JPR1, 03PHMD617). Especially, isothiazol-3(2H)-ones **1** are potent industrial microbiocides because of their antifungal and antibacterial activities (02SUR79).

Saccharin **2**, 1,2-benzisothiazol-3(2H)-one 1,1-dioxide, is one of the most well-known isothiazolone derivatives, which was first prepared by Remsen and Fahlberg in 1879 by an oxidative cyclization of *ortho*-toluene-sulfonamide (1879CB469, 73AHC233, 85AHC105).

Furthermore, the development of a new and efficient method for the synthesis of functionalized monocyclic sultams[1] **3** (R = Ar) (99HCA685, 03ZN(B)111,

---

[1]For cyclic sulfin- and sulfonamides, the IUPAC approved names sultime and sultame were also used in this review.

ADVANCES IN HETEROCYCLIC CHEMISTRY
VOLUME 94 ISSN: 0065-2725 DOI: 10.1016/S0065-2725(06)94004-5

05JEIMC341) and **4** (97T17795, 00T2523) have attracted much attention due to their importance in biological and pharmacological research. The first monocyclic 2, 3-dihydroisothiazole 1,1-dioxides **4**, without the 3-oxogroup but with anti-HIV-1 activity have recently been synthesized (99T7625, (Scheme 1).

The best method for the synthesis of 3-oxosultams **3** is the oxidation of mono- and bicyclic isothiazolium salts **5** and **7** (96T783, 99HCA685). In contrast to the isothiazoles (84CHEC131, 93HOU668, 02SCIS507), the isothiazol-3(2*H*)-ones (02SUR79) and the 1,1-dioxides (97JPR1, 02AHC71, 03H639) as well as the benzoannulated derivatives (85AHC105, 93HOU799, 02SUR279, 02H693, 02SCIS573), the synthesis, reactions, properties and rearrangement of the large number of published papers on the chemistry of isothiazolium salts were not reviewed until now. There is some information on these isothiazolium salts in reviews on isothiazoles (93HOU668, 02SCIS507). This article presents a complete picture of the chemistry of monocyclic, bicyclic, benzocyclic and heterocyclic annulated isothiazolium salts **5–15** (Scheme 2). Several new alternatives to the classical methods of synthesis have been developed. Patents are included, if they reveal new synthetic aspects or interesting applications of formed novel S,N-heterocycles.

II. Synthesis

A. SYNTHESIS OF MONOCYCLIC ISOTHIAZOLIUM SALTS

The general syntheses of monocyclic isothiazolium salts are:

(1) Oxidative cyclization of 3-amino-thioacrylic acid-amides, -esters or 2-amino-1-alkenyl-thioketones;
(2) Ring transformations of 1,2-dithiolium salts with primary amines and oxidative cyclizations;
(3) Cyclocondensation of 3-thiocyanoalkenal-imines and isomerization of 4,5-substituted isothiazolium salts;
(4) N-Alkylation of isothiazoles to quaternary salts;
(5) O/S-Alkylation and oxidative O/S-elimination of isothiazolones and -thiones.

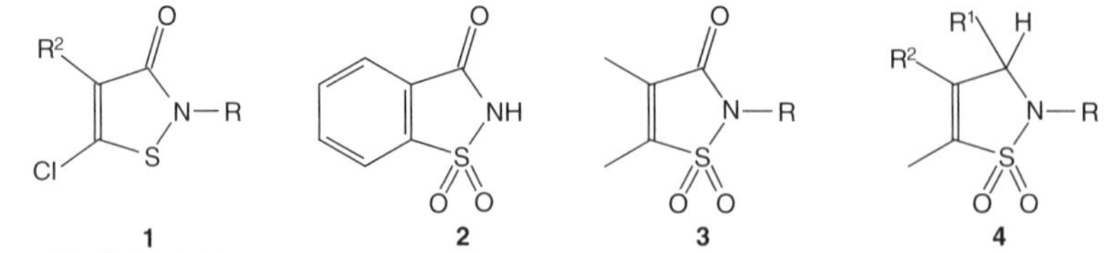

1
R = H, alkyl, CH₂Ph, Ar
R² = H, Cl

2

3
R = alkyl, Ar,
CH₂CO₂R

4
R = H, Me, CH₂Ph
R¹ = Et, *i*-Pr
R² = OH, NH₂

Scheme 1

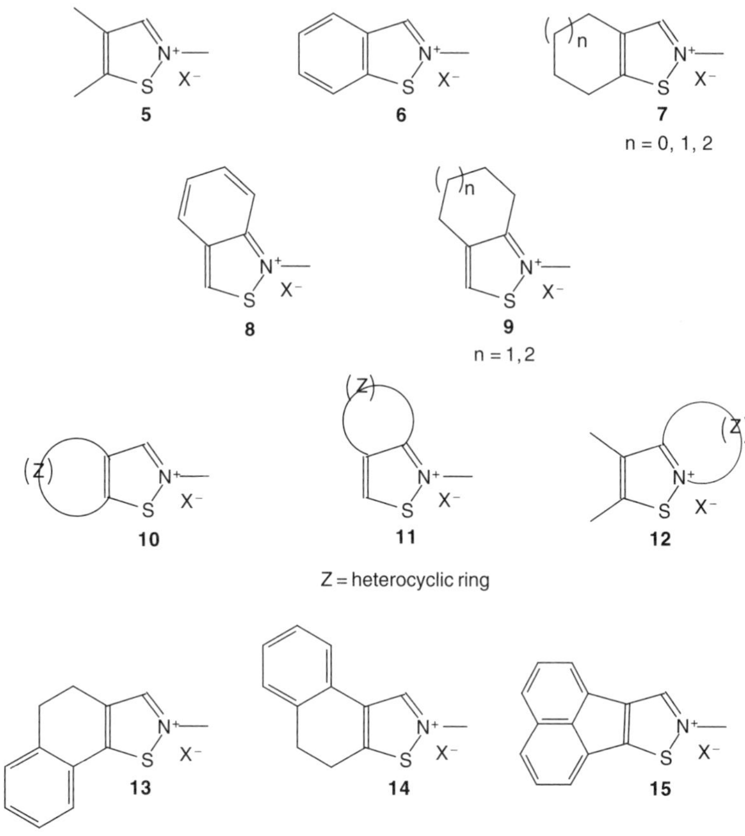

Scheme 2

1. Oxidative Cyclization of 3-Amino-Thioacrylic Acid-Amides, -Esters or 2-Amino-1-Alkenyl-Thioketones

The first synthesis of stable isothiazolium halogenides **17** (65CB1531) was done by oxidative cyclization of the enamine-mustard oil adducts **16** by treatment with Br$_2$/CHCl$_3$ (see also 61CB2950, 65CB1531, see Scheme 3). 4-Unsubstituted salts **18** were also prepared from N-isopropylamino-thiocrotonamides **16** (R^2 = H) by cyclization with I$_2$/pyridine/EtOH (77CB285). The reaction of **17** and **18** with base gave the 5-arylimino-2,5-dihydro-isothiazoles **17** and **18-HX**. The oxidative cyclization of N-morpholinoamino-thiocrotonamides **16** (R^2 = CO$_2$Et) yielded the hydrobromides **19a,b** (82LA884, Scheme 3).

The oxidation of N-monosubstituted 3-amino-2-aryl-thioacrylic acid-morpholides **20** using hydrogen peroxide in perchloric acid gave 5-morpholino-isothiazolium perchlorates[2] **21a,c–g** (83JPR689) in good yields (Scheme 4). The electrochemical oxidation of 3-amino-thioacryl acid-amides **20** (R = CH$_2$CO$_2$Me) is a useful

---

[2]Caution: Perchlorates are potential explosives!

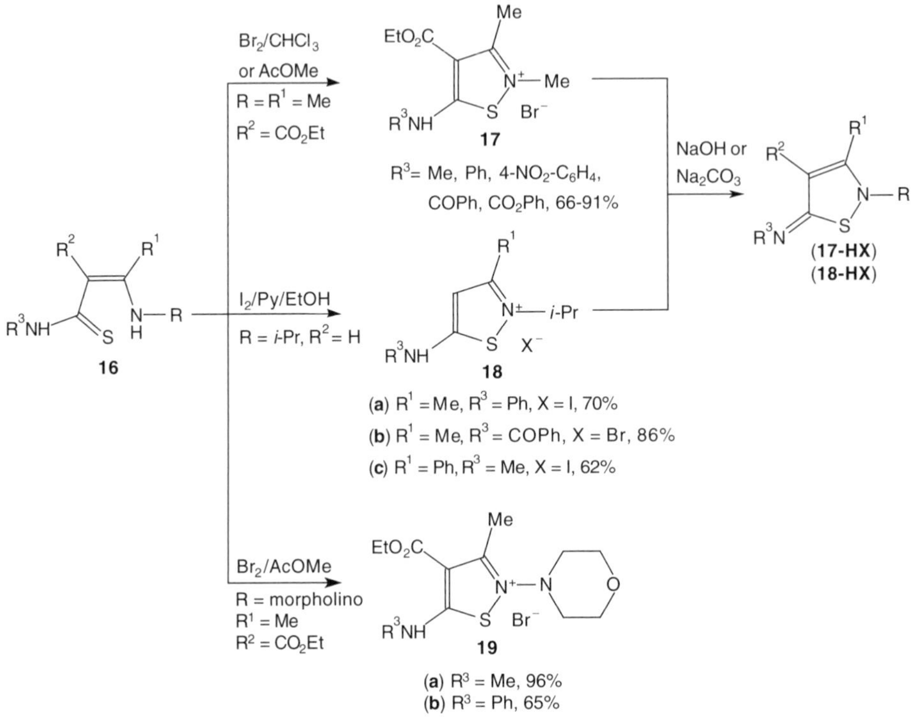

Scheme 3

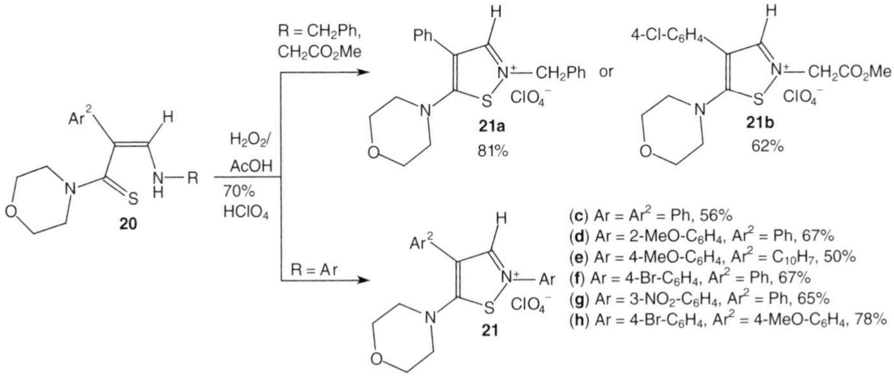

Scheme 4

alternative to the application of chemical oxidizing reagents for the synthesis of the isothiazolium salts, e.g. **21b** (95JPR310); see also the direct synthesis of 3-amino-pyrrole-2-carboxylates without isolating the isothiazolium salts **21** (93AG797, see also Scheme 105).

3-Amino-dithioacrylic-esters **22** prepared from 3-alkylthio-1,2-dithiolium salts with primary amines reacted by oxidative S,N-ring closure with I₂/EtOH

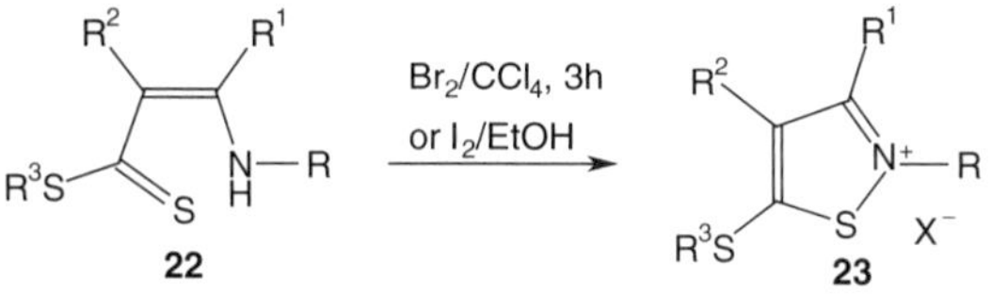

(a) R = R^1 = R^2 = H, R^3 = Me, X = Br, Br$_3$
(b) R = R^2 = Ph, R^1 = H, R^3 = Me, X = ClO$_4$, 34%
(c) R = R^3 = Me, R^1 = Ph, R^2 = H, X = ClO$_4$, 83%
(d) R = R^1 = Ph, R^2 = H, R^3 = Et, X = ClO$_4$, 70%

Scheme 5

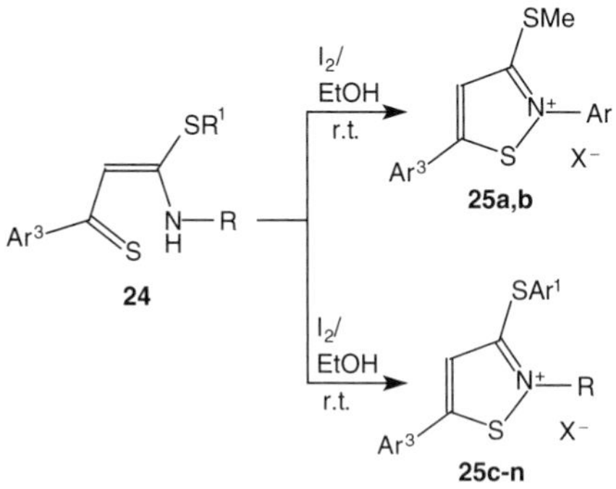

(a) Ar = Ph, Ar^3 = Ph, X = I,30%
(b) Ar = 4-Me-C$_6$H$_4$, Ar^3 = Ph, X= I,33%
(c) R = Me,R^1 = Ph, Ar^3 = Ph, X = I$_3$, 80%
(d) R = Me,R^1 = 4-Me-C$_6$H$_4$, Ar^3 = Ph, X = I$_3$, 87%
(e) R = Me,R^1 = 4-MeO-C$_6$H$_4$, Ar^3 = Ph, X = I$_3$, 90%
(f) R = Me,R^1 = 4-Br-C$_6$H$_4$, Ar^3 = Ph, X = I$_3$, 75%
(g) R = Me,R^1 = 2-MeO-C$_6$H$_4$, Ar^3 = Ph,X = I$_3$, 63%
(h) R = Me,R^1 = 2-Cl-C$_6$H$_4$, Ar^3 = Ph, X = I$_3$, 69%
(i) R = Me, R^1 = 2-Br-C$_6$H$_4$, Ar^3 = Ph, X = I$_3$, 85%
(j) R = Me,R^1 = naphthyl, Ar^3 = Ph, X = I$_3$, 85%
(k) R = Me,R^1 = Ph, Ar^3 = 4-Cl-C$_6$H$_4$, X = I$_3$, 86%
(l) R = Me, R^1 = 2-Br-C$_6$H$_4$, Ar^3 = 4-Cl-C$_6$H$_4$, X = I$_3$, 79%
(m) R = Et, R^1 = Ph, Ar^3 = 4-MeO-C$_6$H$_4$, X = I$_3$, 88%
(n) R = Et, R^1 = 2-Br-C$_6$H$_4$, Ar^3 = 4-MeO-C$_6$H$_4$, X = I$_3$, 70%

Scheme 6

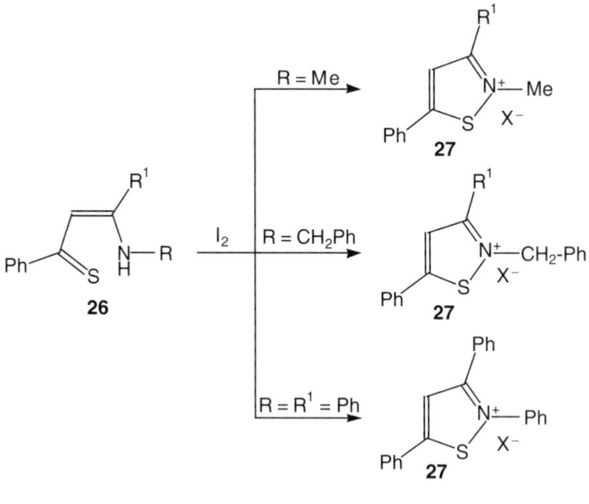

(a) R^1 = H, X = ClO$_4$, 19%
(b) R^1 = Ph, X = ClO$_4$, 40%, X = I$_3$, 45%

(c) R^1 = H, X = ClO$_4$, 48%
(d) R^1 = Ph, X = I$_3$, 54%

(e) X = ClO$_4$, 53%

Scheme 7

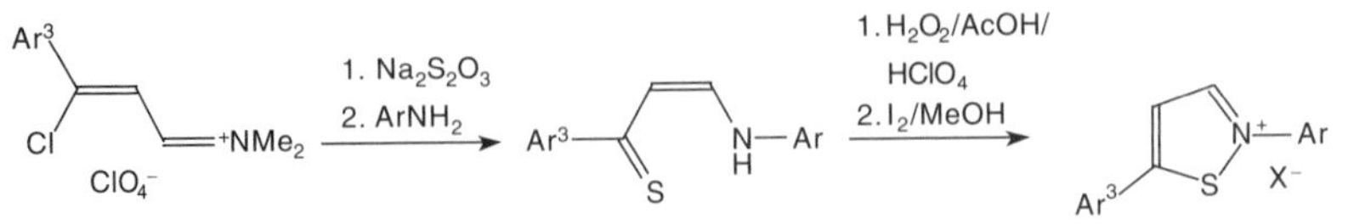

28 29 30

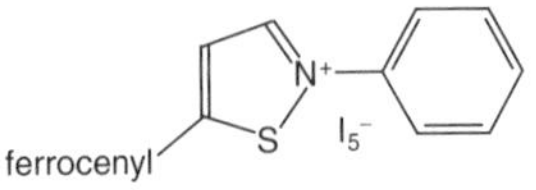

30h
91%

(**a**) Ar = Ar³ = Ph, X = ClO_4, 39%, I_3, 52%, I_8^-, 85%
(**b**) Ar = 4-MeO-C_6H_4, Ar³ = 4-Me-C_6H_4, X = ClO_4, 45%
(**c**) Ar = Ph, Ar³ = 4-CF_3-C_6H_4, X = ClO_4, 68%
(**d**) Ar = 4-MeO-C_6H_4, Ar³ = 4-CF_3-C_6H_4, X = ClO_4, 31%
(**e**) Ar = Ph, Ar³ = 3-CF_3-C_6H_4, X = ClO_4, 76%
(**f**) Ar = Ph, Ar³ = 4-F-C_6H_4, X = ClO_4, 80%
(**g**) Ar = 4-Cl-C_6H_4, Ar³ = 4-MeO-C_6H_4, X = ClO_4, 76%

Scheme 8

(72CJC2568, 73CJC3081) or Br_2/CCl_4 (67ZC306) producing directly the salts **23** (**23a**: 67ZC306; **23b,d**: 72CJC2568; **23c**: 73CJC3081, Scheme 5).

3-Ethoxy-/3-methyl-amino-thioacrylic-*O*-ethylester ($R = Me$, $R^2 = H$) reacted with I_2 or Br_2 in pyridine to give 3,5-diethoxy-2-methyl-isothiazolium triiodide (85%) and tribromide (22%) (88AP863, see also Scheme 11).

3-Arythio-3-alkenylamino-thiopropenones **24c–n** were oxidized to novel 3-arylthio-isothiazolium salts **25c–n** with iodine in ethanol at room temperature (02JOC5375). The 3-alkylthio salts **25a,b** were accordingly prepared (73LA256, Scheme 6).

The oxidative cyclization of 3-alkylamino- and 3-arylamino-thiopropenones **26a–e** with I_2 in ethanol and perchloric acid yielded the 4-unsubstituted salts **27a,b,e** (68CJC1855), **27c,d** (72JCS(P1)2305, Scheme 7).

2,5-Diaryl-isothiazolium salts **30a–g** were prepared by oxidative cyclization of enaminothioketones **29** (88ZC345, 94JPR434) with H_2O_2/AcOH and perchloric acid, which were obtained from the dimethyliminium perchlorates **28** (Scheme 8) by reaction with sodium thiosulfate followed by transamination with the corresponding aniline (**30a**: 68CJC1855, 72JCS(P1)2305; **30a,b,i**: 74ZC189; **30c–g**: 94ZOR1379).

The salts **30c,e** were also prepared from the β-thiocyanatovinylaldehydes, aniline, glacial acetic acid and perchloric acid (94ZOR1379, see Scheme 16).

The treatment of 3-phenyl- and 3-ferrocenyl-3-chloropropenal with sodium sulfide nonahydrate gave rise to intermediate 3-phenyl- and 3-ferrocenyl-3-thiopropenal, which produced with aniline the 1-phenyl- (**29a**) and 1-ferrocenyl-3-phenylaminothiopropenon **29h**. The reaction of **29a,h** with iodine in methanol yielded the oxidized bis(2,5-diphenylisothiazolium) octaiodide **30a** (85%) and 2-phenyl-5-ferrocenylisothiazolium pentaiodide **30h** (91%) (99ZAAC511). The structures of these iodides **30a** ($X = I_8^-$) and **30h** were confirmed by X-ray diffraction (99ZAAC511, Scheme 8). **30a** (I_8^-) also formed a layer structure with isothiazolium cations and polyiodide anions.

2. *Ring Transformations of 1,2-Dithiolium Salts with Primary Amines and Oxidative Cyclizations*

4-Hydroxy-isothiazoles **33** were obtained through nucleophilic substitution of the 3,5-disubstituted 1,2-dithiolium-4-oleates **31** or their perchlorates **32** by ammonia in

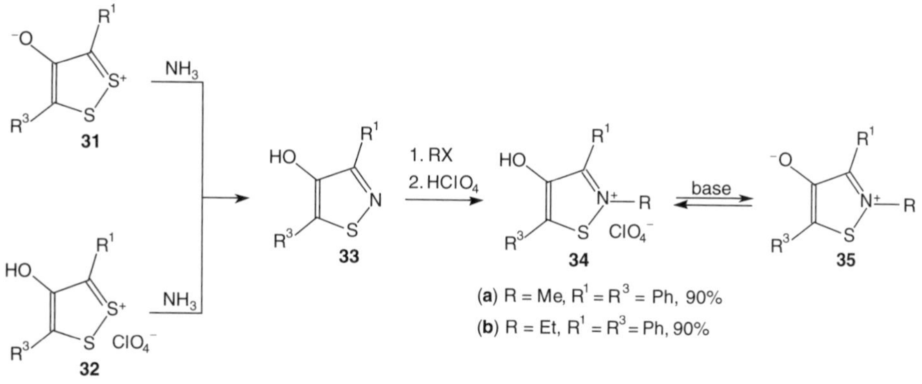
NH3
NH3
ClO4−
1. RX
2. HClO4
ClO4−
base
(a) R = Me, R1 = R3 = Ph, 90%
(b) R = Et, R1 = R3 = Ph, 90%

Scheme 9

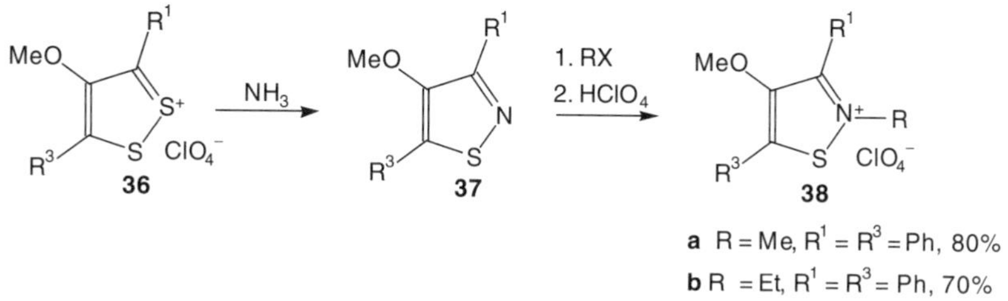
NH3
1. RX
2. HClO4
a R = Me, R1 = R3 = Ph, 80%
b R = Et, R1 = R3 = Ph, 70%

Scheme 10

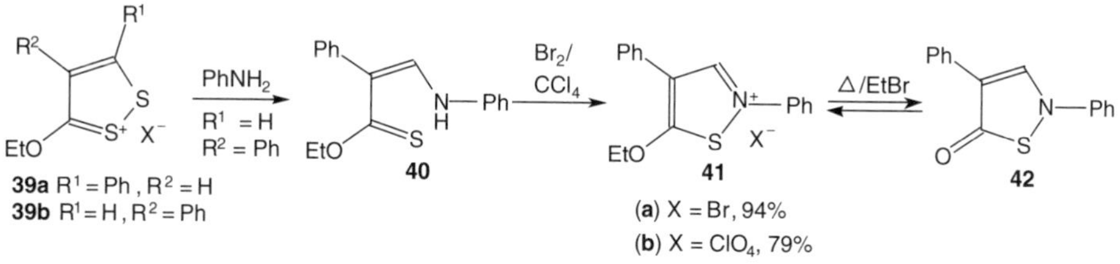
PhNH2
R1 = H
R2 = Ph
Br2/
CCl4
Δ/EtBr
39a R1 = Ph, R2 = H
39b R1 = H, R2 = Ph
(a) X = Br, 94%
(b) X = ClO4, 79%

Scheme 11

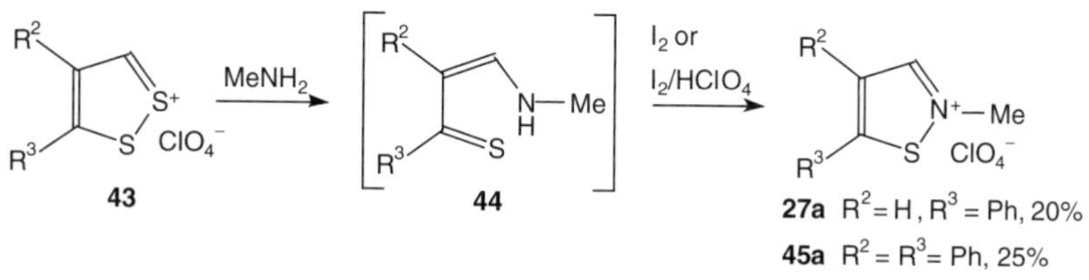
MeNH2
I2 or
I2/HClO4
27a R2 = H, R3 = Ph, 20%
45a R2 = R3 = Ph, 25%

Scheme 12

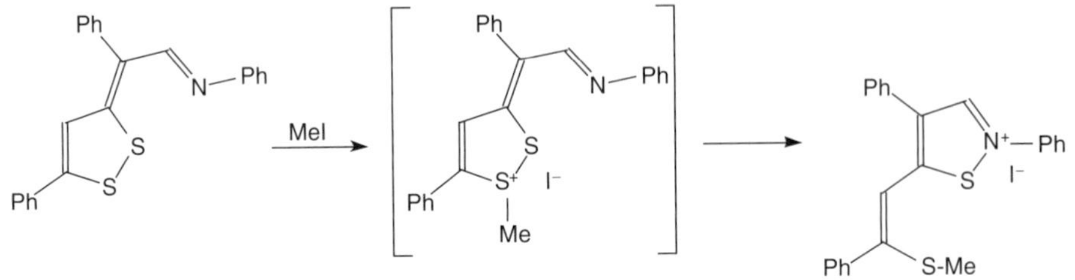

Scheme 13

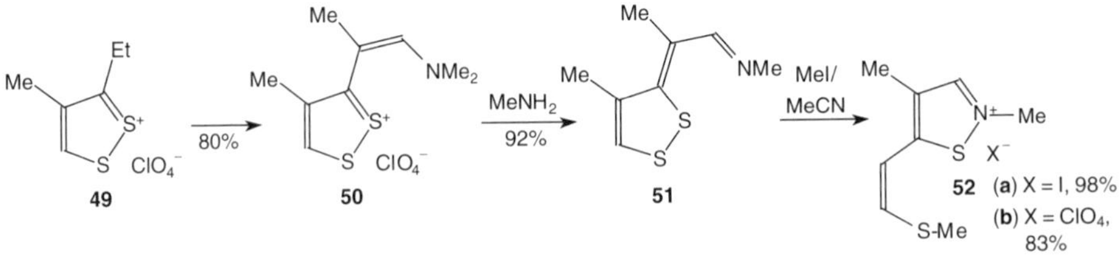

Scheme 14

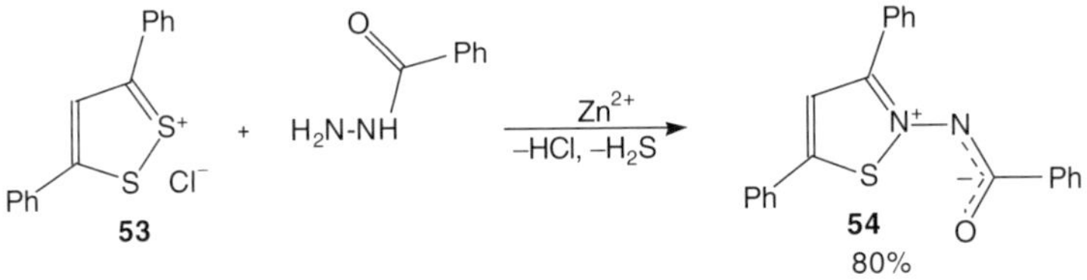

Scheme 15

EtOH. The yields of **33** were between 35 and 50%. The alkylation of the isothiazole **33** gave the isothiazolium salts **34a,b** in 90% yield (79BSF26, Scheme 9). By reaction of the perchlorates **34** with a base, the isothiazolium-4-oleates **35** were obtained (79BSF26, 80PS79).

The formation of the 4-methoxy-3,5-diphenyl-isothiazolium salts **38a,b** also took place by nucleophilic attack of ammonia on **36** followed by alkylation of the iso-thiazole **37** (79BSF26, Scheme 10). Secondly, the isothiazole-4-oleates **35a,b** (Scheme 9) could be transformed by an electrophilic attack of dimethyl sulfate to give also the 4-methoxy salts **38**.

The specific reaction pathway of 1,2-dithiolium salts **39** was determined by the substituent in the 5-position. Primary aromatic amines reacted with monocyclic 5-aryl-3-ethoxy-1,2-dithiolium salts **39a** but did not give 3-phenylamino-thioacrylicacid-*O*-ethylesters because the attack of the amines on 5-position in **39a** is hindered. In contrast, the 4-aryl salts **39b** reacted very easily to produce thioesters **40**, which were converted to salts **41** by cyclization with Br$_2$ in CCl$_4$ (75ZC478). After treatment of salt **41** with a special base as well as by dry heating, the isothiazolone **42** was obtained (Scheme 11).

The treatment of 1,2-dithiolium salts **43** with methylamine gave 3-methylaminothiopropenones **44**, which were oxidized by iodine to form the *N*-methylisothiazolium salts **27a,45a**, isolated as their perchlorates (73CJC3081). The yields of products were rather poor but the reaction is a very quick synthesis of *N*-methylisothiazolium salts **27a,45a** from the 1,2-dithiolium salts **43** (Scheme 12).

The methylation of the 1,2-dithiol **46** occurred at the sulfur atom with formation of the dithiolium salt **47**, which could not be isolated and immediately reacted to form the isothiazolium iodide **48** (84JHC627, Scheme 13).

3-Ethyl-4-methyl-1,2-dithiolium perchlorate **49**, which crystallized from AcOH as colorless plates, was prepared and converted into the Vilsmeier salt **50** in 80% yield. The addition of aqueous methylamine to a solution of salt **50** in dimethylformamide at room temperature afforded the methylimine **51** in 92% as deep yellow needles. After heating with methyl iodide in acetonitrile, the dithiol **51** gave the S-methylated isothiazolium iodide **52a**, which was treated with 70% perchloric acid to form the corresponding perchlorate **52b** as yellow needles was received (69CC1314, Scheme 14).

The reaction of the 3,5-diphenyl-1,2-dithiolium chloride **53** in the presence of zinc acetate and benzoylhydrazine gave a yellow substance in 80% yield. After the X-ray analysis, the structure was established as the 3,5-diphenylisothiazolium-2-benzoylimine **54** (97ZN(B)1139) (Scheme 15).

3. *Cyclocondensation of 3-Thiocyanoalkenal-Imines and Isomerization of 4, 5-Substituted Isothiazolium Salts*

Z/*E*-3-Thiocyanato-2-alkenals **55** are important C_3S-synthons for the synthesis of isothiazolium salts **30a,c,e,i** and **57–64**. The alkenals **55a–h** reacted with primary

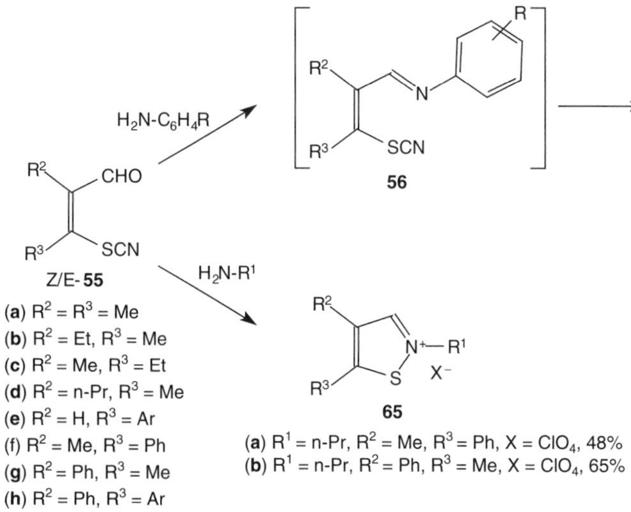
H2N-C6H4R
R2
CHO
R3
SCN
Z/E-55
(a) R2 = R3 = Me
(b) R2 = Et, R3 = Me
(c) R2 = Me, R3 = Et
(d) R2 = n-Pr, R3 = Me
(e) R2 = H, R3 = Ar
(f) R2 = Me, R3 = Ph
(g) R2 = Ph, R3 = Me
(h) R2 = Ph, R3 = Ar
N
R
56
N+
S
X-
30a,c,e,i,
57-64
H2N-R1
N+—R1
65
(a) R1 = n-Pr, R2 = Me, R3 = Ph, X = ClO4, 48%
(b) R1 = n-Pr, R2 = Ph, R3 = Me, X = ClO4, 65%

Scheme 16

Table 1. 2-Arylisothiazolium salts **30a,c,e,i** and **57-64**

	R^2	R^3	R	Reference
57	Me	Me	H, 4-Me, 2-Cl, 4-Cl	92JPR25
			4-MeO, 4-Br, 4-SO$_2$Me	95JPR175
			4-CF$_3$	96MOL142
			3-Me	98JPR361
			4-NH$_2$, 4-CO$_2$H, 4-CO$_2$Me, 4-CO$_2$Et, 2,6-Cl$_2$, 4-NO$_2$	02HCA183
			2,4-(NO$_2$)$_2$, 2,5-Cl$_2$-4-O-i-Pr	03ZN(B)111
58	Et	Me	H, 2-Cl	92JPR25
			4-MeO	98JPR361
59	Me	Et	H, 2-Me, 4-Me, 4-t-Bu, 2-MeO, 3-MeO, 4-MeO, 2-Cl, 3-Cl, 4-Cl, 4-Br, 4-NO$_2$	97SUL35
60	n-Pr	Me	H, 2-Cl	92JPR25
30a	H	Ph	H	74ZC189
30i	H	Ph	2,5-Cl$_2$-4-O-i-Pr	03ZN(B)111
30c	H	4-CF$_3$-C$_6$H$_4$	H	94ZOR1379
30e	H	3-CF$_3$-C$_6$H$_4$	H	94ZOR1379
61	Me	Ph	4-NO$_2$, 2,4-(NO$_2$)$_2$, 2,5-Cl$_2$, 2,5-Cl$_2$-4-O-i-Pr	03ZN(B)111
62	Ph	Me	H, 4-Me, 2-Cl	92JPR25
			4-MeO	01PS29
			4-NO$_2$, 2,4-(NO$_2$)$_2$, 2,5-Cl$_2$-4-O-i-Pr	03ZN(B)111
			2-Cl	05SUC211
63	Ph	Ph	4-Cl, 4-CF$_3$	01PS29
			4-NO$_2$, 2,4-(NO$_2$)$_2$, 2,5-Cl$_2$-4-O-i-Pr	03ZN(B)111
			3-NO$_2$, 2-CF$_3$-4-NO$_2$, 2-Cl-4-NO$_2$	05JEIMC341
			H, 4-MeO, 2-Cl-4-NO$_2$	05SUC211
64	Ph	4-MeO-C$_6$H$_4$	3-NO$_2$, 4-NO$_2$, 4-CF$_3$, 2-Cl-4-NO$_2$, 2,4-(NO$_2$)$_2$, 2,5-Cl$_2$-4-O-i-Pr	05JEIMC341

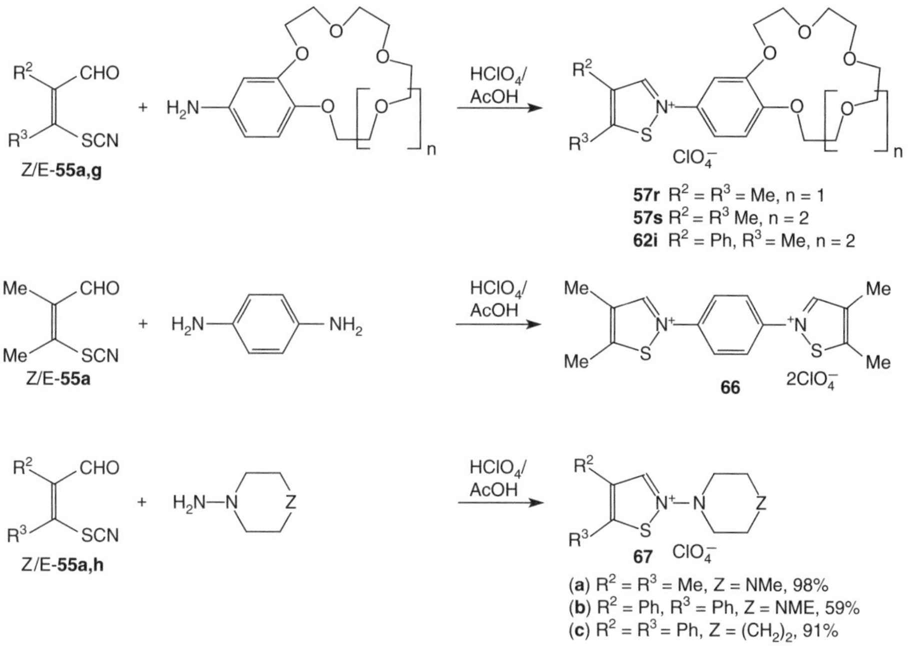

57r R^2 = R^3 = Me, n = 1
57s R^2 = R^3 Me, n = 2
62i R^2 = Ph, R^3 = Me, n = 2

66   2ClO$_4^-$

67 ClO$_4^-$

(**a**) R^2 = R^3 = Me, Z = NMe, 98%
(**b**) R^2 = Ph, R^3 = Ph, Z = NME, 59%
(**c**) R^2 = R^3 = Ph, Z = (CH$_2$)$_2$, 91%

Scheme 17

57 R^2 = Me
58 R^2 = Et

68

(**a**) R = H, R^2 = Me, 77%
(**b**) R = 3-Me, R^2 = Me, 87%
(**c**) R = 4-Me, R^2 = Me, 77%
(**d**) R = 4-MeO, R^2 = Me, 87%
(**e**) R = 4-Br, R^2 = Me, 44%
(**f**) R = 4-Cl, R^2 = Me, 38%
(**g**) R = H, R^2 = Et, 85%
(**h**) R = 4-MeO, R^2 = Et, 97%

69a-h
1-16%

69`a-h
1-17%

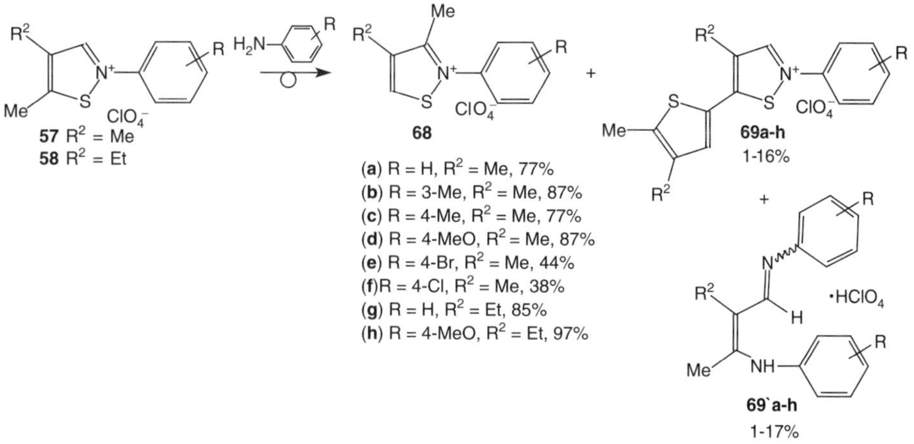

Scheme 18

aromatic or aliphatic amines in the presence of hydrochloric acid in ethanol, tetra-fluoroboric acid and perchloric acid in glacial acetic acid to yield N-aryl **30a,c,e,i** or **57–64** and N-alkylisothiazolium salts **65** in good-to-very-good yields (88ZC287, 90DDP275459, 92JPR25, 95JPR175). The isothiazolium iodides were prepared by anion exchange of perchlorates with KI in ethanol for iodides. Furthermore,

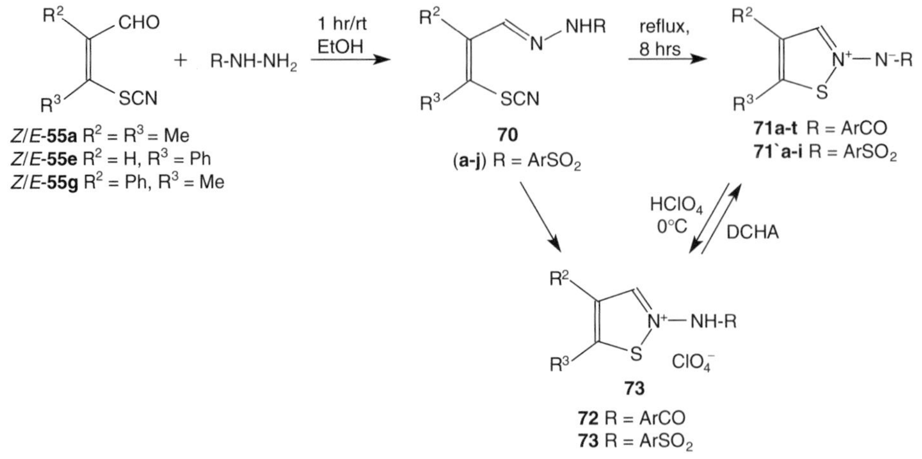

Scheme 19

Table 2. Synthesis of *N*-benzoyl- or sulfonyl- isothiazolium-2-imines **71,71′** and 2-amino-isothiazolium perchlorates **72,73** from β-thiocyanato-vinyl aldehydes **55a,e,g**

	R^a	R^2	R^3	**71,71′** (%)	**72,73** (%)	Reference
71,72a	PhCO	Me	Me	60	–	05H2705
b	4-Me-C$_6$H$_4$-CO	Me	Me	18	–	05H2705
f	4-Cl-C$_6$H$_4$-CO	Me	Me	32	–	05H2705
h	4-NO$_2$-C$_6$H$_4$-CO	Me	Me	42	–	05H2705
q	PhCO	Ph	Me	55	–	89DDP289269
		H	Ph	53	95	89DDP289269
s	4-Cl-C$_6$H$_4$-CO	H	Ph	82	–	89DDP289269
t	3-NO$_2$-C$_6$H$_4$-CO	H	Ph	65	40	89DDP289269
71′,73a	PhSO$_2$	Me	Me	27	–	96ZOR1745
b	4-Me-C$_6$H$_4$-SO$_2$	Me	Me	90	57	96ZOR1745
d	4-Br-C$_6$H$_4$-SO$_2$	Me	Me	32	67	96ZOR1745
e	3-NO$_2$-C$_6$H$_4$-SO$_2$	Me	Me	35	57	99JHC1081
f	4-NO$_2$-C$_6$H$_4$-SO$_2$	Me	Me	54	59	99JHC1081
g	2,4,6-Me$_3$-C$_6$H$_2$-SO$_2$	Me	Me	65	58	96ZOR1745
h	PhSO$_2$	Ph	Me	72	77	96ZOR1745
i	4-Me-C$_6$H$_4$-SO$_2$	H	Ph	48	42	89DEP289270

[a]See also Table 8.

Z/E-thiocyanates **55a–h** reacted with aniline hydrochloride in ethanol to give hydrochlorides **57–64** (X = Cl) (92JPR25).

In all cases, intermediate 3-thiocyanato-2-alkenal arylimine **56** was not isolated, it reacted further by intramolecular cyclocondensation to salts **30a,c,e,i** and **57–64** (90DDP275459, 92JPR25, Scheme 16). In order to establish the structure of salts **57**, an X-ray analysis was performed for salt **57** (R = 4-MeO) (98ZK331). In Table 1 are presented all 2-aryl isothiazolium salts **30a,c,e,i** and **57–64** (20–95%) with the corresponding reference obtained by this approach.

In particular, *N*-phenylisothiazolium salts **57r,s** and **62i** bearing a benzo crown ether substituent were synthesized in 53–60% yield by treatment of the β-thiocyanatovinyl aldehydes **55a,g** with the crown aniline (01PS29). Furthermore, the 2, 2'-(benzene-1,4-diyl)bis(4,5-dimethylisothiazol-2-ium) diperchlorate **66** was prepared in 88% yield as yellow crystals by the reaction of thiocyanate **55a** with benzene-1, 4-diamine (02HCA183).

Thiocyanates **55a,h** with *N*-amino heterocycles produced N,N'-linked isothiazolium salts **67a–c** (06ZN(B)464, Scheme 17, see also Scheme 19).

4,5-Substituted monocyclic isothiazolium salts **57,58** gave with anilines 3,4-dialkyl isothiazolium salts **68a–h** by isomerization, which took place after nucleophilic attack of aniline at the carbon atom in the 5-position of **57** following migration of a sulfur atom to 3-position (98JPR361). The structure of 3,4-dimethyl-2-phenylisothiazolium perchlorate **68a** was confirmed by X-ray analysis (98JPR361). This ring transformation resulted in two by-products, the 5-thienyl salts **69a–h** in only 1–16% yield and salts of vinamidines **69'a–h** in 1–17% yields. The yields of **69** and **69'** became lower on increasing the basicity of the corresponding anilines (98JPR361, Scheme 18).

β-Thiocyanatovinyl aldehydes **55a,e,g** reacted with the acceptor-substituted hydrazines *via* 3-thiocyanato-2-alkenalhydrazones **70** to obtain *N*-aroyl- and *N*-arylsulfonyl isothiazolium-2-imines **71** and **71'**, which served as educts for the synthesis of 2-aminoisothiazolium perchlorates **72** and **73** by treatment with perchloric acid (96ZOR1745, 98H587, 02SCIS287, 05H2705, Scheme 19). The benzhydrazones **70** (R = ArCO) were not isolated (05H2705), they reacted in ethanol under 8 h refluxing to imines **71a–t** (R = ArCO) (Table 2) and treatment with perchloric acid produced the salts **72**. The structure of benzoylimine **71a** was determined by X-ray analysis (05H2705).

In contrast, salts **73b–i** (R = ArSO₂) were obtained from the sulfonylhydrazones **70** (R = ArSO₂) or *N*-arylsulfonylisothiazolium-2-imines **71'** by stirring 70% perchloric acid cooled to 0 °C for 1 h (96ZOR1745). In Table 2 are given the isothiazolium-2-imines **71,71'** and their salts **72,73**, synthesized by this method.

4. N-Alkylation of Isothiazoles to N-Alkylisothiazolium Salts

2-Alkylisothiazolium salts **27,45** and **75** were prepared by N-alkylation of isothiazoles **74** with a very wide range of alkyl bromides and iodides, methyl *p*-toluenesulfonate, methyl trifluoromethansulfonate, methyl fluorosulfonate, dimethylsulfate

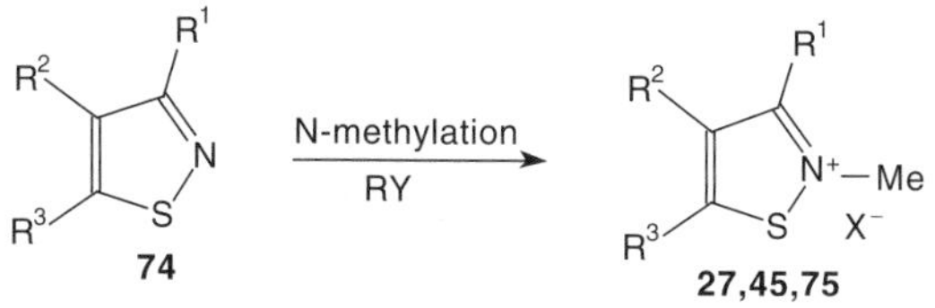

Scheme 20

J. WOLF AND B. SCHULZE

Table 3. Methylation of isothiazoles **74** to 2-methyl-isothiazolium salts **27,45** and **75**

	R^1	R^2	R^3	RY	X	Yield (%)	Reference
75a	H	H	H	MeI	I	43	65JCS(C)4577
				4-Me-C$_6$H$_4$-SO$_3$Me	Ts	70	72JCS(P1)2305
				4-Me-C$_6$H$_4$-SO$_3$Me	FSO$_3$	71	72JCS(P1)2305
				4-Me-C$_6$H$_4$-SO$_3$Me/(MeO)$_2$SO$_2$	ClO$_4$	65/61	75JCS(P2)1620/73CJC3081
b	Me	H	H	MeI	I	43	65JCS(C)4577
				4-Me-C$_6$H$_4$-SO$_3$Me	ClO$_4$	83	75JCS(P2)1620
				4-Me-C$_6$H$_4$-SO$_3$Me	Ts	46	65JCS(C)4577
c	H	Me	H	MeI	I	31	65JCS(C)4577
d	H	H	Me	MeI	I	23	65JCS(C)4577
				4-Me-C$_6$H$_4$-SO$_3$Me	ClO$_4$	48	75JCS(P2)1620
e	H	Me	Me	MeI	I	27	92JPR25
f	H	Ph	Me	MeI	I	20	92JPR25
g	Me	H	Me	MeI	ClO$_4$	85	65JCS(C)4577
				4-Me-C$_6$H$_4$-SO$_3$Me	Ts	85	71JCS(B)2365
h	Me	NO$_2$	H	4-Me-C$_6$H$_4$-SO$_3$Me	ClO$_4$	63	75JCS(P2)1620
i	Me	NO$_2$	Me	CF$_3$SO$_3$Me	ClO$_4$	94	71JCS(B)2365
45a	H	Ph	Ph	(MeO)$_2$SO$_2$	ClO$_4$	80	72CJC2568
75j	H	Ph	H	(MeO)$_2$SO$_2$	ClO$_4$	85	72CJC2568
27a	H	H	Ph	(MeO)$_2$SO$_2$	ClO$_4$	73	68CJC1855
				FSO$_3$Me	FSO$_3$	64	72JCS(P1)2305
27b	Ph	H	Ph	FSO$_3$Me	FSO$_3$	68	72JCS(P1)2305

Compound				Reagent	Anion	Yield (%)	Reference
75k	Cl	Cl	Cl	$(MeO)_2SO_2$	$MeOSO_3$	46	78USP4281136
				Me_3OBF_4	$SbCl_3$	73	78USP4281136
				Me_3OBF_4	PF_6	61	78USP4281136
				CF_3SO_3Me	CF_3SO_3	26	78USP4281136
				FSO_3Me	FSO_3	96	78USP4281136
l	Cl	Cl	H	FSO_3Me	FSO_3	63	78USP4281136
m	Cl	H	H	FSO_3Me	Cl	92	77USP4292430/79USP4267341
				FSO_3Me	NO_3	98	78USP4281136
n	Cl	Cl	Me	FSO_3Me	FSO_3	92	78USP4281136
o	Cl	CN	Cl	FSO_3Me	FSO_3	92	78USP4281136
p	Cl	CN	Ph	FSO_3Me	FSO_3	92	96PHA638
q	Cl	Cl	CN	FSO_3Me	FSO_3	77	78USP4281136
r	Cl	Cl	CO_2Et	FSO_3Me	FSO_3	99	78USP4281136
s	Br	Br	Br	FSO_3Me	FSO_3	88	78USP4281136
t	NH_2	H	H	FSO_3Me	Cl	34	77USP4292430/79USP4267341/79JOC1118
u	NHPh	H	H	FSO_3Me	FSO_3	72	77USP4292430/79USP4267341
v	NHMe	H	H	FSO_3Me	FSO_3	—	79USP4267341
w	MeO	H	H	FSO_3Me	FSO_3	89	79USP4262127
x	EtO	H	H	FSO_3Me	FSO_3	50	79USP4262127
y	BuO	H	H	FSO_3Me	FSO_3	—	79USP4262127
z_1	OctO	H	H	FSO_3Me	FSO_3	—	79USP4262127
z_2	allylO	H	H	FSO_3Me	FSO_3	89	79USP4262127
z_3	BnO	H	H	FSO_3Me	FSO_3	—	79USP4262127
z_4	MeO	Br	H	FSO_3Me	FSO_3	81	79USP4262127
z_5	MeO	Cl	H	FSO_3Me	FSO_3	—	79USP4262127
z_6	MeO	Cl	Cl	FSO_3Me	FSO_3	—	79USP4262127

Table 4.   Ethylation of isothiazoles **74** to 2-ethyl-isothiazolium salts **76**

76	R^1	R^2	R^3	RY	X	Yield (%)	Reference
a	H	H	H	EtI	I		71BSF4373
b	Cl	Cl	Cl	$(EtO)_2SO_2$	$EtOSO_3$		78USP4281136
				Et_3OBF_4	BF_4	50	78USP4281136
				FSO_3Et	FSO_3		78USP4281136
c	Ph	H	H	Et_3OBF_4	BF_4	36	66T2135
d	H	Ph	H	Et_3OBF_4	BF_4	90	66T2135
e	H	H	Ph	Et_3OBF_4	BF_4	89	66T2135
f	D	H	Ph	Et_3OBF_4	BF_4	48	66T2135
g	H	Br	Ph	Et_3OBF_4	BF_4	80	68ZC170
h	Cl	H	H	FSO_3Et	FSO_3	95	77USP4292430/79USP4267341
i	Et	H	H	FSO_3Et	Cl		77USP4292430/79USP4267341
j	Me	H	Me	FSO_3Et	BF_4		88SC1847
k	Me	I	Me	FSO_3Et	BF_4		88SC1847
l	Me	CN	Me	FSO_3Et	BF_4		88SC1847
m	Me	NO_2	Me	FSO_3Et	BF_4		88SC1847
n	Me	CO_2Et	Me	FSO_3Et	BF_4		88SC1847

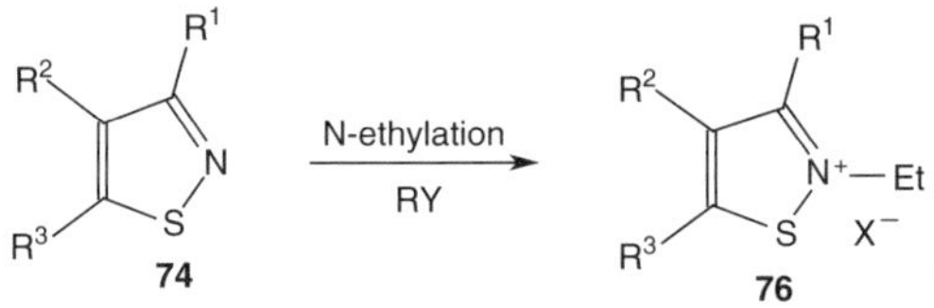

Scheme 21

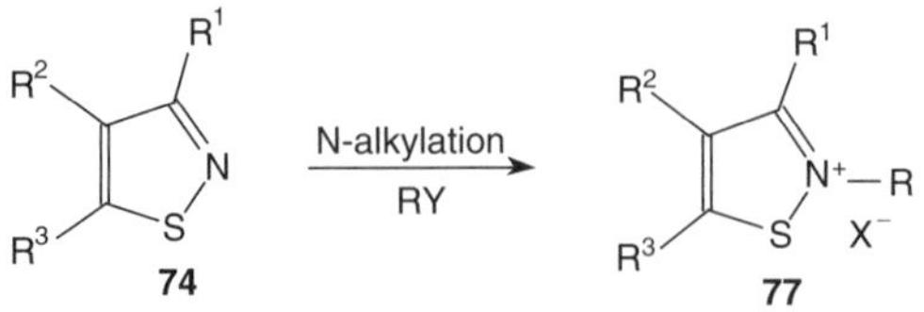

(a) R = n-Pr, $R^1 = R^2 = R^3 =$ Cl, X = CF_3SO_3, 28%
(b) R = allyl, $R^1 = R^2 = R^3 =$ Cl, X = CF_3SO_3, 30%
(c) R = n-Oct, $R^1 = R^2 = R^3 =$ Cl, X = CF_3SO_3, 31%

Scheme 22

and trimethyloxonium tetrafluoroborate (Meerwein's salt). In Table 3 are represented the N-methylated salts **27,45** and **75** (Scheme 20).

2-Ethylisothiazolium salts **76** were prepared by N-alkylation of isothiazoles **74** with alkyl iodide, sulfonate esters, diethyl sulfate and triethyloxonium tetrafluoroborate. In Table 4 are given N-ethylated salts **76** (Scheme 21).

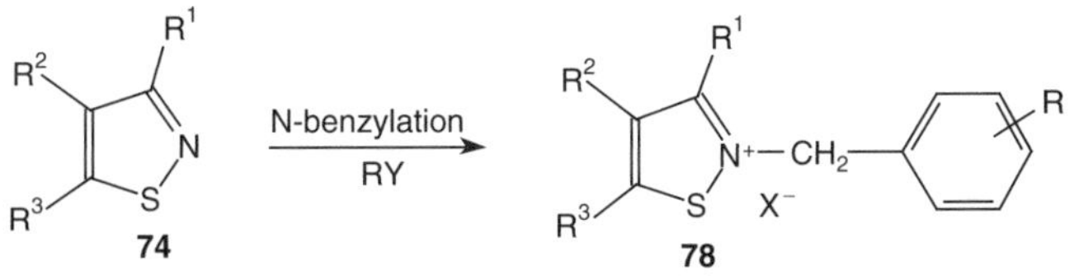

(**78a**) R = R^1 = R^2 = R^3 = H, X = Br, 60%,
X = 4-Me-$C_6H_4SO_3$, 72%, X = NO_3, 73%, X = I, 20%
(**27c**) R = R^1 = R^2 = H, R^3 = Ph, X = ClO_4, 10%
(**78b**) R = R^2 = R^3 = H, R^1 = Me, X = Br, 52%
(**c**) R = 2-NO_2, R^1 = Me, R^2 = R^3 = H, X = Br, 33%
(**d**) R = 3-NO_2, R^1 = Me, R^2 = R^3 = H, X = Br, 35%
(**e**) R = R^1 = H, R^2 = Ph, R^3 = Me, X = ClO_4, 45%

Scheme 23

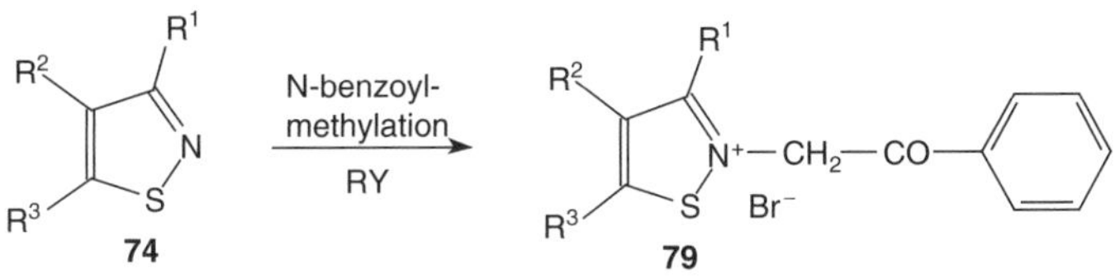

(**a**) R^1 = R^2 = R^3 = H, 53%,
(**b**) R^1 = R^2 = H, R^3 = Ph, 10%
(**c**) R^1 = R^3 = H, R^2 = Ph
(**d**) R^1 = R^3 = H, R^2 = 4-Me-C_6H_4

Scheme 24

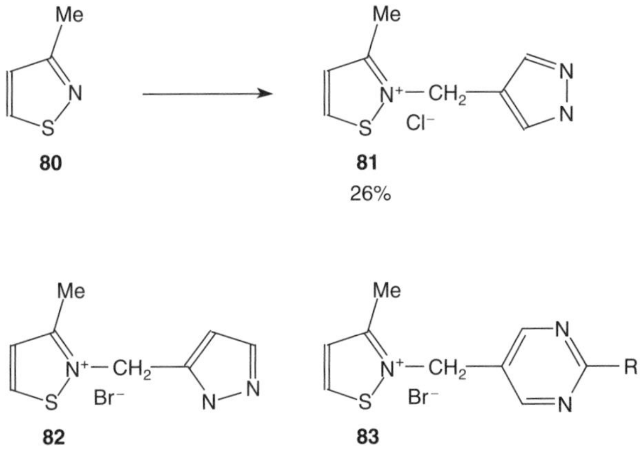

(**a**) R = Me, 62%
(**b**) R = n-Pr, 69%

Scheme 25

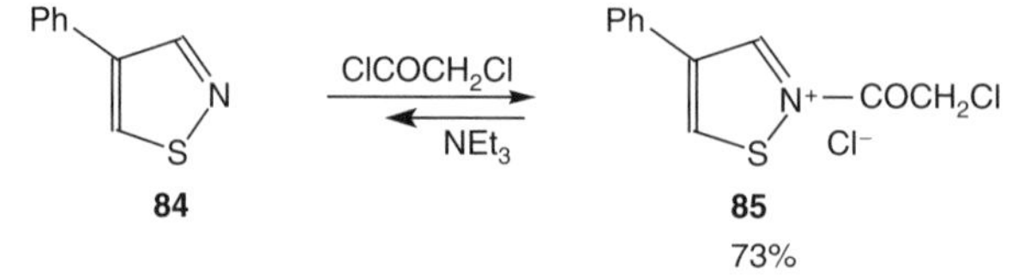

Scheme 26

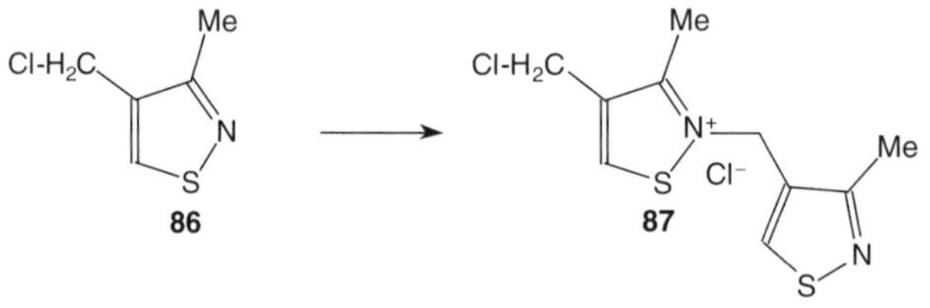

Scheme 27

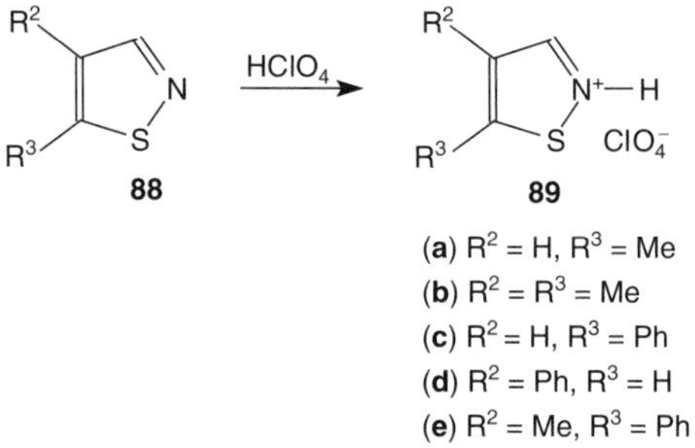

(a) R^2 = H, R^3 = Me
(b) R^2 = R^3 = Me
(c) R^2 = H, R^3 = Ph
(d) R^2 = Ph, R^3 = H
(e) R^2 = Me, R^3 = Ph

Scheme 28

The 2-alkyl-3,4,5-trichloroisothiazolium trifluoromethyl sulfonates **77a–c** were also obtained by N-alkylation of the corresponding isothiazole **74** but only in poor yields (28–31%) (78USP4281136, Scheme 22).

The 2-benzyl-sustituted salts **78** were prepared by quaternization of isothiazole **74** with benzyl bromide, iodide, fluorosulfonate and tosylate. 5-Phenylisothiazole **74** ($R^1 = R^2$ = H, R^3 = Ph) reacted difficultly to salt **27c** (10%) but 3,5-diphenyl-isothiazole are resisted quaternization by benzyl halides (**27c,78a,b**: 72JCS(P1)2305; **78b–d**: 65JCS(C)4577; **78e**: 92JPR25, Scheme 23).

2-Phenacylisothiazolium salts **79** were synthesized by N-alkylation of isothiazoles **74** with phenacyl bromide (**79a,b**: 72JCS(P1)2305; **79b–d**: 85BSB149, Scheme 24).

The 2-(imidazol-4-ylmethyl)- **81** and 2-(imidazol-3-ylmethyl)-3-methyl-isothiazolium salts **82** were prepared by N-alkylation of 3-methylisothiazole **80** in poor yields (26–27%). Similarly, the 2-(pyrimid-5yl-methyl)-3-methyl-isothiazolium bromides **83a** and **83b** were obtained (65JCS(C)4577, Scheme 25).

When the reaction was carried out in the absence of a tertiary amine, N-alkylation of the isothiazole **84** took place and the 2-(chloroacetyl)isothiazolium chloride **85** was precipitated by the addition of ether (85BSB149, Scheme 26). The treatment of salt **85** with triethylamine at room temperature gave back the starting isothiazole **84** by a dealkylation process.

The 4-chloromethyl isothiazole **86**, which could be obtained from the isothiazolyl carbinol by treatment with thionyl chloride, was very reactive towards nucleophilic reagents. Thus 4-chloromethylisothiazole **86** underwent a self-quaternization to form the 4-chloromethyl-3-methyl-2-(3-methylisothiazol-4-ylmethyl)isothiazolium chloride **87** (68JCS(C)611, Scheme 27).

The treatment of β-thiocyanatovinyl aldehyde (76ZC49) with NH$_4$SCN in acetone over 50 °C yielded directly the isothiazole **88**. The 4,5-disubstituted isothiazolium salts **89a–e** were formed by protonation of isothiazole **88** with perchloric acid in ethanol (76JPR507, 76DDP122249, 79ZC41, Scheme 28).

5. *O/S-Alkylation and Oxidative S/O-Elimination of Isothiazolones and -Thiones*

Reaction of 2-alkyl/aryl isothiazol-3(2*H*)-thiones **90** with various alkyl-halogenides gave the 3-*S*-alkyl-substituted isothiazolium salts **25a,91** in good-to-very-good yields (70BSF3076, 73CJC3081, 93JHC929, Scheme 29). Several isothiazolium iodides **25a,91a–n**, perchlorate **25a** and chlorides **91o–v** were also synthesized. The structure of 2-methyl-5-aryl-isothiazolium chloride **91q**, isolated in 60% yield, was confirmed by X-ray analysis. The *S*-alkylated isothiazolium salts **25a,91** with yields and references are represented in Table 5.

The *N*-phenyl/-cyclohexyl isothiazol-5(2*H*)-thiones **92a,b** formed with methyl iodide the stable *S*-methylisothiazolium halogenides **23b** and **93a,b** in very good yields (**23b**: 73CJC3081; **93a,b**: 66JPR312, Scheme 30).

The synthesis of isothiazolium salts **27a** and **45a** in good yields (75–80%) is represented in Scheme 31 by converting the *N*-alkylisothiazol-3(2*H*)-thione **94** with hydrogen peroxide in acetic acid (73CJC3081).

The *N*-phenyl/alkyl-isothiazolium salts **75j** and **96a–c** were also obtained by the oxidative elimination of the isothiazol-5(2*H*)-thione **95** in 67–72% yield (**96a,b**: 72CJC2568; **75j,96c**: 74CJC3021, Scheme 31).

2-Methyl-isothiazol-3(2*H*)-thiones **97** were prepared from the corresponding 2-methyl-isothiazol-3(2*H*)-ones in toluene and oxalylchloride (81USP4281136, see also Schemes 55, 66). They reacted with sodium *N*-chlorobenzenesulfonamide in ethanol and chloroform to produce sulfimides **98**, sulfonamides **99** as well as sulfinamidinates **100** (96PS203, Scheme 32). The stability of the inner salts **98–100** increased from **98 < 99 < 100**.

3-Methoxy isothiazolium salts **75w** and **102a–c** were prepared in high yields by treating the isothiazolone **101** with an excess of methyl fluorosulfonate. In some

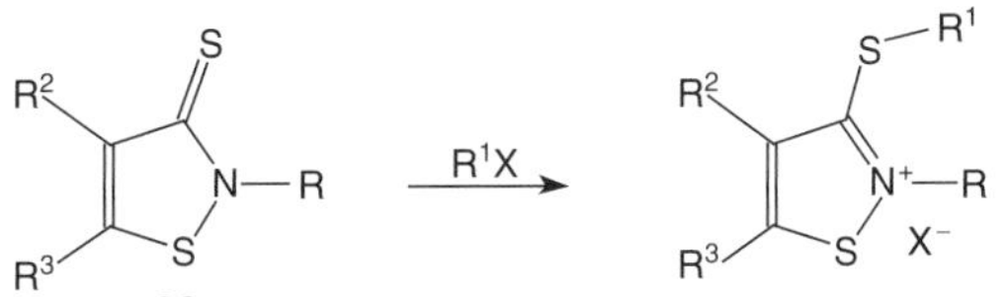

Scheme 29

Table 5. S-Alkylation of isothiazol-3(2H)-thiones **90** to 3-alkylthio-isothiazolium iodides **25a,91a–n** and chlorides **91o–v**

	R	R^1	R^2	R^3	X	Yield (%)	Reference
91a	Me	Me	H	Ph	I	42	68CJC1855
						78	70BSF3076
						35	73CJC3081
b	Me	Et	H	Ph	I	74	70BSF3076
c	Et	Me	H	Ph	I	84	70BSF3076
d	Me	Me	H	4-Me-C$_6$H$_4$	I	85	70BSF3076
e	CH$_2$Ph	Me	H	4-Me-C$_6$H$_4$	I	80	70BSF3076
f	Me	Me	H	4-MeO-C$_6$H$_4$	I	85	70BSF3076
g	Me	Et	H	4-MeO-C$_6$H$_4$	I	80	70BSF3076
h	Et	Me	H	4-MeO-C$_6$H$_4$	I	88	70BSF3076
i	Et	Et	H	4-MeO-C$_6$H$_4$	I	92	70BSF3076
j	CH(CH$_2$)$_5$	Me	H	4-MeO-C$_6$H$_4$	I	90	70BSF3076
k	CH$_2$Ph	Me	H	4-MeO-C$_6$H$_4$	I	92	70BSF3076
l	CH$_2$CH$_2$Ph	Me	H	4-MeO-C$_6$H$_4$	I	78	70BSF3076
m	CH$_2$Ph	Me	H	Thienyl	I	73	70BSF3076
25a	Ph	Me	H	Ph	I	23	68CJC1855
91n	Me	Me	Ph	Ph	I	95	77CJC1123
o	Me	MeCOCHCO$_2$Et	H	Ph	Cl	81	93JHC929
p	Me	MeCH$_2$COCHCO$_2$Me	H	Ph	Cl	70	93JHC929
q	Me	MeCOCHCMeO	H	Ph	Cl	62	93JHC929
r	Me	-CH$_2$-COCHCO-(CH$_2$)$_2$-	H	Ph	Cl	60	93JHC929
s	Et	MeCOCHCO$_2$Et	H	Ph	Cl	71	93JHC929
t	Et	MeCH$_2$COCHCO$_2$Me	H	Ph	Cl	64	93JHC929
u	Et	MeCOCHCMeO	H	Ph	Cl	58	93JHC929
v	Et	-CH$_2$-COCHCO-(CH$_2$)$_2$-	H	Ph	Cl	64	93JHC929

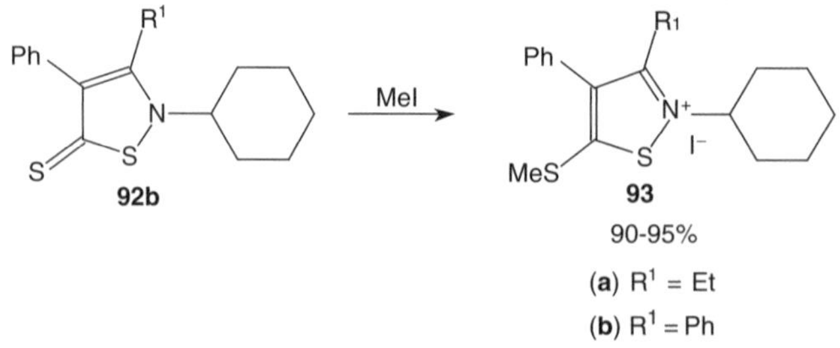

Scheme 30

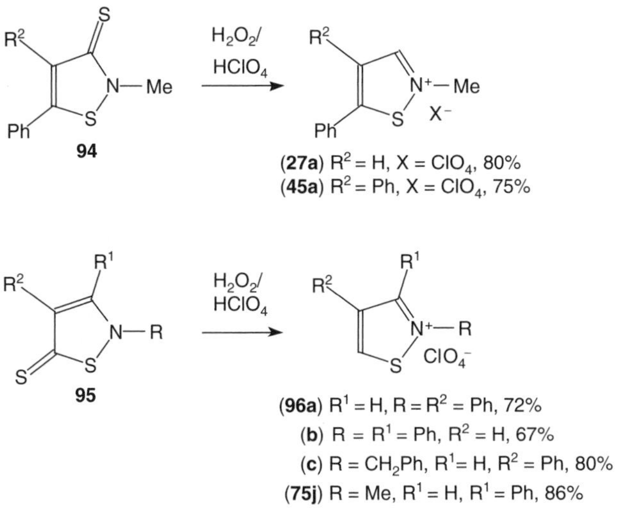
H2O2/
HClO4
94
(27a) R2 = H, X = ClO4, 80%
(45a) R2 = Ph, X = ClO4, 75%
H2O2/
HClO4
95
(96a) R1 = H, R = R2 = Ph, 72%
(b) R = R1 = Ph, R2 = H, 67%
(c) R = CH2Ph, R1= H, R2 = Ph, 80%
(75j) R = Me, R1 = H, R1 = Ph, 86%

Scheme 31

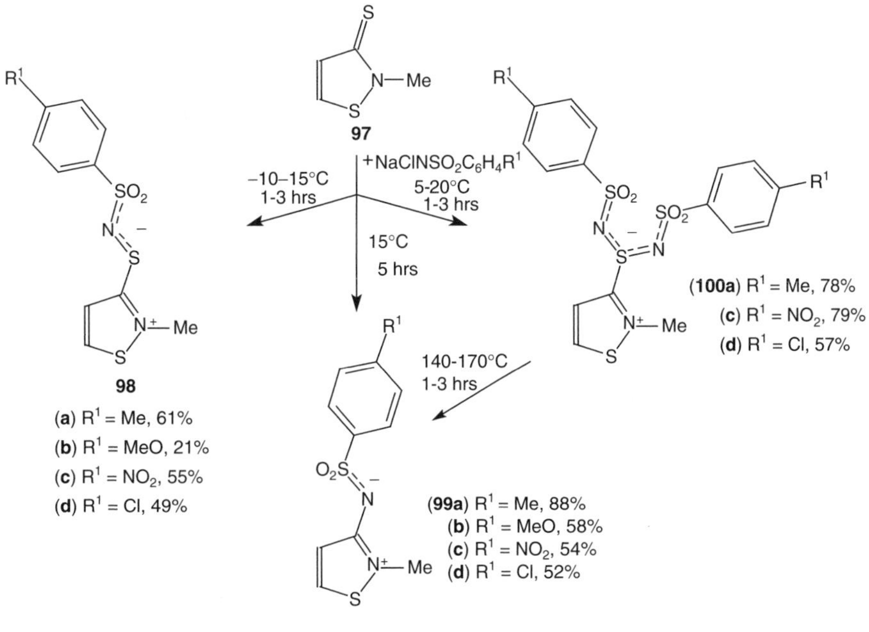
97
+NaClNSO2C6H4R1
−10–15°C
1-3 hrs
5-20°C
1-3 hrs
15°C
5 hrs
140-170°C
1-3 hrs
98
(a) R1 = Me, 61%
(b) R1 = MeO, 21%
(c) R1 = NO2, 55%
(d) R1 = Cl, 49%
(100a) R1 = Me, 78%
(c) R1 = NO2, 79%
(d) R1 = Cl, 57%
(99a) R1 = Me, 88%
(b) R1 = MeO, 58%
(c) R1 = NO2, 54%
(d) R1 = Cl, 52%

Scheme 32

cases, CH_2Cl_2 has been used as solvent. The salts **75w,102a–c** are generally hygroscopic and are decomposed on standing (79JOC1118, Scheme 33).

The quaternary salts **75m,76h** and **104a–d** were formed by oxidative elimination of a carbonyl oxygen with $POCl_3$ from isothiazol-3(2*H*)-one **101** (**75m** (X = Cl),

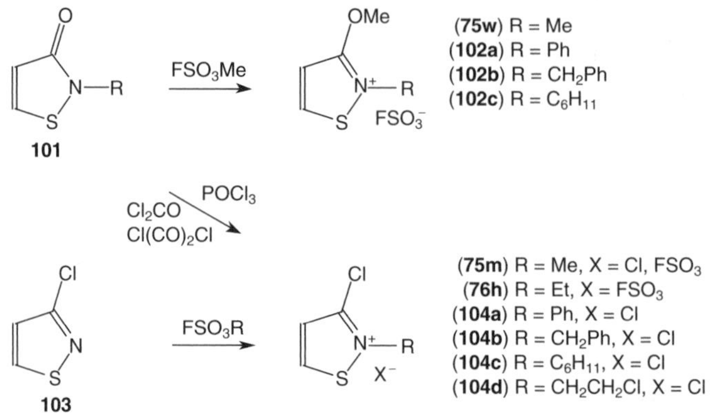

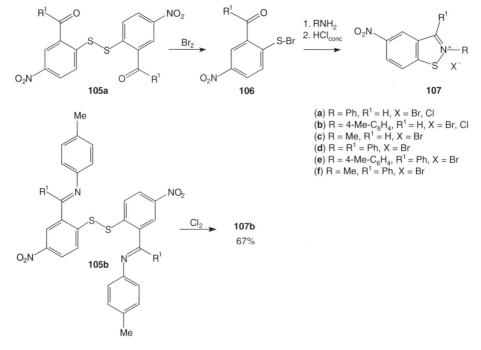

Scheme 33

104a: 68ZC170; **75m,76h** and **104a–d**: 79JOC1118, Scheme 33). The treatment of 3-chloro-isothiazole **103** with alkyl or aryl fluorosulfonates gave the 3-chloroisothia-zolium salts **75m,76h** and **104a–d** (77USP4292430, 79USP4267341, 79JOC1118).

B. 1,2-BENZISOTHIAZOLIUM SALTS

The preparation of the first 1,2-benzisothiazolium salts were reported by Fries et al. (23CB1630, 27LA264, see **107**) and later reviewed by Bambas (52CHE227) as well as

Scheme 34

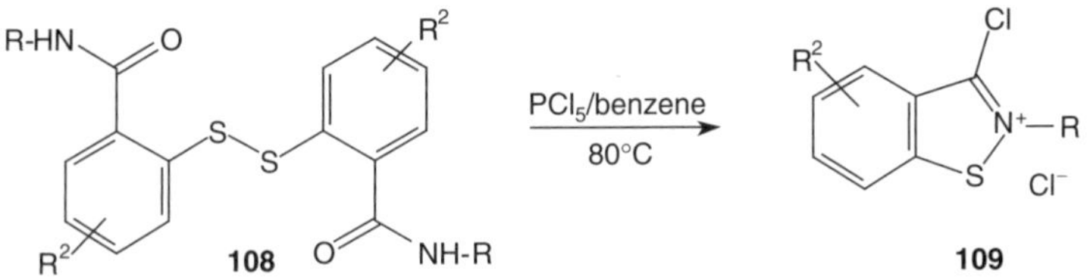

Scheme 35

Table 6. Ring closure reaction from 2,2'-dithiobis(benzamides) **108** to 3-chloro-1,2-benzisothiazolium chlorides **109**

109	R	R^2	Reference
a	Me	H	66CB2566
b	Et	H	66CB2566
c	*n*-Pr	H	66CB2566
d	CH$_2$-*i*-Pr	H	66CB2566
e	*i*-Pr	H	66CB2566
f	*n*-Bu	H	66CB2566
g	CH$_2$CH$_2$Cl	H	66CB2566
h	Allyl	H	66CB2566
i	Ph	H	66CB2566
j	4-Cl-C$_6$H$_4$	H	66CB2566
k	Morpholino	H	66CB2566
l	Et	6-MeO	66CB2566
m	Et	6-Cl	66CB2566

described in isothiazole reports (93HOU799, 02SCIS573). The general synthesis of 1,2-benzisothiazolium salts consists of cyclization of 2,2'-dithiobis(benzoic acid) derivatives *via* arensulfenylhalogenides and of 2-alkylthio- and 2-thiocyanato-benzaldehyde-phenylimines, as well as N-alkylation of the corresponding 1,2-benzisothiazoles, S-alkylation and O/S-elimination of 1,2-benzisothiazolones and -thiones.

1. *Cyclization of 2,2'-Dithiobis(Benzaldehyde) and -(Benzamide) Derivatives via Sulfenylhalides*

The 4-nitro sulfenylbromides **106**, obtained from the disulfides **105a**, reacted with primary amines in the presence of concentrated hydrochloric acid to form the corresponding 1,2-benzisothiazolium salts **107**. The salts were prepared as bromides **107a–f** and chlorides **107a,b** (**107a**: 23CB1630; **107b–f**: 27LA264, Scheme 34).

The 5-nitro-2-(4-methylphenyl)-1,2-benzisothiazolium chloride **107b** was produced from the disulfide **105b** by treatment with chlorine in chloroform in 67% yield (27LA264, Scheme 34).

The synthesis of the 3-chloro-1,2-benzisothiazolium chlorides **109** succeeded on heating *N,N'*-substituted 2,2'-dithiobis(benzamides) **108** with phosphorus pentachloride in benzene (66CB2566, Scheme 35). In this way, a series of 3-chloro

1,2-benzisothiazolium salts **109** was synthesized, represented in Table 6. Perhaps, initial cleavage of the disulfide-bond occurred followed by formation of the 1,2-benzisothiazol-3(2*H*)-one and chlorination to salt **109**.

2. Ring Closure of 2-Alkylthio- and 2-Thiocyanato-Benzaldehyde-Phenylimines and of 4,5,6,7-Tetrahydroderivatives

The *ortho*-benzylthio derivatives **110** were treated with sulfuryl chloride in dichloroethane and intermediate compounds sulfenyl chlorides **111** were apparently formed, which underwent intramolecular cyclization due to the interaction of S–Cl fragment with the C = N bond. This pathway results in the formation of 2-aryl-4,6-dinitro-1,2-benzisothiazolium chlorides **112a–d** even at room temperature. The structure of salts **112** was established by an NOE method (04MC207, Scheme 36).

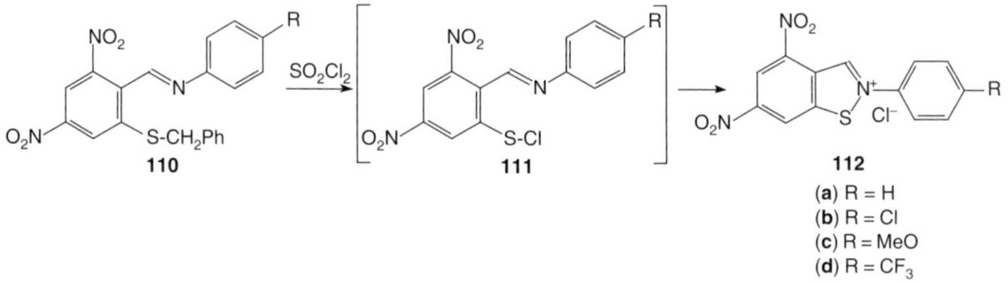

Scheme 36

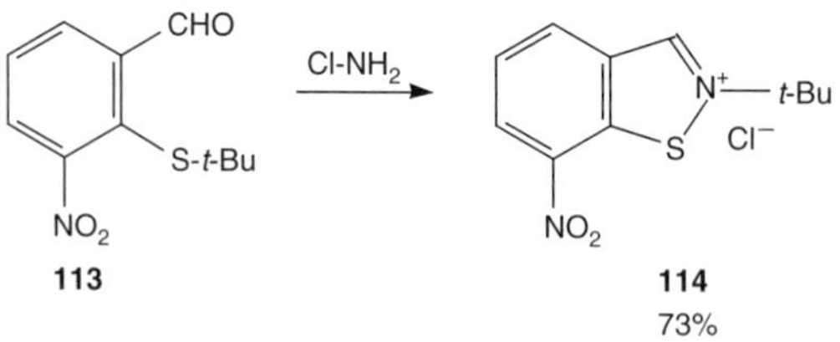

(a) R = H
(b) R = Cl
(c) R = MeO
(d) R = CF$_3$

Scheme 37

114
73%

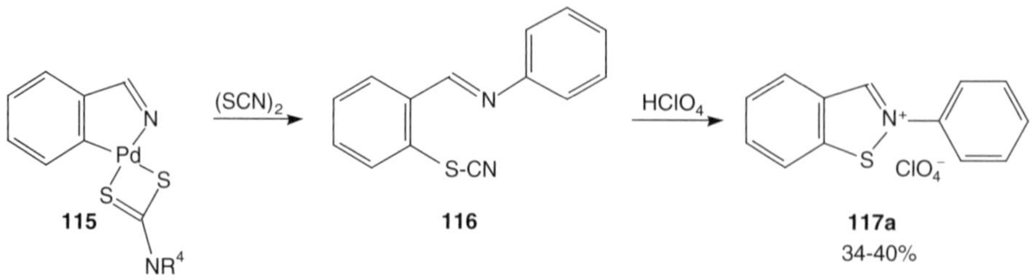

Scheme 38

117a
34-40%

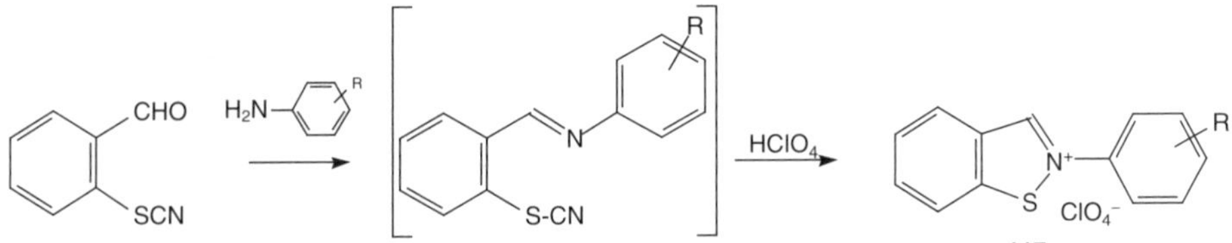

(a) R = H, 65%
(b) R = 4-MeO, 55%
(c) R = 4-CO$_2$H, 84%
(d) R = 4-CO$_2$Me, 78%
(e) R = 2-Cl, 42%
(f) R = 4-Cl, 20%
(g) R = 4-CF$_3$, 44%
(h) R = 4-NO$_2$, 42%
(i) R = 4-NH$_2$, 68%

Scheme 39

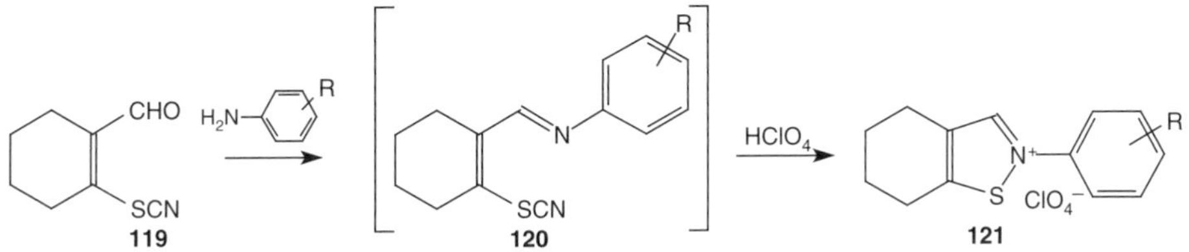

Scheme 40

The treatment of 3-nitro-2-*t*-butylthiobenzaldehyde **113** with chloramine in the usual way gave after cyclization and recrystallization from ethanol 7-nitro-2-*t*-butyl-1,2-benzisothiazolium chloride **114** in good yield (73%) (84JCS(P1)385, Scheme 37).

The reaction of the *N,N*-dimethyldithiocarbamato-palladium(II) complex **115** with dithiocyan in chloroform resulted in *N*-(2-thiocyanatobenzylidene)aniline **116**, which yielded in the presence of perchloric acid the corresponding 2-phenyl-1,2-benzisothiazolium perchlorate **117a** (79TL3339: 34%; 90JCS(P1)2881: 40%, Scheme 38).

Another general new route for the synthesis of 2-aryl-1,2-benzisothiazolium salts **117a–i** consists of cyclocondensation *via* **116**, which were easily generated by reaction of the 2-thiocyanatobenzaldehyde **118** with different anilines (05UP1, Scheme 39).

2-Aryl-4,5,6,7-tetrahydro-1,2-benzisothiazolium salts **121** were synthesized by intramolecular cyclocondensation of unstable intermediates 2-thiocyanato-4,5,6,7-tetrahydro-1,2-benzaldehyde-phenylimines **120**. This is a convenient novel method for their production (95JPR175, 96JPR424, 96T783, 99HCA685, Scheme 40). The list of salts **121** is given in Table 7.

The homologous and benzoannulated 4,5,6,7-tetrahydro-1,2-benzisothiazolium salts **122–125** were prepared similarly by cyclocondensation of the corresponding 2-thiocyanatovinyl aldehydes with aniline in the presence of perchloric acid (**122,123**: R = H, 4-Me, 4-Cl: 00JPR675, **123**: R = 2,5-Cl$_2$-4-*O-i*-Pr: 02JOC8400; **124,125**: 96UP1, Scheme 41).

Table 7.   2-Aryl-4,5,6,7-tetrahydro-1,2-benzisothiazolium salts **121**

	R	Reference
121	H, 4-MeO, 4-Cl, 4-Br, 4-SO$_2$Me	95JPR175
	3-Me, 4-Me, 3-MeO, 2-Cl, 3-Cl, 2,6-Cl$_2$, 2-CF$_3$	96JPR424
	2-Me, 2,6-Me$_2$	96T783
	4-CO$_2$H, 4-CO$_2$Me, 2-Br, 2-F, 3-Br	99HCA685
	4-t-Bu, 2,5-Cl$_2$-4-O-i-Pr	02JOC8400
	2,4-(NO$_2$)$_2$, 2,5-Cl$_2$	03ZN(B)111
	2-MeO, 2-NO$_2$, 4-NO$_2$	04ZN(B)478

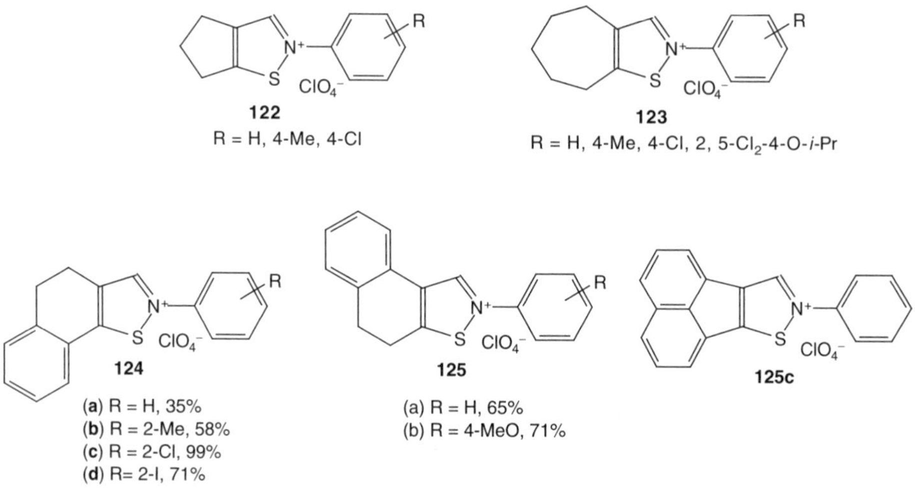

122
R = H, 4-Me, 4-Cl

123
R = H, 4-Me, 4-Cl, 2, 5-Cl$_2$-4-O-i-Pr

124
(a) R = H, 35%
(b) R = 2-Me, 58%
(c) R = 2-Cl, 99%
(d) R= 2-I, 71%

125
(a) R = H, 65%
(b) R = 4-MeO, 71%

125c

Scheme 41

The isomerization of 2-aryl-4,5,6,7-tetrahydro-1,2-benzisothiazolium perchlorates **121** and their homologous **123** with substituted anilines yielded 1-aryl-4,5,6,7-tetra-hydro-2,1-benzisothiazolium perchlorates **126a–c** (2–64%) and their homologous **126d–f** (4–89%) while the spirocyclic salts **127** (1–8%, see also Section IV.D.1) and the vinamidines **128** (5–13%) were only by-products in this process (00JPR675; see also for **127**: $n = 1$: 96JPR424, Scheme 42). The reaction of 2-aryl-5,6-dihydro-4 H-cyclopenta[d]isothiazolium perchlorates **122** ($n = 0$) with anilines produced in nearly quantitative yields (95–99%) only vinamidines **128** ($n = 0$) as the main products of sufficient purity (00JPR675).

β-Thiocyanatovinylaldehydes **119** were also versatile C$_3$S building blocks in the synthesis of acceptor substituted 2-amino-isothiazolium salts **131** and **132** by intramolecular cyclocondensation of thiocyanatovinylaldehyde hydrazones **129** (93TL1909). The alternative cyclization route to 1,2,3-thiadiazines was not observed. Alicyclic aldehydes **119** reacted with benzhydrazides (R = ArCO) to unstable benzhydrazones **129** (R = ArCO) that cyclized spontaneously to **130** (R = ArCO). The imines **130**

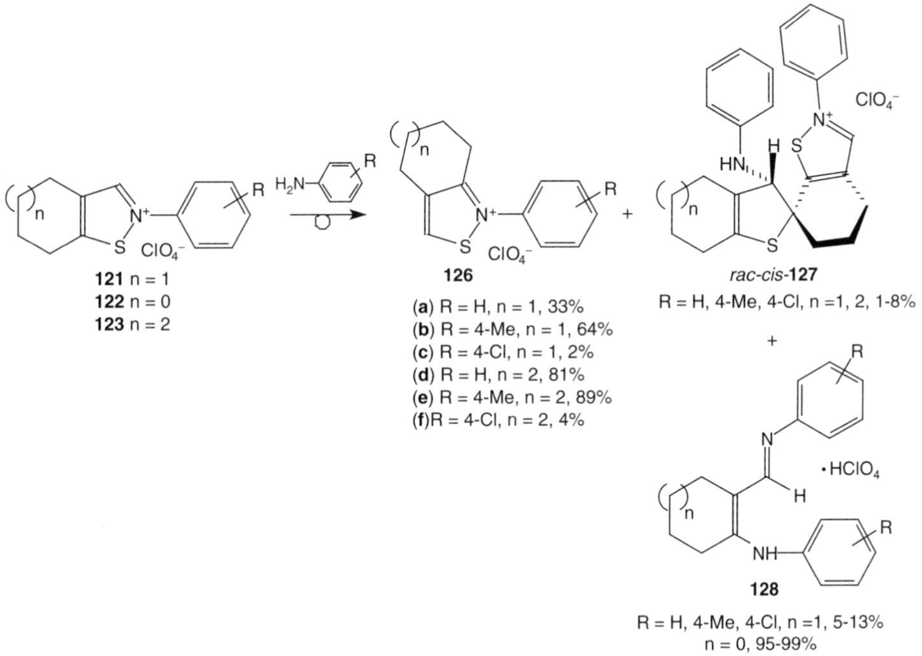

Scheme 42

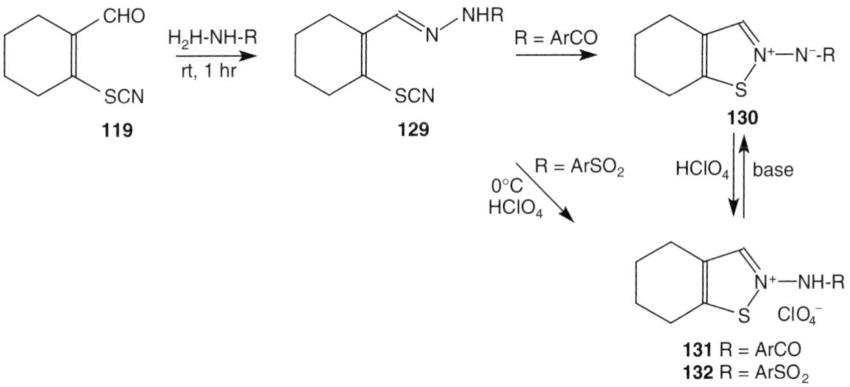

Scheme 43

reacted at 0 °C with 70% perchloric acid giving rise to salts **131** (R = ArCO) (92ZK281, 94JPR115, 98H587, 04SCIS731, Scheme 43).

In contrast, the arylsulfonylhydrazones **129** (R = ArSO$_2$) were thermally stable, and they could be recrystallized from ethanol or acetonitrile. Treatment of these hydrazones **129** with 70% perchloric acid at 0 °C for a short time gave the isothiazolium salts **132a–h** (R = ArSO$_2$) as primary cyclization products (96ZOR1745, 99JHC1081, Scheme 43). The structure of salt **132a** was determined by X-ray analysis (96ZOR1745).

Table 8. Synthesis of isothiazolium salts **131,132**

	R	Yield (%)	Reference
131a	PhCO	75	94JPR115
b	4-Me-C$_6$H$_4$-CO	70	94JPR115
c	4-MeO-C$_6$H$_4$-CO	50	94JPR115
d	4-OH-C$_6$H$_4$-CO	95	94JPR115
e	3-Cl-C$_6$H$_4$-CO	80	00JPR291
f	4-Cl-C$_6$H$_4$-CO	90	94JPR115
g	3-NO$_2$-C$_6$H$_4$-CO		00JPR291
h	4-NO$_2$-C$_6$H$_4$-CO	95	94JPR115
i	3-CF$_3$-C$_6$H$_4$-CO	75	00JPR291
j	2-Cl-C$_6$H$_4$-CO		00JPR291
k	2-NO$_2$-C$_6$H$_4$-CO	30	94JPR115
l	2-Me-C$_6$H$_4$-CO	82	94JPR115
132a	PhSO$_2$	75	96ZOR1745
b	4-Me-C$_6$H$_4$-SO$_2$	89	96ZOR1745
c	4-MeO-C$_6$H$_4$-SO$_2$	67	96ZOR1745
d	4-Br-C$_6$H$_4$-SO$_2$	52	96ZOR1745
e	3-NO$_2$-C$_6$H$_4$-SO$_2$	53	99JHC1081
f	4-NO$_2$-C$_6$H$_4$-SO$_2$	95	96ZOR1745
g	2,4,6-Me$_3$-C$_6$H$_2$-SO$_2$	45	96ZOR1745
h	2,4,6-i-Pr$_3$-C$_6$H$_2$-SO$_2$	47	96ZOR1745

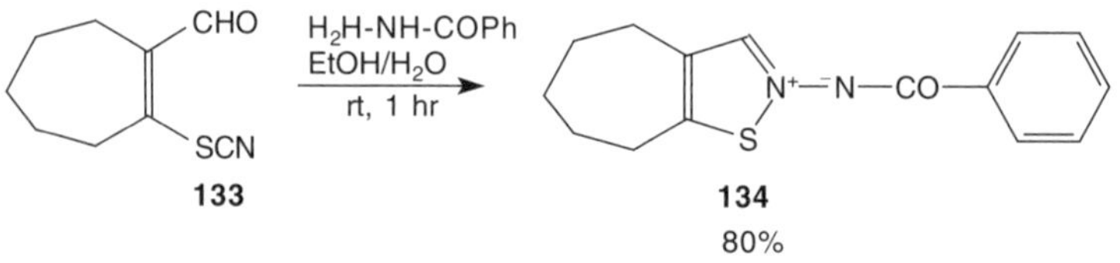

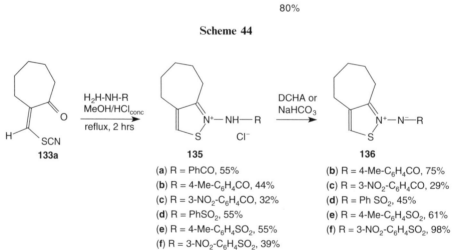

Scheme 44

Scheme 45

The yields and related references for the salts **131,132** are represented in Table 8.

2-Thiocyanatocyclohepten-1-carbaldehyde **133** reacted with benzhydrazides by the same cyclocondensation pathway to give *N*-benzoyl-isothiazolium-2-imines **134** in 80% yield (94JPR115, Scheme 44).

The cyclocondensation of 2-thiocyanatomethylenecycloheptan-2-one **133a** (77JPR305, 91ACSA302) with benzhydrazides produced hydrochlorides **135a–c** and the reaction with DCHA yielded imines **136b,c** (94JPR115, 00JPR291, Scheme 45).

The reaction of **133a** with benzenesulfonylhydrazines also gave by ring closure the hydrochlorides **135d–f** (00SUL109) while the treatment with sodium bicarbonate resulted in the formation of the imines **136d–f**. The structure of sulfonyl-imines **136f** was determined by X-ray crystallography (00SUL109, Scheme 45).

3. N/S-Alkylation, Protonation and O/S-Elimination

The 2-substituted 1,2-benzisothiazolium salts **138** were obtained by N-alkylation of 1,2-benzisothiazole **137** with triethyloxonium tetrafluoroborate as well as with dimethylsulfate. These alkylations afforded a useful method for the preparation of quaternary salts **138** with different anions such as perchlorate **138a** (91JHC749), iodide **138c** (74AJC1221, 74JHC1011) and tetrafluoroborate **138b–e** (68ZC170, 71LA46, Scheme 46).

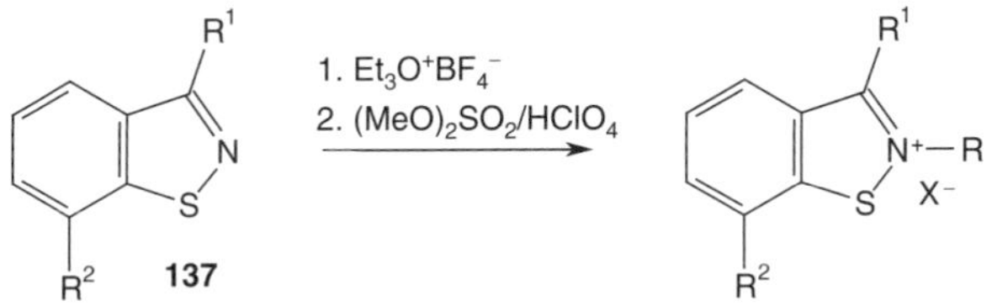

(**138a**) R = R^1 = Me, R^2 = CO-Me, X = ClO$_4$, 93%
(**b**) R = Me, R^1 = R^2 = H, X = BF$_4$, I
(**c**) R = Et, R^1 = Br, R^2 = H, X = BF$_4$
(**109a**) R = Me, R^1 = Cl, R^2 = H, X = BF$_4$, 90%
(**b**) R = Et, R^1 = Cl, R^2 = H, X = BF$_4$

Scheme 46

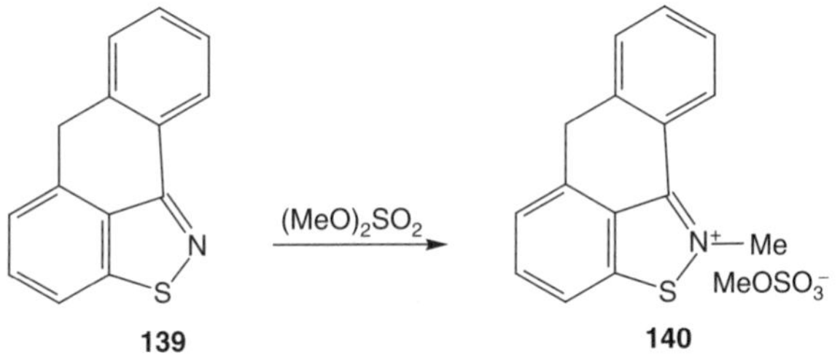

Scheme 47

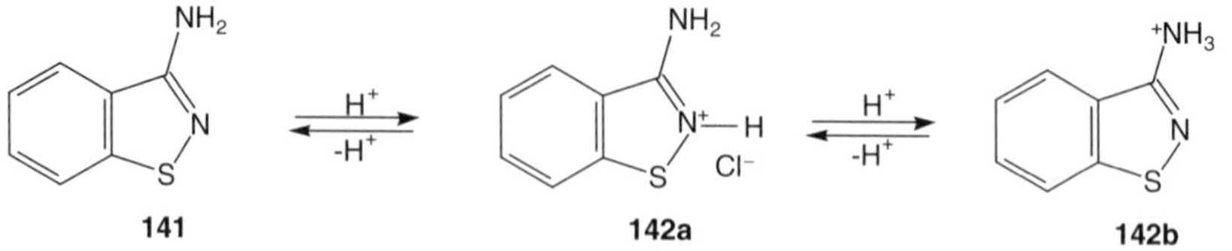

Scheme 48

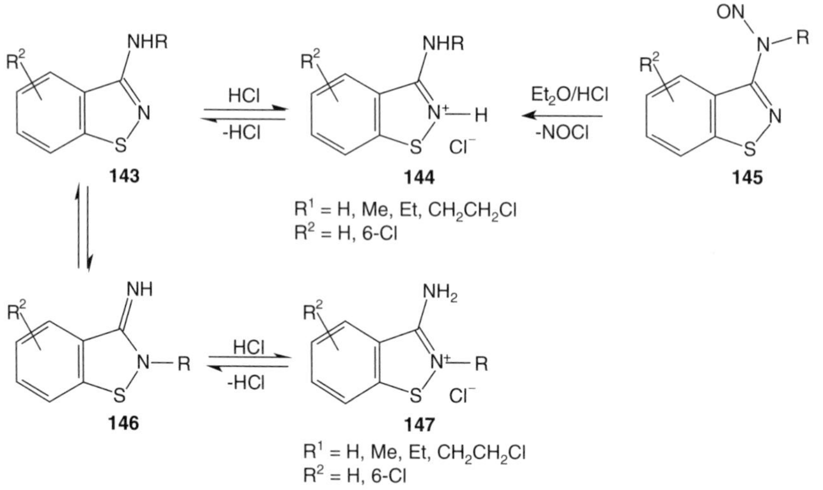

R¹ = H, Me, Et, CH₂CH₂Cl
R² = H, 6-Cl

R¹ = H, Me, Et, CH₂CH₂Cl
R² = H, 6-Cl

Scheme 49

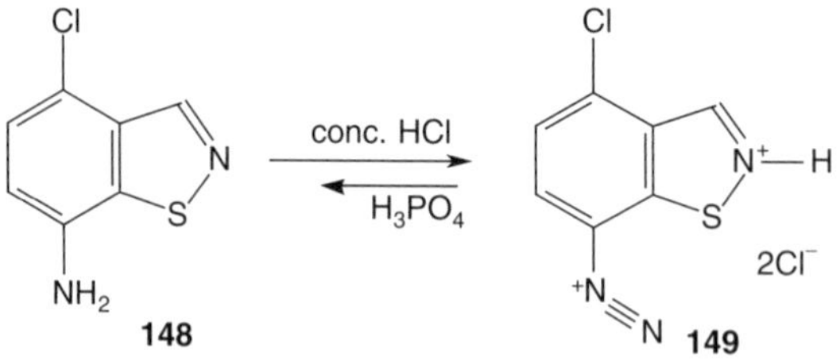

Scheme 50

6*H*-Anthra[9,1-*cd*]isothiazole **139** was N-methylated by dimethyl sulfate to the isothiazolo-anthrenium methylsulfate **140** (72IJC361, Scheme 47). The salt **140** was synthesized to study the structural changes in their absorption spectra, e.g., cross conjugation effects as well as steric inhibitions.

The 3-amino-1,2-benzisothiazole **141** could be protonated to form quaternary 1, 2-benzisothiazolium cation **142a** (81G71, Scheme 48). This tautomeric process gave two protonated forms: cations **142a** and **142b**. The latter is formed by protonation of the exocyclic nitrogen of the 3-amino group.

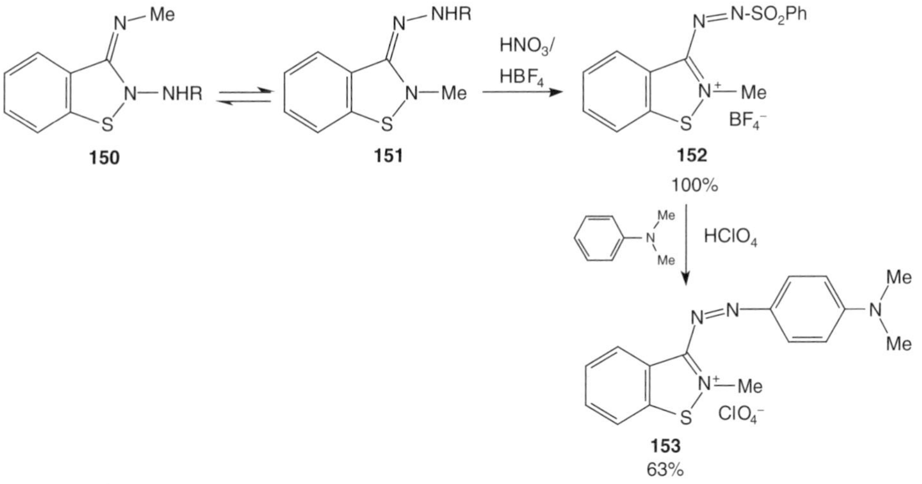

Scheme 51

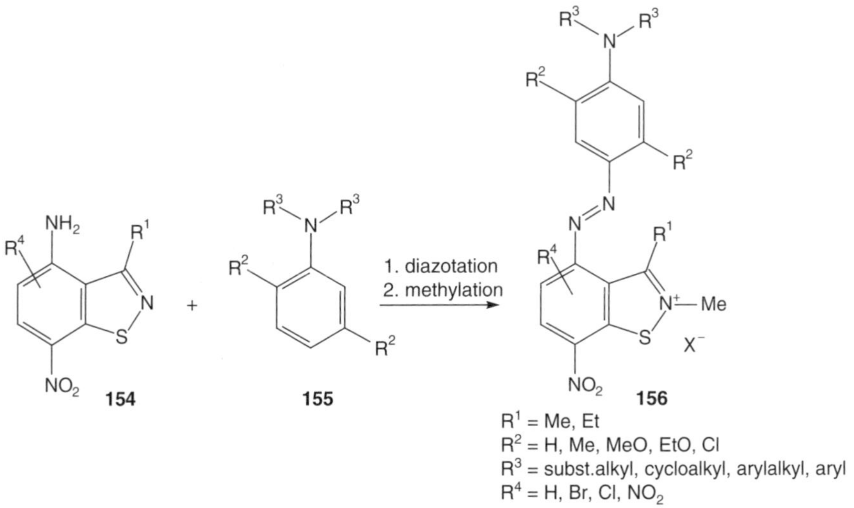

R^1 = Me, Et
R^2 = H, Me, MeO, EtO, Cl
R^3 = subst.alkyl, cycloalkyl, arylalkyl, aryl
R^4 = H, Br, Cl, NO$_2$

Scheme 52

From 3-amino 1,2-benzisothiazole **143**, salts of 3-amino-1,2-benzisothiazole hydrochloride **144** were obtained by addition of hydrochloric acid followed by heating until equilibrium between the 3-imino-1,2-benzisothiazolium chloride **146** and 3-amino **147** was established. The equilibrium was affected by the solvent, pH and the nature of the substituents R and R^2. The 3-nitrosamino benzisothiazole **145** could be easily converted in ether solution with hydrochloric acid at room temperature into hydrochlorides **144** (69CB1961, Scheme 49).

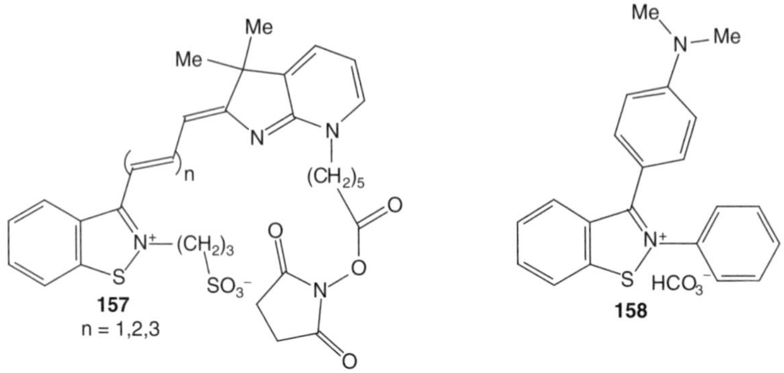

157
n = 1,2,3

158

Scheme 53

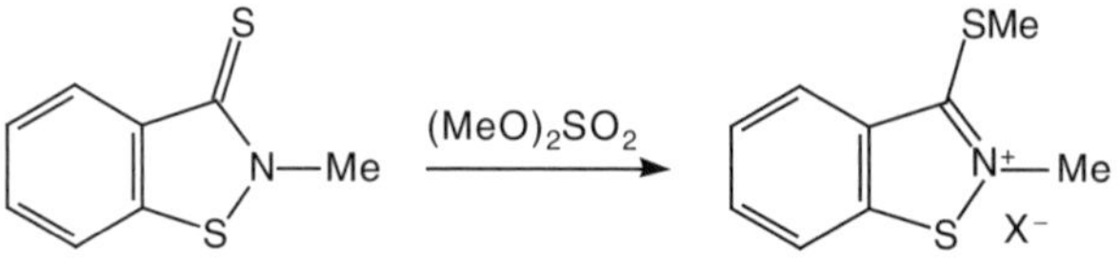

159

160

(a) X = ClO$_4$, 82%
(b) X = MeOSO$_3$, 84%

Scheme 54

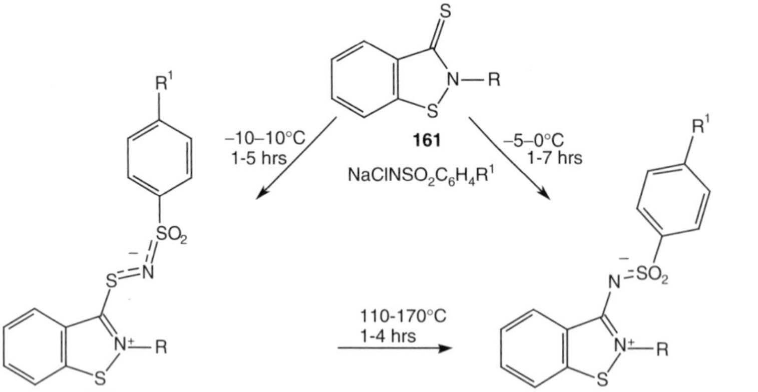

−10–10°C
1-5 hrs

161

−5–0°C
1-7 hrs

NaClNSO$_2$C$_6$H$_4$R^1

110-170°C
1-4 hrs

162

(a) R = R^1= Me, 5%
(b) R = Me, R^1= MeO, 52%
(c) R = Me, R^1 = NO$_2$, 52%
(d) R = Me, R^1= Cl, 34%
(e) R = CH$_2$Ph, R^1 = NO$_2$, 51%
(f) R = CH$_2$Ph, R^1 = Cl, 76%

163

(a) R = R^1= Me, 7%
(b) R = Me, R^1 = MeO, 82%
(c) R = Me, R^1 = NO$_2$, 55%
(d) R= Me, R^1= Cl, 55%
(e) R = CH$_2$Ph, R^1 = Me, 35%
(f) R = CH$_2$Ph, R^1 = MeO, 3%
(g) R = CH$_2$Ph, R^1 = NO$_2$, 54%
(h) R = CH$_2$Ph, R^1 = Cl, 66%

Scheme 55

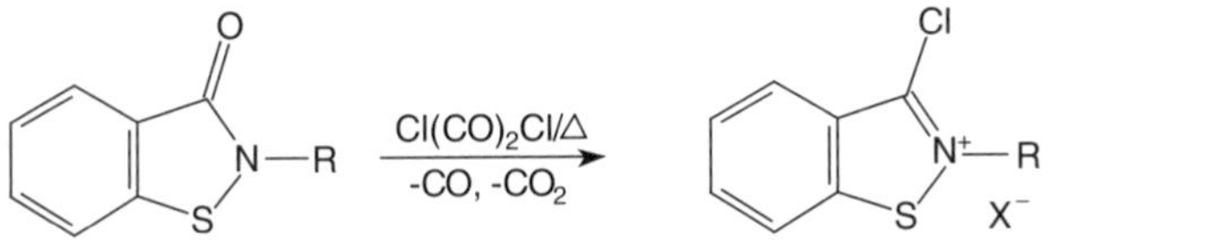

164

(**109a**) R = Me, X = Cl,ClO$_4$, ReO$_4$
(**109i**) R = Ph, X=Cl, ClO$_4$
(**165a**) R = 4-MeO-C$_6$H$_4$, X = Cl, ClO$_4$

Scheme 56

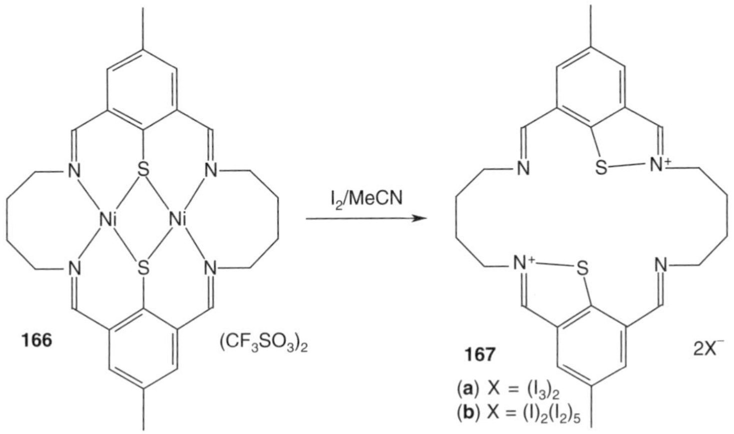

166   (CF$_3$SO$_3$)$_2$

167   2X$^-$

(**a**) X = (I$_3$)$_2$
(**b**) X = (I)$_2$(I$_2$)$_5$

Scheme 57

The 7-aminobenzisothiazole **148**, prepared by reduction of the 4-chloro-7-nitro-benzisothiazole, was diazotized in concentrated hydrochloric acid and the 4-chloro benzisothiazole-7-diazonium chloride **149** was obtained (71JCS(C)3994, Scheme 50).

The 3-methylimino 1,2-benzisothiazole **150**, which in solution is in equilibrium with 2-methyl benzisothiazole **151**, reacted with concentrated tetrafluoric and nitric acids to produce 3-benzensulfonylazo-2-methyl-1,2-benzisothiazolium tetrafluoroborate **152** in quantitative yield. This salt **152** was treated with N,N'-dimethyl-aniline to give 2-methyl-3-[4-(1-dimethylaminophenyl)azo]-1,2-benzisothiazolium perchlorate **153** in 63% yield (71LA46, Scheme 51).

Azo dyes **156** could be prepared by interaction of aniline **154** with secondary amine **155** followed by N-alkylation. Because of their good water solubility, salts **156** were used for dyeing or printing textiles, leather or synthetic materials (71DEP2020479, Scheme 52).

Inner salts **157**, obtained by N-alkylation of the isothiazole, are azomethine dye-based fluorescent nucleotides. They were used as the labels for nucleic acid (03JP2003034696, Scheme 53).

The 3-[4-(dimethylamino)phenyl]-2-phenyl carbonate **158** is a photosensitive salt that has a high sensitivity to semiconductor laser light and good storage stability

(02JP2002202598, Scheme 53). This dye is polymerized by an exposing it to 450 nm laser light.

2-Methyl-3-methylthio-1,2-benzisothiazolium salts **160** were formed on treatment of 2-methyl-1,2-benzisothiazol-3(2H)-thione **159** with dimethyl sulfate to give **160b** followed by conversion to its perchlorate (**160a**) in acetic acid with very good yields (**160a**: 71LA46, 77CJC1123; **160b**: 71LA201, Scheme 54).

Boberg et al. (96PS203) described the preparation of the sulfimides **162** and sulfonamides **163** from 2-methyl- or 2-benzyl-1,2-benzisothiazol-3(2H)-thiones **161** by reacting with sodium N-chlorobenzensulfonamide (Scheme 55). Some sulfimides **162** decomposed at room temperature in the solid state, quickly in solution and especially in polar solvents. Their instability depends on the heterocyclic part of the molecule and also on the substituents in the aryl ring, e.g., electron withdrawing substituents like NO_2 or Cl stabilized sulfimides **162**.

3-Chloro-1,2-benzisothiazolium chloride **109a,i,165a** could be obtained by treatment of the 1,2-benzisothiazol-3(2H)-one **164** with oxalylchloride in xylene solvent. At the temperatures of 120–140 °C, elimination of CO as well as of CO_2 occurred and the colorless crystals of the salt **109a,i,165a** were isolated by oxidative elimination. The chloride anions **109a,i,165a** could be changed in aqueous solution into the perrhenate **109a** with NH_4ReO_4-solution and/or into perchlorates **109a,i,165a** with 70% aqueous perchloric acid (65ZN(B)712, Scheme 56).

The reaction of $[Ni_2L_3](CF_3SO_3)_2$-complex **166** with an excess of I_2 in MeCN resulted in ligand oxidation to form the metal free macrocycle (ligand)$^{2+}$ **167** by slow evaporation of complex **166** that contains 2 five-membered isothiazole rings (00JCS(D)3113, Scheme 57). Over approximately one day small brown and red single crystals developed. These were filtered off, washed and dried in poor yield (13%) and confirmed by X-ray determinations of (ligand)(I$_3$)$_2$ **167a** as well as of (ligand)(I)$_2$(I$_2$)$_5$ **167b**.

C. 2,1-Benzisothiazolium Salts

The general synthesis of 2,1-benzisothiazolium salts consists mainly in the N-alkylation of the corresponding 2,1-benzisothiazole as well in the S-alkylation of 2,1-benzisothiazol-3(2H)-ones and thiones as well as in O/S-elimination.

The 1-methyl 2,1-benzisothiazolium salts **169a–m** were prepared by methylation of 2,1-benzisothiazole **168**. The 1-alkylated isothiazolium salts **169** are presented with their yields and references in Table 9. These alkylation afforded a very simple

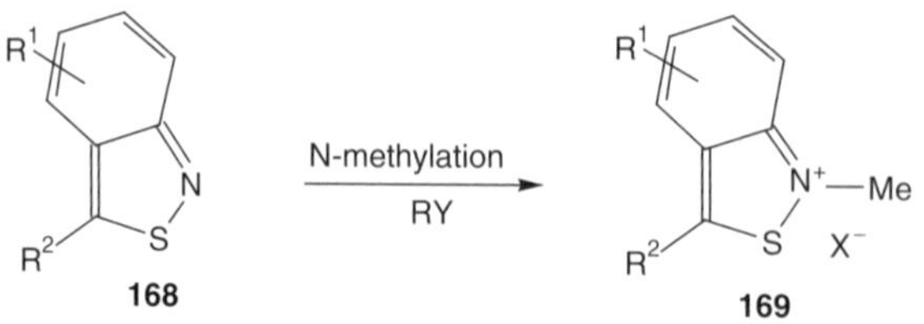

Scheme 58

Table 9. N-Alkylation of 2,1-benzisothiazoles **168** to 2-alkyl-2,1-benzisothiazolium salts **169**

169	R^1	R^2	RY	X	Yield (%)	Reference
a	H	H	$(MeO)_2SO_2$	$MeOSO_3$	~100	73JCS(P1)1863
			MeI	I		74AJC1221/74JHC1011
			4-Me-C_6H_4-SO_3Me	I	~100	73JCS(P1)1863
b	6-Cl	H	$PhCH_2Br$	I	~100	73JCS(P1)1863
			$CH(MeO)_3$ + Lewis acid	BF_4	47	72CPB2372
			4-Me-C_6H_4-SO_3Me	I	~100	73JCS(P1)1863
c	5-Cl	H	4-Me-C_6H_4-SO_3Me	I	~100	73JCS(P1)1863
d	5-NO_2	H	4-Me-C_6H_4-SO_3Me	Ts	~100	73JCS(P1)1863
e	H	Styryl	FSO_3Me	ClO_4	92	78JOC1233
			$(MeO)_2SO_2$	ClO_4	59	75AJC129
f	H	Me	FSO_3Me	ClO_4	83	78JOC1233
g	H	MeO	$(MeO)_2SO_2$	ClO_4	59	75AJC129
h	H	Cl	$PhCH_2Br$	I	~100	73JCS(P1)1863
			$(MeO)_2SO_2$	ClO_4	84	75AJC129
i	H	$CH(CF_3)_2$	FSO_3Me	FSO_3		78JOC2500
j	H	NH_2	MeI	I	88	65JMC515
k	5,6-$(MeO)_2$	NHEt	MeI	I	78	65JMC515
l	6-Cl	NMe_2	MeI	I	92	65JMC515
m	5-Cl	Ph	$CH(MeO)_3$ + Lewis acid	BF_4	60	72CPB2372
			$CH(MeO)_3$ + Lewis acid	$SbCl_6$	57	72CPB2372
n	5-Cl	4-Cl-C_6H_4	$CH(MeO)_3$ + Lewis acid	BF_4	9	72CPB2372

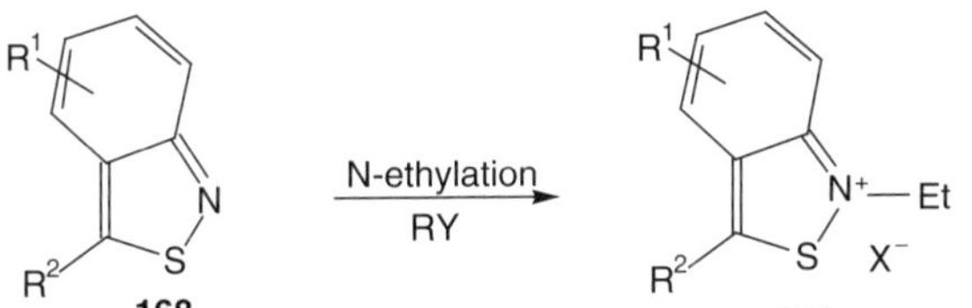

Scheme 59

(a) $R^1 = R^2 = H$, X = I
(b) $R^1 = 5$-Cl, $R^2 = Ph$, X = BF_4, 37%

way to obtain quaternary salts in good yields (**169a–l**: 47–100%, except **169n**: 9%, Scheme 58) from isothiazole **168** on direct heating in the presence of Lewis acids with alkyl iodides and bromides, dialkyl sulfates, methyl fluorosulfonate, alkyl toluene-4-sulfonates as well as trialkyl orthoformates.

2,1-Benzisothiazole **168a** slowly reacted to form 1-ethyl 2,1-benzisothiazolium iodide **170a** (76AJC1745). The alkylation of isothiazole **168b** with triethyl orthoformates in the presence of Lewis acids readily gave the 1-ethyl-2,1-benzisothiazolium tetrafluoroborate **170b** in 37% yield (72CPB2372, Scheme 59).

Stirring the 2,1-benzisothiazoles **168** with dialkyl sulfates, alkyl bromides or alkyl toluene-4-sulfonates gave the 1-N-alkyl-substituted isothiazolium salts **171** in very good yields (90–100%) (**171a,b,e–i**: 73JCS(P1)1863; **171c**: 83JHC1707; **171d**: 78JOC1233,

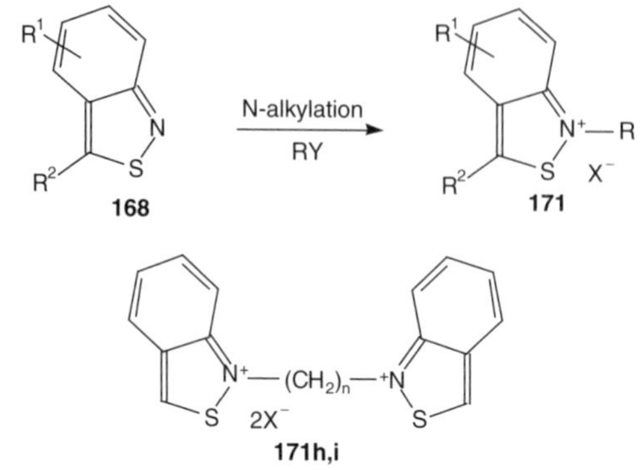

(a) R = CH$_2$Ph, R^1 = R^2 = H, X = Br
(b) R = CH$_2$Ph, R^1 = 5-Cl, R^2 = H, X = Br
(c) R = CH$_2$Ph, R^1 = 6-Cl, R^2 = H, X = Br
(d) R = COMe, R^1 = H, R^2 = styryl, X = ClO$_4$
(e) R = CH$_2$CO$_2$H, R^1 = R^2 = H, X = Br, I
(f) R = CH$_2$CO$_2$Et, R^1 = R^2 = H, X = Br
(g) R = CH$_2$CH$_2$SO$_3$Ph, R^1 = R^2 = H, X = I
(h) n = 3, X = I,
(i) n = 4, X = Br

Scheme 60

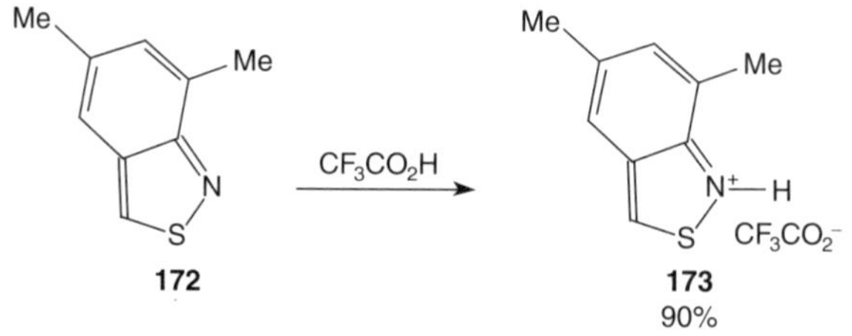

Scheme 61

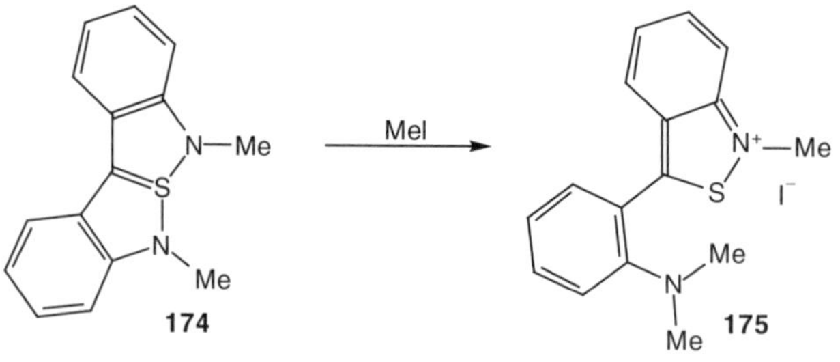

Scheme 62

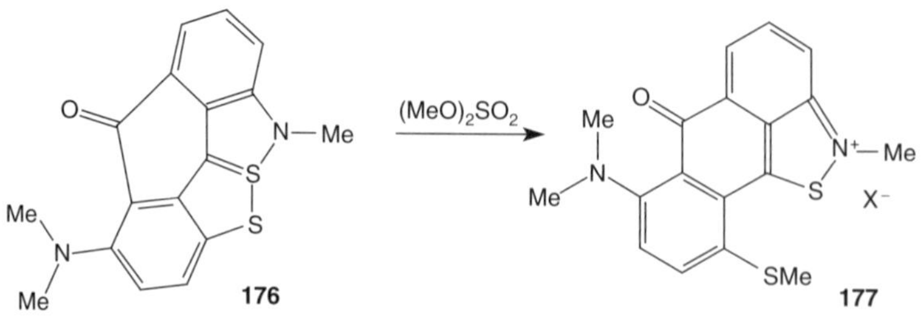

X = MeOSO$_3$, ClO$_4$

Scheme 63

Table 10.   N-Alkylated 3-arylazo-1-methyl-2,1-benzisothiazolium salts **178**

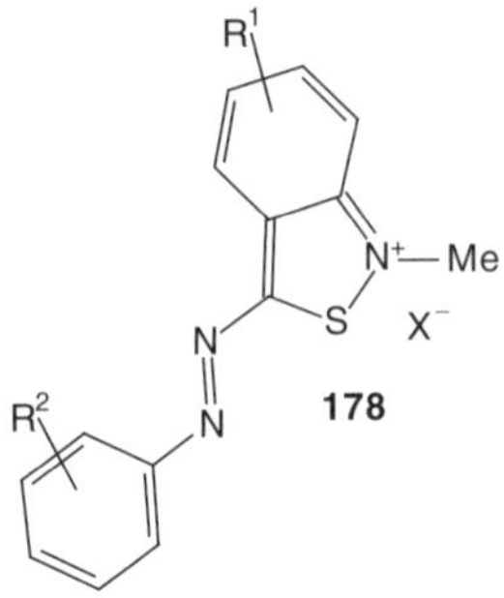

178	R^1	R^2	X	Reference
a	H	4-N(Et)(CH$_2$)$_2$NMe$_3$	MeOSO$_3$	68DEP19661015
b	H	4-NMe$_2$	MeOSO$_3$	66DEP19650610
c	H	2-SO$_2$Me, 4-NMe$_2$	Cl	72DEP2155694
d	6-Cl	2-NHAc, 4-Et$_2$	MeOSO$_3$	68DEP19661015
e	5-NO$_2$	4-NMe$_2$	MeOSO$_3$	68DEP19661015
f	5-NO$_2$	4-NEt$_2$	MeOSO$_3$	68DEP19661015
g	5-NO$_2$	2-SO$_2$Me, 4-NMe$_2$	Cl	72DEP2155694
h	7-Cl	2-SO$_2$NH$_2$, 4-NMe$_2$	Cl	72DEP2155694
i	4-MeO	2-SO$_2$NH$_2$, 4-NEt$_2$	Cl	72DEP2155694
j	5-Cl	2-SO$_2$NMe$_2$, 4-NMe$_2$	Cl	72DEP2155694
k	6-Cl	2-SO$_2$NMe$_2$, 4-NMe$_2$	Cl	72DEP2155694
l	5,7-Cl$_2$	2-SO$_2$NMe$_2$, 4-NMe$_2$	Cl	72DEP2155694
m	5-NO$_2$	4-N(Me)(CH$_2$CH$_2$OH)	MeOSO$_3$	99COL21
n	5-NO$_2$	2-NHAc, 4-N((CH$_2$)$_2$OH)$_2$	MeOSO$_3$	99COL21
o	5-NO$_2$	4-N((CH$_2$)$_2$OH)((CH$_2$)$_2$CN)	MeOSO$_3$	99COL21
p	5-NO$_2$	2-MeO, 4-N((CH$_2$)$_2$OH)-((CH$_2$)$_2$CN)	MeOSO$_3$	99COL21
q	5-NO$_2$	4-NAc$_2$	MeOSO$_3$	99COL21
r	5-NO$_2$	4-N((CH$_2$)$_2$OAc)((CH$_2$)$_2$CN)	MeOSO$_3$	99COL21
s	5-NO$_2$	2-NHAc,3-MeO, 4-N(Et)((CH$_2$)$_2$CN)	MeOSO$_3$	99COL21
t	5-NO$_2$	2-NHAc,3-MeO, 4-N((CH$_2$)$_2$OH)((CH$_2$)$_2$MeO)	MeOSO$_3$	99COL21

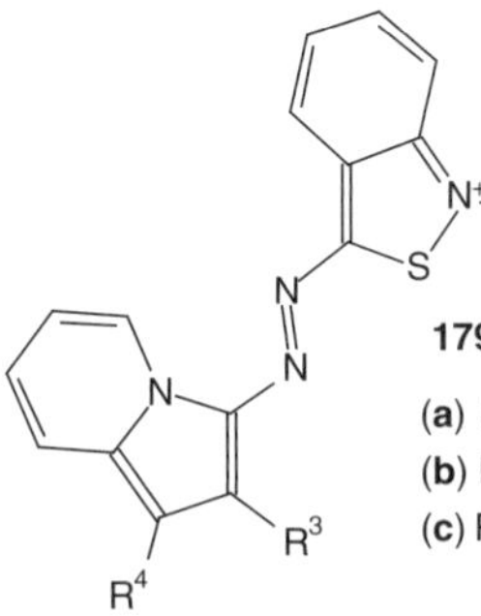

(a) R^3 = Me, R^4 = H
(b) R^3 = R^4 = Me
(c) R^3 = Ph, R^4 = Me

Scheme 64

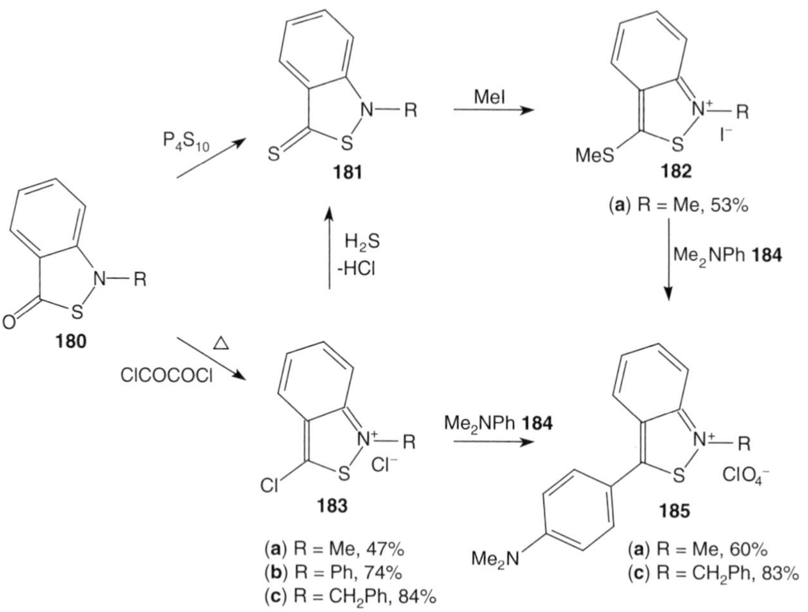

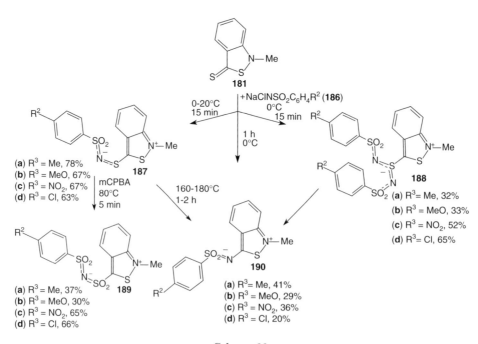

Scheme 65

Scheme 66

Scheme 60). Davis et al. (73JCS(P1)1863) found that quaternary salts **171** rapidly decomposed in aqueous acid or base yielding 2-aminobenzaldehydes.

The 2,1-benzisothiazole **172** was protonated with trifluoroacetic acid to give 5, 7-dimethyl-2,1-benzisothiazole trifluoroacetate **173** in very high yield (79ZN(B)123, Scheme 61). Protonation occurred at the N-atom according to ^{1}H NMR analysis.

1-Methyl-3-(2-dimethylaminophenyl)-2,1-benzisothiazolium iodide **175** was only obtained in small amount by the reaction of 2,3,4,5-dibenzothiadiazapentalene **174** with methyl iodide (85CJC882, Scheme 62).

The methylation of **176** with dimethyl sulfate afforded in quantitative yield the 2-methyl-7-dimethylamino-10-methylthio-6-oxo-6*H*-anthra[1,9-*cd*]isothiazolium methylsulfate **177**, which could be converted into the perchlorate with perchloric acid (92KGS256, Scheme 63).

The 3-arylazo-1-methyl 2,1-benzisothiazolium methyl sulfates **178a–f** are blue green dyes for polyacrylonitrile fibers. They can be prepared by quaternization of an arylazobenzisothiazole with dimethyl sulfate (**178b**: 66DEP19650610; **178a,d–f**: 68DEP19661015; **178m–t**: 99COL21, Table 10). By anion exchange or treatment with salts or acids, the anion of dyes **178** can be replaced by chloride **178c,g–l** (72DEP2155694), bromide, sulfate, perchlorate as well as acetate and other anions. Shah et al. (99COL21) synthesized eight cationic dyes **178m–t** derived from 3-amino-5-nitro-2,1-benzisothiazole with various *N,N*-disubstituted aniline derivatives that then were applied to polyacrylnitrile and acid modified polyester fibers and gave yellow to orange to green hues (99COL21). All these dyes **178m–t** were brighter and deeper shade with high tinctorial strength.

Furthermore, cationic dyes **179** of the diazahemicyanine class are derived from indolizines. This invention related to water-soluble azo dyestuffs is useful for the colorations of synthetic polymeric materials particularly polymers and copolymers of acrylnitrile and dicyanoethylene as well as modified polyesters and polyamides (81GBP2075540, Scheme 64).

A series of compounds with a 2,1-benzisothiazol-basis were easily available from 2, 1-benzisothiazol-3(2*H*)-one **180**. The treatment of 1-methyl-2,1-benzisothiazol-3 (2*H*)-one **180** with phosphorus pentasulfide in pyridine gave the 1-methyl 2,1-benzisothiazol-3(2*H*)-thione **181** in 28% yield. The preparation of 1-methyl-3-methylthio-2,1-benzisothiazolium iodide **182** resulted from the reaction of thione **181** with methyl iodide in benzene in very good yield (78JHC529, Scheme 65).

After heating of **180** in inert solvents and with oxalylchloride, 1-substituted 3-chloro-2,1-benzisothiazolium chlorides **183** were obtained (76JPR161); they could be transformed into their perchlorates. Activated aromatic amines like dimethylaniline reacted with the salts **183a,c** to give deeply colored cyanines **185a,c** (60–83%) (76JPR161, Scheme 65). This color also could be obtained with 3-methylthio-2,1-benzisothiazolium salts **182a**, formed through methylation of their thiones **181a**.

The 2,1-benzothiones **181** were mainly synthesized from 3-chloro-2,1-benzisothiazolium chlorides **183** with H$_2$S in inert solvents and HCl-elimination (76JPR161, Scheme 65).

The 1-methyl-2,1-benzisothiazol-3(2*H*)-thiones **181** reacted with sodium *N*-chlorobenzensulfonamide **186** to sulfimides **187**, sulfinamidinates **188** and to sulfonamides **190** (20–78%). Sulfonamidates **189** were synthesized by an oxidation of sulfimides

187. The sulfimides **187** and sulfinamidinates **188** also could be converted into sulfonamides **190** (96PS203, Scheme 66).

D. HETEROCYCLIC ANNULATED ISOTHIAZOLIUM SALTS

The first described method involves the synthesis of thieno[2,3-*d*] and [3,2-*c*] iso-thiazolium salts **192** by cyclization of the 2-mercaptothiophen-cyclohexylimine complexes **191**. Most heterocyclic annulated isothiazolium salts were generally isolated with [2,3-*a*] and [1,2-*b*] annulation.

Solutions of complexes **191** in chloroform were added quickly to bromine, sodium iodide, *N*-bromosuccinimide or picrate and heated for 1 h at 50–60 °C to give 2-cyclohexylthieno[3,2-*d*]- **192a** and 2-cyclohexylthieno[2,3-*d*]-isothiazolium salts **192b** (65KGS220, 67KGS1022, 69KGS475, Scheme 67).

Only two 3,4-heterocyclic annulated isothiazolium salts **194** and **197** were synthesized. Since the isothiazolium salts were synthesized from β-aminopropenone derivatives by thionation, followed by oxidation (65JCS32, 68CJC1855), it was of interest to learn if a 2-acyl-3-aminothiophenone could be converted to the thieno [3,2-*c*]isothiazolium system. Accordingly, crude thiones **193** treated with iodine in

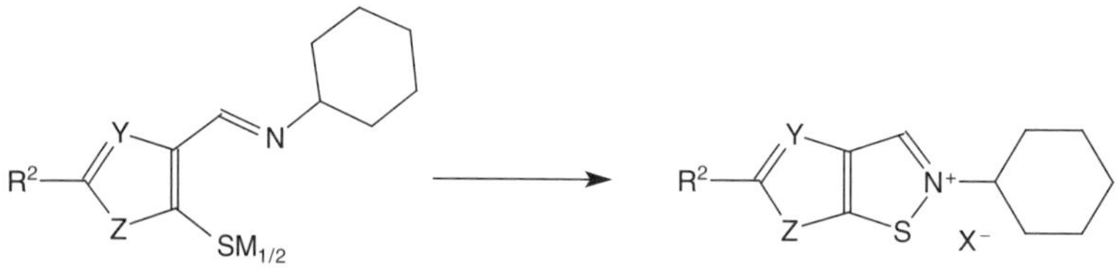

191
M = Ni, Zn

192

(**a**) R^2 = H, Et, Y = C, Z = S, X = Br, Br$_3$, ZnBr$_4^-$, picric acid
(**b**) R^2 = H, Y = S, Z = C, X = Br, Br$_3$, I, ZnBr$_4^-$, picric acid

Scheme 67

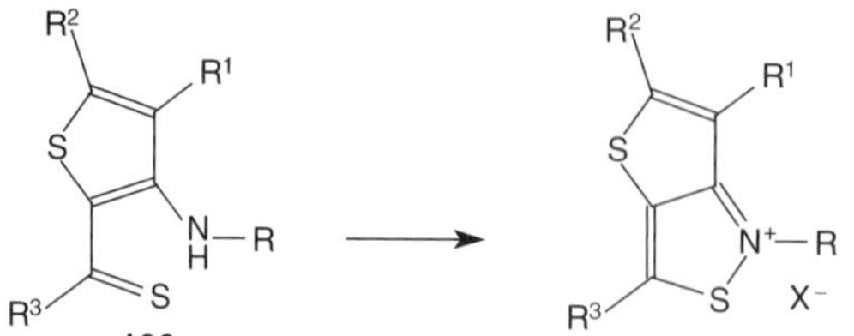

193

194

(**a**) R = R^2 = R^3 = Ph, R^1 = H, X = ClO$_4$, 38%
(**b**) R = R^1 = R^3 = Ph, R^2 = H, X = ClO$_4$, 41%
(**c**) R = Me, R^1 = R^2 = H, R^3 = Ph, X = ClO$_4$, 30%
(**d**) R = R^2 = Ph, R^1 = H, R^3 = Me, X = ClO$_4$, 35%
(**e**) R = Me, R^1 = R^2 = (CH = CH)$_2$, R^3 = Ph, X = I$_3$, 32%

Scheme 68

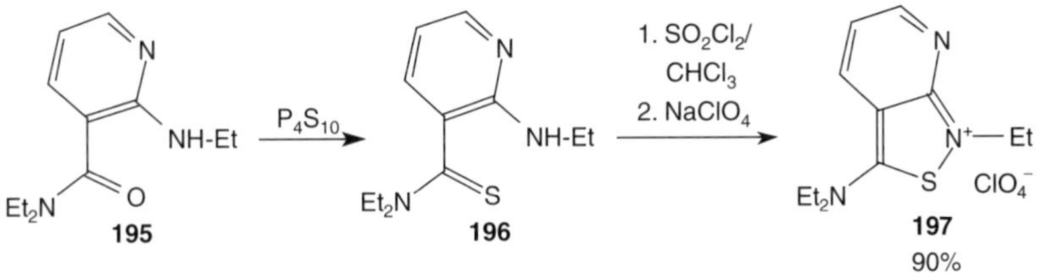

Scheme 69

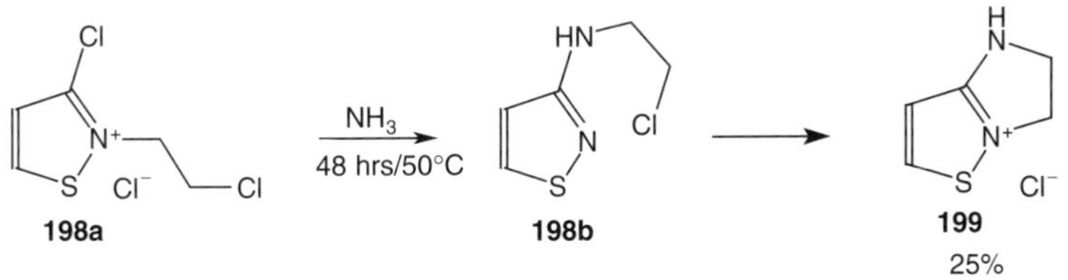

Scheme 70

ethanol afforded the isothiazolium salts **194** isolated as its perchlorates **194a–d** and triiodide **194e** (77CJC1123, Scheme 68).

The preparation of the 3-diethylamino-1-ethylisothiazolo[3,4-*b*]pyridinium perchlorate **197** smoothly proceeded on treatment of the 2-ethylamino-*N,N*-diethyl-thionicotinamide **196** with sulfuryl chloride in chloroform in 90% yield (82S972, Scheme 69). The thionicotinamide **196** could be obtained from nicotinamide **195** by heating in pyridine with P_4S_{10}.

The 2,3-heterocyclic annulated isothiazolium chloride **199** was obtained by intramolecular quaternization of salt **198b** in poor yield (79JOC1118, Scheme 70).

Isothiazole **201a**, synthesized from the isothiazol-5-thione **200** by dimethyl sulfate, reacted to the stable 7-cyano-6-methylthio-thiazolo[3,2-*b*]-isothiazolium perchlorate **202** in 77% yield in the presence of perchloric acid by cyclization (83M999, Scheme 71). In an aqueous base, the isothiazole ring of salt **202** was opened and **201a** was regenerated.

The 4-cyano isothiazole **201b** was prepared by stepwise S-alkylation of **200** and treatment of **201b** with concentrated sulfuric acid gave the new condensed salt **203** (72IJS328, Scheme 71).

The disulfide **204** immediately reacted with bromine and base to sulfenylbromide **205** and then the 1-methyl-2,3-dihydroimidazolo[1,2-*b*]1,2-benzisothiazolium bromide **206** was generated. Salt **206** is very hygroscopic and decomposed in the presence of moisture with hydrolytic disproportionation (78CB2716, Scheme 72).

The 1,2-benzisothiazolo[2,3-*b*]1,2-benzisothiazolium triiodide **208** was prepared by treatment of 3-(2-thiophenyl)-1,2-benzisothiazole **207** (R = H) with iodine; a dark solid was obtained, which was recrystallized from nitromethane as a fine red prism (88CJC1405, Scheme 73). The attempt to convert triiodide **208** into the perchlorate in acetic acid failed. The reaction of 3-(2-thiocyanatophenyl)-1,2-benzisothiazole **207** (R = CN) with perchloric acid did not yield an isolable salt (90JCS(P1)2881).

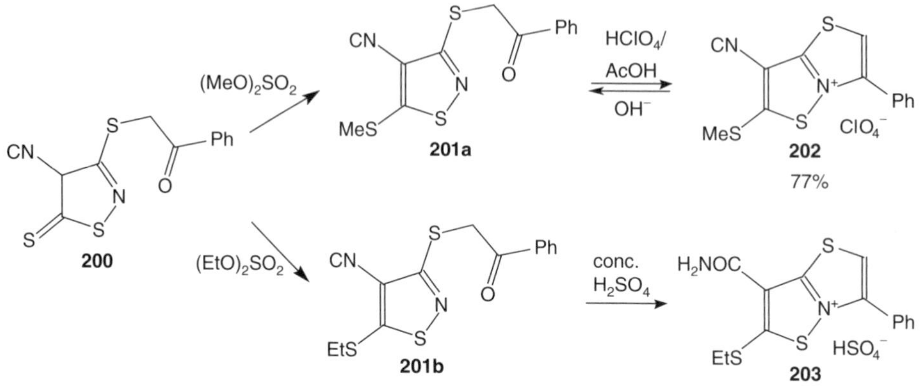

Scheme 71

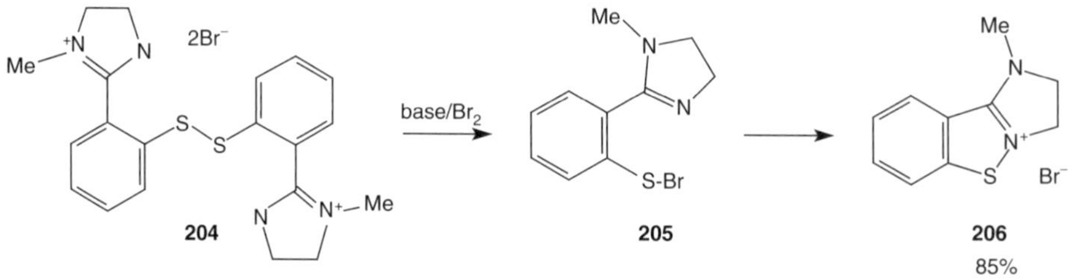

Scheme 72

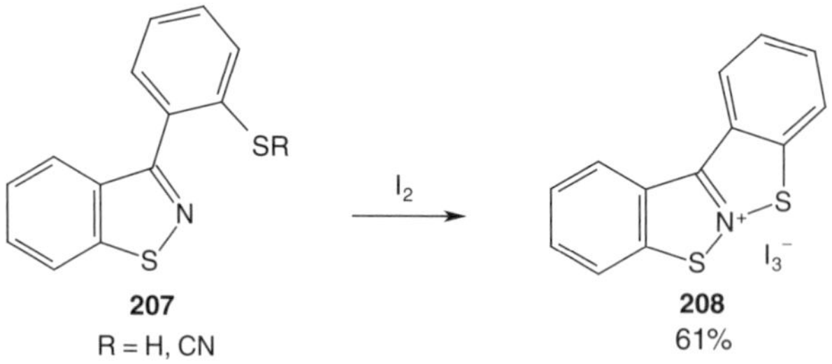

Scheme 73

First attempts were directed to the ethoxybutyrylisothiazole **209** prepared from 3-cyano isothiazole with 3-ethoxypropylmagnesium bromide in poor yield (30%). Cleavage of the ether with hydrobromic acid, followed by evaporation to dryness of the acidic solution, gave in a nearly quantitative yield the 4,5,6,7-tetrahydro-4-oxoisothiazolo[2,3-*a*]pyridinium bromide **210** (97%) (69JCS(C)707, Scheme 74).

Interestingly, the fused 1,2,5-thiadiazole **211** rearranged to adduct **212** in a very high yield. When this isothiazole **212** (R^1 = CN, CO_2H, CO_2Me) was dissolved in mineral acids such as concentrated perchloric, hydrochloric or sulfuric acid, the addition of water then afforded a series of pentacyclic isothiazolium salts **213** (88JCS(P1)2141, Scheme 75).

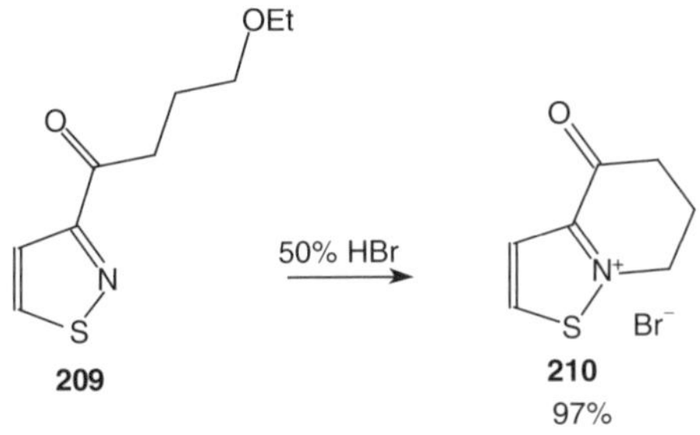

Scheme 74

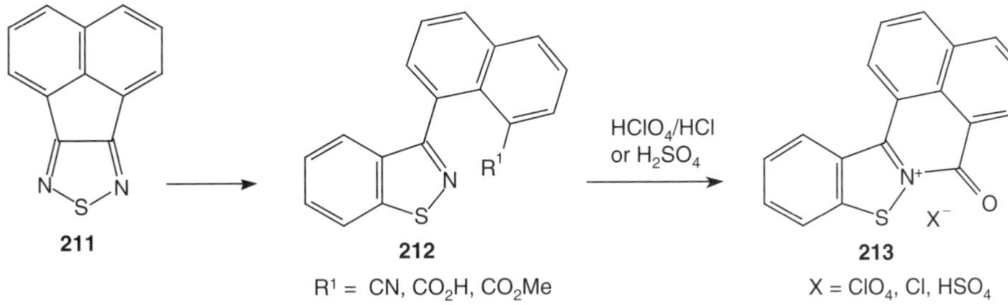

$R^1 = CN, CO_2H, CO_2Me$

Scheme 75

(a) $Ar^3 = Ph$, 42%
(b) $Ar^3 = $ thienyl, 26%

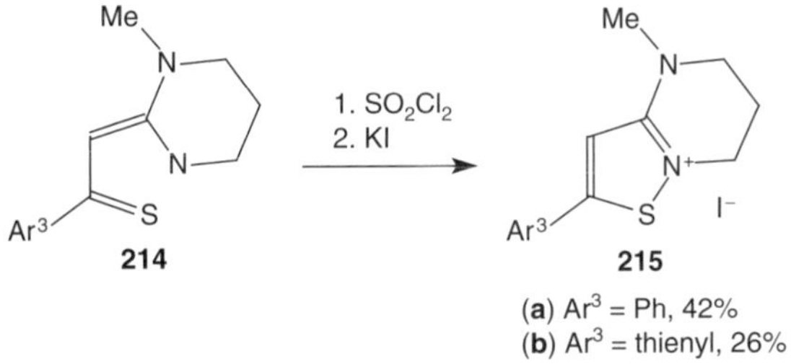

Scheme 76

1,1-Diamino-3-thiono-1-alkene **214** reacted with sulfuryl chloride to give iso-thiazolo[2,3-a]-4,5,6,7-tetrahydro pyrimidinium iodide **215** by an oxidative cycliza-tion in 26–42% yield (82T1673, Scheme 76).

The oxidation of the thioacetamides **216c,d** with hydrogen peroxide as well as in the presence of perchloric acid gave the 2-dimethylamino salt **217c** and oxidation with bromine afforded the corresponding 3-bromoderivatives **217d**. In a similar procedure, using iodine as oxidant, the 2-isopropyl salt **217b** was generated. The isothiazolo[2,3-a]pyridinium salts **217a,e–g** were obtained by base-catalyzed addition of **218** and bromine in 25–35% yield (73JCS(CC)150, Scheme 77).

Treatment of mercaptan **219** with N-chlorosuccinimide (NCS) in benzene under dry nitrogen and at 0–5 °C converted it to sulfenylchloride **220**, which could only be

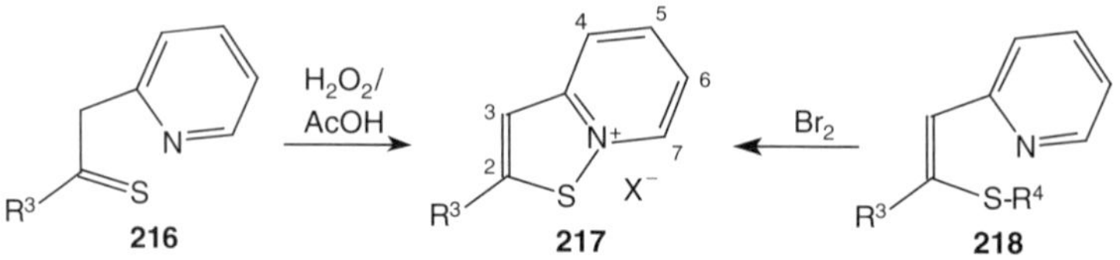

(a) R³ = H, X = ClO₄
(b) R³ = i-Pr, X = I₃
(c) R³ = NMe₂, X = ClO₄
(d) R³ = NMe₂, 3-Br, X = ClO₄
(e) R³ = Me, X = ClO₄
(f) R³ = Ph, X = ClO₄
(g) R³ = H, 7-Me, X = ClO₄

Scheme 77

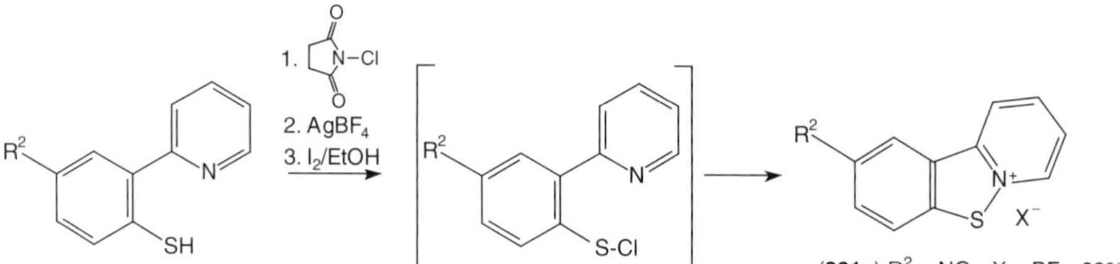

(221a) R² = NO₂, X = BF₄, 60%
(221b) R² = CN, X = BF₄, 57%
(222a) R² = H, X = I₃, 67%

Scheme 78

isolated in a very crude form. This sulfenyl chloride **220** reacted directly *in situ* to 7-nitrobenzisothiazolo[2,3-*a*]pyridinium tetrafluoroborate **221a** in 60% yield by addition of an equivalent of silver tetrafluoroborate in acetonitrile. Similarly, 2-(5-cyano-2-thiophenyl)pyridine **219b** was converted to the sulfenylchloride **220b** and then in the corresponding tetrafluoroborate **221b** (57%) (82JHC509, Scheme 78).

The crude 2-(2-thiophenyl)pyridine **219** (R² = H) in ethanol was treated with a saturated solution of iodine in ethanol until the oxidation and precipitation of the crystalline 1,2-benzisothiazolo[2,3-*a*]pyridinium triiodide **222a** was complete. This salt **222a** was filtered and recrystallized from nitromethane as brown needles (67%) (85CJC882, Scheme 78).

The thiocyanates **224** and **226** were obtained as analytically pure crystalline solids and the former was converted into 1,2-benzisothiazolo[2,3-*a*]pyridinium perchlorate **225** (79TL3339: 25%; 90JCS(P1)2881: 51%) and 1,2-benzisothiazolo[4,3,2-*hij*] quinolinium perchlorate **227** (79TL3339: 58%; 90JCS(P1)2881: 85%) by reaction with bromine followed by perchloric acid (Scheme 79).

3-Amino-1,2-benzisothiazole **228** smoothly reacted with acetylacetone in the presence of either perchloric acid to obtain **229a** (X = ClO₄) in 55% yield or tetrafluoroboric acid to give the salt **229a** (X = BF₄) in 58% yield [**229a**: ClO₄, CF₃SO₃: 83ZOR1134; ClO₄, BF₄: 88JCR(S)46; **229b**: 83ZOR1134, Scheme 80].

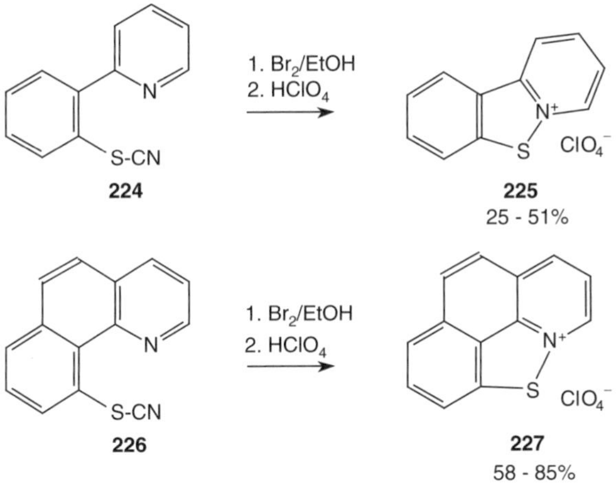

Scheme 79

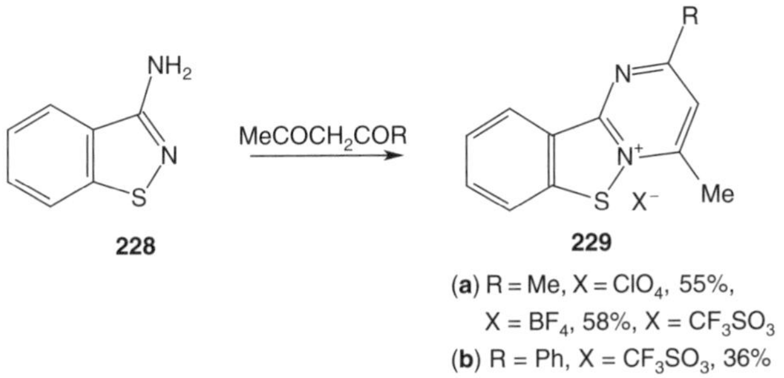

Scheme 80

In the absence of the strong acid, the acyclic bis-condensation product was derived in poor yield (16%).

III. Structure

A. X-RAY DIFFRACTION

The structure of several isothiazolium salts was confirmed by X-ray analysis (72CJC324, 92JCS(CC)571, 93JHC929, 95ZK73, 96JPR424, 97ZN(B)1139, 00JCS (D)3113). Crystal and molecular structures of substituted isothiazolium salts were investigated to show the conformation of the isothiazole ring and their substituents.

The salt 2-(4-methoxyphenyl)-4,5-dimethylisothiazolium perchlorate **57g** (R = 4-MeO) was analyzed by X-ray diffraction and displays an approximately planar conformation of the isothiazolium ring and the C6–C7–N–S torsion angle is 42.7° (98ZK331).

Similarly, *N*-phenylsulfonylamino-4,5,6,7-hexahydro-1,2-benzisothiazolium perchlorate **131i** has a planar isothiazolium ring and not a 1,2,3-thiadiazine ring (96ZOR1745). The bond lengths of this salt are intermediate between those of single and double bonds. The endocyclic C7–N1 (1.695 Å) and C1–S1 (1.695 Å) bonds and exocyclic N1–N2 bond (1.402 Å) are analogous to the corresponding bonds in *N*-aroylisothiazolium-2-imines (94JPR115). The S–O distances (1.420 and 1.426 Å) in the sulfonyl group are consistent with analogous distances in 5-methylsulfonyl-isothiazoles (1.429 Å) (77JCS(P2)1332).

Accordance to other named isothiazolium salts **57** (R = H), offered the inspection of the X-ray data of 3,4-dimethyl-2-phenylisothiazolium perchlorate **68a**, received after an isomerization of 4,5-dimethyl-2-phenylisothiazolium perchlorate in ethanol, the typical salt structure with S–N bond of 1.682 Å and 2-phenyl ring is 86.9° out of plane of the isothiazole ring (98JPR361).

Richter et al. demonstrated that the crystal structure of 2-phenyl-5-ferrocenyl-isothiazolium pentaiodide **30h** is a layer structure arranged toward *z*-axis and bis (2,5-diphenylisothiazolium) octaiodide **30a** toward *x*-axis with changed layer of isothiazolium cation and polyiodide anion (99ZAAC511).

B. MOLECULAR SPECTRA

The earliest studies of isothiazolium salts dealt with UV and IR spectra and elemental analyses. In later investigations, ^{1}H and ^{13}C NMR spectroscopy and mass spectrometry became practical tools for characterizing various substituted isothiazolium salts such as monocyclic, bicyclic and benzoannulated salts. Today, the most useful methods are NMR and IR spectroscopy as well as mass spectrometry.

1. *^{1}H NMR*

^{1}H NMR spectra of isothiazolium salts show a clear singlet for the proton in the 3-position between 8.54 and 10.17 ppm. Typical values of ^{1}H NMR characteristics for some isothiazolium salts are presented in ppm in Table 11.

The H-3 proton of 4,5-diphenyl isothiazolium salt **63a** is shifted to 10.00 ppm compared with 4,5-dimethyl- **57f,g** or 4,5,6,7-tetrahydro perchlorates **121l**, which are below 9.50 ppm. The salt **73n** is an exception with a chemical shift of the H-3 proton at 8.54 ppm. After an isomerization of 2-(4-methoxyphenyl)-4,5-dimethylisothiazolium perchlorate **57f** to 2-(4-methoxyphenyl)-3,4-dimethylisothiazolium perchlorate **68d**, described in this chapter, the 5-methyl group of **57g** was transferred to the 3-position of the isothiazole ring and the H-5 proton of **68d** is located at 9.25 ppm in contrast to the H-3 proton of **57g** at 9.39 ppm.

2. *^{13}C NMR and ^{15}N NMR*

The structures of isothiazolium salts were also identified by their ^{13}C NMR spectra with the help of the proton-coupled ^{13}C signal. The characteristic signal of the C-3

Table 11. ¹H NMR data of isothiazolium salts **57f,g,63a,68d,73n,121l**

Salt	C_6H_5R	R^1	R^2	R^3	Reference
4,5-Dimethyl-2-phenylisothiazolium perchlorate **57f**	7.63 (m, 2H), 7.56 (m, 3H)	9.05 (s, 1H)	2.41 (s, 3H)	2.73 (s, 3H)	92JPR25
2-(4-methoxyphenyl)-4,5-Dimethylisothiazolium perchlorate **57g**	3.87 (s, 3H), 7.22, 7.76 (dd, 4H)	9.39 (s, 1H)	2.36 (s, 3H)	2.78 (s, 3H)	95JPR175
2-(4-methoxyphenyl)-3,4-Dimethylisothiazolium perchlorate **68d**	3.84 (s, 3H), 7.18, 7.61 (dd, 4H)	2.37 (s, 3H)	2.34 (s, 3H)	9.25 (s, 1H)	98JPR361
2-(4-nitrophenyl)-4,5-Diphenylisothiazolium perchlorate **63a**	8.33, 8.64 (dd, 4H)	10.17 (s, 1H)	7.51–7.70 (m, 10H)		03ZN(B)111
2-Phenyl-4,5,6,7-tetra-hydro-1,2-benzisothiazolium perchlorate **121l**	7.59 (m, 3H), 7.62 (m, 2H)	9.28 (s, 1H)	1.83 (m, 4H), 2.88 (t, 2H), 3.46 (t, 2H)		95JPR175
4,5-Dimethyl-N-(4-bromo-phenyl)-sulfonylaminoisothiazolium perchlorate **73n**	7.73 (m, 4H)	8.54 (s, 1H)	2.28 (s, 3H)	2.65 (s, 3H)	96ZOR1745

atom could easily be identified by its intensive shielding of 3–5 ppm in contrast to the corresponding isothiazole.

The ^{15}N signal assigned to 4,5-dimethyl-2-phenylisothiazolium perchlorate **57f** was identified also by its ^{13}C–^{1}H-hetero-correlated 2D NMR spectrum. The ^{15}N signal is located at −173 ppm and the estimated coupled constant to the C-3 atom (12.2 Hz), C_{ipso} Ph (13.3 Hz) and H-3 (3.3 Hz) could be identified (92JPR25).

Special ^{13}C NMR spectra data of various substituted salts are given in ppm in Table 12.

The characteristic C-3 atom of salts **57f,g,63a,68d,73n,121l** are identified at 152.3–158.0 and 169.5 ppm. The exception is the perchlorate **68d** with a chemical shift at 169.5 ppm, which is shifted to lower field because of the unsubstituted C-atom.

3. Ultraviolet Spectra

Generally, these spectra are usefully applied to the elucidation of the 2-aryl ring system. All UV spectra of isothiazolium salts contain two absorption bands, at intervals of 212–267 and 268–321 nm, whereas the absorptions of salts with electron acceptor substituents were shifted hypochromically (92JPR25). Some UV absorption λ_{max} values (in nm) and the corresponding absorption coefficients ε of isothiazolium salts are displayed in Table 13.

The first absorption bands of 4-ethyl-5-methyl- **68h** and 4,5-diphenyl-substituted **63a** salts are shifted more hypochromically than 4,5-dimethyl salts **57c,g** and **121l**.

4. Infrared Spectra

IR spectra of monocyclic, bicyclic and benzisothiazolium salts have been used mainly for the identification of nitro- sulfonyl-, carbonyl- and ester-groups as well as for the characterization of the typical perchlorate anion absorption. For instance, the absence of the typical absorption of the SCN- and CHO-groups in the spectra is proof that these functional groups took part in the intramolecular cyclization of β-thiocyanatovinyl aldehydes with aromatic or aliphatic amines to the corresponding salt (88ZC287, 92JPR25, 95JPR175, 96ZOR1745, 03ZN(B)111). Several characteristic IR absorption maxima $\tilde{v}_{max}$ of isothiazolium salts are collected in Table 14. The absorption are given in wave numbers.

The typical perchlorate absorptions are located at 1090–1112 cm^{-1} and the characteristic double NO_2-absorption are given at about 1345 and 1535 cm^{-1}.

The IR spectra serve as useful and easy method to identify perchlorate and functional groups and to differentiate between the β-thiocyanatovinyl aldehydes and the salts.

5. Mass Spectrometry

The information contained in the mass spectra of isothiazolium salts strongly depends on the ionisation methods. The electron bombardment (EI) of non-volatile salts afforded thermal reaction in the ion source. The thermal reactions included the formation of anhydrobases [(M-H)$^{+\bullet}$-HX] dimerization and decomposition

Table 12. ^{13}C NMR data of isothiazolium salts **57f,g,63a,68d,73n,121l**

Salt	R^2	R^3	C-3	C-4	C-5	Reference
4,5-Dimethyl-2-phenylisothiazolium perchlorate **57f**	11.5	15.3	153.5	133.5	169.3	92JPR25
2-(4-methoxyphenyl)-4,5-Dimethylisothiazolium perchlorate **57g**	11.0	13.1	156.4	133.5	166.9	95JPR175
2-(4-methoxyphenyl)-3,4-Dimethylisothiazolium perchlorate **68d**	14.4	16.0	169.5	134.5	149.4	98JPR361
2-(4-nitrophenyl)-4,5-Diphenylisothiazolium perchlorate **63a**	121.6–148.6		158.0	135.5	167.0	03ZN(B)111
2-Phenyl-4,5,6,7-tetra-hydro-1,2-benzisothiazolium perchlorate **121l**	20.5, 22.4, 21.2, 25.3		152.3	134.6	170.5	95JPR175
4,5-Dimethyl-N-(4-bromo-phenyl)-sulfonylaminoisothiazolium perchlorate **73n**	10.5	12.4	154.1	128.5	158.9	96ZOR1745

Table 13.   UV data of isothiazolium salts **57c,g,68h,63a,121l**

Salt	λ_{max}	ε	λ_{max}	ε	Reference
2-(4-methoxyphenyl)-4,5-Dimethylisothiazolium perchlorate **57g**	266	3.76	317	3.61	95JPR175
2-[4-(methoxycarbonyl)-phenyl]-4,5-Dimethylisothiazolium perchlorate **57c**	256	3.87	295	3.99	02HCA183
4-Ethyl-2-(4-methoxy-phenyl)-3-methylisothiazolium perchlorate **68h**	224	4.06	268	3.99	98JPR361
2-(4-nitrophenyl)-4,5-Diphenylisothiazolium perchlorate **63a**	212	4.26	321	3.92	03ZN(B)111
2-Phenyl-4,5,6,7-tetrahydro-1,2-benzisothia-zolium perchlorate **121l**	256	3.79	300	3.89	95JPR175

Table 14.   IR data of isothiazolium salts **30c,57c,58h,63a,68d**

Salt	$\tilde{v}_{max}$	Reference
2-(4-methoxyphenyl)-3,4-Dimethylisothiazolium perchlorate **68d**	1090 ClO_4 1260 MeO	98JPR361
2-(4-nitrophenyl)-4,5-Diphenylisothiazolium perchlorate **63a**	1100 ClO_4 1345 NO_2 1535 NO_2	03ZN(B)111
2-[4-(methoxycarbonyl)-phenyl]-4,5-Dimethylisothiazolium perchlorate **57c**	1112 ClO_4 1722 C = O	02HCA183
4-Ethyl-2-(4-methoxy-phenyl)-5-methylisothiazolium perchlorate **58h**	1092 ClO_4 1249 MeO	98JPR361
2-Phenyl-5-(4-trifluoro-methylphenyl)isothiazolium perchlorate **30c**	1100 ClO_4 1330 C-F	94ZOR1379

products depending on the temperature of the ion source, the chemical structure of cations and the nature of anions (94ZOR1379, 02HCA183, 03ZN(B)111). The EI mass spectra of salts could be interpreted in the following way (Scheme 81a).

The ESI MS method yielded results to identify isothiazolium salts **62** and **63** (05SUC211). The aqueous methanolic solution of the salts gave by application of this spray method intensive signals of the cation (ESI pos.) and accordingly the anions (ESI neg.). The determination of the molar mass and the total formula are possible. Information about the structure of the cations could be obtained by using the ion-molecule-push-process. The cations decomposed in the typical fragmentation pathway are formed in Scheme 81b.

6. Elemental Analysis

Elemental analyses (EA) are performed to validate the conclusion of other measurements such as NMR, IR, UV spectroscopy and MS spectrometry. In the earliest

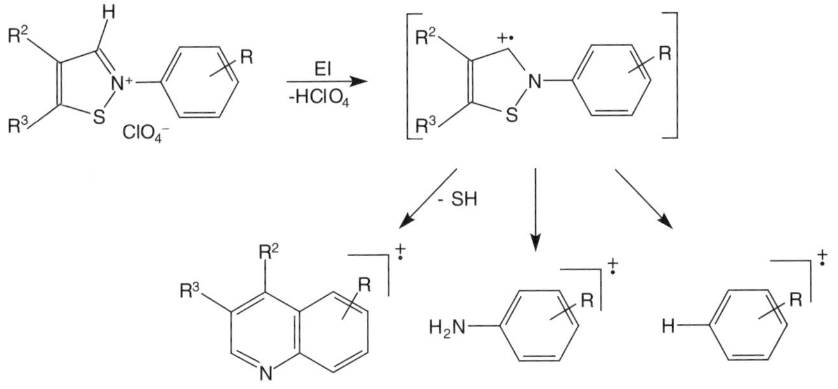

Scheme 81a

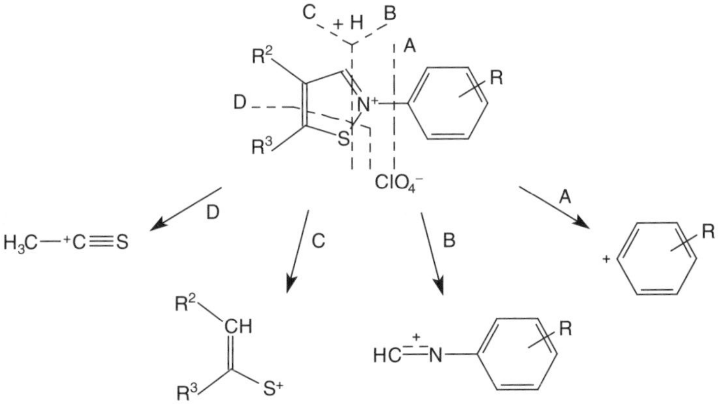

Scheme 81b

investigation, EA was used only with IR and UV spectra to determine the structure of salts (94ZOR1379). Today, the elemental analyses of C, H, N, O and S atoms are generally calculated and analysed with error less than 1% (03ZN(B)111, 05SUC211).

C. Other Analytical Methods

Especially, 2-phenyl-4,5,6,7-tetrahydro-1,2-benz- (**121**) and 5-methyl-2,4-diphenyl-isothiazolium salts **62** (R = H, 4-MeO, 2-Cl, 2,6-Cl$_2$) were studied with the HPLC-MS(MS) method to monitor the oxidation of isothiazolium salts with H$_2$O$_2$/acetic acid (96%) (03CG147). The strongly acidic reaction mixture was separated on a RP-18 column without any sample pretreatment and included intermediates, which were identified by API-MS(MS)-techniques. The aim of this work was to establish the reaction mechanism using several N-functionalized salts.

IV. Reactivity

A. General Survey

Isothiazolium salts react by S-oxidation and C-3 oxyfunctionalization with H_2O_2 (30%)/acetic acid or magnesium monoperoxyphthalate (MMPP) to stable mono- and bicyclic 3-hydroperoxysultims, -sultams, 3-hydroxysultams and 3-oxosultams, retaining the heterocyclic ring system.

Furthermore, isothiazolium salts are masked enaminothioketones and thioacrylic acid derivatives, which can be obtained by reductive ring cleavage.

The reactivity of isothiazolium salts toward nucleophiles is higher than that of isothiazoles. Therefore, on quaternization of isothiazoles to isothiazolium salts, the tendency of nucleophilic ring opening increases.

Isothiazolium salts react with N-nucleophiles such as ammonia, primary amines, hydrazines and hydroxylamines by ring transformation to isothiazoles, pyrazoles and oxazoles. The synthesis of 3-aminopyrroles occurs by base-catalyzed ring transformation and desulfuration of substituted 5-amino-2-methyl-isothiazolium salts. N-Aryl isothiazolium salts with an active methyl- or methylene group in the 5-position rearrange in a base-induced reaction with secondary amines by deprotonation and oxidative dimerization to thieno-annulated N-aryl-6aλ^4-thia-1,6-diazapentalenes, spirocyclic isothiazolium salts and thianthene derivatives. In contrast, weaker bases, such as substituted anilines, compete due to their basicity and nucleophilicity in the reaction with N-aryl-4,5-dimethylisothiazolium salts. Thus, ring transformation occurs by nucleophilic attack of aniline at the 5-position following by an apparent migration of the sulfur atom to the 3-position and elimination of aniline. The reaction of 3-chloro-1,2-benzisothiazolium salts with amines as N-nucleophiles gives rearranged 3-amino-substituted salts including 3-amino-1,2-benzisothiazoles.

Nucleophilic attack with C-nucleophiles of monocyclic and 1,2-benzisothiazolium salts at the sulfur atom are carried out with ring cleavage and recyclization to thiophenes.

There are also ring transformations of monocyclic and benzisothiazolium salts by ring extension to 1,3-thiazines, quinolines, 1,4-benzothiazepines and 1,2,3-thiadiazine 1,1-dioxides.

Many heterocyclic annulated isothiazolium salts are prepared by intramolecular quaternization of special 3-substituted salts.

B. Oxidation to Sultims and Sultams

The oxidation of monocyclic **57,61–63** and bicyclic isothiazolium salts **121,123** is classified by two principle oxidation methods. First, these salts can be oxidized with H_2O_2 in acetic acid to yield sultims *rac-cis-/-trans-***232–235**, 3-hydroperoxysultams **239–243** and 3-oxosultams **251–257** (method A). Furthermore, these isothiazolium salts can react with MMPP in water or alcohol to yield sultims *rac-cis-/-trans-***236–238** as well as 3-hydroxysultams **244–248**, 3-alkoxysultams **249–250** and **251–257** (method B). Thus, many oxidation products were obtained. In Scheme 82, the general

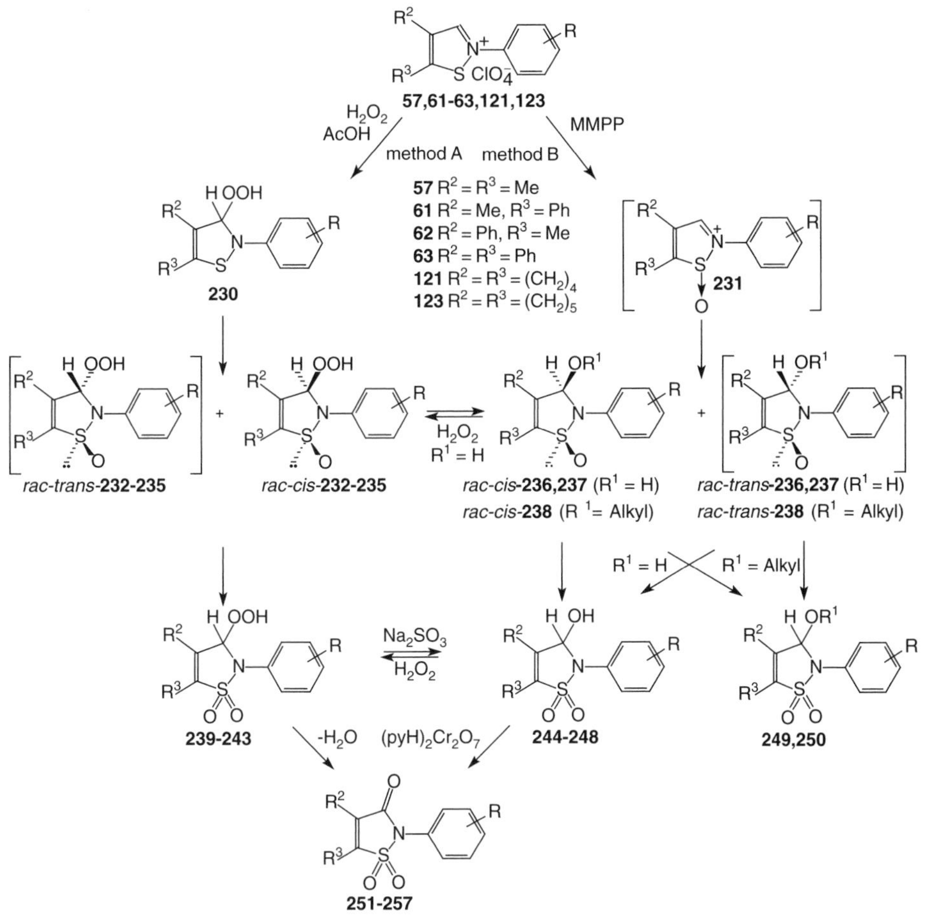

Scheme 82

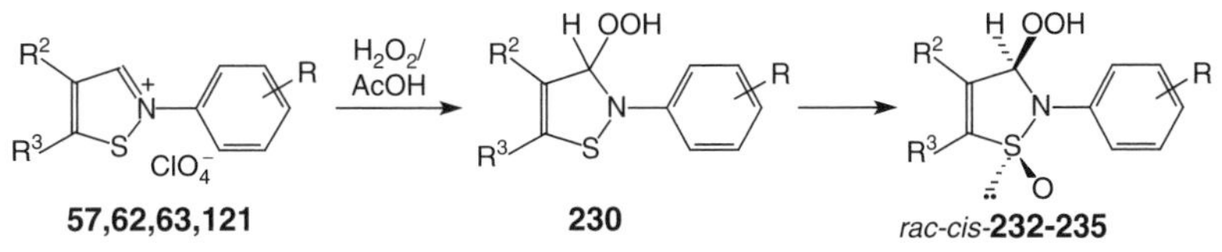

Scheme 83

reaction pathway for the oxidation of isothiazolium salts **57,61–63** and **121,123** is represented (see Tables 15 and 16).

The first step in the oxidation of salts **57,62,63** and **121** with H_2O_2 in glacial acetic acid at room temperature gave the stable 3-hydroperoxysultims *rac-cis-***232a–k** (96MOL139, 99HCA685), **233a–e** (02HCA183), **234a–d** and **235a–c** (05SUC211) (10–97%) *via* the unstable 3-hydroperoxyisothiazoles **230** by nucleophilic attack of

Table 15.   Synthesis of 3-hydroperoxysultims *rac-cis*-**232–235**

rac-cis-	R^2	R^3	R	Yield (%)	Reference
232a	-(CH$_2$)$_4$-		2-Cl	70	99HCA685
b	(CH$_2$)$_4$		2,6-Cl$_2$	42	99HCA685
c	(CH$_2$)$_4$		2-Br	60	99HCA685
d	(CH$_2$)$_4$		2-CF$_3$	50	99HCA685
e	(CH$_2$)$_4$		2-F	43	99HCA685
f	(CH$_2$)$_4$		3-Br	44	99HCA685
g	(CH$_2$)$_4$		4-CO$_2$H	30	99HCA685
h	(CH$_2$)$_4$		4-CO$_2$Me	45	99HCA685
i	(CH$_2$)$_4$		2-NO$_2$	59	04ZN(B)478
j	(CH$_2$)$_4$		4-NO$_2$	23[a]	04ZN(B)478
k	(CH$_2$)$_4$		4-Cl	36	04HCA376
233a	Me	Me	2,6-Cl$_2$	55[a]	02HCA183
b	Me	Me	4-CO$_2$H	10	02HCA183
c	Me	Me	4-CO$_2$Me	20	02HCA183
d	Me	Me	4-CO$_2$Et	15	02HCA183
e	Me	Me	4-NO$_2$	12	02HCA183
234a	Ph	Me	H	21	05SUC211
b	Ph	Me	2-Cl	68	05SUC211
c	Ph	Me	4-NO$_2$	29[a]	05SUC211
d	Ph	Me	2,4-(NO$_2$)$_2$	20[a]	05SUC211
235a	Ph	Ph	4-NO$_2$	74[a]	05SUC211
b	Ph	Ph	2,4-(NO$_2$)$_2$	86	05SUC211
c	Ph	Ph	2-Cl,4-NO$_2$	97[a]	05SUC211

[a]Mixture of sultims and sultams or salts.

H$_2$O$_2$ (Scheme 83). A HPLC-MS(MS) method was applied to monitor the oxidation of isothiazolium salts **62** (R^2 = Ph, R^3 = Me) and **121** (R^2 = R^3 = (CH$_2$)$_4$) with H$_2$O$_2$/acetic acid supported by API-MS(MS) techniques and a transient intermediate was assigned as the primary oxidation product **230** (R^1 = R^2 = (CH$_2$)$_4$) (03CG147). Only, the isothiazole **230** (R = 2,4-(NO$_2$)$_2$, R^2 = R^3 = Ph) could be isolated from 4,5-disubstituted salts **63** (05SUC211). The sultims *rac-trans*-**232–235** were not separated, but they could be identified by HPLC-MS(MS) monitoring. The structures *rac-cis*-**232b** and **233c** were confirmed by X-ray analysis (99HCA685, 02HCA183, 02ZN(B)383). In Table 15, 3-hydroperoxysultims *rac-cis*-**232–235** are given.

The reduction of 3-hydroperoxysultims *rac-cis*-**232,233** with dimethylsulfoxide very easily yielded 3-hydroxysultims *rac-cis*-**236a,h** and **237c,d** (**236a**: 04HCA376; **236h**: 02ZN(B)383; **237c,d**: 02HCA183, Scheme 84). The isothiazol-3(2H)-one 1-oxides **258** were synthesized by elimination of water from 3-hydroperoxyisothiazole *rac-cis*-**232,233** (method A) or by oxidation of the isothiazol-3(2H)-one **259** (method B) (04HCA376, Scheme 84). The structures *rac-cis*-**236h,237c** (02ZN(B)383, 02HCA183) and **258h** (04HCA376) were determined by X-ray analysis.

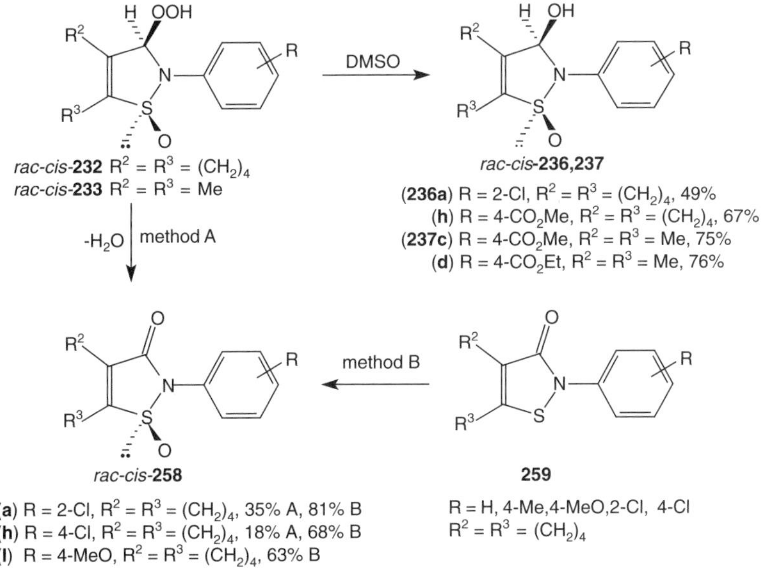
DMSO
-H2O method A
method B
rac-cis-232 R2 = R3 = (CH2)4
rac-cis-233 R2 = R3 = Me
rac-cis-236,237
(236a) R = 2-Cl, R2 = R3 = (CH2)4, 49%
(h) R = 4-CO2Me, R2 = R3 = (CH2)4, 67%
(237c) R = 4-CO2Me, R2 = R3 = Me, 75%
(d) R = 4-CO2Et, R2 = R3 = Me, 76%
rac-cis-258
(a) R = 2-Cl, R2 = R3 = (CH2)4, 35% A, 81% B
(h) R = 4-Cl, R2 = R3 = (CH2)4, 18% A, 68% B
(l) R = 4-MeO, R2 = R3 = (CH2)4, 63% B
259
R = H, 4-Me,4-MeO,2-Cl, 4-Cl
R2 = R3 = (CH2)4

Scheme 84

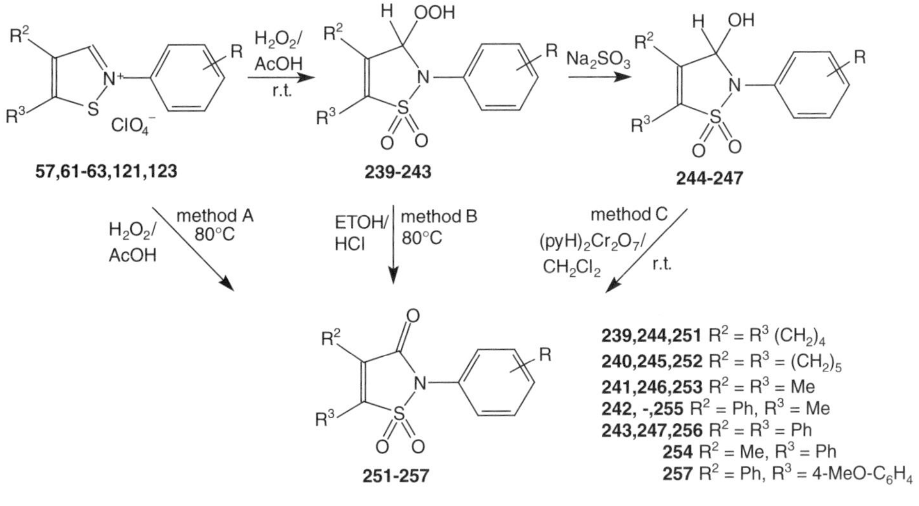
H2O2/ AcOH r.t.
Na2SO3
57,61-63,121,123
239-243
244-247
method A 80°C H2O2/ AcOH
ETOH/ HCl method B 80°C
method C (pyH)2Cr2O7/ CH2Cl2 r.t.
251-257
239,244,251 R2 = R3 (CH2)4
240,245,252 R2 = R3 = (CH2)5
241,246,253 R2 = R3 = Me
242, -,255 R2 = Ph, R3 = Me
243,247,256 R2 = R3 = Ph
254 R2 = Me, R3 = Ph
257 R2 = Ph, R3 = 4-MeO-C6H4

Scheme 85

The oxidation of salts **57,61–63** and **121,123** with H_2O_2/AcOH at room temperature gave the 3-hydroperoxysultams **239–243** and at 80 °C they directly reacted to the isothiazol-3(2*H*)-one 1,1-dioxides **251–257** (method A). The elimination of water using ethanol and concentrated hydrochloric acid gave 3-oxosultams **251–257** (method B) as well as by reduction of **239–243** with Na_2SO_3 *via* **244–247** (method C), respectively (Scheme 85). In Table 16, 3-hydroperoxysultams **239–243** and 3-oxosultams **251–257**

Table 16. 3-Hydroperoxysultams **239–243** and 3-oxosultams **251–257**

	R^2	R^3	R	Reference
239		-(CH$_2$)$_4$-	H, 2-Me, 2,6-Me$_2$, 4-Br	96T783
			3-Me, 4-Me, 4-MeO, 4-CO$_2$H, 4-CO$_2$Me, 2-Br, 3-Br, 2-Cl, 2,6-Cl$_2$, 2-F, 2-CF$_3$	99HCA685
			4-t-Bu, 2,5-Cl$_2$-4-O-i-Pr	02JOC8400
			2-MeO, 4-Cl, 2-NO$_2$, 4-NO$_2$	04ZN(B)478
240		-(CH$_2$)$_5$-	2,5-Cl$_2$-4-O-i-Pr	02JOC8400
241	Me	Me	H, 4-MeO, 4-CO$_2$H, 4-CO$_2$Me, 4-CO$_2$Et, 4-SO$_2$Me, 2-Cl, 2,6-Cl$_2$, 4-NO$_2$	02HCA183
242	Ph	Me	H, 4-MeO, 4-Cl, 4-NO$_2$, 2,4-(NO$_2$)$_2$	05SUC211
243	Ph	Ph	H, 4-MeO, 2-Cl-4-NO$_2$, 4-NO$_2$, 2,4-(NO$_2$)$_2$	05SUC211
251		-(CH$_2$)$_4$-	H, 2-Me, 2,6-Me$_2$, 4-MeO, 4-Br	96T783
			3-Me, 4-Me, 4-CO$_2$H, 4-CO$_2$Me, 2-Br, 3-Br, 2-Cl, 2,6-Cl$_2$, 2-F, 2-CF$_3$	99HCA685
			2,4-(NO$_2$)$_2$, 2,5-Cl$_2$	03ZN(B)111
			4-Cl, 4-NO$_2$	04ZN(B)478
252		-(CH$_2$)$_5$-	2,5-Cl$_2$-4-O-i-Pr	02JOC8400
253	Me	Me	H, 4-MeO, 4-CO$_2$H, 4-CO$_2$Me, 4-CO$_2$Et, 4-SO$_2$Me, 2-Cl, 2,6-Cl$_2$, 4-NO$_2$	02HCA183
			2,5-Cl$_2$-4-O-i-Pr, 2,4-(NO$_2$)$_2$	03ZN(B)111
254	Me	Ph	2,5-Cl$_2$, 2,5-Cl$_2$-4-O-i-Pr, 4-NO$_2$, 2,4-(NO$_2$)$_2$	03ZN(B)111
255	Ph	Me	2,5-Cl$_2$-4-O-i-Pr, 4-NO$_2$, 2,4-(NO$_2$)$_2$	03ZN(B)111
			2-OH, 2,5-Cl$_2$, 4-CF$_3$	05ZN(B)41
			H, 4-MeO, 2-Cl	05SUC211
256	Ph	Ph	2,5-Cl$_2$-4-O-i-Pr, 4-NO$_2$, 2,4-(NO$_2$)$_2$	03ZN(B)111
			H, 4-MeO, 2-Cl-4-NO$_2$	05SUC211
257	Ph	4-MeO-C$_6$H$_4$	2-Cl-4-NO$_2$, 2-CF$_3$-4-NO$_2$, 4-CF$_3$, 3-NO$_2$	05JEIMC341
			2,5-Cl$_2$-4-O-i-Pr, 2-Cl-4-NO$_2$, 3-NO$_2$, 4-NO$_2$, 2,4-(NO$_2$)$_2$	05JEIMC341

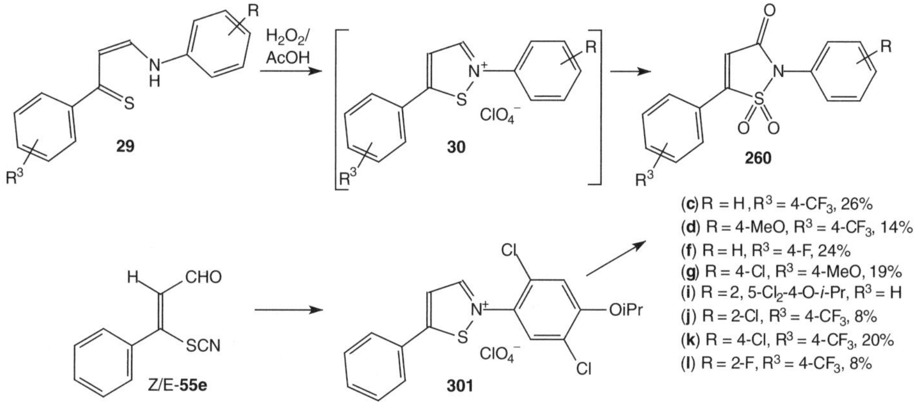

H2O2/
AcOH
29
30
ClO4−
260
Z/E-55e
301
OiPr
(c) R = H, R3 = 4-CF3, 26%
(d) R = 4-MeO, R3 = 4-CF3, 14%
(f) R = H, R3 = 4-F, 24%
(g) R = 4-Cl, R3 = 4-MeO, 19%
(i) R = 2, 5-Cl2-4-O-i-Pr, R3 = H
(j) R = 2-Cl, R3 = 4-CF3, 8%
(k) R = 4-Cl, R3 = 4-CF3, 20%
(l) R = 2-F, R3 = 4-CF3, 8%

Scheme 86

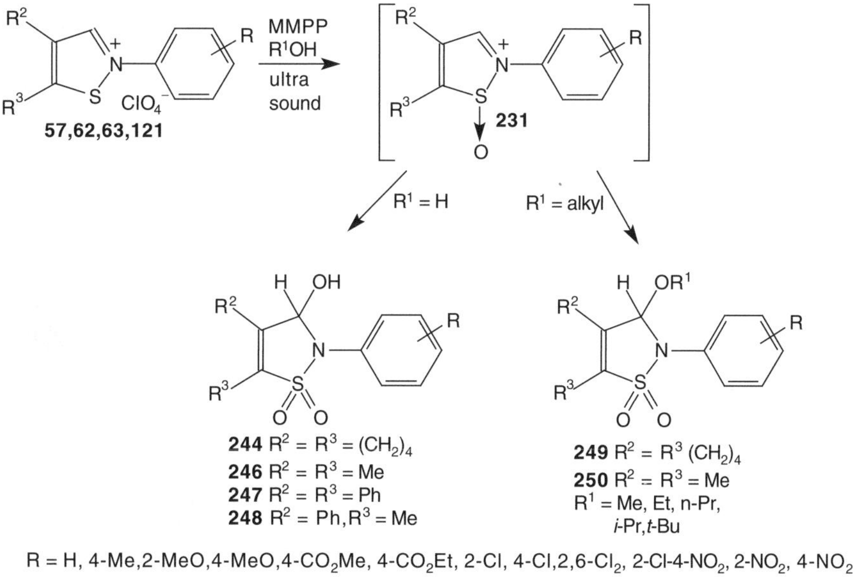

MMPP
R1OH
ultra
sound
57,62,63,121
231
R1 = H
R1 = alkyl
244 R2 = R3 = (CH2)4
246 R2 = R3 = Me
247 R2 = R3 = Ph
248 R2 = Ph, R3 = Me
249 R2 = R3 (CH2)4
250 R2 = R3 = Me
R1 = Me, Et, n-Pr,
i-Pr, t-Bu
R = H, 4-Me, 2-MeO, 4-MeO, 4-CO2Me, 4-CO2Et, 2-Cl, 4-Cl, 2,6-Cl2, 2-Cl-4-NO2, 2-NO2, 4-NO2

Scheme 87

are presented. Further, the synthesis of novel hydroperoxysultams **239** (R = 2,5-Cl_2-4-*O*-*i*-Pr), **240** and their potential as renewable chemoselective electrophilic oxidants for a wide range of nitrogen, sulfur and phosphorus heteroatoms in nonaqueous media is described (02JOC8400).

The 3-hydroxysultams **244–247** are variously substituted in the 2-phenyl ring (R = H, 2-Me, 2,6-Me_2, 4-CO_2Me, 4-CO_2Et, 2-Cl, 2,5-Cl_2-4-*O*-*i*-Pr, 4-Br, 2-Cl-4-NO_2: 96T783, 99HCA685, 02HCA183, 05SUC211). The structures of the 3-hydroperoxysultams **239** [R = 4-Br, 96T783, 02ZN(B)383], **241** [R = 4-CO_2Me, 02ZN(B)383], of the 3-hydroxysultams **245** [R = 2,5-Cl_2-4-*O*-*i*-Pr, 02ZN(B)383], **246**

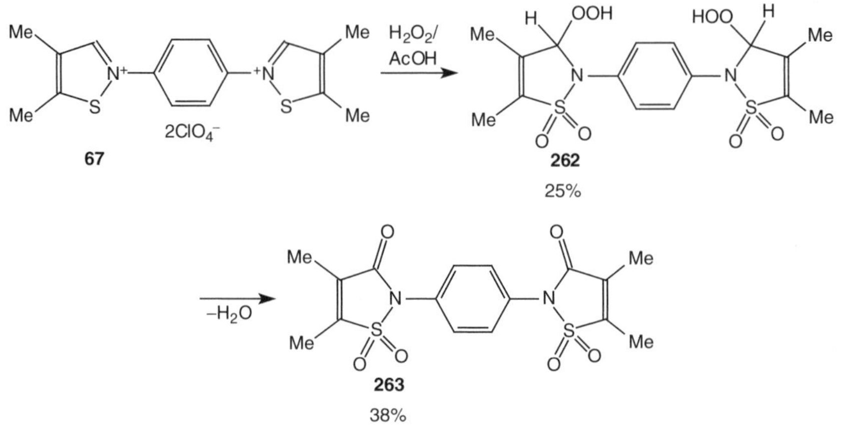

Scheme 88

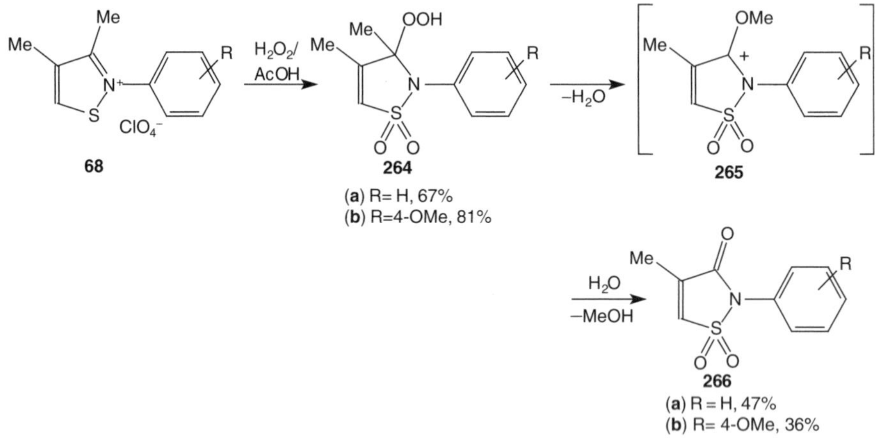

Scheme 89

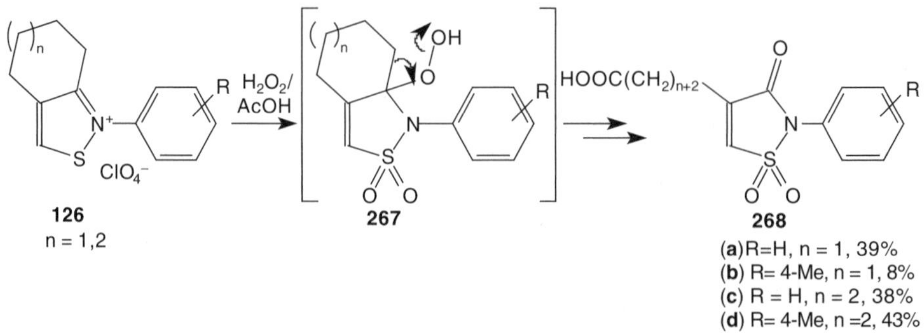

Scheme 90

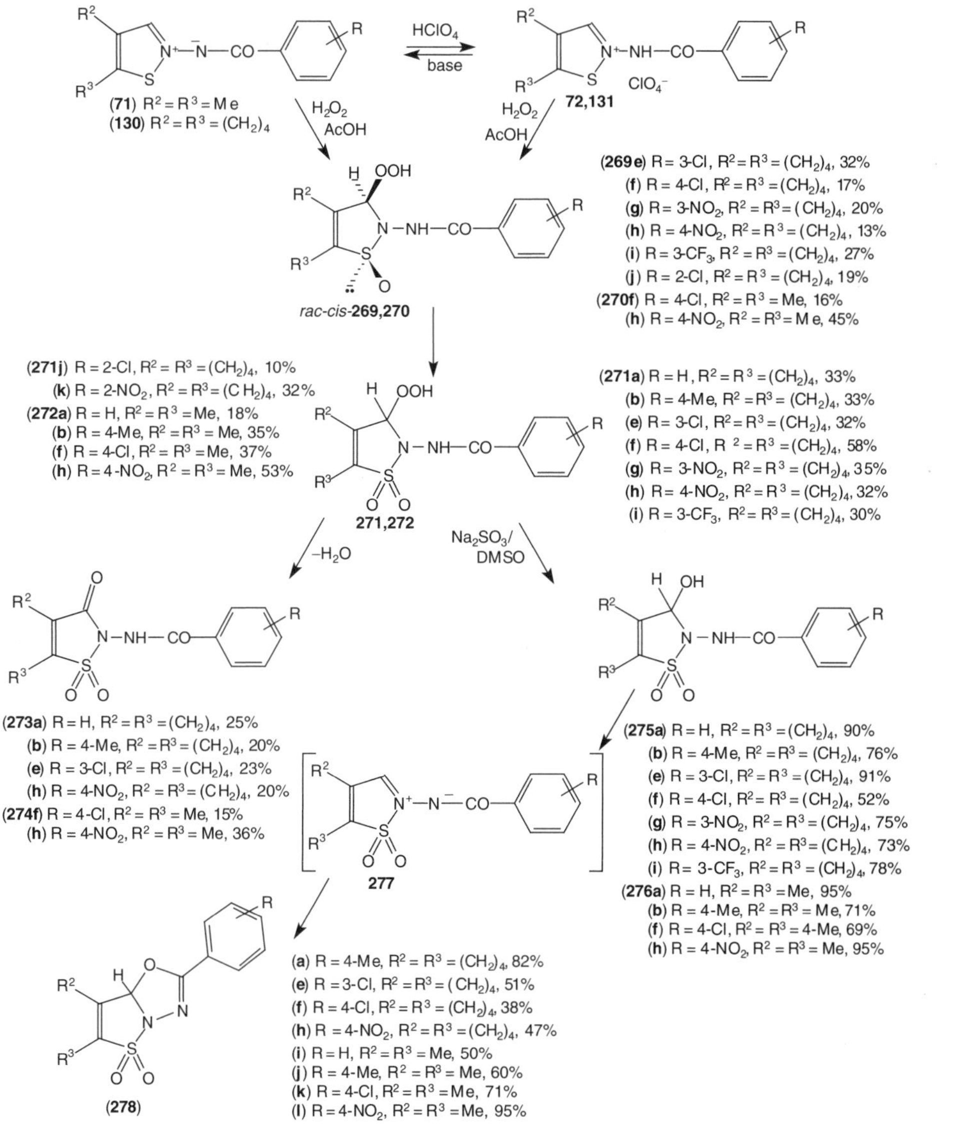
(71) R² = R³ = Me
(130) R² = R³ = (CH₂)₄
HClO₄
base
72,131
H₂O₂
AcOH
rac-cis-269,270
(269e) R = 3-Cl, R² = R³ = (CH₂)₄, 32%
(f) R = 4-Cl, R² = R³ = (CH₂)₄, 17%
(g) R = 3-NO₂, R² = R³ = (CH₂)₄, 20%
(h) R = 4-NO₂, R² = R³ = (CH₂)₄, 13%
(i) R = 3-CF₃, R² = R³ = (CH₂)₄, 27%
(j) R = 2-Cl, R² = R³ = (CH₂)₄, 19%
(270f) R = 4-Cl, R² = R³ = Me, 16%
(h) R = 4-NO₂, R² = R³ = Me, 45%
(271j) R = 2-Cl, R² = R³ = (CH₂)₄, 10%
(k) R = 2-NO₂, R² = R³ = (CH₂)₄, 32%
(272a) R = H, R² = R³ = Me, 18%
(b) R = 4-Me, R² = R³ = Me, 35%
(f) R = 4-Cl, R² = R³ = Me, 37%
(h) R = 4-NO₂, R² = R³ = Me, 53%
271,272
(271a) R = H, R² = R³ = (CH₂)₄, 33%
(b) R = 4-Me, R² = R³ = (CH₂)₄, 33%
(e) R = 3-Cl, R² = R³ = (CH₂)₄, 32%
(f) R = 4-Cl, R² = R³ = (CH₂)₄, 58%
(g) R = 3-NO₂, R² = R³ = (CH₂)₄, 35%
(h) R = 4-NO₂, R² = R³ = (CH₂)₄, 32%
(i) R = 3-CF₃, R² = R³ = (CH₂)₄, 30%
-H₂O
Na₂SO₃/
DMSO
(273a) R = H, R² = R³ = (CH₂)₄, 25%
(b) R = 4-Me, R² = R³ = (CH₂)₄, 20%
(e) R = 3-Cl, R² = R³ = (CH₂)₄, 23%
(h) R = 4-NO₂, R² = R³ = (CH₂)₄, 20%
(274f) R = 4-Cl, R² = R³ = Me, 15%
(h) R = 4-NO₂, R² = R³ = Me, 36%
277
(275a) R = H, R² = R³ = (CH₂)₄, 90%
(b) R = 4-Me, R² = R³ = (CH₂)₄, 76%
(e) R = 3-Cl, R² = R³ = (CH₂)₄, 91%
(f) R = 4-Cl, R² = R³ = (CH₂)₄, 52%
(g) R = 3-NO₂, R² = R³ = (CH₂)₄, 75%
(h) R = 4-NO₂, R² = R³ = (CH₂)₄, 73%
(i) R = 3-CF₃, R² = R³ = (CH₂)₄, 78%
(276a) R = H, R² = R³ = Me, 95%
(b) R = 4-Me, R² = R³ = Me, 71%
(f) R = 4-Cl, R² = R³ = 4-Me, 69%
(h) R = 4-NO₂, R² = R³ = Me, 95%
(278)
(a) R = 4-Me, R² = R³ = (CH₂)₄, 82%
(e) R = 3-Cl, R² = R³ = (CH₂)₄, 51%
(f) R = 4-Cl, R² = R³ = (CH₂)₄, 38%
(h) R = 4-NO₂, R² = R³ = (CH₂)₄, 47%
(i) R = H, R² = R³ = Me, 50%
(j) R = 4-Me, R² = R³ = Me, 60%
(k) R = 4-Cl, R² = R³ = Me, 71%
(l) R = 4-NO₂, R² = R³ = Me, 95%

Scheme 91

[R = 4-CO_2Me, 02ZN(B)383], of the 3-oxosultams **253** [R = 2-Cl, 05ZN(B)41] and **255** [R = 2-OH, 4-CF_3, 2,5-Cl_2, 05ZN(B)41] were confirmed by X-ray analysis.

The isothiazol-3(2*H*)-one 1,1-dioxides **254–257** with stabilizing aryl substituents in the 2-, 4- and/or 5-position are potential inhibitors toward human leukocyte elastase (HLE) (03ZN(B)111, 05JEIMC341). HLE is a serine protease implicated in several inflammatory diseases and represents a major target for the development of low-molecular weight inhibitors.

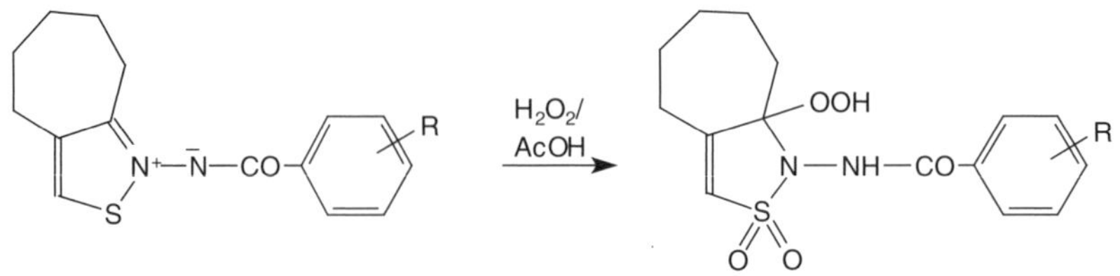

136   **279**   (a) R = H, 48%
                                                      (b) R = 3-NO$_2$, 41%
                                                      (c) R = 4-NO$_2$, 25%

Scheme 92

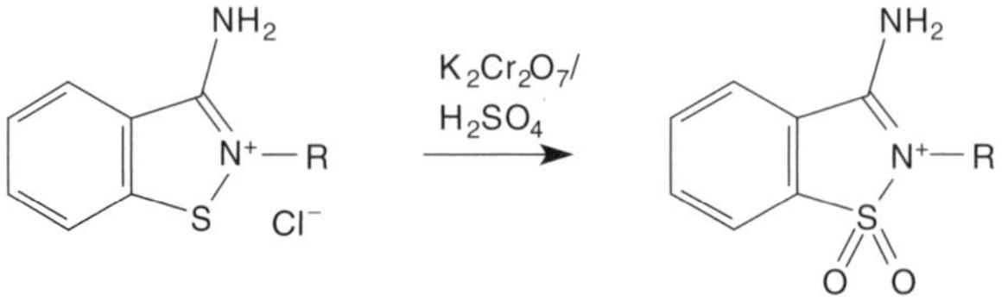

147   **280**

                                 (a) R = Me, 46%
                                 (b) R = Et, 26%

Scheme 93

Surprisingly, the 5-arylisothiazole 1,1-dioxides **260c,d,f,g,i–e** also were directly obtained by oxidation of the enaminothioketones **29** with H$_2$O$_2$/AcOH at room temperature *via* salts **30** in poor yields. In one case, the 3-oxosultam **260i** could be directly isolated from the salt **30i** (94ZOR1379, Scheme 86).

Furthermore, salts **57,62,63** and **121** were oxidized with MMPP in an ultrasonic bath (method B, see Scheme 82) to obtain directly 3-hydroxysultams **244,246–248** and 3-alkoxy sultams **249,250** as an efficient and simple method. At first, there occurred electrophilic attack of MMPP at the sulfur atom of the salts in contact to H$_2$O$_2$/AcOH to give the very highly polarized and reactive S-oxide **231** followed by nucleophilic addition of water or alcohol to sultims *rac-***236,237** (R^1 = H) and *rac-***238** (R^1 = alkyl) (see Scheme 82), which could not be isolated. Investigation by HPLC-MS(MS) was also realized with salts and MMPP/EtOH. Thus, 3-ethoxysultims *rac-cis/trans-***250** (R = CO$_2$Me) could be identified (03CG147). A series of 2-phenyl and 4,5-disubstituted sultams **244** and **246–250** could be obtained in good yields (03S2265, 04ZN(B)478, 05SUC211, Scheme 87)]. The solid-state sulfonamide structures of 3-alkoxy sultams **249** [R = 4-Cl, R^1 = Et: 04ZN(B)478] and **250** (R = 4-Me, R^1 = Me: 03S2265) were revealed by X-ray crystallography.

The oxidation of bis(isothiazolium) salt **67** gave bis(3-hydroperoxide) **262** in 25% yield and bis(3-oxosultam) **263** in 38% yield, respectively (02HCA183, Scheme 88).

The oxidation of 3,4-dimethyl isothiazolium salts **68** with H$_2$O$_2$/AcOH proceeded with rearrangement to 3-oxosultams **266** *via* hydroperoxide **264** and **265**. The structure **266b** was confirmed by X-ray analysis (98JPR361, Scheme 89).

The isomeric salts **126** are valuable starting compounds for the synthesis of bicyclic 3-hydroperoxides **267**, which surprisingly reacted in heated H_2O_2 in acetic acid, to isothiazol-3(2H)-one 1,1-dioxides **268a–d** by a *Criegee-type* rearrangement (00JPR675, Scheme 90).

Furthermore, the oxidation of imines **71,130** and salts **72,131** with H_2O_2 in acetic acid at 0 °C gave the 3-hydroperoxysultims *rac-cis-***269e–j** and **270f,h** in 13–45% yield (**269e–j**: 00JPR291; **270f,h**: 05H2705, Scheme 91) as the first isolable products. The sultims *rac-trans-***269** and **270** could not be isolated. In contrast, oxidation of **71,130** and **72,131** at room temperature yielded the corresponding 3-hydroperoxysultams **271a,b,e–j** and **272a,b,f,h** in 10–58% yield (**271a,b,e–j**: 00JPR291; **271a,b,f,h**: 05H2705, Scheme 91). The 3-hydroperoxides **271,272** were converted into the 3-hydroxysultams **275a,b,e–i** and **276b,f** by reduction with sodium sulfite in water or with dimethylsulfoxide to **276a,h** in 52–95% yield (**275a,b,e–i**: 00JPR291; **276a,b,f,h**: 05H2705, Scheme 91).

The 3-oxosultams **273a,b,e,h** and **274f,h** were directly isolated by oxidation of 2-imines **72,130** and salts **73,131** or by elimination of water from the 3-hydroxperoxide **271,272** in 15–36% yield (**273a,b,e,h**: 00JPR291; **274f,h**: 05H2705, Scheme 91). The structure **274h** was confirmed by X-ray analysis (05H2705). Surprisingly, the heteropentalene isothiazolo[3,2-*b*]-1,3,4-oxadiazole 5,5-dioxides **278a,e–h** and **278m–p** were obtained by dehydration of 3-hydroxides **275** and **276** in toluene and 1,5-electrocyclization of the azomethinimine **277** as crystalline products in 38–95% yield (05H2705, Scheme 91).

3-Hydroperoxysultams **279a–c** (25–48%) resulted from the oxidation with 30% H_2O_2 in acetic acid at room temperature of the isothiazolium-1-imines **136** containing substituents in the 3- and 4-position of the isothiazole ring (00JPR291, Scheme 92). The structure of **279a** was confirmed by X-ray diffraction (01ZK309).

The 3-amino-1,2-benzisothiazolium salt **147** can be oxidized with potassium dichromate to salt **280a,b** in moderate yields (70CB3166, Scheme 93).

C. Reduction and Ring Opening

Isothiazolium salts **23c,27b** and **76j–n,281** are masked 2-amino-1-alkenyl-thioketones and can be converted with thioether (A) and metal hydrides (B, C), in a simple method, by reductive ring opening to β-enaminothioketones **22c,26b** and **282a–g** (Scheme 94), which are useful intermediates for the synthesis of a wide range of heterocyclic compounds. Isothiazoles are not ring-opened by complex

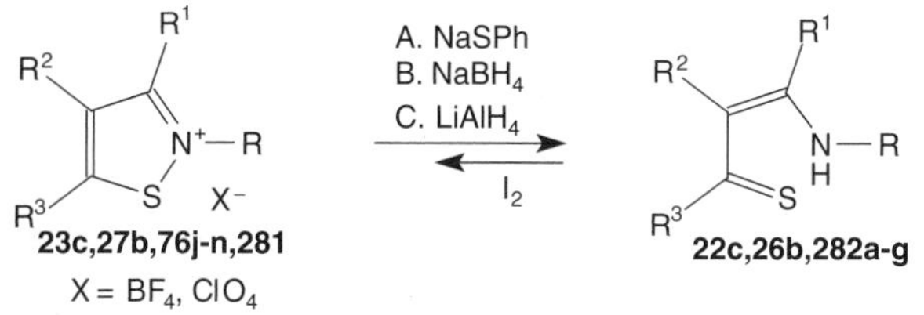

Scheme 94

Table 17. Synthesis of β-enaminothioketones **22c,26b,282a–g**

	R	R^1	R^2	R^3	A	B	C	Reference
					Method (%)			
282a	Me	H	Ph	MeS	60			84CJC1580
22c	Me	Ph	H	MeS	59			73CJC3081
26b	Me	Ph	H	Ph	55			72JCS(P1)2305
282b	Me	H	H	Ph	35			72JCS(P1)2305
c	Et	Me	H	Me		90	86	88SC1847
d	Et	Me	CO$_2$Et	Me		91	74	88SC1847
e	Et	Me	CN	Me		93	75	88SC1847
f	Et	Me	I	Me		85	71	88SC1847
g	Et	Me	NO$_2$	Me		82	69	88SC1847

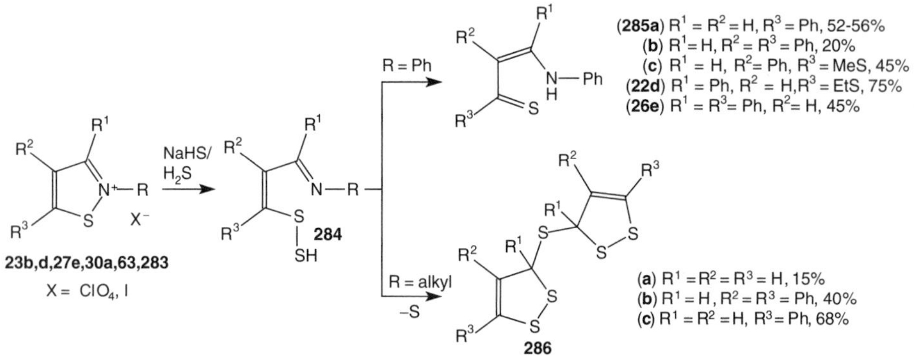

Scheme 95

metal hydrides and therefore, these hydrides reduced the functional groups (Table 17).

N-Aryl-isothiazolium salts **23b,d,27e,30a** and **63** were treated with equivalent quantities of ethanolic sodium hydrosulfide or hydrogen sulfide in ethanol to yield enaminothioketones **285a,b,22d,26e** and esters **285c,d** (73CJC3081, 72CJC2568). In contrast, N-alkyl salts **283** were reduced to bis(1,2-dithiol-3-yl)sulfane **286a–c** (72JCS(P1)2305, 73CJC3081, Scheme 95).

The 2-methylisothiazolium salts **27a,b,91a** and **287** afforded a mixture of acyclic thiones **288a–c** (85T1885) and **26a,b,289a,b** by treatment with benzylamine. Only in one case, the pure **289b** was obtained from the salt **91a** (72JCS(P1)2305, 85T1885, Scheme 96).

2-Methyl-3-phenyl-isothiazolium salts **27a,b** were attacked at the N-atom with the N,S-binucleophile 2-aminoethanthiol to obtain thioethylaminothiones **290a,b** in 48–57% yield (72JCS(P1)2305, Scheme 96).

Quaternization of isothiazole increases the tendency for nucleophilic ring cleavage. Accordingly, 2-phenylisothiazolium bromides **291** reacted with benzylamine or

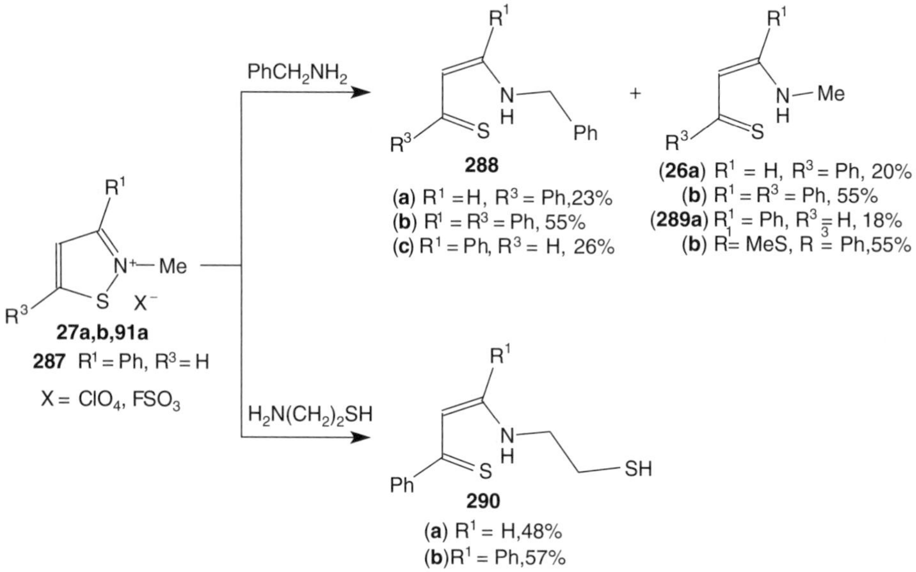
PhCH2NH2
H2N(CH2)2SH
27a,b,91a
287 R1 = Ph, R3 = H
X = ClO4, FSO3
288
(a) R1 = H, R3 = Ph, 23%
(b) R1 = R3 = Ph, 55%
(c) R1 = Ph, R3 = H, 26%
(26a) R1 = H, R3 = Ph, 20%
(b) R1 = R3 = Ph, 55%
(289a) R1 = Ph, R3 = H, 18%
(b) R1 = MeS, R3 = Ph, 55%
290
(a) R1 = H, 48%
(b) R1 = Ph, 57%

Scheme 96

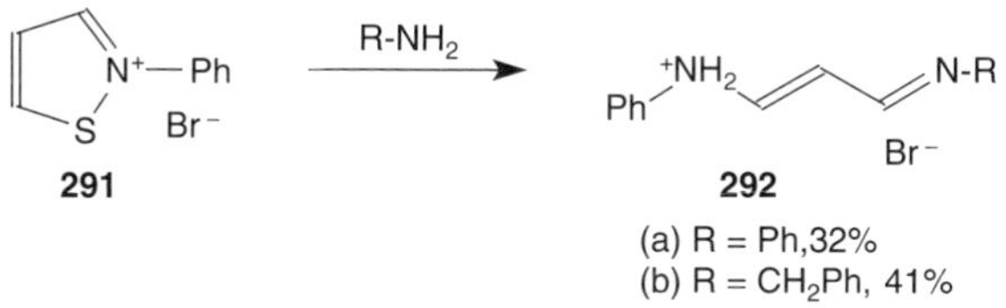
R-NH2
291
292
(a) R = Ph, 32%
(b) R = CH2Ph, 41%

Scheme 97

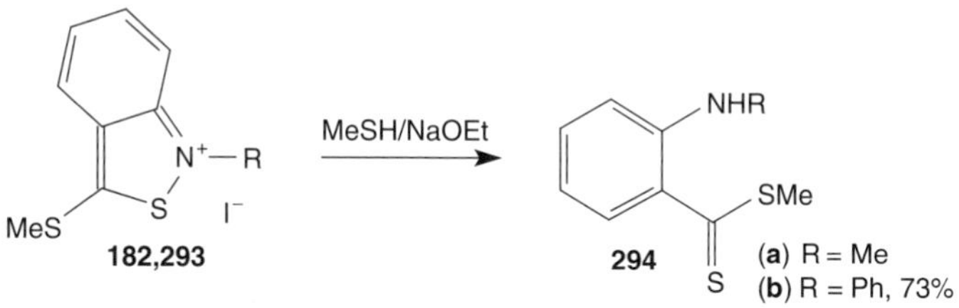
MeSH/NaOEt
182,293
294
(a) R = Me
(b) R = Ph, 73%

Scheme 98

aniline to acyclic acrolein-phenylimine **292a** and -benzylimine **292b** bromides in poor yields (72JCS(P1)2305, Scheme 97).

2,1-Benzisothiazolium iodides **182,293** reacted reductively with methanethiol to 2-methylamino- (**294a**) and 2-phenylamino-methyl dithiobenzoate **294b** (82CJC440, Scheme 98).

3-Piperidino-1,2-benzisothiazolium chloride **295** was heated with thiophenol in acetonitrile to form 2-thio-benzamidine **296** by ring opening of the isothiazole (82LA14, Scheme 99).

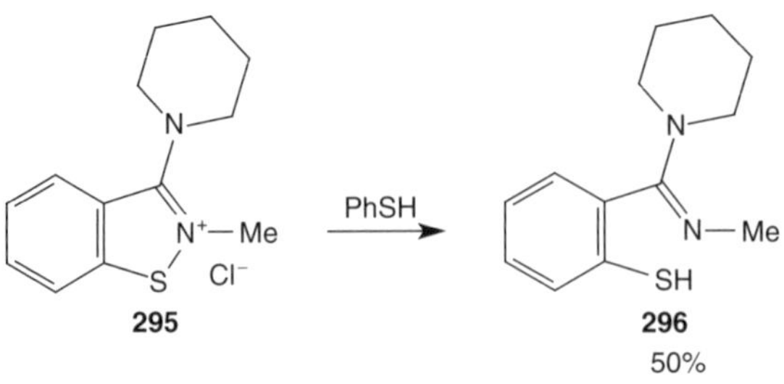

Scheme 99

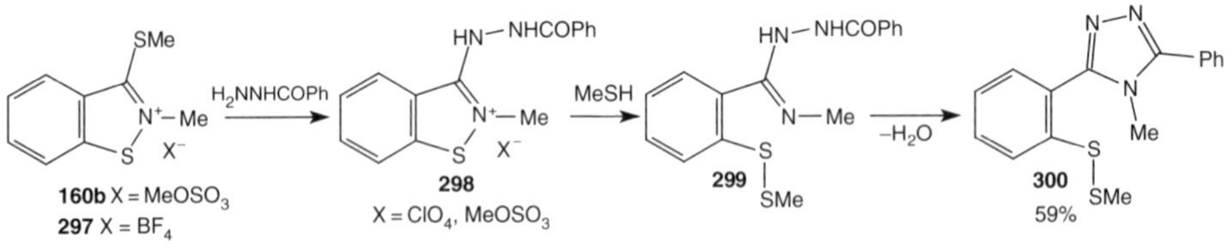

Scheme 100

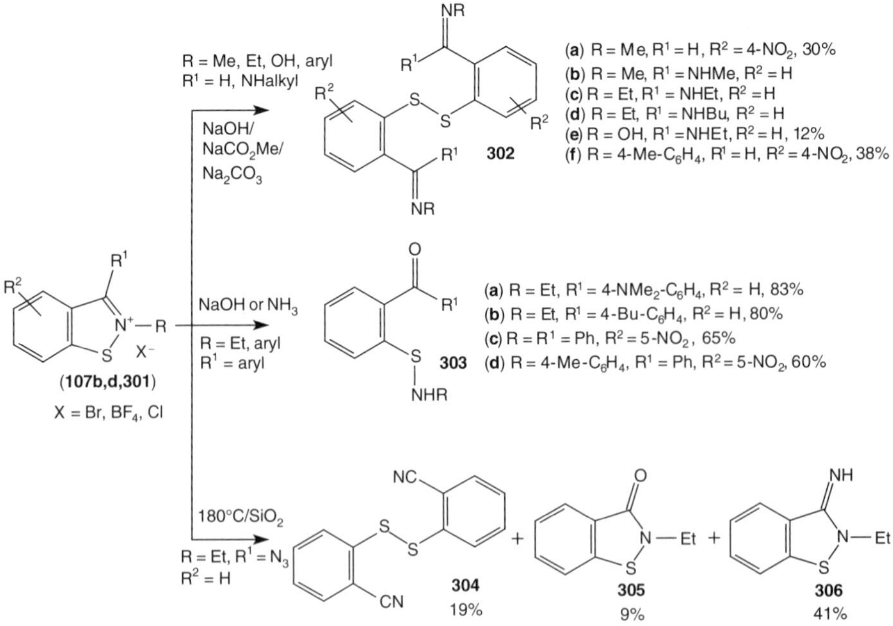

Scheme 101

The reaction of quaternary salts **160b** and **297** with benzhydrazide gave salts **298**, which were converted with methanethiol to **299**. After dehydration, 1,2,4-triazole **300** was produced (71LA46, 71LA201, Scheme 100).

The 1,2-benzisothiazolium salts **107** and **301** reacted with sodium hydroxide, sodium acetate or sodium carbonate to give 2,2′-dithiobenzamidines **302b–e**

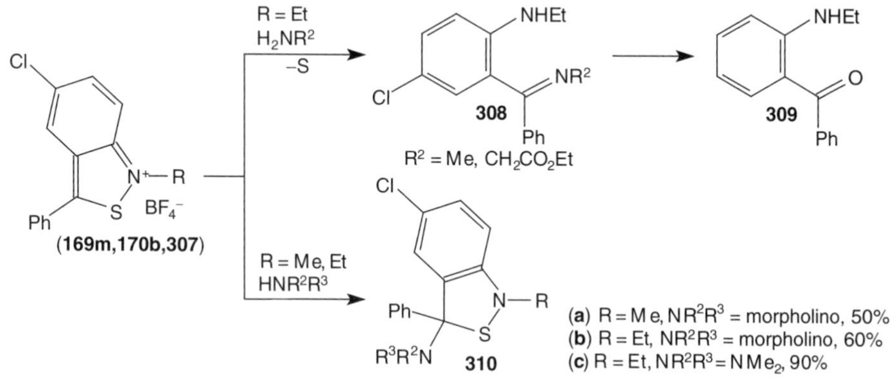

Scheme 102

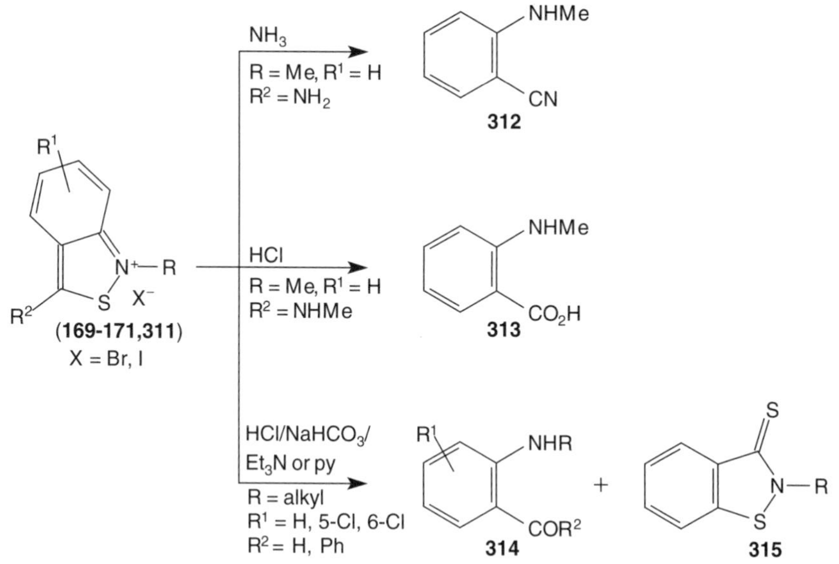

Scheme 103

and -benzimides **302a,f** by ring cleavage in poor yields (**302a,f**: 27LA264; **302b–d**: 76CB659; **302f**: 70CB3166, Scheme 101). 3-Aryl and also 3-ethyl salts **301** were cleaved to sulfenamides **303a–d** (**303a,b**: 79CB3286; **303c,d**: 27LA264). On thermolysis of salt **301** with dry SiO$_2$ at 180 °C, 2,2′-dithiobenzonitrile **304**, 1,2 -benzisothiazol-3(2*H*)-one **305** and 3-imino-1,2-benzisothiazole **306** were obtained (90HCA1679, Scheme 101).

1-Alkyl 2,1-benzisothiazolium salts **169m,170b** and **307** could be transformed with primary amine to 2-ethylaminobenzophenon-methylimines **308**, which reacted to benzophenon derivative **309** by acid hydrolysis. Furthermore, the reaction of salts **169m,170b** and **307** with secondary amines offered the stable 2,1-benzisothiazoles **310a–c** in 50–90% yield (72CPB2372, Scheme 102).

2,1-Benzisothiazolium salts **169–171,311** were cleaved with concentrated hydrochloric acid or aqueous ammonia to 2-aminobenzonitrile **312** and 2-aminobenzoic acid **313** (65JMC515, Scheme 103). Therefore, salts **169–171,311** reacted with acidic or basic sodium hydrogen carbonate, triethylamine or pyridine to 2-aminobenzaldehyde **314** ($R^2 = H$) and to 2,1-benzisothiazol-3(2H)-thiones **315** (73JCS(P1)1863, 82CJC440). 3-Phenyl salts **169–171,311** gave 2-methylaminobenzophenon **314** ($R^2 = Ph$), respectively (82CJC440, Scheme 103).

D. RING TRANSFORMATION IN OTHER HETEROCYCLIC COMPOUNDS

1. *Ring Transformation with Retention of the Ring Size*

Ring transformation of the isothiazolium salts can be converted into N-, O- and S-heterocycles, isothiazole derivatives, annulated isothiazoles and rearrangement products such as thiadiazapentalenes, spirocyclic isothiazolium salts and thianthrenes.

a. N-Heterocyclic Compounds. The reaction of monocyclic isothiazolium salts **27a,b,76c,78a,91a** with *N*-nucleophiles hydrazine and phenylhydrazine gave pyrazoles **316a–i** in good yields (**316a,b,f,g**: 72JCS(P1)2305; **316c**: 66T2135; **316d,h,i**: 85T1885, Scheme 104).

The salts **21b** and **317** ($R = CO_2$alkyl, CN) spontaneously reacted in the presence of triethylamine in methanol with sulfur extrusion to 2-alkoxycarbonyl (**319a–j**) or 2-cyanopyrroles **319k** (93AG797, Scheme 105). Electron donor substituents in the

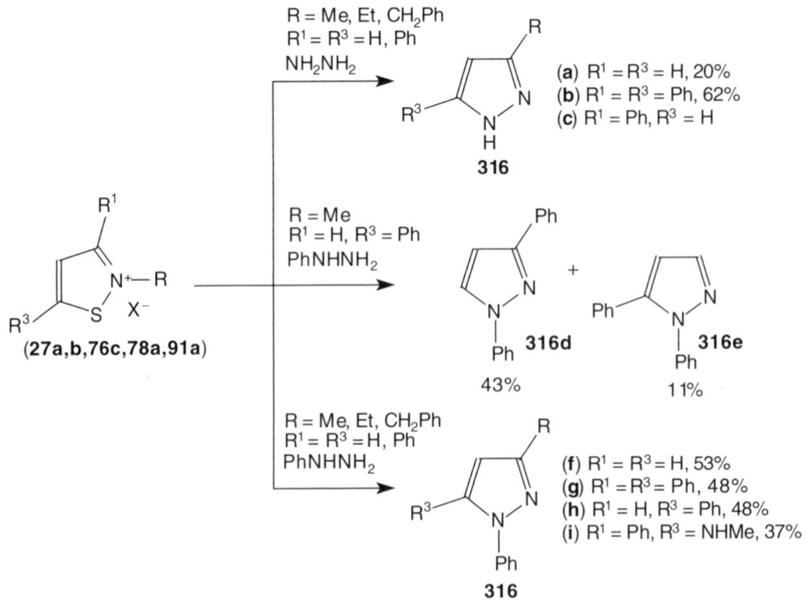

Scheme 104

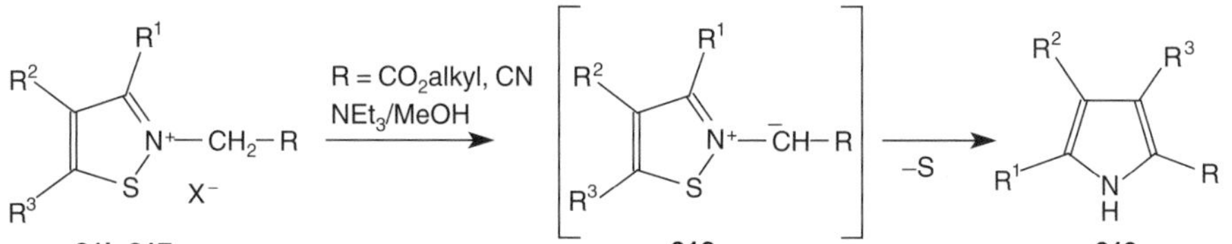

Scheme 105

Table 18. 2-Alkoxycarbonyl and 2-cyano pyrroles **319**

319	R	R^1	R^2	R^3	Yield (%)
a	CO_2Me	H	Ph	Ph	87
b	CO_2Me	Ph	H	Ph	85
c	CO_2Me	H	2-Me-C_6H_4	Morpholino	91
d	CO_2Me	H	4-F-C_6H_4	Morpholino	81
e	CO_2Me	H	4-Cl-C_6H_4	Morpholino	92
f	CO_2Me	Me	Ph	NHPh	100
g	CO_2Me	Me	CO_2Et	NHPh	73
h	CO_2Me	H	4-Cl-C_6H_4	MeS	72
i	CO_2Me	H	H	4-Cl-C_6H_4	68
j	CO_2Et	H	4-Cl-C_6H_4	NMe-piperazino	80
k	CN	Me	Ph	NHPh	82
l	CO_2Met	H	Ph	Morpholino	82
m	CO_2Et	H	4-Cl-C_6H_4	Piperidino	76
n	CO_2-i-Bu	H	4-Cl-C_6H_4	Morpholino	80

5-position and electron acceptor substituents R promoted the ring transformation. The direct electrochemical synthesis of 3-aminopyrrole-2-carboxylates **319e–g,l–n** was carried out by anodic oxidation, for example, of the thioacrylamides **20** without isolating isothiazolium salts **317** (95JPR310). This result demonstrated that electrochemical oxidation of 3-amino-thioacrylamides is a useful alternative to the application of chemical oxidizing reagents. The pyrrole derivatives **319** are represented in Table 18.

The ring transformation to *N*-methylpyrrole **320** as a minor product occurred from the 2-methyl salt **75j** on reaction with dimethyl-(2-ethoxycarbonylmethylene)sulfurane (84CJC1580, Schemes 106, 108).

b. O-Heterocycles. Isoxazoles **321** and **322** as well as isothiazole *N*-oxide **323** were formed on treatment of salts **27a** and **b** with hydroxylamine and sodium alcoholate (**321–323a,b**: 72JCS(P1)2305, 85T1885, Scheme 107).

c. S-Heterocycles. The initial nucleophilic attack of a carbanion at the sulfur atom of monocyclic isothiazolium salts **23b,c,27a,30a** and **75a,j,96a,324** was carried out under ring cleavage and recyclization condition to thiophenes **325–328**

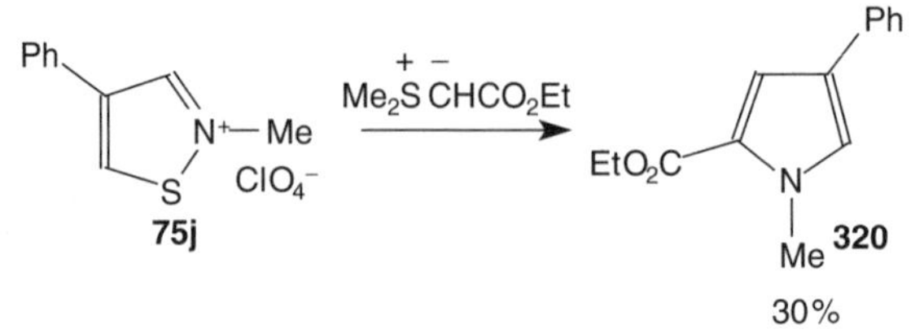

Scheme 106

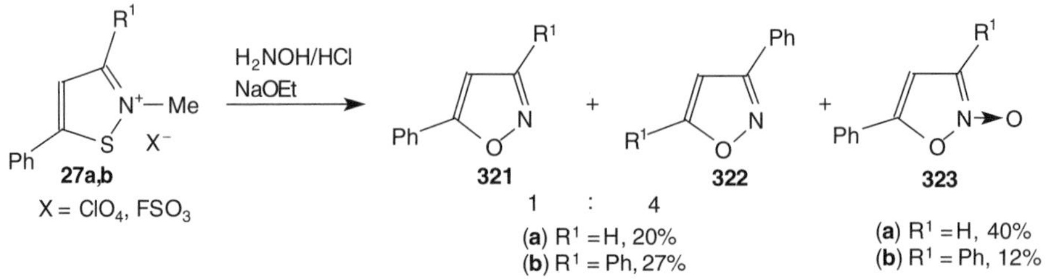

Scheme 107

(**325a–c**: 73CJC3081; **325d,e,326**: 77CJC1123; **327a–c,328**: 84CJC1580, Scheme 108). The carbanion sodium-3-oxo-3-phenyl-propiolate and stable S-ylides were utilized.

3-Chloro-1,2-benzisothiazolium chlorides **109a,b,i,h** and **329** reacted with methyl ketones or N-heterocycles with an activated methyl group to give 3-amino-1-benzo-thiophenes **330a–l** and **331a–j** (**330a–l**: 72LA58; **331a–j**: 79S442, Scheme 109) (Table 19).

N-Aryl 3-chloro-1,2-benzisothiazolium chlorides **109** were converted with thio-acetic acid to 3-aryl-imino-1,2-benzdithioles **332**. In contrast, N-alkyl salts **109** reacted to form 2-alkyl-1,2-benzisothiazol-3(2H)-thiones **161**, which are in equilib-rium with **332** (67CB2435, Scheme 110).

The 3-thiono-3H-1,2-dithioles **334a,b** in poor yields and the dealkylated iso-thiazoles **335** resulted from the treatment of salts **27a,b,30a,45a** and **75j,333** (R = Me, Ph) with sulfur in pyridine (72CJC2568, Scheme 111). The reaction of salts **27a,333** (R = t-Bu, $R^2 = R^3 = H$) with toluene yielded isothiazole **335a** (83PS119).

d. Isothiazole Derivatives. The reaction of 3-chloro- (**75,76,104**) or 3-methoxy monocyclic salts **102** with ammonia gave 3-aminoisothiazole **337a–f** (53–95%) and 3-unsubstituted or 3-phenyl isothiazolium salts **75,76** afforded isothiazole **338a–d** (67–89%) (**337a,b**: 79JOC1118; **337a,c–f**: 78JHC695; **338a,d**: 72JCS(P1)2305; **338a,b**: 66T2135; **338c**: 85T1885, Scheme 112). Ammonia made a nucleophilic attack at the sulfur atom to yield S-aminoiminoyl **336**, which efficiently recyclized to **337** and by loss of amine to **338**.

The treatment of 3-chloro 1,2-benzisothiazolium salts **109** with ammonia gave the rearranged 3-aylkl/aryl-amino-isothiazole **339**, which could be converted by proto-nation to cation **144**. Only in the presence of strongly acid solution were salts **147**

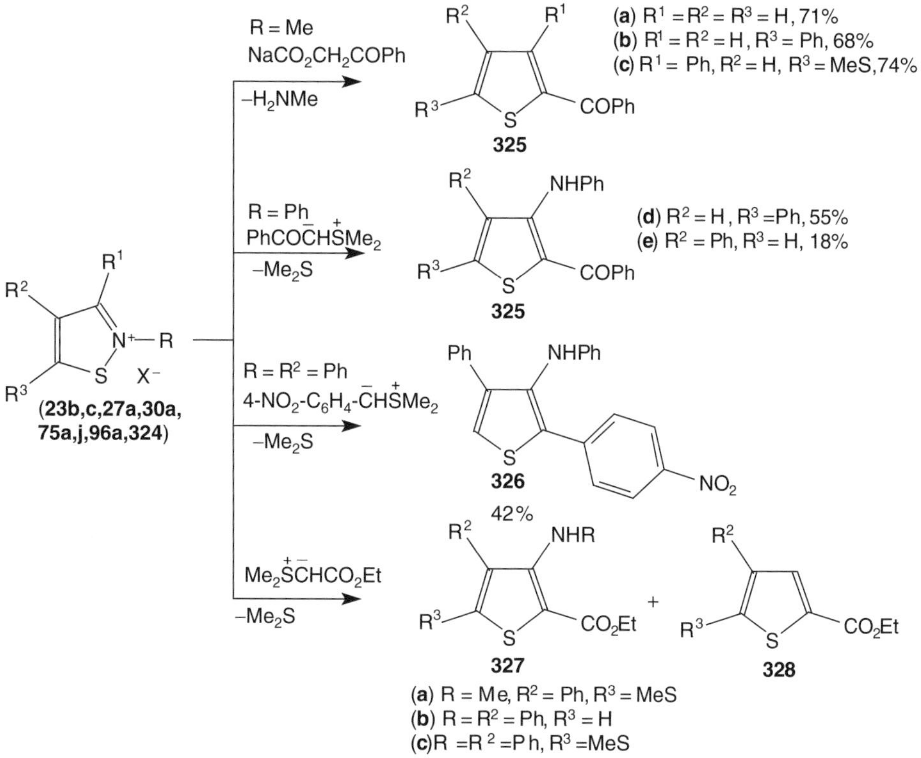

R = Me
NaCO₂CH₂COPh

–H₂NMe

(a) R¹ = R² = R³ = H, 71%
(b) R¹ = R² = H, R³ = Ph, 68%
(c) R¹ = Ph, R² = H, R³ = MeS, 74%

325

R = Ph
PhCOCHSMe₂

–Me₂S

(d) R² = H, R³ = Ph, 55%
(e) R² = Ph, R³ = H, 18%

325

R = R² = Ph
4-NO₂-C₆H₄-CHSMe₂

–Me₂S

326

42%

Me₂SCHCO₂Et

–Me₂S

327   **328**

(a) R = Me, R² = Ph, R³ = MeS
(b) R = R² = Ph, R³ = H
(c) R = R² = Ph, R³ = MeS

Scheme 108

MeCOR¹/py
100°C/30 min

330

(a) R = R¹ = Me
(b) R = Et, R¹ = Me
(c) R = Ph, R¹ = Me
(d) R = 2-Me-C₆H₄, R¹ = Me
(e) R = 4-F-C₆H₄, R¹ = Me
(f) R = 3-Cl-C₆H₄, R¹ = Me
(g) R = allyl, R¹ = Me
(h) R = Et, R¹ = 2-thienyl
(i) R = Et, R¹ = Ph
(j) R = Et, R¹ = 3-NO₂-C₆H₄
(k) R = Et, R¹ = t-Bu
(l) R = Et, R¹ = i-Bu

329 (a) R = 2-Me-C₆H₄
(b) R = 3-Cl-C₆H₄
(c) R = 4-F-C₆H₄

Het-Me/py
20-100°C/1-15 hrs

331

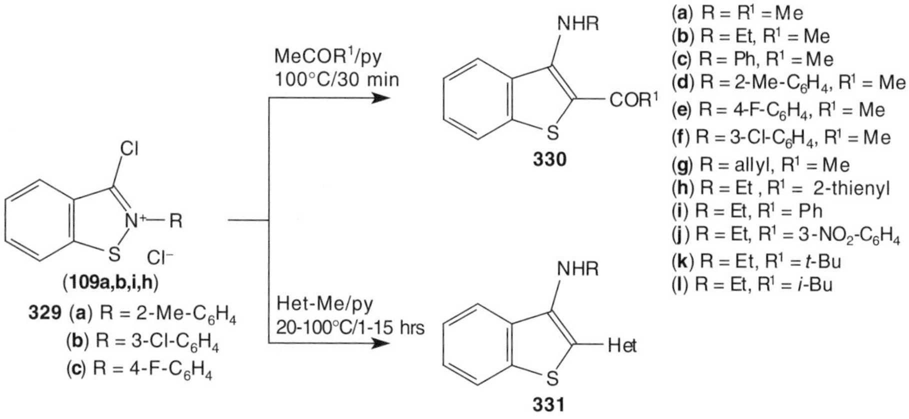

Scheme 109

formed. They are in equilibrium with **144** (**147**: R = alkyl: 66CB2566, 69CB1961; **147**: R = Ph: 80CB2490, Scheme 113).

3-Piperidino-1,2-benzisothiazolium chloride **295** could be obtained by reaction of salt **109a** with piperidine in ethanol at 5 °C in 35% yield (82LA14, Scheme 113).

Table 19. N-heterocyclic thiophenes **331**

331	R	Het	Yield (%)	Reference
a	Et		65	79S442
b	Et		55	79S442
c	Et		16	79S442
d	Et		40	79S442
e	Et		53	79S442
f	Et		65	79S442
g	Et		66	79S442
h	Et		42	79S442
i	Me		62	79S442
j	Me		87	79S442

The 1,2-benzisothiazolium salts **109a–c** reacted with hydroxylamine to give 3-alkylamino-1,2-benzisothiazole 2-oxides **340a–c** in equilibrium with salt **341** (70CB3166, Scheme 113).

By contrast, 5-nitro-1,2-benzisothiazolium salts **107** reacted with ammonia in ethanol to isothiazole **343** by 2-dealkylation (23AG159, Scheme 114). The salts **109b** and **344** are dealkylated in 1,2-dichlorobenzene or by dry distillation to **137** in 78–95% yield (66CB2566, 68ZC170, Scheme 114).

For the first time, 5-aryl-3-arylthioisothiazoles **345c–n** were prepared by dealkylation of the corresponding salts **25** with KI in dimethylsulfoxide and 3–5 h at 120 °C (02JOC5375, Scheme 115).

A Dimroth rearrangement took place by the reaction of a base with 3-amino-2-methylisothiazolium fluorosulfonate **75t** to **337a** (79JOC1118, Scheme 116).

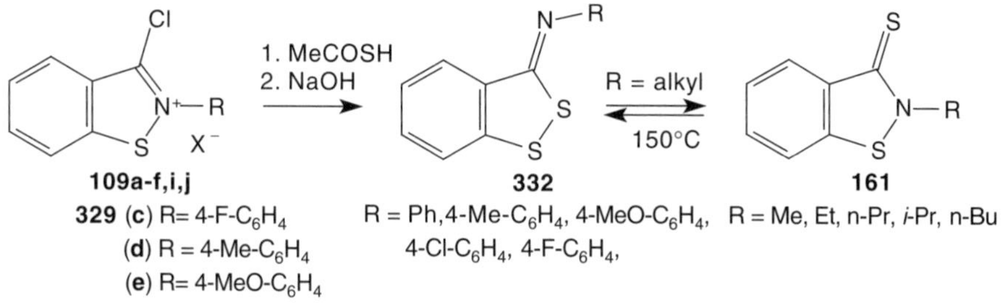
1. MeCOSH
2. NaOH
R = alkyl
150°C
109a-f,i,j
329 (c) R= 4-F-C6H4
(d) R = 4-Me-C6H4
(e) R= 4-MeO-C6H4
332
R = Ph,4-Me-C6H4, 4-MeO-C6H4, 4-Cl-C6H4, 4-F-C6H4,
161
R = Me, Et, n-Pr, i-Pr, n-Bu

Scheme 110

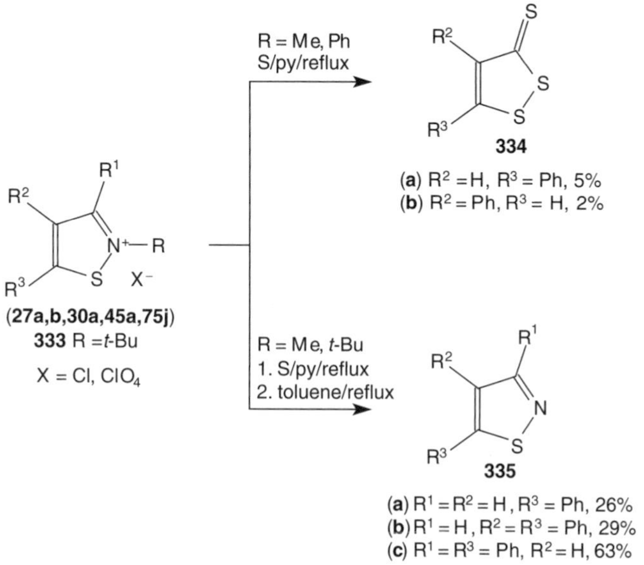
R = Me, Ph
S/py/reflux
334
(a) R2 = H, R3 = Ph, 5%
(b) R2 = Ph, R3 = H, 2%
(27a,b,30a,45a,75j)
333 R = t-Bu
X = Cl, ClO4
R = Me, t-Bu
1. S/py/reflux
2. toluene/reflux
335
(a) R1 = R2 = H, R3 = Ph, 26%
(b) R1 = H, R2 = R3 = Ph, 29%
(c) R1 = R3 = Ph, R2 = H, 63%

Scheme 111

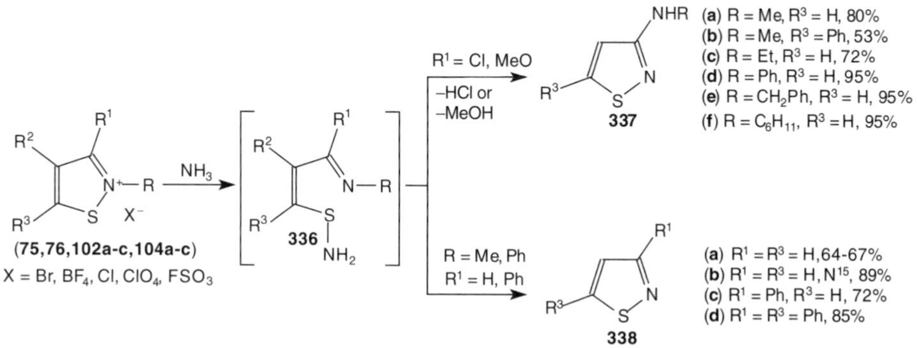
(75,76,102a-c,104a-c)
X = Br, BF4, Cl, ClO4, FSO3
NH3
336
R1 = Cl, MeO
–HCl or
–MeOH
337
(a) R = Me, R3 = H, 80%
(b) R = Me, R3 = Ph, 53%
(c) R = Et, R3 = H, 72%
(d) R = Ph, R3 = H, 95%
(e) R = CH2Ph, R3 = H, 95%
(f) R = C6H11, R3 = H, 95%
R = Me, Ph
R1 = H, Ph
338
(a) R1 = R3 = H, 64-67%
(b) R1 = R3 = H, N15, 89%
(c) R1 = Ph, R3 = H, 72%
(d) R1 = R3 = Ph, 85%

Scheme 112

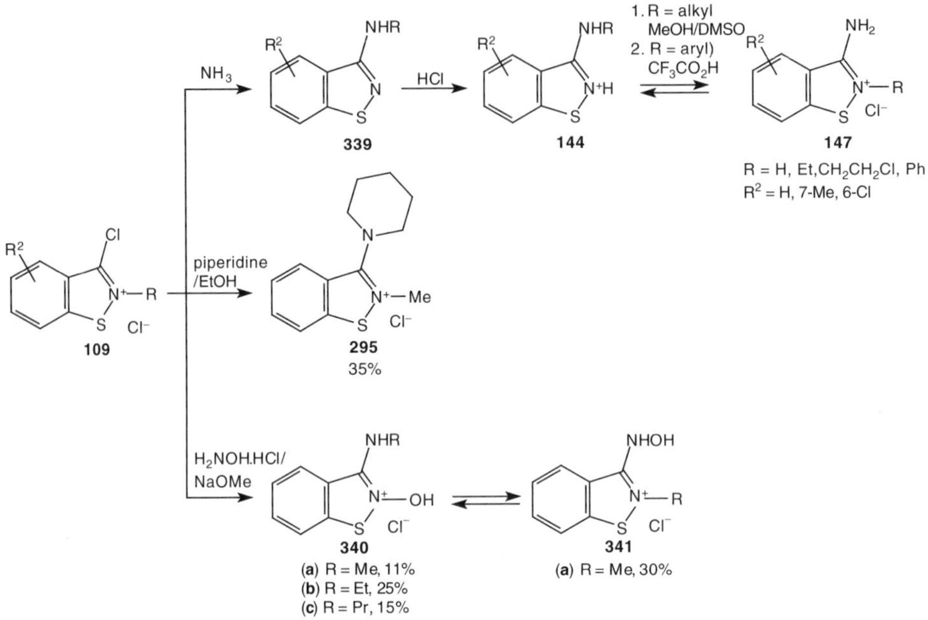

Scheme 113

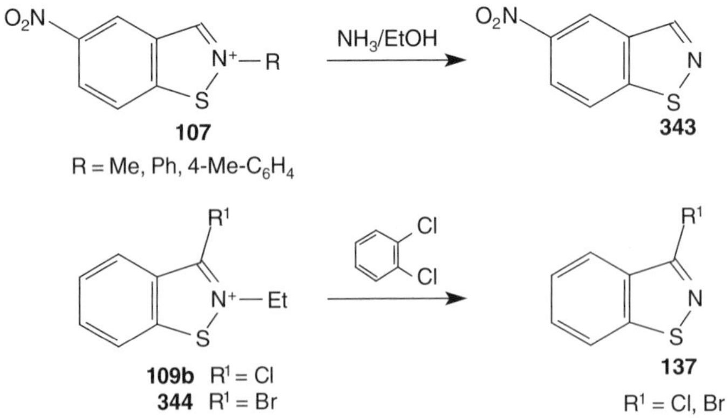

Scheme 114

The reaction of 3-phenacylthio- (**347**) and 5-phenacylthio-isothiazolium salts **350** with triethylamine afforded 3-phenacylidenisothiazole **349** (39%) and isothiazol-5-thione **352** (50%) by deprotonation of the exocyclic methylene group (85BSB149, Scheme 117).

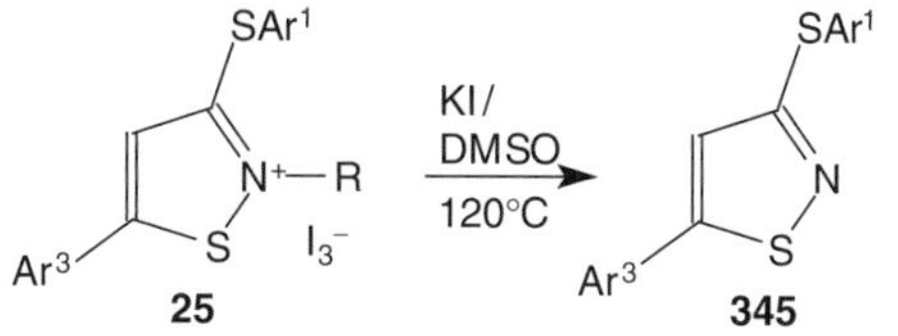

(c) R = Me, R^1 = Ph, 76%
(d) R = Me, R^1 = 4-Me-C$_6$H$_4$, 91%
(e) R = Me, R^1 = 4-MeO-C$_6$H$_4$, 92%
(f) R = Me, R^1 = 4-Br-C$_6$H$_4$, 86%
(g) R = Me, R^1 = 2-MeO-C$_6$H$_4$, 87%
(h) R = Me, R^1 = 2-Cl-C$_6$H$_4$, 67%
(i) R = Me, R^1 = 2-Br-C$_6$H$_4$, 84%
(j) R = Me, R^1 = naphthyl, 85%
(k) R = Me, R^1 = Ph, Ar^3 = 4-Cl-C$_6$H$_4$, 86%
(l) R = Me, R^1 = 2-Br-C$_6$H$_4$, Ar^3 = 4-Cl-C$_6$H$_4$, 79%
(m) R = Et, R^1 = Ph, Ar^3 = 4-MeO-C$_6$H$_4$, 74%
(n) R = Et, R^1 = 2-Br-C$_6$H$_4$, A r^3 = 4-MeO-C$_6$H$_4$, 74%

Scheme 115

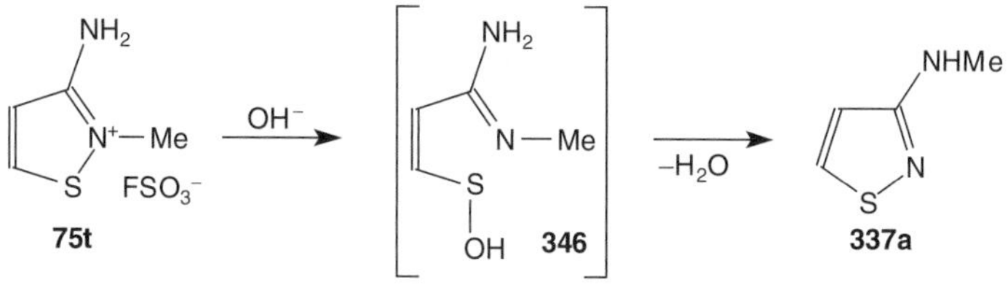

Scheme 116

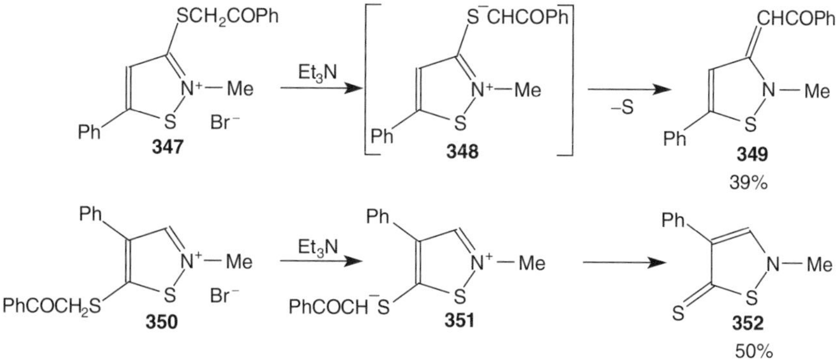

Scheme 117

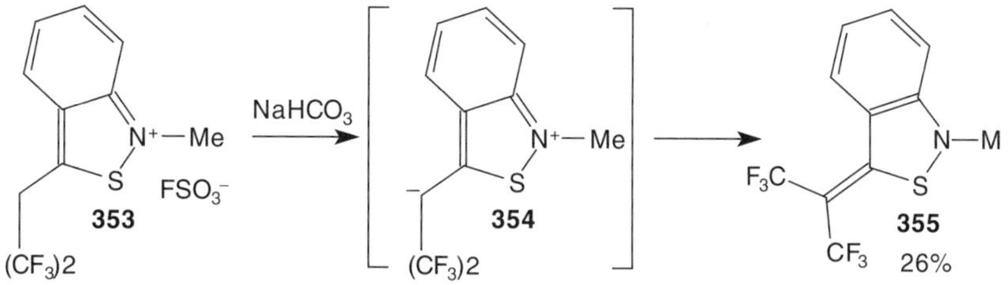

Scheme 118

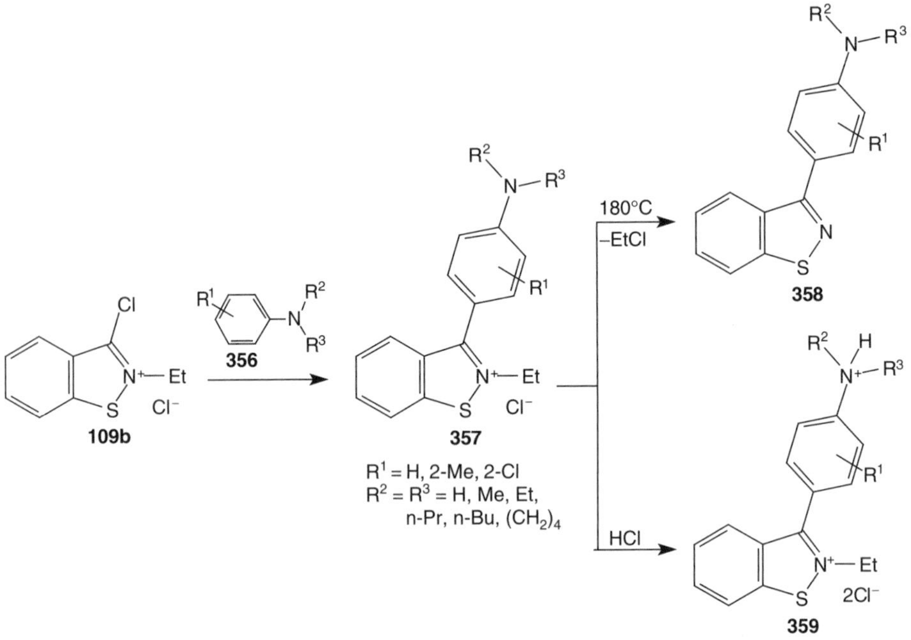

Scheme 119

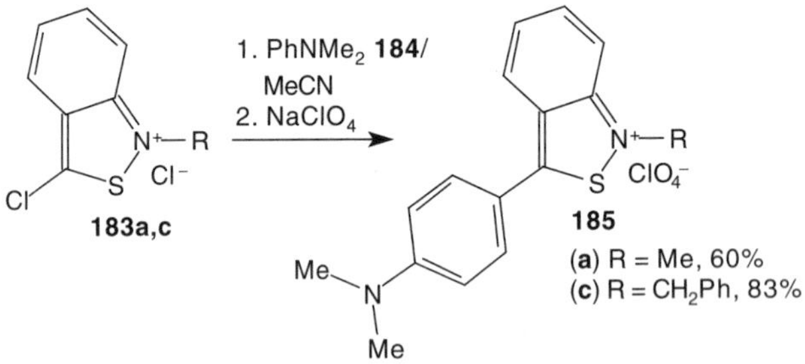

Scheme 120

The 2-methyl 2,1-benzisothiazolium fluorosulfonate **353** was treated with sodium bicarbonate to obtain 2,1-benzisothiazole **355** in 26% yield (78JOC2500, Scheme 118).

3-Chloro-2-ethyl-1,2-benzisothiazolium chloride **109b** reacted with *N*-mono- and *N*,*N*-dialkylanilines **356** to give 3-phenyl 1,2-benzisothiazolium salts **357**, which were converted by thermal treatment to isothiazole **358** and with hydrochloric acid to salts **359** in 70–80% yield (79CB3286, Scheme 119). The isothiazole **358** can be oxidized to 2-phenyl 1,2-benzisothiazole 1,1-dioxides.

In contrast, 3-chloro-2,1-benzisothiazolium salts **183a,c** reacted with *N*,*N*-dimethylaniline **184** to give 3-anilino-2,1-benzisothiazolium salts **185** in good yield (76JPR161, Scheme 120).

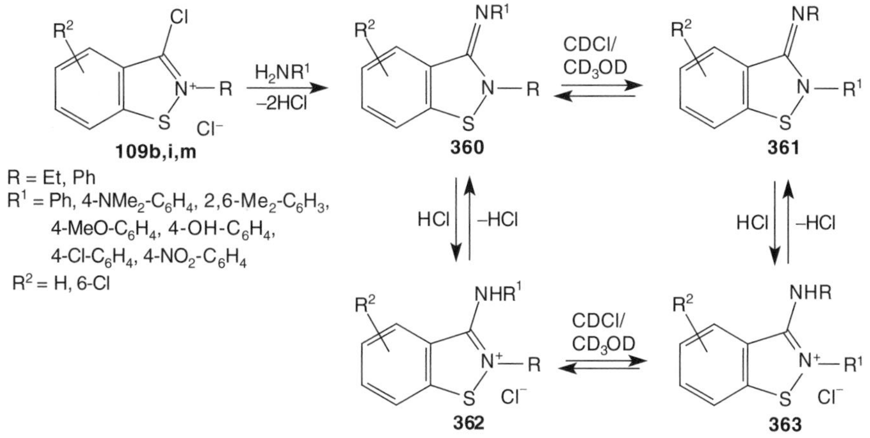

Scheme 121

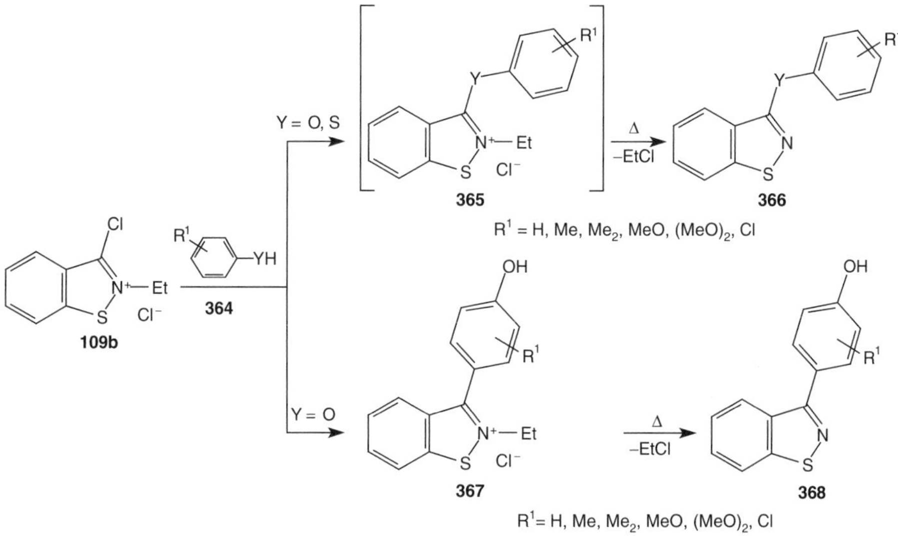

Scheme 122

The treatment of salts **109** with aromatic amines gave 3-arylimino isothiazoles **360** (59–84%) and **361** (31–58%), which can be rearranged to 3-amino salts **362,363** by protonation (76CB659, 80CB2490, Scheme 121).

Furthermore, chloride **109b** reacted with substituted phenols and thiophenols **364** to yield primarily the 3-phenoxy- or the corresponding 3-phenylthio-1, 2-benzisothiazoles **366**. When the attack of salts **109b** occurred at the aromatic ring of phenol, the 3-phenyl-1,2-benzisothiazoles **368** were produced (74CB1667, Scheme 122).

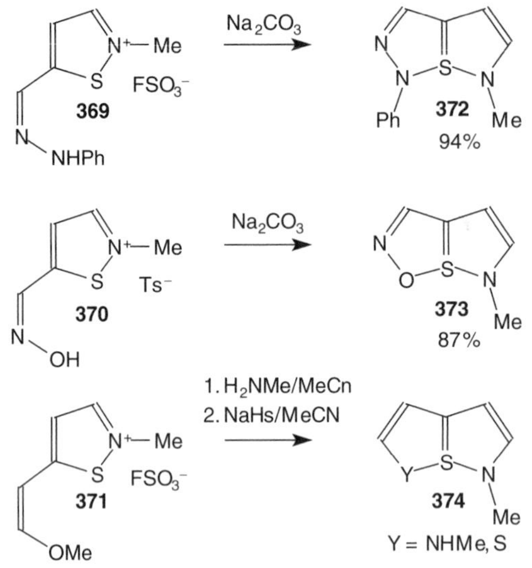
Na2CO3
N+—Me
S FSO3−
369
N
NHPh
N
N
S
N
Ph
372
Me
94%
Na2CO3
N+—Me
S Ts−
370
N
OH
N
O
S
N
373
Me
87%
1. H2NMe/MeCn
2. NaHs/MeCN
N+—Me
S FSO3−
371
OMe
Y
S
N
374
Me
Y = NHMe, S

Scheme 123

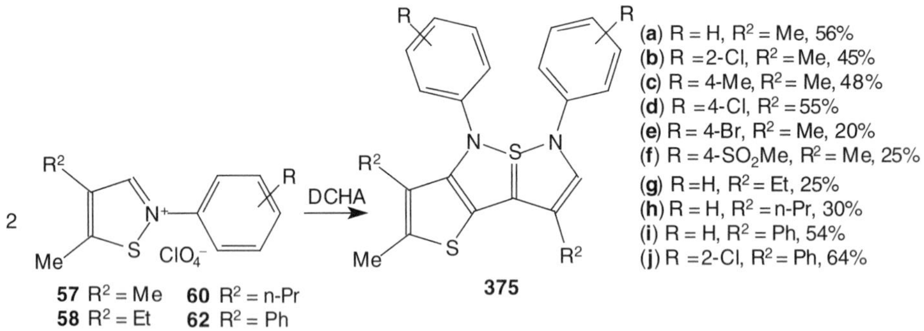
2
R2
Me
N+
S ClO4−
R
57 R2 = Me 60 R2 = n-Pr
58 R2 = Et 62 R2 = Ph
DCHA
R
R
N
S
N
R2
Me
S
R2
375
(a) R = H, R2 = Me, 56%
(b) R =2-Cl, R2 =Me, 45%
(c) R = 4-Me, R2 = Me, 48%
(d) R =4-Cl, R2 =55%
(e) R = 4-Br, R2 = Me, 20%
(f) R = 4-SO2Me, R2 = Me, 25%
(g) R =H, R2 = Et, 25%
(h) R = H, R2 =n-Pr, 30%
(i) R = H, R2 = Ph, 54%
(j) R =2-Cl, R2 = Ph, 64%

Scheme 124

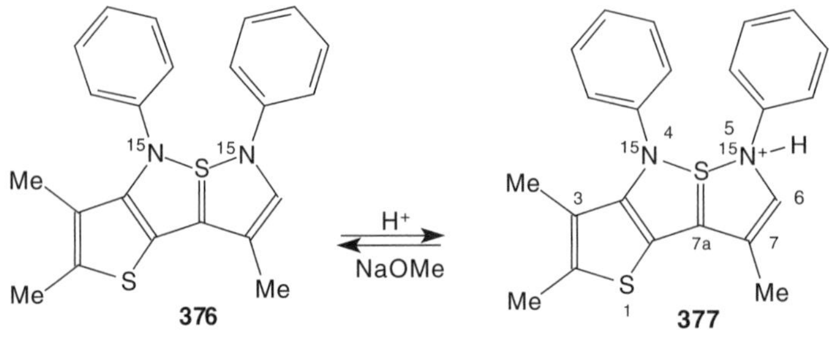
15N
S
15N
Me
Me
S
Me
376
H+
NaOMe
15N 4
5
15N+—H
Me
3
6
7a
7
Me
S
1
Me
377

Scheme 125

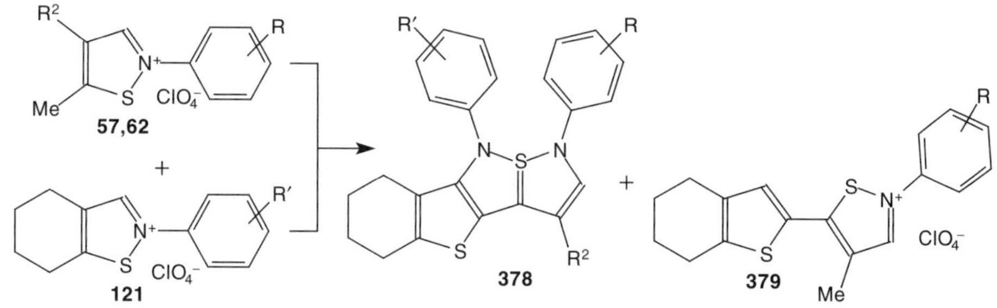

R = R'= H,4-MeO,-4-Br, -4-Cl,4-SO$_2$Me, 4-CF$_3$
R^2 = Me, Ph

Scheme 126

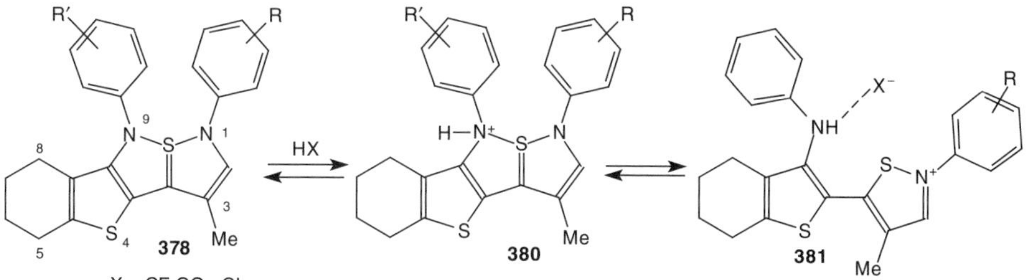

X = CF$_3$CO$_2$, Cl

Scheme 127

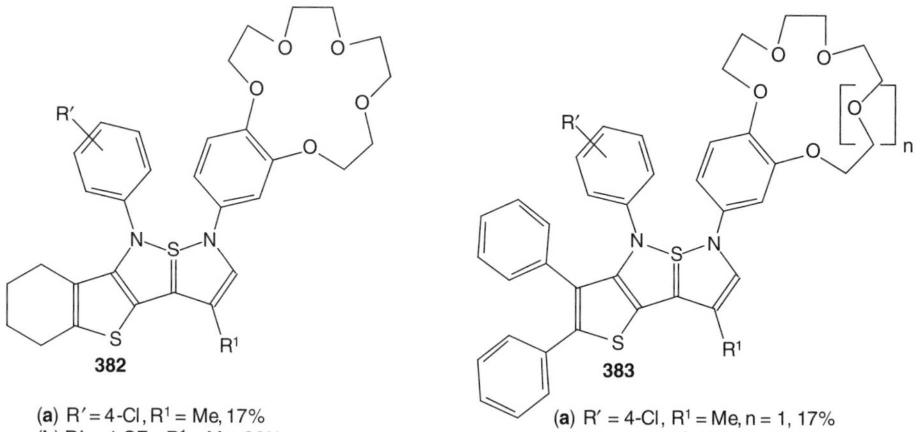

382

(a) R' = 4-Cl, R^1 = Me, 17%
(b) R' = 4-CF$_3$, R^1 = Me, 88%
(c) R' = 4-Cl, R^1 = Ph, 30%
(d) R' = 4-CF$_3$, R^1 = Ph, 45%
(e) R' = H, R^1 = Ph, 54%

383

(a) R' = 4-Cl, R^1 = Me, n = 1, 17%
(b) R' = 4-Cl, R^1 = Me, n = 2, 34%
(c) R' = 4-Cl, R^1 = Ph, n = 1, 40%
(d) R' = 4-CF$_3$, R^1 = Ph, n = 1, 20%

Scheme 128

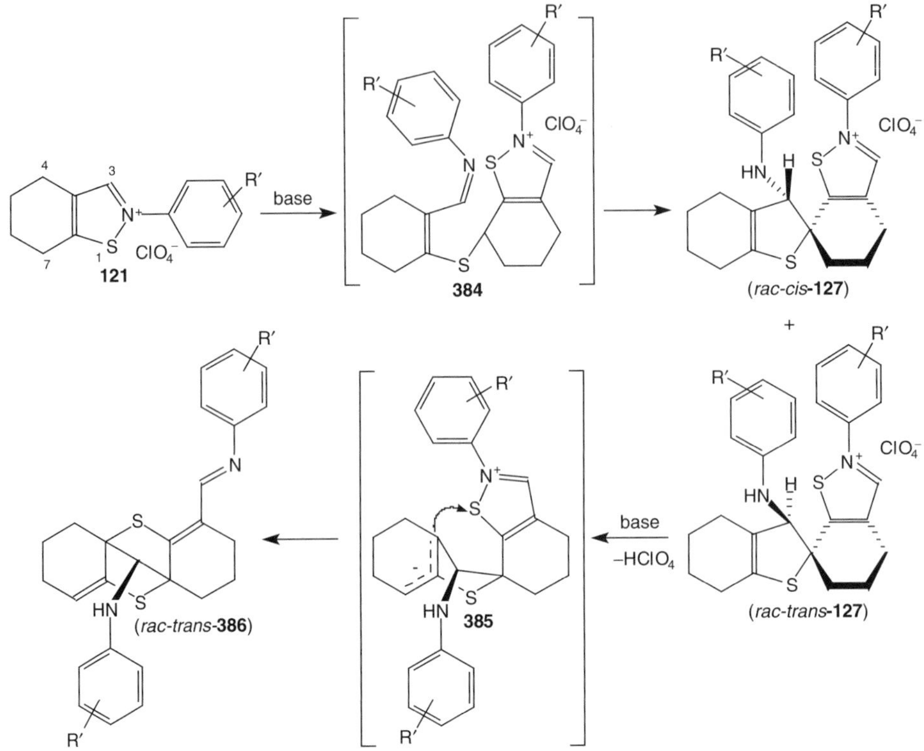

R′ = (a) H,(b) 4-Me, (c) 4-MeO,(d) 4-Cl,(e) 4-SO₂Me, (f) 3-Me,
(g) 3-MeO,(h) 3-Cl,(i) 2-Me,(j) 2-Cl, (k) 2,6-Cl₂, (l) 2-CF₃

Scheme 129

e. Heterocyclic Annulated Isothiazoles. The deprotonation of the salts **369–371** with sodium carbonate yielded 1,2-thiazolo[5,1-*e*]-1,2,3-thiadiazole **372** (94%), 1,2-thiazolo[1,5-*b*]-1,2,5-oxathiazole **373** (87%) and annulated products **374** (23–61%) (76IEC268, 79JCS(P1)2340, Scheme 123).

f. Thiadiazapentalenes as Rearrangement Products. The base-induced dimerization of 5-methyl-isothiazolium salts **57,58,60,62** surprisingly afforded 6aλ⁴-thia-1,6 -diazapentalenes **375a–j** (**375a,c,i**: 88ZC287, 90MRC419; **375b,d,g,h,j**: 92JPR25; **375e,f**: 95JPR175, Scheme 124).

Doubly ¹⁵N labeled thiadiazapentalene **376** was protonated with trifluoroacetic acid at N-5 to give **377**, maintaining the N–S–N bond with a coupling constant ^2J (N-4, N-5) of 3.5 Hz (92JPR25, Scheme 125).

The isothiazolium salts with an active 5-methyl **57** and **62** or a 7-methylene group **121** reacted with DCHA depending on substituents R and R′ to thiadiazapentalenes **378**. Salts **379** were obtained only as by-products (95JPR175, 95ZK73, Scheme 126).

The structure of the basic skeleton of the thiadiazapentalenes was confirmed by X-ray analysis of **378** (R = 4-MeO, R′ = H) and *ab initio* MO calculations. Some mechanistic aspects are supported by the MO results (95JPR175).

The ring opening of thiadiazapentalenes **378–381** was made possible by protonation at the basic center N-9 *via* **380** when R′ is an electron acceptor substituent (R′ = Cl, CF₃) (96MOL142, Scheme 127).

The thiadiazapentalenes **375** and **378** were studied in terms of the influence of several substituents toward Ag(I), Hg(II) and Na(I) in solution. The combination of thiadiazapentalenes **382** and **383** with benzo crown ether substituents resulted in a simultaneous binding of one hard and one soft metal ion by one molecule (01PS29, Scheme 128).

N-Phenyl isothiazolium salts **121** with an active 7-methylene group reacted under the influence of base to isolable spirocyclic isothiazolium salts *rac-cis*-**127a–h** (36–82%) and *rac-trans*-**127**. The salts *rac-trans*-**127** were not stable and rearranged to thianthrene derivatives *rac-trans*-**386a–h** (11–93%). The *o*-substituted salts **121i–l** also reacted with sodium acetate only to thianthrenes *rac-trans*-**386i–l** (58–95%) (90DDP295385, 94ZOR1404, 96JPR424, 96ZK761, Scheme 129).

Thus, a very simple separation of diastereoisomeric salts *rac-cis*- and *rac-trans*-**127** was possible (Scheme 129). The structures of *rac-cis*-**127** (R′ = 3-Me) and *rac-trans*-**386** (R′ = 2,6-Cl₂) were confirmed by X-ray analysis (96JPR424, 96ZK761).

The synthesis of 2,3-dihydrothiophen derivatives by base-catalyzed reaction of 5-ethyl-4-methylisothiazolium salts **59** is reported (97SUL35).

2. Ring Transformation with Expansion of the Ring

The 2-methyl-3-methylaminoisothiazolium chloride **75v** was treated with cyanide in water and a nucleophilic attack on the sulfur occurred to yield the intermediate

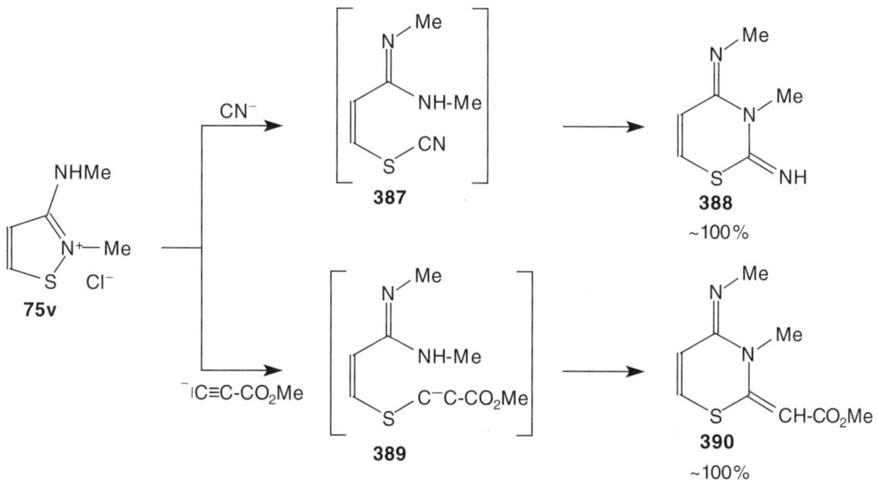

Scheme 130

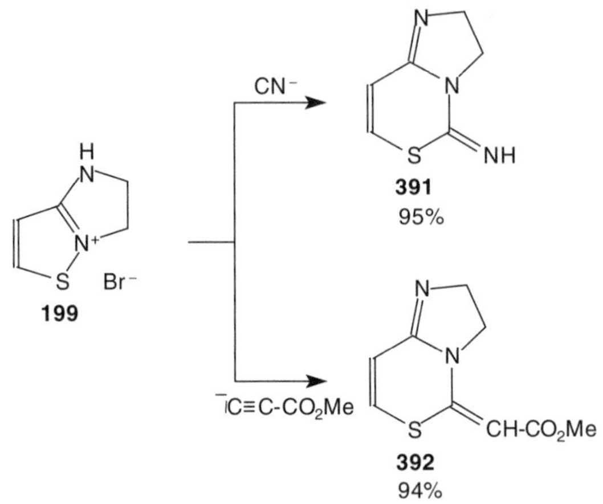
CN⁻
NH
391
95%
199
Br⁻
⁻C≡C-CO₂Me
CH-CO₂Me
392
94%

Scheme 131

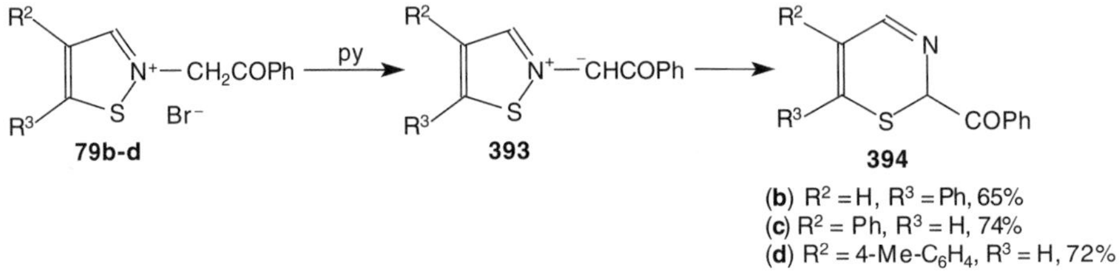
N⁺—CH₂COPh
Br⁻
py
79b-d
N⁺—⁻CHCOPh
393
COPh
394
(b) R² = H, R³ = Ph, 65%
(c) R² = Ph, R³ = H, 74%
(d) R² = 4-Me-C₆H₄, R³ = H, 72%

Scheme 132

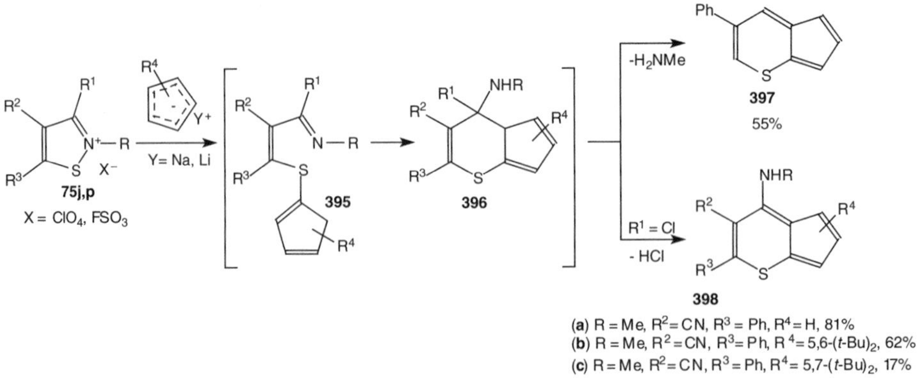
Y= Na, Li
75j,p
X = ClO₄, FSO₃
395
396
-H₂NMe
397
55%
R¹ = Cl
- HCl
398
(a) R = Me, R² = CN, R³ = Ph, R⁴ = H, 81%
(b) R = Me, R² = CN, R³ = Ph, R⁴ = 5,6-(t-Bu)₂, 62%
(c) R = Me, R² = CN, R³ = Ph, R⁴ = 5,7-(t-Bu)₂, 17%

Scheme 133

open-chain amidine **387**, which recyclized to give ring expanded 2-imino-3-methyl-4*H*-1,3-thiazine **388** in high yield (79TL1281). In another case, the isothiazole ring was attacked with methyl propynolate as nucleophile to ring expands 2-(methoxy-carbonyl-methylene)-1,3-thiazine **390** (79TL1281, 83JCS(P1)1953, Scheme 130).

The bicyclic bromide **199** reacted with cyanide and methyl propynolate in quantitative yields to expand to bicyclic 1,3-thiazines **391** and **392**, respectively (79TL1281, Scheme 131).

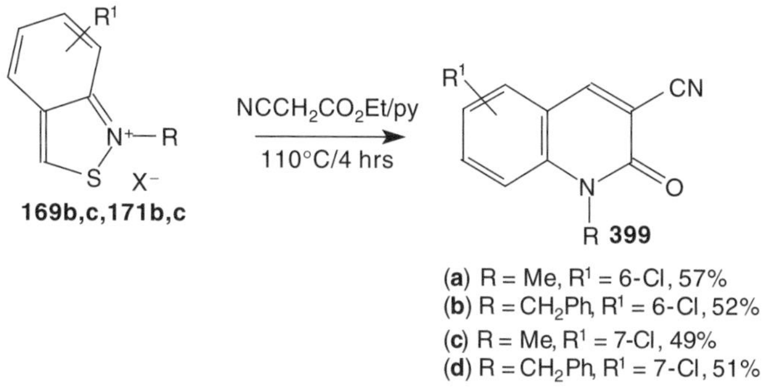

(a) R = Me, R¹ = 6-Cl, 57%
(b) R = CH₂Ph, R¹ = 6-Cl, 52%
(c) R = Me, R¹ = 7-Cl, 49%
(d) R = CH₂Ph, R¹ = 7-Cl, 51%

Scheme 134

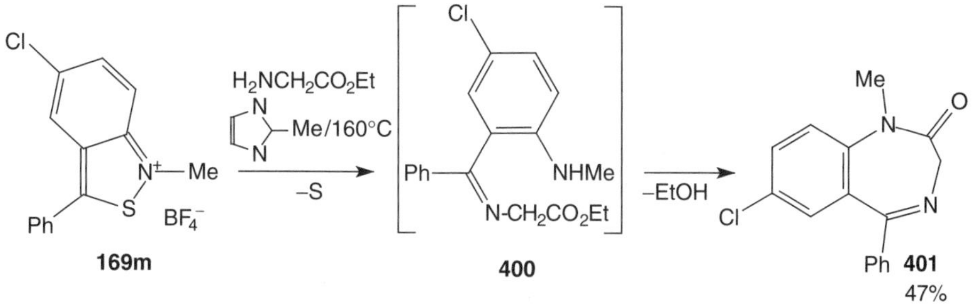

Scheme 135

The 2-phenacylisothiazolium bromides **79b–d**, synthesized by reaction of the corresponding isothiazole with phenacyl bromide (see Scheme 24), were treated with pyridine and afforded 2-benzoyl-2H-1,3-thiazines **394b–d** *via* deprotonation of the exocyclic methylene group to **393** followed by intramolecular nucleophilic attack on sulfur atom (85BSB149, Scheme 132).

The salts **75** reacted with substituted cyclopentadienyl anion in hexane at room temperature to cyclopenta[*b*]thiin **397** and **398a–c** in good yields (**397**: 73CJC3081; **398a–c**: 96PHA638, Scheme 133).

1-Methyl- (**399a,c**) or 1-benzyl-2-quinolones **399b,d** were synthesized in one-step reaction of 2,1-benzisothiazolium salts **169b,c** and **171b,c** with ethyl cyanoacetate in hot pyridine in 49–57% yield (83JHC1707, Scheme 134). In this case, the salts **169b,c** and **171b,c** are the equivalent of 2-aminobenzaldehyde. The ethyl cyanoacetate anion attacks carbon atom, and closure of the quinolone ring occurs by loss of the sulfur atom.

5-Chloro-1-methyl-3-phenyl-2,1-benzisothiazolium chloride **169m** was treated with ethyl glycinate to yield in one-step reaction 7-chloro-1-methyl-2-oxo-5-phenyl-1,3-dihydro-2H-1,4-benzodiazepine **401** in 47%, which is the famous tranquilizer Valium® (72CPB2372, Scheme 135).

The oxidation of 2-benzenesulfonylisothiazol-2-imines **71′,130** and their perchlorates **73,132** with hydrogen peroxide gave 1,2,3-thiadiazine 1-oxides **402** and **403** in 26–52% yield; they were then converted into the corresponding 1,2,3-thiadiazine 1,1-dioxides **406** and **407** in 24–90% yield using *m*-chloroperoxybenzoic acid.

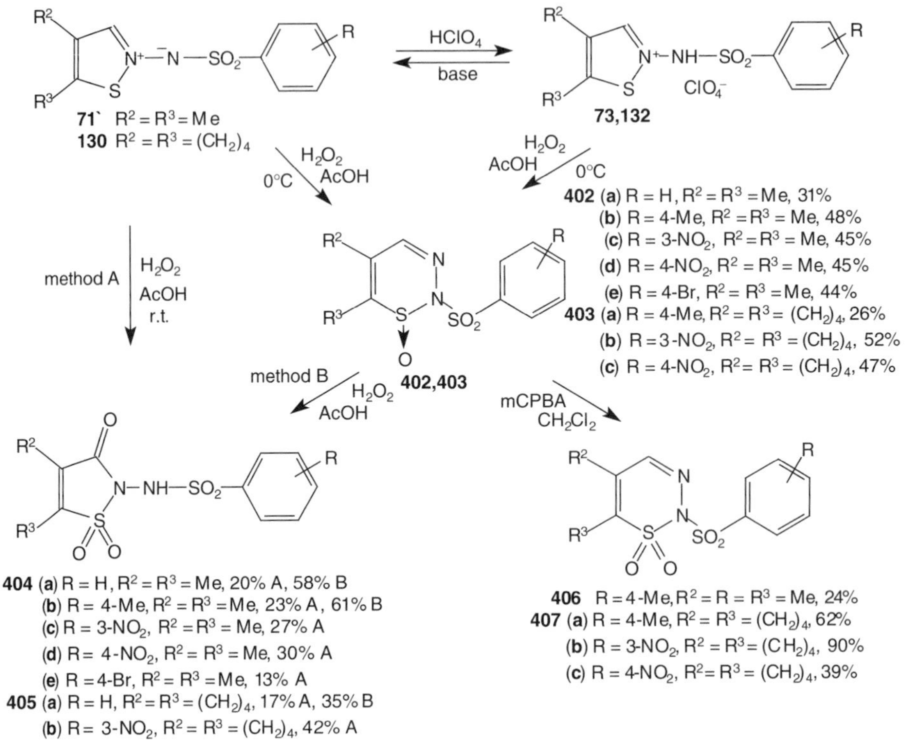

Scheme 136

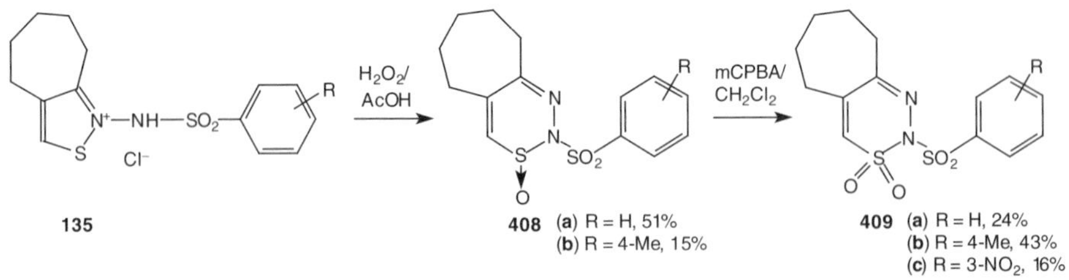

Scheme 137

Furthermore, oxidation of **402** and **403** with hydrogen peroxide furnished isothiazol-3(2*H*)-one 1,1-dioxides **404** and **405** (method B) as ring contraction products, which can be obtained directly by oxidation of imines **71′** and **130** or perchlorates **73** and **132** (method A), respectively. The structures **402e** (R = 4-Br), **404c** (R = 3-NO₂) and **407b** (R = 3-NO₂) were confirmed by X-ray diffraction (99JHC1081, Scheme 136).

The oxidation of isothiazolium chlorides **135** with hydrogen peroxide in acetic acid at room temperature gave the 1,2,3-thiadiazine 1-oxides **408a,b** in 15–51% yield. They were also converted into 1,2,3-thiadiazine 1,1-dioxides **409a–c** in 16–43% yield. The structure **408b** was confirmed by X-ray diffraction (00SUL109, Scheme 137).

E. Condensation

3-Chloro-2-methyl-isothiazolium salts **75k,m,q,r** were converted with malonodi-nitrile to 2-alkyl-3-(dicyanomethylen)-2,3-dihydroisothiazoles **410a–d** by a condensation reaction (77DEP2851023, 78USP4281136, Scheme 138).

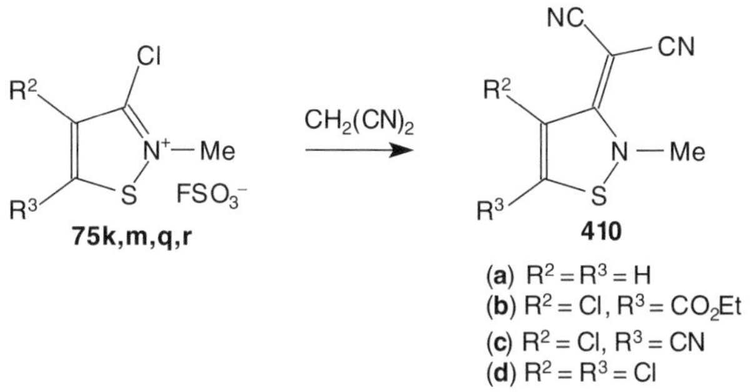

(a) $R^2 = R^3 = H$
(b) $R^2 = Cl$, $R^3 = CO_2Et$
(c) $R^2 = Cl$, $R^3 = CN$
(d) $R^2 = R^3 = Cl$

Scheme 138

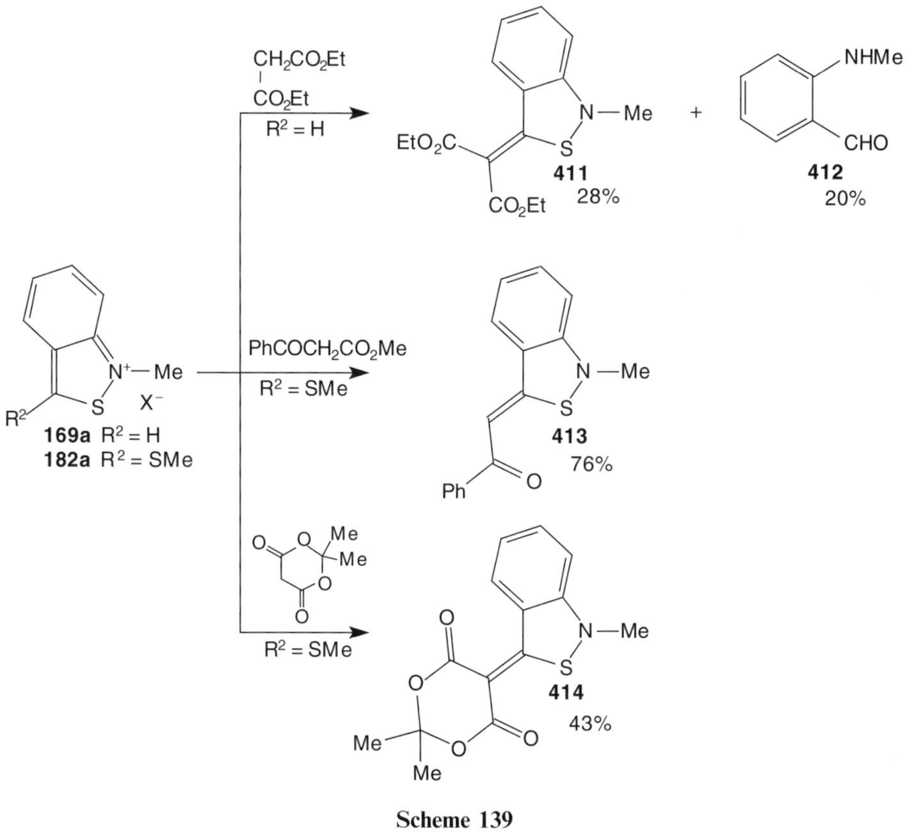

Scheme 139

2,1-Benzisothiazolium salts **169a** and **182a** reacted with carbanions in contrast to isothiazolium and 1,2-benzisothiazolium salts **301** not at the sulfur atom but by attack on the C-3 carbon atom. Therefore, 2-methyl 2,1-benzisothiazolium salts **169a** and the diethyl malonate anion yielded a mixture of two products, 2-methylaminobenzaldehyde **412** (20%) and diethyl 1-methyl-2,1-benzisothiazol-3-ylidenmaloneate **411** (28%). The treatment of salt **182a** with sodium benzoylacetate afforded 2,1-benzisothiazole derivative **413** (76%). Furthermore, the reaction of salt **182a** with the anion of 2,2-dimethyl-1,3-dioxane-4,6-dione (Meldrum's acid) gave condensation product **414** (43%) (82CJC440, Scheme 139).

REFERENCES

1879CB469	I. Remsen and C. Fahlberg, *Chem. Ber.*, **12**, 469 (1879).
23CB1630	K. Fries and G. Brothuhn, *Chem. Ber.*, **56**, 1630 (1923).
23AG159	R. Stolle and W. Geisel, *Angew. Chem.*, **36**, 159 (1923).
27LA264	K. Fries, K. Eishold, and B. Vahlberg, *Liebigs Ann. Chem.*, **454**, 264 (1927).
52CHE227	L. L. Bambas, *Chem. Heterocycl. Compd.*, **4**, 227 (1952).
61CB2950	J. Goerdeler and H. W. Pohland, *Chem. Ber.*, **94**, 2950 (1961).
65CB1531	J. Goerdeler and J. Gnad, *Chem. Ber.*, **98**, 1531 (1965).
65JCS32	D. Leaver, D. M. McKinnon, and W. A. H. Robertson, *J. Chem. Soc.*, 32 (1965).
65JCS(C)4577	P. Chaplen, R. Slack, and K. R. H. Wooldridge, *J. Chem. Soc. (C)*, 4577 (1965).
65JMC515	R. F. Meyer, B. L. Cummings, P. Bass, and H. O. J. Collier, *J. Med. Chem.*, **8**, 515 (1965).
65KGS220	Y. L. Gol'dfarb, M. A. Kalik, and M. L. Kirmalova, *Khim. Geterosikl. Soedin.*, **2**, 220 (1965), [CA, **63**, 71918 (1965)].
65ZN(B)712	J. Faust and R. Mayer, *Z. Naturforsch. B*, **20**, 712 (1965).
66CB2566]	H. Böshagen, *Chem. Ber.*, **99**, 2566 (1966).
66DEP19650610	BASF, DE Pat. 19650610 (1966), [CA, **66**, 65467g (1967)].
66JPR312	R. Mayer, H. J. Hartmann, and J. Jentzsch, *J. Prakt. Chem.*, **31**, 312 (1966).
66T2135	J. M. Landesberg and R. A. Olofson, *Tetrahedron*, **22**, 2135 (1966).
67CB2435	H. Böshagen, H. Feltkamp, and W. Geiger, *Chem. Ber.*, **100**, 2435 (1967).
67KGS1022	Y. L. Gol'dfarb, M. A. Kalik, *Khim. Geterosikl. Soedin.*, **6**, 1022 (1967), [CA, **69**, 86887 (1968)].
67ZC306	J. Faust, *Z. Chem.*, **7**, 306 (1967).
68CJC1855	D. M. McKinnon and E. A. Robak, *Can. J. Chem.*, **46**, 1855 (1968).
68DEP19661015	BASF, DE Pat. 19661015 (1968), [CA, **72**, 80338 (1970)].
68JCS(C)611	A. J. Layton and E. Lunt, *J. Chem. Soc (C)*, 611 (1968).
68ZC170	J. Faust, *Z. Chem.*, **8**, 170 (1968).
69CB1961	W. Geiger, H. Böshagen, and H. Medenwald, *Chem. Ber.*, **102**, 1961 (1969).
69CC1314	D. H. Reid and J. D. Symon, *Chem. Comm.*, **22**, 1314 (1969).
69JCS(C)707	D. G. Jones and G. Jones, *J. Chem. Soc (C)*, 707 (1969).
69KGS475	Y. L. Gol'dfarb and M. A. Kalik, *Khim. Geterosikl. Soedin.*, **3**, 475 (1969), [CA, **72**, 21549 (1970)].
70BSF3076	G. Le Coustumer and Y. Mollier, *Bull. Soc. Chim. Fr.*, 3076 (1970).

70CB3166   H. Böshagen, W. Geiger, and H. Medenwald, *Chem. Ber.*, **103**, 3166 (1970).

71BSF4373   J. C. Poite, S. Coen, and J. Roggero, *Bull. Soc. Chim. Fr.*, 4373 (1971).

71DEP2020479   H. Hagen and G. Hansen, DE Pat. 2020479 (1971), [CA, **76**, 87134 (1972)].

71JCS(B)2365   A. G. Burton, P. P. Forsythe, C. D. Johnson, and A. R. Katritzky, *J. Chem. Soc. (B)*, 2365 (1971).

71JCS(C)3994   E. Haddock, P. Kirby, and A. W. Johnson, *J. Chem. Soc. (C)*, 3994 (1971).

71LA46   S. Hünig, G. Kießlich, K. H. Oette, and H. Quast, *Liebigs Ann. Chem.*, **754**, 46 (1971).

71LA201   S. Hünig, G. Kießlich, and H. Quast, *Liebigs Ann. Chem.*, **748**, 201 (1971).

72CJC324   F. Leung and S. C. Nyburg, *Can. J. Chem.*, **50**, 324 (1972).

72CJC2568   G. E. Bachers, D. M. McKinnon, and J. M. Buchshriber, *Can. J. Chem.*, **50**, 2568 (1972).

72CPB2372   O. Aki, Y. Nakagawa, and K. Sirakawa, *Chem. Pharm. Bull.*, **20**, 2372 (1972).

72DEP2155694   P. Moser, DE Pat. 2155694 (1972), [CA, **77**, 103290 (1972)].

72IJC361   A. L. Cherian, P. Y. Pandit, and S. Seshadri, *Indian J. Chem.*, **10**, 361 (1972), [CA, **78**, 43340 (1973)].

72IJS328   A. Joos, *Int. J. Sulfur Chem.*, **2**, 328 (1972).

72JCS(P1)2305   P. Sykes and H. Ullah, *J. Chem. Soc., Perkin Trans*, **1**, 2305 (1972).

72LA58   H. Böshagen and W. Geiger, *Liebigs Ann. Chem.*, **764**, 58 (1972).

73AHC233   H. Hettler, *Adv. Heterocycl. Chem.*, **15**, 233 (1973).

73CJC3081   D. M. McKinnon and M. E. Hassan, *Can. J. Chem.*, **51**, 3081 (1973).

73JCS(CC)150   G. G. Abott and D. Leaver, *J. Chem. Soc., Chem. Commun.*, 150 (1973).

73JCS(P1)1863   M. Davis, E. Homfeld, and K. S. L. Srivastava, *J. Chem. Soc., Perkin Trans.*, **1**, 1863 (1973).

73LA256   F. Boberg and W. von Gentzkow, *Liebigs Ann. Chem.*, 256 (1973).

74AJC1221   M. Davis, L. W. Deady, and E. Homfeld, *Aust. J. Chem.*, **27**, 1221 (1974).

74CB1667   H. Böshagen and W. Geiger, *Chem. Ber.*, **107**, 1667 (1974).

74CJC3021   J. L. Charlton, S. M. Loosmore, and D. M. McKinnon, *Can. J. Chem.*, **52**, 3021 (1974).

74JHC1011   M. Davis, L. W. Deady, and E. Homfeld, *J. Heterocycl. Chem.*, **11**, 1011 (1974).

74ZC189   J. Liebscher and H. Hartmann, *Z. Chem.*, **14**, 189 (1974).

75AJC129   M. Davis, L. W. Deady, E. Homfeld, and S. Pogany, *Aust. J. Chem.*, **28**, 129 (1975).

75JCS(P2)1620   A. R. Katritzky, H. O. Tarhan, and B. Terem, *J. Chem. Soc., Perkin Trans.*, **2**, 1620 (1975).

75ZC478   J. Faust, *Z. Chem.*, **15**, 479 (1975).

76AJC1745   L. W. Deady and D. C. Stillman, *Aust. J. Chem.*, **29**, 1745 (1976).

76CB659   H. Böshagen and W. Geiger, *Chem. Ber.*, **109**, 659 (1976).

76DDP122249   M. Mühlstädt, R. Brämer, and B. Schulze, DD Pat. 122249 (1976), [CA, **87**, 39468e (1977)].

76IEC268   M. A. Hossain and J. Metzger, *Ind. Eng. Chem.*, **15**, 268 (1976), [CA, **85**, 192618 (1976)].

76JPR161   J. Faust and R. Mayer, *J. Prakt. Chem.*, **318**, 161 (1976).

76JPR507   M. Mühlstädt and B. Schulze, *J. Prakt. Chem.*, **318**, 507 (1976).

76ZC49   M. Mühlstädt, R. Brämer, and B. Schulze, Z. Chem., 16, 49 (1976).
77CB285   J. Goerdeler, R. Büchler, and S. Solyom, Chem. Ber., 110, 285 (1977).
77CJC1123   D. M. McKinnon, M. E. Hassan, and M. S. Chauhan, Can. J. Chem.,
                  55, 1123 (1977).
77DEP2851023   J. A. Virgilio, M. Manowitz, and E. Heilweil, DE Pat. 2851023 A1
                  (1977), [CA, 91, 91637 (1979)].
77JCS(P2)1332   J. L. McVicars, M. F. Mackay, and M. Davis, J. Chem. Soc., Perkin
                  Trans., 2, 1332 (1977).
77JPR305   B. Schulze, S. Herre, R. Brämer, C. Laux, and M. Mühlstädt, J.
                  Prakt. Chem., 319, 305 (1977).
77USP4292430   J. Rokach, C. S. Rooney, and E. J. Cragoe, Jr., US Pat. 4292430
                  (1977), [CA, 96, 35236 (1982)].
78CB2716   R. W. Hoffmann and S. Goldmann, Chem. Ber., 111, 2716 (1978).
78JHC529   A. H. Albert, D. E. O'Brien, and R. K. Robins, J. Heterocycl. Chem.,
                  15, 529 (1978).
78JHC695   J. Rokach and P. Hamel, J. Heterocycl. Chem., 15, 695 (1978).
78JOC1233   N. F. Haley, J. Org. Chem., 43, 1233 (1978).
78JOC2500   M. S. Raasch, J. Org. Chem., 43, 2500 (1978).
78USP4281136   J. A. Virgilio, M. Manowitz, and E. Heilweil, US Pat. 4281136 (1978),
                  [CA, 95, 203937 (1981)].
79BSF26   D. Barillier, Bull. Soc. Chim. Fr., 2, 26 (1979).
79CB3286   H. Böshagen and W. Geiger, Chem. Ber., 112, 3286 (1979).
79JCS(P1)2340   A. G. Briggs, J. Czyzewski, and D. H. Reid, J. Chem. Soc., Perkin
                  Trans., 1, 2340 (1979), [CA, 87, 183791 (1977)].
79JOC1118   J. Rokach, P. Hamel, Y. Girard, and G. Reader, J. Org. Chem., 44,
                  1118 (1979).
79S442   H. Böshagen and W. Geiger, Synthesis, 442 (1979).
79TL1281   J. Rokach, P. Hamel, Y. Girad, and G. Reader, Tetrahedron Lett.,
                  1281 (1979).
79TL3339   R. C. Davis, T. J. Grinter, D. Leaver, and R. M. O'Neil, Tetrahedron
                  Lett., 3339 (1979).
79USP4262127   J. A. Virgilio, M. Manowitz, and E. Heilweil, US Pat. 4262127 (1979),
                  [CA, 95, 62181 (1981)].
79USP4267341   J. Rokach, C. S. Rooney, and E. J. Cragoe, Jr., US Pat. 4267341
                  (1979), [CA, 95, 132864 (1981)].
79ZC41   B. Schulze and M. Mühlstädt, Z. Chem., 19, 41 (1979).
79ZN(B)123   U. Klingebiel and D. Bentmann, Z. Naturforsch. B, 34, 123 (1979).
80CB2490   H. Böshagen and W. Geiger, Chem. Ber., 113, 2490 (1980).
80PS79   D. Barillier, Phosphorus Sulfur, 8, 79 (1980).
81G71   A. Braibanti and M. T. Lugari-Mangia, Gazz. Chim. Ital., 111, 71
                  (1981), [CA, 95, 79873 (1981)].
81GBP2075540   P. Gregory, GB Pat. 2075540 (1981), [CA, 96, 164133d (1982)].
81PMC117   A. De, Prog. Med. Chem., 18, 117 (1981).
81USP4281136   J. A. Virgilio, M. Manowitz, and E. Heilweil, US Pat. 4281136 (1981),
                  [CA, 95, P2039370 (1981)].
82CJC440   D. M. McKinnon, K. A. Duncan, and L. M. Millar, Can. J. Chem.,
                  60, 440 (1982).
82JHC509   R. A. Abramovitch, M. N. Inbasekaran, A. L. Miller, and J. M.
                  Hanna Jr., J. Heterocycl. Chem., 19, 509 (1982).
82LA14   H. Böshagen and W. Geiger, Liebigs Ann. Chem., 1, 14 (1982).
82LA884   K. Grohe and H. Heitzer, Liebigs Ann. Chem., 1, 884 (1982).
82S972   D. J. LeCount and D. J. Dewsbury, Synthesis, 972 (1982).
82T1673   S. Rjappa, M. D. Nair, R. Sreenivasan, and B. G. Advani, Tetra-
                  hedron, 38, 1673 (1982).

83JCS(P1)1953   S. Rajappa and B. G. Advani, *J. Chem. Soc., Perkin Trans*, **1**, 1953 (1983).
83JHC1707   M. Davis and M. J. Hudson, *J. Heterocycl. Chem.*, **20**, 1707 (1983).
83JPR689   J. Liebscher, A. Areda, and B. Abegaz, *J. Prakt. Chem.*, **325**, 689 (1983).
83ZOR1134   V. A. Chuiguk and E. L. Komar, *Khim. Geterotsikl. Soedin.*, 1134 (1983), [CA, **99**, 212491 (1983)].
83M999   K. Gewald and H. Röllig, *Monatsh. Chem.*, **114**, 999 (1983).
83PS119   M. Komatsu, N. Harada, H. Kashiwagi, Y. Ohshiro, and T. Agawa, *Phosphorus Sulfur*, **16**, 119 (1983).
84CHEC131   D. L. Pain, B. J. Pearth, and K. R. H. Wooldrigde, *Comp. Heterocycl. Chem.*, **6**, 131 (1984).
84CJC1580   D. M. McKinnon, K. A. Duncan, and L. M. Millar, *Can. J. Chem.*, **62**, 1580 (1984).
84JCS(P1)385   L. K. A. Rahman and R. M. Scrowston, *J. Chem. Soc., Perkin Trans*, **1**, 385 (1984).
84JHC627   G. L'abbé, *J. Heterocyclic Chem.*, **21**, 627 (1984).
85AHC105   M. Davis, *Adv. Heterocycl. Chem.*, **38**, 105 (1985).
85BSB149   M. E. Hassan, *Bull. Soc. Chim. Belg.*, **94**, 149 (1985).
85CJC882   D. M. McKinnon, K. A. Duncan, A. M. McKinnon, and P. A. Spevack, *Can. J. Chem.*, **63**, 882 (1985).
85T1885   M. E. Hassan, M. A. Magraby, and M. A. Aziz, *Tetrahedron*, **41**, 1885 (1985).
88AP863   K. Hartke and H. G. Müller, *Arch. Pharm.*, **321**, 863 (1988).
88CJC1405   D. M. McKinnon and K. R. Lee, *Can. J. Chem.*, **66**, 1405 (1988).
88JCR(S)46   J. M. Golec and R. M. Scrowston, *J. Chem. Res. (S)*, 46 (1988).
88JCS(P1)2141   M. R. Bryce, T. A. Dansfield, K. A. Kandeel, and J. M. Vernon, *J. Chem. Soc., Perkin Trans.*, **1**, 2141 (1988).
88SC1847   P. Cuadrado, A. M. González, and F. J. Pulido, *Synthetic Comm.*, **18**, 1847 (1988).
88ZC287   B. Schulze, J. Hilbig, L. Weber, K. Rosenbaum, and M. Mühlstädt, *Z. Chem.*, **28**, 287 (1988).
88ZC345   M. Pulst, D. Greif, and E. Kleinpeter, *Z. Chem.*, **28**, 345 (1988).
89DDP289269   B. Schulze, K. Mütze, and M. Mühlstädt, DD Pat. 289269 (1989), [CA, **115**, 114501 (1991)].
89DEP289270   B. Schulze, K. Mütze, and M. Mühlstädt, DE Pat. 289270 (1989), [CA, **115**, 114500 (1991)].
90DDP295385   B. Schulze, S. Wagner, G. Kirsten, E. Kleinpeter, and M. Mühlstädt, DD Pat. 295385 (1990), [CA, **116**, 151750 (1992)].
90DDP275459   B. Schulze, J. Hilbig, K. Rosenbaum, J. Sieler, and M. Mühlstädt, DD Pat. 275459 A1 (1990), [CA, **113**, 152398 (1990)].
90HCA1679   M. Huys-Francotte and H. Balli, *Helv. Chim. Acta*, **73**, 1679 (1990).
90JCS(P1)2881   R. C. Davis, T. J. Grinter, D. Leaver, R. M. O'Neil, and A. Gordon, *J. Chem. Soc., Perkin Trans.*, **1**, 2881 (1990).
90MRC419   L. Weber, R. Szargan, B. Schulze, and M. Mühlstädt, *Magn. Reson. Chem.*, **28**, 419 (1990).
91ACSA302   B. Schulze, M. Kretschmer, M. Mühlstädt, K. Rissanene, K. Laihia, and E. Kleinpeter, *Acta Chem. Scand.*, **45**, 302 (1991).
91JHC749   D. M. McKinnon and A. A. Abouseid, *J. Heterocycl. Chem.*, **28**, 749 (1991).
92JCS(CC)571   D. Szabo, I. Karpovits, A. Kucsman, M. Czugler, G. Argay, and A. Kalman, *J. Chem. Soc., Chem. Comm.*, 571 (1992).
92JPR25   B. Schulze, K. Rosenbaum, J. Hilbig, and L. Weber, *J. Prakt. Chem.*, **334**, 25 (1992).

92KGS256   M. V. Gorelik, V. Y. Shteiman, and R. A. Alimova, *Khim. Get-erotsikl. Soedin.*, **2**, 256 (1992), [CA, **117**, 251264 (1992)].
92ZK281   R. Kempe, J. Sieler, D. Selke, and B. Schulze, *Z. Kristallogr.*, **199**, 281 (1992).
93AG797   A. Rolfs and J. Liebscher, *Angew. Chem.*, **105**, 797 (1993).
93HOU668   B. Schulze, *Houben-Weyl Methoden Org. Chem.*, **1**, 668 (1993).
93HOU799   B. Schulze, *Houben-Weyl Methoden Org. Chem.*, **1**, 799 (1993).
93JHC929   S. H. Kim, K. Kim, J. Kim, K. Kim, and J. H. Kim, *J. Heterocycl. Chem.*, **30**, 929 (1993).
93TL1909   B. Schulze, K. Mütze, S. Selke, and R. Kempe, *Tetrahedron Lett.*, **34**, 1909 (1993).
94JPR115   B. Schulze, D. Selke, S. Kirrbach, and R. Kempe, *J. Prakt. Chem.*, **336**, 115 (1994).
94JPR434   U. Dietrich, A. Feindt, M. Pulst, M. Weißenfels, D. Greif, and W. Coa, *J. Prakt. Chem.*, **336**, 434 (1994).
94ZOR1379   B. Schulze, U. Dietrich, K. Illgen, and J. Sieler, *Z. Org. Khim.*, **30**, 1379 (1994), [CA, **123**, 339853 (1995)].
94ZOR1404   B. Friedrich, A. Fuchs, and B. Schulze, *Z. Org. Khim.*, **30**, 1404 (1994), [CA, **123**, 339854 (1995)].
95JPR175   B. Schulze, U. Obst, G. Zahn, B. Friedrich, R. Cimiraglia, and H. J. Hofman, *J. Prakt. Chem.*, **337**, 175 (1995).
95JPR310   A. Rolfs, H. Brosig, and J. Liebscher, *J. Prakt. Chem.*, **337**, 310 (1995).
95ZK73   U. Obst, B. Schulze, and G. Zahn, *Z. Kristallogr.*, **210**, 73 (1995).
96JPR424   B. Schulze, B. Friedrich, S. Wagner, and P. Fuhrmann, *J. Prakt. Chem.*, **338**, 424 (1996).
96MOL139   K. Illgen, C. Hartung, R. Herzschuh, and B. Schulze, *Molecules*, **1**, 139 (1996).
96MOL142   B. Friedrich, A. Fuchs, M. Findeisen, and B. Schulze, *Molecules*, **1**, 142 (1996).
96PHA638   K. Hartke and C. Ashry, *Pharmazie*, **51**, 638 (1996).
96PS203   F. Boberg, B. Bruchmann, A. Herzberg, and A. Otten, *Phosphorus Sulfur*, **108**, 203 (1996).
96T783   B. Schulze, S. Kirrbach, K. Illgen, and P. Fuhrmann, *Tetrahedron*, **52**, 783 (1996).
96UP1   A. Noack, K. Illgen, and B. Schulze, unpublished results (1996).
96ZK761   B. Friedrich, B. Schulze, and P. Fuhrmann, *Z. Kristallogr.*, **211**, 761 (1996).
96ZOR1745   S. Kirrbach, K. Muetze, R. Kempe, R. Meusinger, A. Kolberg, and B. Schulze, *Z. Org. Khim.*, **32**, 1745 (1996), [CA, **126**, 293292 (1997)].
97JPR1   B. Schulze and K. Illgen, *J. Prakt. Chem.*, **339**, 1 (1997).
97SUL35   B. Friedrich, A. Noack, and B. Schulze, *Sulfur Lett.*, **21**, 35 (1997).
97T17795   S. T. Ingate, J. L. Marco, M. Witvrouw, C. Pannecouque, and E. De Clercq, *Tetrahedron*, **53**, 17795 (1997).
97ZN(B)1139   S. Sawusch, U. Schilde, E. Uhlemann, and F. Weller, *Z. Naturforsch. B*, **52**, 1139 (1997).
98H587   L. L. Rodina, A. Kolberg, and B. Schulze, *Heterocycles*, **49**, 587 (1998).
98JPR361   A. Noack, S. Jelonek, F. B. Somoza Jr., and B. Schulze, *J. Prakt. Chem.*, **340**, 361 (1998).
98ZK331   T. Gelbrich, A. Noack, and B. Schulze, *Z. Kristallogr.*, **213**, 331 (1998).
99COL21   K. H. Shah and R. G. Patel, *Colourage*, 21 (1999).

99HCA685   C. Hartung, K. Illgen, J. Sieler, B. Schneider, and B. Schulze, *Helv. Chim. Acta*, **82**, 685 (1999).

99JHC1081   A. Kolberg, J. Sieler, and B. Schulze, *J. Heterocycl. Chem.*, **36**, 1081 (1999).

99T7625   J. L. Marco, S. T. Ingate, and P. M. Chinchón, *Tetrahedron*, **55**, 7625 (1999).

99ZAAC511   R. Richter, O. Seidelmann, and L. Beyer, *Z. Anorg. Allg. Chem.*, **625**, 511 (1999).

00JCS(D)3113   S. Brooker, G. B. Caygill, P. D. Croucher, T. C. Davidson, D. L. J. Clive, S. R. Magnuson, S. P. Cramer, and C. Y. Ralston, *J. Chem. Soc., Dalton Trans.*, 3113 (2000).

00JPR291   A. Kolberg, S. Kirrbach, D. Selke, B. Schulze, and S. Morozkina, *J. Prakt. Chem.*, **342**, 291 (2000).

00JPR675   A. Noack, I. Röhlig, and B. Schulze, *J. Prakt. Chem.*, **342**, 675 (2000).

00SUL109   A. Kolberg, J. Sieler, and B. Schulze, *Sulfur Lett.*, **24**, 109 (2000).

00T2523   J. L. Marco, S. T. Ingate, C. Jaime, and I. Beá, *Tetrahedron*, **56**, 2523 (2000).

01PS29   M. Wust, B. Zur Linden, K. Gloe, and B. Schulze, *Phosphorus Sulfur*, **170**, 29 (2001).

01ZK309   A. Kolberg, B. Schulze, and J. Sieler, *Z. Kristallogr.*, **216**, 309 (2001).

02AHC71   F. Clerici, *Adv. Heterocycl. Chem.*, **83**, 71 (2002).

02H693   Z. Liu and Y. Takeuchi, *Heterocycles*, **56**, 693 (2002).

02HCA183   K. Taubert, J. Sieler, L. Hennig, M. Findeisen, and B. Schulze, *Helv. Chim. Acta*, **85**, 183 (2002).

02JOC5375   D. J. Lee, B. S. Kim, and K. Kim, *J. Org. Chem.*, **67**, 5375 (2002).

02JOC8400   F. G. Gelalcha and B. Schulze, *J. Org. Chem.*, **67**, 8400 (2002).

02JP2002202598   Y. Murota and T. Sorori, *J. Pat.* 2002202598 (2002), [CA, **137**, 116971 (2002)].

02SCIS287   J. Schatz, *Sci. Synth.*, **9**, 287 (2002).

02SCIS507   D. W. Brown and M. Sainsbury, *Sci. Synth.*, **11**, 507 (2002).

02SCIS573   D. W. Brown and M. Sainsbury, *Sci. Synth.*, **11**, 573 (2002).

02SUR79   K. Taubert, S. Kraus, and B. Schulze, *Sulfur Rep.*, **23**, 79 (2002).

02SUR279   A. Siegemund, K. Taubert, and B. Schulze, *Sulfur Rep.*, **23**, 279 (2002).

02ZN(B)383   B. Schulze, K. Taubert, F. G. Gelalcha, C. Hartung, and J. Sieler, *Z. Naturforsch, B*, **57**, 383 (2002).

03CG147   S. Baumann, M. Möder, R. Herzschuh, and B. Schulze, *Chromatographia*, **57**, 147 (2003).

03H639   B. Schulze, D. Gidon, A. Siegemund, and L. L. Rodina, *Heterocycles*, **61**, 639 (2003).

03JP2003034696   J. Nishigaki, K. Nakamura, K. Takeuchi, H. Inomata, and M. Kojima, *J. Pat.* 2003034696 (2003), [CA, **138**, 170463 (2003)].

03S2265   K. Taubert, A. Siegemund, A. Eilfeld, S. Baumann, J. Sieler, and B. Schulze, *Synthesis*, **14**, 2265 (2003).

03PHMD617   R. Miedzybrodzki, *Post. Hig. Med. Dósw.*, **57**, 617 (2003).

03ZN(B)111   M. Gütschow, M. Pietsch, K. Taubert, T. H. E. Freysoldt, and B. Schulze, *Z. Naturforsch, B*, **58**, 111 (2003).

04HCA376   A. Siegemund, C. Hartung, S. Baumann, and B. Schulze, *Helv. Chim. Acta*, **87**, 376 (2004).

04MC207   O. Y. Sapozhnikov, V. V. Mezhnev, E. V. Smirnova, B. G. Kimel, M. D. Dutov, and S. A. Shevelev, *Mendeleev Comm.*, 207 (2004).

04SCIS731   J. G. Schanle, *Sci. Synth.*, **27**, 731 (2004).

04ZN(B)478   A. Siegemund, C. Hartung, A. Eilfeld, J. Sieler, and B. Schulze, *Z. Naturforsch, B*, **59**, 478 (2004).

05H2705   S. Schmidt, A. Kolberg, L. Hennig, J. Hunger, and B. Schulze, *Heterocycles*, **65**, 2705 (2005).

05JEIMC341   M. Gütschow, M. Pietsch, A. Themann, J. Wolf, and B. Schulze, *J. Enzyme Inhib. Med. Chem.*, **20**, 341 (2005).

05SUC211   J. Wolf, T. H. E. Freysoldt, C. Hartung, J. Sieler, and B. Schulze, *J. Sulfur Chem.*, **26**, 211 (2005).

05UP1   A. Siegemund-Eilfeld and B. Schulze, unpublished results (2005).

05ZN(B)41   B. Schulze, K. Taubert, A. Siegemund, T. H. E. Freysoldt, and J. Sieler, *Z. Naturforsch, B*, **60**, 41 (2005).

06ZN(B)464   V. M. Zakharova, A. Siegemund-Eilfeld, J. Sieler, and B. Schulze, *Z. Naturforsch, B*, **61**, 464 (2006).

Subject Index of Volume 94